Digital Logic Design and Computer Organization

With Computer Architecture for Security

About the Author

Nikrouz Faroughi has a BS in computer engineering, MS in computer science, MS in electrical engineering, and PhD in electrical engineering with a specialization in computer engineering from Michigan State University. He has worked as a systems analyst and currently is a professor and graduate coordinator in the Computer Science Department and a faculty member in the Computer Engineering Program at California State University, Sacramento. As a consultant, he has worked and also served as a technical manager at Intel Corporation.

Digital Logic Design and Computer Organization

With Computer Architecture for Security

Nikrouz Faroughi
California State University, Sacramento

Mc Graw Hill Education

New York Chicago San Francisco
Athens London Madrid
Mexico City Milan New Delhi
Singapore Sydney Toronto

Cataloging-in-Publication Data is on file with the Library of Congress

McGraw-Hill Education books are available at special quantity discounts to use as premiums and sales promotions, or for use in corporate training programs. To contact a representative please visit the Contact Us page at www.mhprofessional.com.

Digital Logic Design and Computer Organization with Computer Architecture for Security

1 2 3 4 5 6 7 8 9 0 QVS/QVS 1 2 0 9 8 7 6 5 4

ISBN 978-0-07-183690-6
MHID 0-07-18-3690-X

The pages within this book were printed on acid-free paper.

Sponsoring Editor
Michael McCabe

Editorial Supervisor
Donna M. Martone

Acquisitions Coordinator
Amy Stonebraker

Project Manager
Dheeraj Chahal, MPS Limited

Copy Editor
Lisa McCoy

Proofreader
Surendra Shivam, MPS Limited

Production Supervisor
Lynn M. Messina

Composition
MPS Limited

Art Director, Cover
Jeff Weeks

Contents

Preface

This book is designed with the goal of providing a comprehensive understanding of digital logic design and computer organization in a single textbook. In addition, the book contains an entire chapter on computer architecture for security.

The book covers both the fundamentals of digital logic design and design with the Verilog hardware description language. Separate chapters are allocated to cover design methodologies of simple and complex combinational and sequential circuits. Modern tools and methodology for circuit design are discussed in general, and Verilog examples illustrate only the basic and synthesizable features of the language. If desired, instructors may choose to use VHDL instead. However, the book does not require using a hardware description language.

The book covers memory organization, processing core and processor organization, and computer security through hardware. As advancements in technologies and demand for high-speed and low-power designs have changed the fundamentals of computer organization, an attempt is made to provide not only simple examples to illustrate basic design concepts, but also provide an understanding of modern computer design objectives.

The book also covers computer architecture concepts from instruction set architecture, including the architecture for secure execution, pipelining and parallelism, and memory hierarchy. An attempt is made to provide numerous examples that illustrate the applications of pipelining and parallelism to increase concurrency and reduce or hide latency (two factors that affect performance). Program code examples are also used to illustrate the link between CPU architecture and a compiler and between programing methodologies and performance.

Overview of Chapters

There are 11 chapters in this book. An overview of digital systems, innovations in computing, number systems, digital logic design, and computer organization/ architecture and security is given in Chap. 1.

Chapters 2 and 3 cover simple and complex combinational circuits, including integer and floating-point arithmetic. In Chap. 2, where design methodologies for small circuits are discussed, it is assumed that when it is necessary to minimize truth tables with more than four input variables, students would be using logic minimization software, such as Espresso, which is available for free download from the Internet. The chapter also provides an introduction to design tools, structural and behavioral design models, and design with programmable logic devices, and includes sample designs in Verilog and presents synthesis and simulation results. Chapter 3 covers methodologies used to design large combinational circuits and introduces integer and floating-point computer arithmetic and also presents design examples.

Chapters 4, 5, and 6 cover simple and complex sequential circuits from basic modules to complex data paths and control to timing constraints, design efficiency, and power usage. Chapter 4 covers latches, flip-flops, and their timing requirements. Chapter 5 covers finite state machine (FSM) design and timing requirements and handling of asynchronous inputs. Chapter 6 covers single-cycle, multicycle, and pipelined data paths and controls. Design examples illustrate data path and FSM-based, microprogrammed, and pipelined control unit organizations. Several data path design examples, including for unsigned and signed multiplication and two-dimensional virtual object rotation, are also presented.

Chapter 7 is dedicated to memory and covers memory technologies, including SDRAM technologies, and memory design, including interleaving and multichannel. Memory communication protocols, performance, and uniform memory access (UMA) and nonuniform memory access (NUMA) organizations are also presented. Examples of programing methodologies to take advantage of a NUMA organization to improve performance are also discussed.

Chapter 8 covers CPU design, from single-cycle and pipeline, to reduced instruction set computer (RISC), deep pipelining, and branch prediction, to static and dynamic instruction-level parallelism (ILP), to multithreading. The chapter includes design and simulation of CPU data path examples and presents program code examples to illustrate compiler optimization to improve performance, branch prediction, ILP, and multithreading.

Chapter 9 is dedicated to microcomputer architecture and covers the history from simple bus-based to integrated to modern point-to-point architectures, and topics from I/O port addressing, to interrupt-driven I/O and direct memory access (DMA), to modern "plug and play" device controller interfaces, such as the USB host controller interface. Interruption and related operating system tasks are also discussed. A data path and instruction set of an interrupt handling CPU is also used as an example to explain the architecture and operations of a simple computer.

Chapter 10 covers the rationale and organization of a memory hierarchy. Cache coherency in a single processor system and an introduction to cache coherency in shared-memory multiprocessor systems are covered. Examples are used to illustrate advantages of different cache-mapping techniques in terms of miss rate, amount of hardware, and power usage. The chapter also presents virtual address translation, the management of page tables, and alternative processor organizations for translating a virtual address.

Chapter 11 starts by providing a general understanding of the security engineering methodology applied to computer architecture. It introduces access control, security policy models, hardware security policy mechanisms, and software/physical attack mechanisms, and presents an introduction to cryptography techniques. The chapter also covers the architecture of a trusted computing base (TCB) either as a secure coprocessor—to implement, for example, secure data storage and communication—or as a secure general purpose processor. The architecture of a secure processor for enforcing program (instructions and data) confidentiality and integrity is also presented in detail.

While the topics of Chap. 11 are compiled into one chapter for the convenience of readers, the topics in this chapter can concurrently be covered with other chapters. For instance, students can design simple cryptography circuits when learning sequential circuit design techniques. Other selected topics from this chapter that may be covered in conjunction with the topics of other chapters are hardware Trojans and hardware security policy mechanisms, memory authentication, secure handling of interrupts, and the architecture of a secure co-processor and processor. In order to provide a sample presentation of this chapter's topics in conjunction with the topics of other chapters, the Exercise sections in Chaps. 1, 3, and 5 to 10 include a list of exercises from Chap. 11 under the title "Computer Security." Instructors may choose exercises from this list in these chapters.

Keywords are bolded, for easy reference, when it is first introduced. The abbreviated keywords are not bolded and occasionally spelled out for readers' convenience. Also, for better student understanding, at times brief texts inside two square brackets ("[]") explain topics that are related (e.g., certain operating system tasks) but are outside the scope of this book or are out of context. The instructors at academic institutions who adopt this as the required textbook in their classes will have access to exercise solutions and PowerPoint presentation slides.

Audience

By most accounts, this textbook intentionally covers both digital design and computer organization in more depth than do existing similar textbooks. For the two subject areas, the objectives are to provide a more balanced depth versus breadth of coverage. In a single semester, instructors can judiciously choose both the topics and the depth versus breadth for each topic they want to emphasize in their classes. The textbook also has enough topics for a two-quarter or two-semester course sequence to cover both the digital logic design and computer organization/architecture subjects in depth and allow more time for students to acquire a good understanding of design practices and tradeoffs. The following is a suggested list of ways the textbook may be used:

1. For undergraduate students with no or limited background in digital logic, the course could cover Chaps. 1 through 5 and selected topics from Chaps. 6 to 9 and exposure to some other topics from the remaining chapters. Some sections and design examples may be skipped.
2. For undergraduate students in computer science and computer engineering who have some digital logic knowledge, the course could cover

Chap. 1, review/cover selective topics from Chaps. 2 to 5, and cover Chaps. 6 to 10 and topics from Chap. 11.

3. Academic departments that offer degree programs for prospective graduate students with no or limited background in digital logic design and computer organization, this may be an ideal textbook for covering both digital logic design and computer organization and architecture in detail in a single book.
4. Professionals who wish to refresh their knowledge in digital logic and/or computer organization and architecture and/or to familiarize themselves with security-related computer architecture concepts would benefit from this book.

Acknowledgment

A number of people have provided valuable inputs to the content of this book. Special thanks go to my colleagues Isaac Ghansah and Thomas Mathews, for their valuable suggestions and contribution to the content of the book; and Martin Nicholes (now at Intel), for his insightful comments on the content of Chap. 11. I would also like to thank the reviewers for their thoughtful comments. The final copy reflects some of the changes they have suggested.

Many of my students have also provided useful feedback on the drafts of this book and have helped identify some bugs in the text. Their detailed analysis of some textbook examples was effective in identifying some typos and mislabeling. I especially would like to thank Kevin Schultz, Andrew Larsen, Branden Garner, Chris Dalisay, Thomas Lee, Robert Carreras, Ian Reif, and Matt Larsen. I would also welcome any corrections that may have been missed and any suggestions to improve the book. I also would like to thank sponsoring editor Michael McCabe, editorial supervisor Donna Martone, production supervisor Lynn Messina, copy editor Lisa McCoy, and art director Jeff Weeks at McGraw-Hill Publishers and project manager Dheeraj Chahal and Surendra Shivam with MPS Limited for their valuable support, cover page design and final preparation of the book.

Last but not least, I would like to thank my wife, Gita, and sons Kian and Ryon for their patience and support, especially for accepting the long hours and my occasional preoccupation with this project.

Digital Logic Design and Computer Organization With Computer Architecture for Security

CHAPTER 1
Introduction

1.1 Introduction

Computers, iPads, cell phones, etc. have created a digital revolution that has changed many aspects of our lives. All forms of data, from numbers and text to audio, image, and video, are represented in a series of digits as 0's and 1's. Digital systems have changed the way we communicate, work, are entertained, and even shop, and are very much in everything we see and use. They are also in cars, in grocery checkout equipment, in utility meters, in set-top boxes, in emergency equipment, in medical devices, in factory control systems, etc. As more people use digital systems, more data is also created, processed, stored, transmitted, and accessed. This has created a demand for more powerful computers, whether personal computers or large systems used in many areas such as e-commerce, banking, search engines, and research.

Innovations in computing, however, are evolutionary and depend on many factors such as integrated chip (IC) technologies and software development, including operating systems. The innovations in the IC technologies have steadily pushed the transistor count to billions in a single chip. **Feature size**, the size of the elements in an IC that determines the size of a transistor as an electronic switch, has become smaller and smaller over the years. Both the reductions in the feature size and increases in the size of a die (rectangular semiconductor material) have increased transistor density by about 35 percent every year. This, in turn, has increased transistor count on a single chip between 40 and 55 percent every 18 to 24 months [1]. This rate of increase in the number of transistors is commonly known as Moore's law.

Over the years, microprocessor designers have used Moore's law as a guide to design future processors. They have used the increasing number of available transistors to design high-performance processors, revolutionizing personal computers.

Innovations in application developments have also revolutionized the way digital systems are designed. Today, innovations in computer-aided design (CAD) tools for IC have enabled chip designers to use a hardware description language (HDL) to describe the behavior of a digital circuit. The description can then be simulated, debugged, evaluated, and even automatically mapped to hardware, creating a circuit. CAD tools for circuit design are now commonly used in industry as well as in educational settings.

In a digital world, there is also the possibility of unauthorized access to data and information. Personal information, as well as intellectual properties of various organizations, may be stolen, modified, or erased. Malicious software could gain access to private computer systems or disrupt computer operations. However, because digital systems are made of both hardware and software and hardware is more secure than software, hardware can play an important role in keeping digital information secure.

This chapter provides an introduction and an overview of the topics covered in the remaining chapters. In this book, we discuss the hardware aspect of digital systems, from basic circuits to circuit modules that perform computations to the design of a processing element, commonly called a **processing core** or central processing unit (CPU). We also discuss memory, memory system design, and computer systems that contain multiple cores, either as a multicore processor or multiprocessor system. The book also covers an introduction to computer architecture for security.

1.1.1 Data Representation

Digital systems all contain circuits that input and output logic values as either true or false. A voltage range defines each logic value. For example, using a 5 V power source, any voltage value between 2.4 and 5 V is considered true, and value between 0 and 0.8 V is considered false. Battery-powered handheld digital devices typically use a lower voltage source to save power. True and false logic values are interpreted as 1's and 0's, forming binary numbers.

Binary numbers are used to represent characters to create text, pixels to create an image, digital audio and video data, and integer and real numbers used in computations. Characters are typically represented either in 8-bit American Standard Code for Information Interchange (ASCII) codes or 16-bit Unicode. While only 256 (2^8) different characters (letters, decimal digits, and symbols) can be represented with ASCII codes, over 65,000 (2^{16}) different representations are possible with Unicode, thus making Unicode suitable for word-based languages such as Asian languages.

Images are made of thousands or millions of pixels as dots seen on the screen. Each pixel on a color monitor is composed of three dots (red, green, and blue) that converge into a single dot on the screen, creating a dot with different colors or shades of gray. For example, a monitor that operates in true color mode uses 8 bits to represent each of the red, blue, and green colors, creating a 24-bit color code capable of displaying over 16 million (2^{24}) colors. There is also deep color, where 30 or more bits are used to represent billions of colors.

Digital audio and video data are created by digitizing (i.e., converting) analog and continuous electrical signals into a stream of numbers. For instance, a microphone converts a continuous sound wave that travels through the air into an analog electrical signal. A digitizer then samples the electrical signal at fixed intervals to create a stream of integer numbers representing the sound. The interval is determined from a sampling rate. For example, a sampling rate of 44.1 kilohertz(kHz) will generate 44,100 samples per second, producing a compact disk (CD) quality sound [2].

The higher the sampling rate, the more closely the sampled data represents the actual sound. Each sample value indicates the signal strength at the sampling time. Using 8 bits to represent each sampled value implies that signal strength is divided into 256 levels, with 0 being the lowest and 255 the highest. With 16 bits, the signal strength can be divided into 65,536 (2^{16}) different levels. Therefore, with more bits, one is able to more accurately capture the sound wave in digital form, but more data is also generated. Stereo sound systems have two independent audio channels. The sound from each channel is sampled, producing a sound file that is twice the size of the file of a single channel (mono) sound. However, which is better, mono or stereo, depends on the sampling rate and bits used to represent each sampled value.

Representation of Integer Numbers

Binary numbers are either signed or unsigned. The range for a 3-bit unsigned binary number is 0 to 7, or in binary $(000)_2$ to $(111)_2$, where the subscript 2 is used here to indicate binary. For computer arithmetic, signed numbers are represented as 2's complement numbers. A negative binary number is converted to its equivalent 2's complement representation by flipping (inverting) each bit and then adding 1 to the inverted result. For example, the 4-bit 2's complement representation of $-3 = -(11)_2$ is determined as follows:

1. Write $-(11)_2$ as a 4-bit binary number; that is, $-(0011)_2$.
2. Invert each bit; that is, 1100.
3. Add 1 to the inverted number to get the 4-bit 2's complement representation of −3 as $(1101)_{2s}$, where "2s" is used here to indicate a 2's complement number.

One-half of all 2's complement numbers stored in a computer are positive, and the other half are negative. The representation for a positive 2's complement number is the same as binary. The 4-bit 2's complement representation of $+3 = (11)_2$ is $(0011)_{2s}$. The most significant bit (MSB) of a 2's complement number indicates the sign of the representation; 1 indicates a negative representation and 0 a positive representation. A similar procedure is used to convert a negative 2's complement number, such as $(1101)_{2s}$, to its equivalent binary representation $-(0011)_2$, as illustrated next:

1. Invert the bits of the negative 2's complement number $(1101)_{2s}$; that is, 0010.
2. Add 1 to the inverted result to get the 4-bit magnitude of the 2's complement number as $(0011)_2$.
3. Include the negative sign; that is, $-(0011)_2$, or $-(11)_2 = -3$ in decimal.

3-Bit Binary	Decimal Equivalent if Unsigned	Decimal Equivalent if 2's Complement	Decimal Equivalent if Signed Magnitude
000	0	0	+0
001	1	1	1
010	2	2	2
011	3	3	3
100	4	−4	−0
101	5	−3	−1
110	6	−2	−2
111	7	−1	−3

Table 1.1 Equivalent Decimal Numbers for 3-Bit Unsigned, 2's Complement, and Signed-Magnitude Numbers

Negative numbers are also represented as signed-magnitude (sm) numbers. For example, $(0011)_{sm} = +3$ and $(1011)_{sm} = -3$, where "sm" is used here to indicate a signed-magnitude number. In this case, the MSB indicates if the number is negative (MSB = 1) or positive (MSB = 0), and the rest of the bits represent the magnitude of the number as $(011)_2 = 3$.

Table 1.1 shows decimal numbers that are equivalent to 3-bit unsigned, 2's complement, and signed-magnitude representations. While arithmetic operations, such as addition, can be performed on 2's complement numbers, signed-magnitude numbers are only used in the representation of real numbers; no arithmetic operations are directly performed on signed-magnitude numbers. Computer arithmetic is covered in Chap. 3; multiplication is also covered in Chap. 6.

Table 1.2 lists 4- and 8-bit unsigned binary, 2's complement, and signed-magnitude representations of +5 and −5. An m-bit 2's complement number can be converted to an n-bit 2's complement representation, where $n > m$, by simply repeating the sign bit $n - m$ times. This is called 2's complement **sign extension.**

Decimal	4 Bits	8 Bits	
+5	$(0101)_2$ $(0101)_{2s}$ $(0101)_{sm}$	$(00000101)_2$ $(00000101)_{2s}$ $(00000101)_{sm}$	Extend the number with 0s Apply sign extension, sign = 0 Sign bit = 0, extend with 0s
−5	$-(0101)_2$ $(1011)_{2s}$ $(1101)_{sm}$	$-(00000101)_2$ $(11111011)_{2s}$ $(10000101)_{sm}$	Extend the number with 0s Apply sign extension, sign = 1 Sign bit = 1, extend the magnitude $(101)_2$ with 0s to get the 7-bit magnitude $(0000101)_2$

Table 1.2 Examples of Unsigned, 2's Complement, and Signed-Magnitude Representations

Representation of Real Numbers

Computers also operate on real numbers, such as 2.75. The representation of real numbers in computers is called floating-point (FP) numbers, where each FP number includes three integer parts: sign bit, biased exponent, and unsigned fraction. The combination of the sign bit and the unsigned fraction creates a signed-magnitude number. As we will see in Chap. 3, FP arithmetic takes several steps and requires operating on the exponent and fraction values, both integer numbers, separately.

The exponent of an FP number is an unsigned number and represents a **biased exponent**. A fixed value, called bias, is used to convert a biased exponent to a negative or positive exponent. Consider an FP number representation that uses 4-bit biased exponents with bias = 7. In this case, biased exponent values 0, 15, and the range 1 to 14 represent various FP numbers. Equation (1.1) shows the relationship between an exponent and its equivalent biased exponent value.

$$\begin{aligned} \text{exponent} &= \text{biased\_exponent} - \text{bias} \\ \text{biased\_exponent} &= \text{exponent} + \text{bias} \end{aligned} \tag{1.1}$$

Example 1.1. Representation of real number 2.75 as a 16-bit FP number using a 4-bit biased exponent with *bias* = 7 and 11-bit fraction.

$$\begin{aligned} 2.75 &= 2 + 0.5 + 0.25 \\ &= 2 + \frac{1}{2} + \frac{1}{4} \\ &= (10)_2 + (0.1)_2 + (0.01)_2 \\ &= (10.11)_2 \\ &= (10.11)_2 * 2^0\text{; exponent} = 0 \\ &= (1.011)_2 * 2^1\text{; exponent} = 1\text{, decimal point moved one position to the left} \\ &= (1.011)_2 * 2^8\text{; biased-exponent} = 8\text{, bias} = 7 \\ &=> (0, 1000, 01100000000)_2\text{; 16-bit FP number representation} \\ &= \text{0x4300; where “0x” indicates a hex number} \end{aligned}$$

The 16-bit FP representing 2.75 has a sign bit = 0 (positive), 4-bit biased exponent = $(1000)_{2,}$ and 11-bit unsigned fraction = $(01100000000)_2$. The implicit decimal point is to the left of the fraction. While the 1 before the decimal point in $(1.011)_2$ is a part of the FP number, it is not included in the 16-bit representation stored in memory. Likewise, the 16-bit FP representation of −2.75 is $(1, 1000, 01100000000)_2$, or 0xC300, where “0x” indicates a hex number.

Using a k-bit biased exponent, if biased exponent is 0, the FP number represents 0.0 if the fraction is also 0. If the biased exponent is 0 and the fraction is not 0, the number represents an extremely small real number known as a **denormal**. If the biased exponent is between 1 and $k - 2$, it represents very small to very large real numbers, called **normal** FP numbers. If the biased exponent is $k - 1$, the FP number is considered either infinity (∞) if the fraction is 0, or an invalid number, such as $\sqrt{-1}$, if the fraction is not 0.

Biased Exponent		Exponent		Unsigned Fraction	Meaning
Decimal	Binary	Bias = 7	Bias = 8		
0	0000	0	0	0	Represents FP number zero (0.0)
0	0000			≠ 0	Represents a very small FP number called denormal, not typically stored in memory
15	1111			0	Represents infinity (e.g., the result of 1.0 divided by 0.0)
15	1111			≠ 0	Represents an invalid FP number (e.g., the result of computing $\sqrt{-1}$),
1–14	0001–1110	–6 to 7	–7 to 6	Any	Represents a normal FP number

Table 1.3 4-Bit Biased Exponent versus Exponent

The bias value determines which set of real numbers is represented in the computer. As illustrated in Table 1.3, with 4-bit biased exponents, the exponent range for normal FP numbers is between –6 and +7 when *bias* = 7, and between –7 and +6 when *bias* = 8. This means that with *bias* = 7, the 16-bit format can be used to represent more of the large real numbers: the largest *exponent* = 7 and the smallest *exponent* = –6. However, with *bias* = 8, the format can represent more of the small real numbers: the largest *exponent* = 6 and the smallest *exponent* = –7. Modern computers implement 32- and 64-bit Institute of Electrical and Electronics Engineers (IEEE) FP standards, discussed in Chap. 3.

1.1.2 Data Path

Whether we are dealing with unsigned, signed, or FP numbers, the inputs and outputs of digital circuits are in binary. A small digital circuit implements a simple function and generates a single-bit output. A complex circuit, on the other hand, generates results that are multiple bits and may implement one or more functions. A complex digital circuit is often made of a data path and a control unit, as illustrated in Fig. 1.1. For the moment, much of the details in the figure are not shown. However, note that there are multiple paths for data to travel.

Specifically, a data path contains various circuits as modules and collectively performs one or more functions. A module may be an arithmetic type, such as an adder that generates the sum of its two numbers; a selector that outputs one of its several inputs; a register that retains a number, etc. In Fig. 1.1, the data path contains a register file consisting of several registers, an adder, a multiplier, and a selector. It generates either the sum or product of two register values, labeled *value*1 and *value*2, as a result and stores the result in a register.

A controller (or control unit) generates a set of signals, each 1 or 0, that controls the functions of a data path. For example, in the figure, the selector

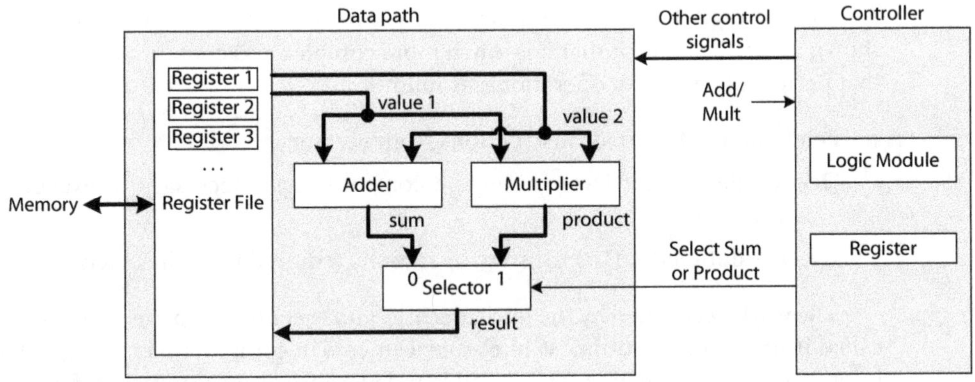

FIGURE 1.1 Block diagram of a complex digital circuit with data path and controller.

outputs either the sum when its control signal is 0 or the product when the signal is 1. A register control signal decides the exact timing when the register loads the value available at its input. In the figure, the initial contents of registers are read from some external module such as memory, and a result may also be stored in memory.

1.1.3 Computer Systems

Figure 1.2 illustrates the block diagram of a computer system known as the **Von Neumann machine**, the basic architecture of virtually every computer ever built. The program instructions and data are stored in memory, and CPU is responsible for accessing instructions and data from memory and executing the instructions.

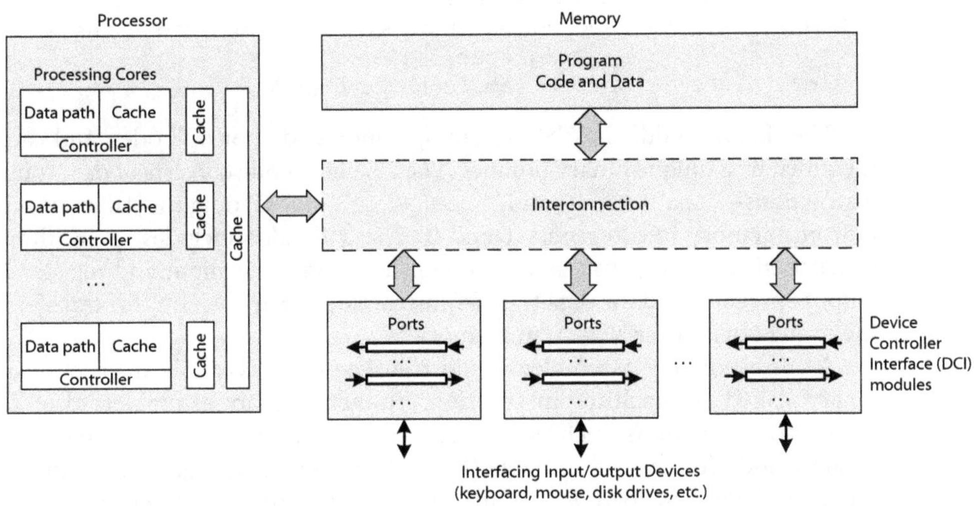

FIGURE 1.2 Block diagram of a computer system known as a Von Neumann machine.

A CPU is a digital circuit with a data path and a control unit similar to the one shown in Fig. 1.1, except that it is much more complex and contains subdata paths that perform three main operations, as follows:

Fetch data path: Loads instructions from memory

Decode data path: Determines the control signals necessary to execute an instruction

Execute data path: Performs the operations required by an instruction

The work performed by the fetch, decode, and execute data paths is collectively called instruction execution. While advancements in computer technologies have improved the performance of both CPUs and memory over the years, performance of CPUs has increased at a faster rate than that of memory. Therefore, the Von Neumann architecture presents a communication bottleneck between a faster CPU and a slower memory.

The execute data path can execute a set of unique instructions, including data transfer instructions that transfer data between a CPU register and memory or an input/output (I/O) device. The set also includes other instructions, such as arithmetic and instructions needed to execute a for-loop, while-loop, subroutine call and return, etc.

A compiler translates high-level program statements to their equivalent assembly instructions. Consider the high-level language program statement "`A = B + C;`", where A, B, and C are variables and represent values stored in memory. Using the data path in Fig. 1.1, a compiler would translate the statement to its equivalent assembly instructions, such as those listed here using an arbitrary syntax:

```
Load R1, B      //Load value of B into register 1 (R1)
Load R2, C      //Load value of C into register 2 (R2)
Add R3, R1, R2 //Add contents of R1 and R2 and store the sum
                //in register 3 (R3)
Store A, R3     //Store sum in memory as A
```

The "Load," "Add," and "Store" are operation codes, or op-codes, and each is identified by a unique binary number. The assembly code consists of data transfer instructions—"`Load R1, B`" and "`Load R2, C`"—for loading the values of B and C from memory into registers 1 and 2. The code also includes the arithmetic instruction "`Add R3, R1, R2`," with registers 1 and 2 as input and register 3 as output operands, and the data transfer instruction "`Store A, R3`" for transferring the content of register 3, as A, in memory.

A corresponding assembler would translate each assembly instruction into binary called a **machine instruction**. In general, an assembler also links static library functions, such as C language routines "`strcpy()`," "`sqrt()`," etc. (if any), and creates an executable (binary) file (e.g., myprogram.exe). Program execution begins by loading the program into memory, and then the processor fetches each instruction from memory to decode the op-code and generate the necessary control signals and to execute the instruction.

The most frequently accessed instructions and data are also kept inside cache memories to increase performance. Cache memories minimize the number of times instructions and data are accessed from the slower memory, reducing average processor wait time to receive instructions and data from the memory.

A **device controller interface** (DCI) contains I/O ports for the processor to communicate with a device such as keyboard or a disk drive. A DCI may also contain other modules, including internal memory to temporarily store a device's data before transferring it to memory, or vice versa, to receive data from memory before transferring it to the device. Finally, an interconnection infrastructure interconnects processor, memory, various DCIs, and other modules designed to facilitate communications with memory and improve the overall performance of the system.

A computer system may also include special and dedicated processors, such as a graphic processing unit (GPU) in personal computers and the digital signal processor (DSP) used in many embedded systems. Both GPUs and DSPs have specialized data paths and controllers and are designed, respectively, for computer graphic and gaming operations and for efficient processing of digital audio and video data.

1.1.4 Embedded Systems

An embedded system is a complete system with both hardware and software built as single or multiple ICs. A simple embedded system is typically called a **microcontroller** and is used in the design of simple devices such as a computer keyboard. A complex embedded system designed as a single chip is known as a system-on-chip (SoC). Handheld devices such as cell phones, digital camcorders, etc. all use embedded systems. These systems are also used in the design of modern DCIs, such as a Universal Serial Bus (USB) **Host Controller Interface** that can interface with multiple different devices.

Depending on the application, an embedded system, in addition to one or more processing units, may contain digital data transmitter/receiver modules and signal conversion modules, such as analog-to-digital (A/D) and digital-to-analog (D/A) converters. An A/D converter converts analog signals—for example, from a microphone—to digital data for digital communication or storage. A D/A converter, on the other hand, converts, for example, digital audio data back to an analog signal before it is fed to speakers.

An embedded system may be implemented as a custom application-specific IC (ASIC), or sometimes for fast prototyping, as a field programmable gate array (FPGA). An FPGA chip contains uncommitted but configurable circuit modules. Modern FPGAs contain CPU, memory, and configurable circuit modules to build an SoC without requiring fabrication labs [3].

1.2 Logic Design

A digital or logic circuit is the implementation of one or more Boolean expressions, where each defines a logical relationship between one or more inputs and a single output. Input and outputs are named by Boolean variables and are called signals, each being either true (T) as 1 or false (F) as 0. Equation (1.2) defines a Boolean or logic expression for a logic circuit with three input signals a, b, and c and one output signal f. The circuit is illustrated as a block diagram in Fig. 1.3(a).

$$f = ((\text{NOT } a) \text{ AND } b) \text{ OR } c \tag{1.2}$$

The AND, OR, and NOT in the expression refer to Boolean logic operators. Transistors are used to implement each Boolean operator as a logic gate. A modern IC uses millions of gates to implement a complex logic circuit such as a processor. An AND gate produces 1 (true) if both of its inputs are 1, an OR produces 1 if either or both of its inputs are 1, and a NOT produces a logic value opposite to that of its input; it outputs 1 if the input is 0 (F) and 0 if the input is 1 (T). For example, when $a = 0$, $b = 1$, and $c = 0$, the logic value of f is determined as follows in Eq. (1.3):

$$\begin{aligned} f &= ((\text{NOT } 0) \text{ AND } 1) \text{ OR } 0 \\ &= (1 \text{ AND } 1) \text{ OR } 0 \\ &= 1 \text{ OR } 0 \\ &= 1 \end{aligned} \tag{1.3}$$

The operators, with the exception of the NOT, can be extended to more than two Boolean variables, and their equivalent logic gates would be implemented with two or more (up to a maximum number) distinct input connections. There are also other gates, such as NAND and NOR, that require fewer transistors to implement. A NAND is logically equivalent to an AND followed by a NOT (AND-NOT logic), and a NOR is equivalent to an OR-NOT. In general, a logic circuit is implemented with either NAND or NOR gates. Section 1.2.2 presents the implementations of NOT, NAND, and NOR gates using transistors. While Fig. 1.3(b) illustrates the implementation of expression f with NOT, AND, and OR gates (called an AND-OR circuit), Fig. 1.3(c) illustrates an equivalent circuit with NAND gates.

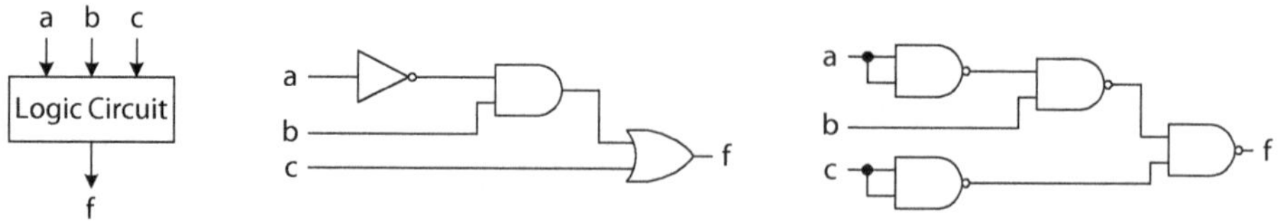

(a) Circuit block diagram (b) AND-OR implementation (c) NAND implementation

FIGURE 1.3 Logic circuit block diagram AND-OR and NAND circuits. vsd.

(a) Logic Values as True or False					(b) Logic Values as 1 (T) or 0 (F)				
a	*b*	*c*	*f*	*g*	*a*	*b*	*c*	*f*	*g*
F	F	F	F	F	0	0	0	0	0
F	F	T	T	T	0	0	1	1	1
F	T	F	T	T	0	1	0	1	1
F	T	T	T	T	0	1	1	1	1
T	F	F	F	F	1	0	0	0	0
T	F	T	T	T	1	0	1	1	1
T	T	F	F	F	1	1	0	0	0
T	T	T	T	T	1	1	1	1	1

Table 1.4 Input Logic Combinations and Corresponding Output Values of Eqs. (1.2) and (1.4)

As illustrated in Table 1.4, there are eight possible combinations for the input values *a*, *b*, and *c* in Fig. 1.3. With *a*, *b*, and *c* concatenated to form a 3-bit number $(abc)_2$, the five numbers $abc = (001)_2$, $(010)_2$, $(011)_2$, $(101)_2$, and $(111)_2$ for which $f = 1$ correspond to five prime numbers 1, 2, 3, 5, and 7, respectively. The remaining three numbers $abc = (000)_2$, $(100)_2$, and $(110)_2$ for which $f = 0$ are not prime numbers. Therefore, Eq. (1.2) defines a logic circuit that outputs 1 each time its 3-bit input as *abc* represents a prime number and 0 otherwise. Table 1.4(*a*) and Table 1.4(*b*) are two different truth table representations for logic expression *f*. The equivalent expression *g* is discussed next.

1.2.1 Circuit Minimization

While a logical relationship may be represented by many equivalent Boolean expressions, the goal is to determine a minimized expression that (1) requires less hardware (i.e., fewer gates, fewer gate inputs, and shorter and fewer connection wires) to implement the expression, and (2) the resultant circuit requires less time to generate results. Boolean algebra is used to minimize complex expressions.

Equation (1.4) describes a logic expression for g, which is equivalent to *f* in Eq. (1.2). More hardware would be required to implement expression *g* than *f*. However, as shown in Table 1.4, the columns associated with the outputs *f* and *g* are identical. This proves Eqs. (1.2) and (1.4) are equivalent; both describe the same function, but Eq. (1.2) is minimized. A circuit that implements Eq. (1.2) would use a lot less hardware and would operate faster than the circuit that implements Eq. (1.4).

$$\begin{aligned} g = {} & ((\text{NOT } a) \text{ AND } (\text{NOT } b) \text{ AND } c) \text{ OR} \\ & ((\text{NOT } a) \text{ AND } b \text{ AND } (\text{NOT } c)) \text{ OR} \\ & ((\text{NOT } a) \text{ AND } b \text{ AND } c) \text{ OR} \\ & (a \text{ AND } (\text{NOT } b) \text{ AND } c) \text{ OR} \\ & (a \text{ AND } b \text{ AND } c) \end{aligned} \tag{1.4}$$

1.2.2 Implementation

Figure 1.4 illustrates the circuit schematic of a NOT gate using a *p*-type and an *n*-type metal-oxide semiconductor field effect (MOSFET) transistor. The schematic is called a CMOS ("C" for complement) circuit because the pMOS and nMOS transistors complement one another; when one transistor is in the OFF position (not conducting), the other is in the ON position (conducting). As illustrated in Fig. 1.4(*a*), when input x is logic 0 (0 V), the pMOS transistor turns ON and the nMOS transistor turns OFF. This changes the value of the output signal f, as expected, to logic 1, as indicated in the schematic by the light showing turned on and the voltmeter reading showing 4.999 V.

In Fig. 1.4(*b*), when input x is logic 1 (5.0 V), the opposite happens. The pMOS turns OFF, the nMOS turns ON, and the output f becomes logic 0, as indicated by the light showing turned off and the voltmeter reading showing 0 V (0.05 microvolts ≈ 0 V).

Compared to old mainframe computers that were built using less power-efficient gates, modern chips are designed using power-efficient CMOS gates. As illustrated in Fig. 1.4, because one of the transistors is always OFF, the CMOS NOT gate primarily uses power only when one transistor is turning ON and the other is turning OFF. This happens each time input x switches from logic 1 to logic 0 or from logic 0 to logic 1 voltage. However, if this switching of x values happens frequently, so will the ON and OFF switching of the transistors, which will cause the NOT gate to consume more power.

As both the number of transistors on the chip and their ON and OFF switching speed increase, more power is required to operate the chip, which in turn dissipates more heat that must be removed by cooling the chip. For instance, the Intel 80386 processor used about 2 W, whereas the 3.3-GHz (gigahertz) Intel Core i7 processor consumes about 130 W (65 times more). As opposed to mainframe and supercomputers of yesteryears, modern computer systems operate with fan-cooling

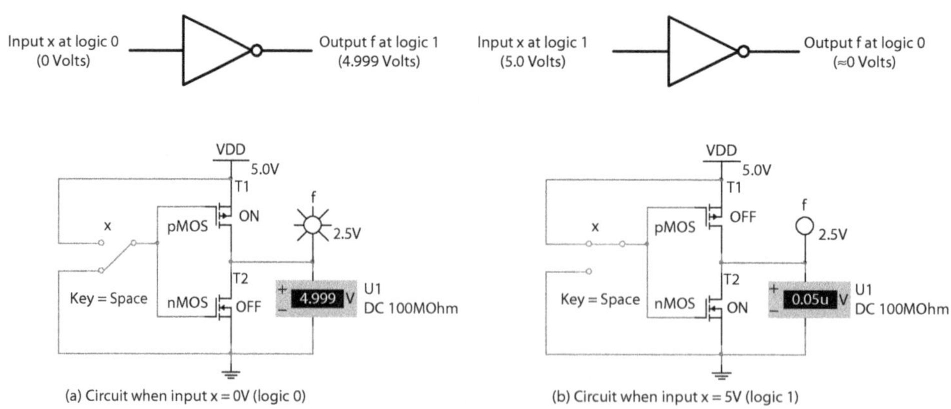

FIGURE 1.4 A CMOS NOT gate circuit schematic and its multisim [4] simulations.

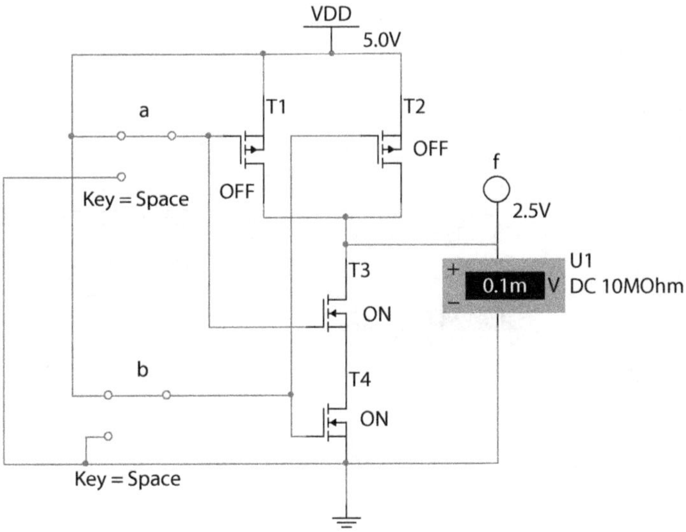

a	b	T1	T2	T3	T4	f
0	0	ON	ON	OFF	OFF	1
0	1	ON	OFF	OFF	ON	1
1	0	OFF	ON	ON	OFF	1
1	**1**	**OFF**	**OFF**	**ON**	**ON**	**0**

Figure 1.5 The circuit schematic of a two-input CMOS NAND gate.

systems. Designers must deal with how much sustained power a chip can consume without exceeding its temperature barrier that can cause malfunction or permanent damage to the chip. Power consumption will be further discussed in Chap. 6.

Figure 1.5 illustrates the circuit schematic of a two-input CMOS NAND gate. Two parallel pMOS transistors and two nMOS transistors in series are connected to the power source and ground, as shown in the figure. The NAND gate outputs logic 0 when both of its inputs a and b are at logic 1, making both the nMOS transistors ON and both the pMOS transistors OFF. A truth table with four possible logic values of inputs a and b, the corresponding logic values of output f, and the transistor operating modes is also shown in the figure.

Likewise, Fig. 1.6 illustrates the circuit schematic of a two-input NOR gate. In this case, two pMOS transistors are connected in series and two nMOS transistors in parallel. The output is logic 0 (or 0) when at least one of the nMOS transistors is ON ($a = 1$, $b = 1$, or both $a = 1$ and $b = 1$). The output is logic 1 (or 1) when both the pMOS transistors are ON and both the nMOS transistors are OFF ($a = 0$ and $b = 0$).

1.2.3 Types of Circuits

Boolean expressions describe both **combinational** and **sequential** circuits. The outputs of a combinational circuit depend only on its current input values. Equations (1.2) and (1.4) both describe a simple combinational circuit. The adder, multiplier, and selector modules in Fig. 1.1, on the other hand, are considered complex combinational circuits; each circuit outputs multiple bits, which are concurrently generated based on their respective current input logic values.

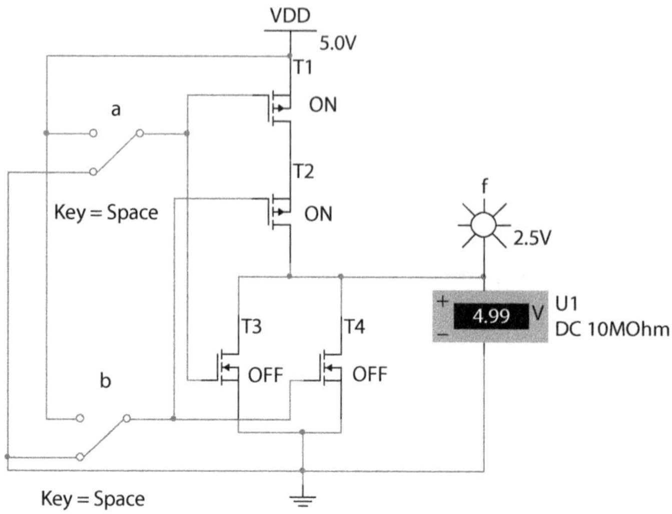

a	b	T1	T2	T3	T4	f
0	**0**	**ON**	**ON**	**OFF**	**OFF**	**1**
0	1	ON	OFF	OFF	ON	0
1	0	OFF	ON	ON	OFF	0
1	1	OFF	OFF	ON	ON	0

FIGURE 1.6 The circuit schematic of a two-input CMOS NOR gate.

In contrast, a sequential circuit retains certain state information based on the previously entered logic values. For example, a counter that outputs 0, 1, 2, 3, etc. is a sequential circuit. It uses a current output (e.g., 2) to generate the next number in the sequence (i.e., 3). The current count is saved internally as the state of the counter. Sometimes, a sequential circuit requires one or more inputs to determine its next state. An up/down counter, for instance, requires a control signal to decide the direction of the count. If the counter's current output is 2, its next output is either 1 if counting down or 3 if counting up.

Other examples of sequential circuits are the registers and control unit in Fig. 1.1. Each register stores a value internally when it is signaled by the controller and retains its current content until the next time that it is signaled. The controller follows a set of states and generates control signals to operate the data path. One or more control signals may be used to control the functions of each module in the data path. In the figure, a control signal is used to operate each of the registers and the selector. If logic 1 is used for register load and 0 for retain, then a register will load the logic values at its input when its control signal is 1, and will retain its current content if the signal is 0.

Suppose the controller in Fig. 1.1 is a three-state controller that performs the following three simple tasks and computes the sum of two numbers entered one at a time:

State 1: Load a value in register 1 from memory.

State 2: Load a value in register 2 from memory.

State 3: Select the sum and store it in register 3.

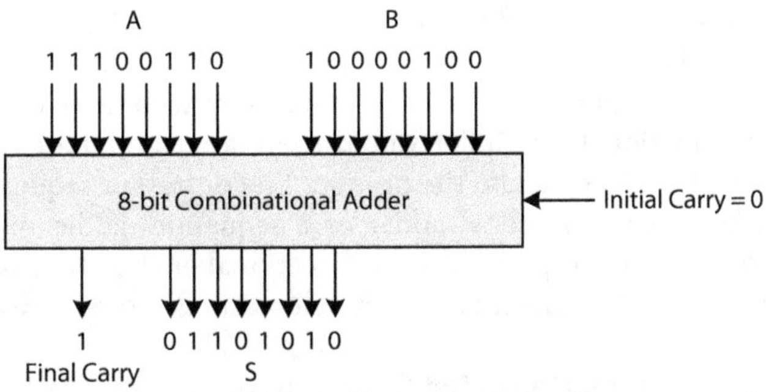

FIGURE 1.7 A combination 8-bit adder.

In general, a circuit that performs, for example, an arithmetic function could be designed either as a combinational or sequential circuit. A combinational arithmetic circuit produces its output bits in one step, with the bits generated in parallel, as illustrated in Fig. 1.7. On the other hand, a sequential arithmetic circuit produces its final output sequentially, using several steps. In each step, the circuit uses the results from a previous step to generate the results of the current step. The process repeats for a fixed number of steps until the final output is produced.

For example, an 8-bit sequential adder could use a single-bit adder repeatedly eight times to generate the final 8-bit sum, as illustrated in Fig. 1.8, much like how we add two numbers by hand, one digit at a time, from right to left. In the figure, during each step, one bit from number *A*, one bit from number *B*, and the carry bit from the previous step are used to generate the next sum bit. The carry bit that is kept internally is used during the next step. The figure also shows the display of one at a time, internally held carry bits. The initial carry bit is assumed to be 0. The clock signal controls the timing of when the next bit of each *A* and *B* numbers

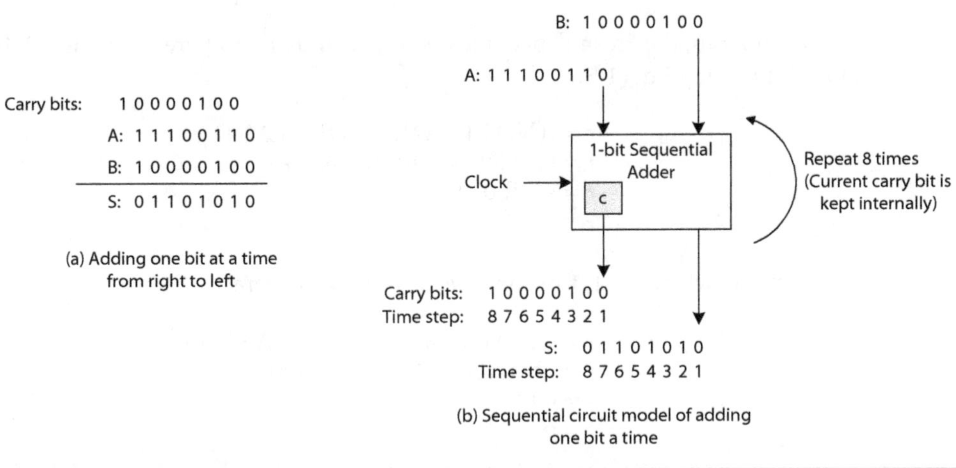

FIGURE 1.8 A sequential 8-bit adder: (*a*) illustrates multibit numbers *A* and *B* are added by hand one bit at a time; (*b*) the multibit addition steps (algorithm) are shown as a sequential circuit.

enters the adder and the next bit of S is captured. After eight steps, the final 8-bit sum S is generated.

A multiplier could also be designed as a combination or sequential circuit. A combinational multiplier would use many combinational adder modules at the same time to generate the product. In contrast, a sequential multiplier could use either a combinational adder or a sequential adder repeatedly to generate the product. In comparison, a combinational arithmetic circuit is always faster than its equivalent sequential circuit, but requires more hardware.

1.2.4 Computer-Aided Design Tools

The rules of Boolean algebra are also implemented in software and incorporated into many logic design CAD tools. Designers typically use an HDL, or more specifically, a register transfer language (RTL), such as Verilog and VHDL, which stands for Very High-Speed Integrated Circuit (VHSIC) HDL, to describe a digital circuit. Although it is possible to enter a Boolean expression with an RTL, designers often use high-level statements, such as "if-else," to describe a circuit. For example, the following statement describes the behavior of the selector module in Fig. 1.1, where x is used to select the output of the adder ("sum") or the multiplier ("product") as "result":

```
if (x == 0)
    result = sum;
else
    result = product;
```

The previous "if-else" statement is equivalent to logic expression in Eq. (1.5), where s, p, and r represent 1 bit of "sum," 1 bit of "product," and 1 bit of "result," respectively:

$$r = ((\text{NOT } x) \text{ AND } s) \text{ OR } (x \text{ AND } p) \tag{1.5}$$

When $x = 0$, the "result" becomes "sum," as illustrated here for 1-bit r, 1-bit s, and 1-bit p using Eq. (1.5).

$$\begin{aligned} r &= ((\text{NOT } 0) \text{ AND } s) \text{ OR } (0 \text{ AND } p) \\ r &= (1 \text{ AND } s) \text{ OR } (0 \text{ AND } p) \\ r &= s \text{ OR } 0 \\ r &= s \end{aligned}$$

Likewise, when $x = 1$, r becomes p, as illustrated here:

$$\begin{aligned} r &= ((\text{NOT } 1) \text{ AND } s) \text{ OR } (1 \text{ AND } p) \\ r &= (0 \text{ AND } s) \text{ OR } (1 \text{ AND } p) \\ r &= 0 \text{ OR } p \\ r &= p \end{aligned}$$

Assuming that *sum*, *value*1, and *value*2 in Fig. 1.1 are each an 8-bit value, the + operator in the HDL statement "sum = value1 + value2" would indicate an 8-bit adder similar to the one shown in Fig. 1.7, where the carry in and out bits are ignored.

An RTL description can be simulated to verify if a circuit is accurately described. The description can then be synthesized (translated) to an equivalent minimized circuit representation called a **net-list** that would be further simulated to verify circuit timing requirements. Finally, using an FPGA, a net-list can be automatically mapped to hardware, creating a circuit.

Combinational and sequential circuits are discussed in Chaps. 2 and 3 and Chaps. 4 through 6, respectively.

1.3 Computer Organization

While logic design deals with circuit description, synthesis, minimization, and simulation, computer organization refers to circuit components and their physical relationship that comprise a processing core (CPU), a processor, memory, and I/O device controller and interface; their interconnection makes up a computer. For example, in Fig. 1.1 a register file, an adder, a multiplier, and a selector are organized into a data path. The control unit and the data path are organized (via a set of specific control signals) to create the desired computing unit that generates either the sum or the product of two numbers. Two CPUs with different internal organizations could execute the same set of instructions. For example, 32-bit Intel and AMD processors execute the same set of instructions, but each has a very different internal organization.

Computer organization is also affected by advancements in computer technologies. The following lists advancements in computer technologies that have changed the organization of a microcomputer (e.g., Fig. 1.2):

- Advancements in data path designs have made CPUs more efficient; modern processing cores (e.g., Intel Core i7) can execute multiple instructions in parallel.
- Advancements in memory technologies and organization, for example, cache and synchronous dynamic random access memory (SDRAM), have reduced the average memory read/write time, allowing a processor to spend its time executing instructions instead of waiting to receive instructions and data from memory.
- Advancements in the I/O device controller interface (e.g., USB 1.0, 2.0, etc.) have simplified personal computing. Almost all devices are now "plug and play" and do not require device installation and system reboot.
- Advancements in system interconnection mechanisms have resulted in even more communication paths among a system's components. A hierarchy of communication paths is used to better organize the interconnection of various components. High-speed communication paths are used between memory and the fastest components, such as a processor and a GPU, and slower communication paths are used to communicate with slower components, such as I/O devices.

However, the limitation on power usage has restricted how fast a processor can operate. For instance, in 2003, the Intel Pentium 4 Xeon processor operated at 3.2 GHz, and in 2010, nearly seven years later, with only a slight increase in speed, the Intel Nehalem Xeon processor operates at 3.33 GHz [1]. Therefore, because of this limitation on power usage, the only way for designers to create more powerful computers, from personal computers to servers to very large-scale systems (e.g., warehouse and cloud computing), is to use multiple processors.

The organization of complex circuits is discussed in Chap. 6, memory design in Chap. 7, CPU design in Chap. 8, and computer design in Chap. 9.

1.4 Computer Architecture

While computer organization deals with how different parts of a computer operate, computer architecture deals with the design of computer arithmetic modules, such as adder and multiplier, and instruction set, and deals with performance improvement concepts, including ways to execute more instructions per second, reduce program's total execution time, and perform more tasks.

1.4.1 Pipelining

The concept of pipelining is similar to a factory assembly line that assembles parts in stages to produce more products (e.g., cars) in less time. The chart in Fig. 1.9 illustrates the working of a simplified car assembly line with three stages. As shown in the figure, once the assembly line is full (one car being worked on in each stage), one car can be produced every 10 minutes, assuming that a car takes 30 minutes to build: 10 minutes to install an engine, 10 minutes to install car doors, and 10 minutes to install car wheels. The more stages and shorter delays per stage there are in the assembly line, the more cars that can be built. For example, suppose the assembly line can be divided into simple stages, each requiring only two minutes work (i.e., 2-minute slots). In this case, in an ideal situation, more than 260,000 cars can be built in 1 year, 1 car every 2 minutes.

When designing a CPU, the pipelining concept is used to organize a CPU's data path into stages to execute programs faster. Consider a program statement "`A = B + C;`", and its equivalent assembly language program as shown:

```
Load R1, B      //Load value of variable B from memory into
                //register R1
Load R2, C      //Load value of variable C from memory into
                //register R2
```

Stage 3: Install wheels			Car1	Car2	Car3	...
Stage 2: Install doors		Car1	Car2	Car3	...	...
Stage 1: Install Engine	Car1	Car2	Car3	...	...	...
Time slots (e.g., 10-minutes slots)	1	2	3	4	5	...

Figure 1.9 A chart showing a simplified car assembly line.

Execute			Load r1, B	Load r2, C	Add r3, r1, r2	Store A, r3
Decode		Load r1, B	Load r2, C	Add r3, r1, r2	Store A, r3	...
Fetch	Load r1, B	Load r2, C	Add r3, r1, r2	Store A, r3	...	...
Time (T)	1	2	3	4	5	6

FIGURE 1.10 A chart illustrating instruction execution in a pipeline fashion.

```
Add R3, R1, R2 //Add the content of R1 and R2 and put the sum
               //in R3
Store A, R3    //Store the sum in memory as variable A
```

The chart in Fig. 1.10 illustrates the execution of the four instructions using a data path consisting of three pipelined stages as follows:

Stage Fetch: Read the next instruction from memory.

Stage Decode: Generate the data path control signals.

Stage Execute: Execute the instruction.

During the time step $T = 1$, instruction "`Load R1, B`" is fetched (read) from memory; during $T = 2$, while the load instruction is being decoded in the Decode stage, "`Load R2, C`" is fetched from memory. During $T = 3$, while "`Load R1, B`" is being executed and "`Load R2, C`" is being decoded, "`Add R3, R1, R2`" is fetched. Starting with $T = 3$, when the pipeline is full and all its three stages are busy operating, one instruction executes every time step between steps 3 through 6, as illustrated in the chart. Each instruction still requires three time steps to complete execution, but Fetch, Decode, and Execute tasks for different instructions are overlapped and operate concurrently. A three-stage pipelined data path consists of three separate and disjointed Fetch, Decode, and Execute data paths.

While a pipelined CPU, in general, can execute more instructions than a nonpipelined CPU can per second, branch instructions and time to access memory can delay the execution of some instructions.

Floating-Point Unit

The pipelining concept also applies to complex arithmetic modules, such as a floating-point unit (FPU) that operates on FP numbers. The execution of an FP instruction requires several arithmetic and shift operations, and if the operations are performed in a pipelined fashion, a program will execute faster. For example, consider the following for-loop statement that operates on arrays of FP numbers:

```
float A [100], B [100], C [100];
int i;
for (i = 0; i < 100; i++)
    C [i] = A [i] + B [i];
```

The for-loop executes 100 FP add instructions. Using a pipelined FPU, these 100 add instructions would require about 100 time-steps to execute. This reduces the total time needed to execute the for-loop, even though each FP ADD instruction, in reality, would require multiple time-steps to execute, as will be discussed in Chap. 3. In this case, the Execute stage shown in Fig. 1.10 would itself be made of several pipelined stages.

1.4.2 Parallelism

Pipelining is applied when tasks are dependent, such as the Fetch, Decode, and Execute tasks required to execute an instruction. Parallelism, on the other hand, is applied when tasks are independent and can be performed in parallel. In addition, both pipelining and parallelism require high availability of inputs and fast handling of outputs. A factory can only produce more products in less time if its assembly line runs efficiently, necessary parts arrive in time, and final products are hauled away quickly. This is similar to data and instructions arriving from memory to the processor faster and computed data quickly stored in memory. This is accomplished by using faster (cache) memory to hold most recently used instructions and data for quick access, while slower but larger and less expensive memory is used to hold different programs and data.

The following sections provide brief descriptions of parallelism techniques when applied to the design of CPUs, processors, and systems.

Single Instruction Multiple Data

As the number of available transistors increased due to advancements in IC technologies, modern processors started to include special instructions that would operate on multiple data items in parallel, thereby increasing performance [5]. An example of these instructions is the streaming single instruction multiple data (SIMD) extension (SSE) instruction set used in Intel processors, or 3DNow instructions used in the AMD processors [6, 7].

Many application areas, such as computer gaming, require multiple arithmetic operations to generate a single result. A virtual game object typically includes thousands of data points, each called a vertex. Moving a virtual game object on a computer screen requires each of its vertices to be repositioned on the screen using a technique called vertex transformation. Each vertex transformation requires several multiplication and addition operations involving the coordinates of the vertex and the rotation angle of the object, as will be illustrated in Chap. 6 using a two-dimensional (2-D) coordinate rotation digital computer (CORDIC) rotation algorithm. The design and simulation of a 2-D CORDIC rotation pipeline data path is also discussed.

Figure 1.11 illustrates an SIMD data path with four multipliers that generate four product terms in parallel. With SIMD architecture, a single instruction operates on multiple data items, thus reducing the total time that would be required to transform a single vertex. With multiple SIMD execution units, which are commonly used in a GPU, it is possible to create more-realistic video games.

Typically, the SIMD capabilities of a general-purpose processor, such as the Pentium IV, are limited to only few data items—not enough for more advanced

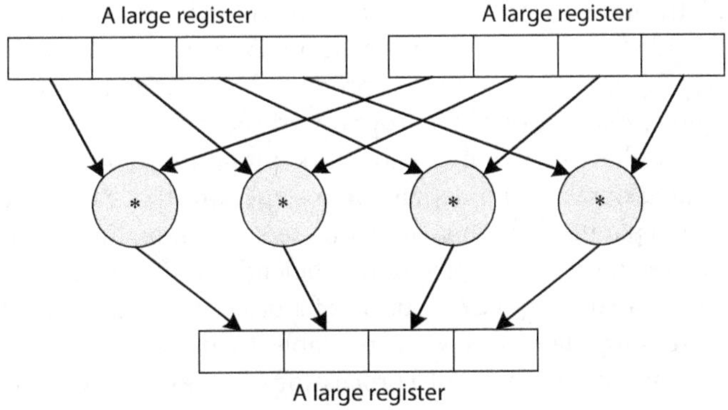

FIGURE 1.11 An SIMD multiply data path.

gaming. GPUs, on the other hand, contain similar and more specialized data paths to increase performance.

Instruction-Level Parallelism

With the increased number of available transistors in an IC, further increase in performance required executing multiple instructions in parallel (i.e., at the same time). In this case, a CPU data path is designed to fetch multiple instructions from memory, decode multiple instructions, and execute multiple independent instructions simultaneously [8].

The list of independent instructions is either determined dynamically by hardware inside the processor, such as with the Intel Core i7, or statically by a compiler, such as in the Intel Itanium-based systems. The assembly instructions of a program for an Itanium-based system are organized by compiler into bundles, with a maximum of three independent instructions in each bundle. The processor fetches, decodes, and executes each bundle of instructions in parallel. However, existing programs must be recompiled to take advantage of Itanium's data path, which many believe was the reason for its demise.

Figure 1.12 illustrates the instruction-level parallelism (ILP) execution of a three-instruction, statically organized instruction bundle. In each time-step three instructions are fetched, decoded, and executed. First, the three instructions

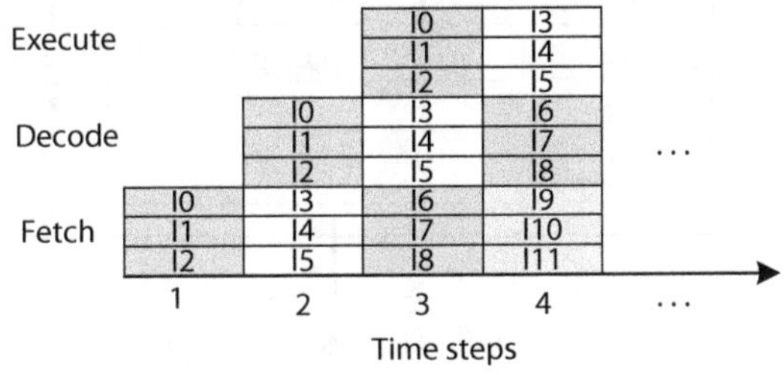

FIGURE 1.12 Instruction-level parallelism illustrating fetching, decoding, and executing three instructions at a time.

I0 through I2 are fetched, and then while these three instructions are being decoded, the next three instructions I3 through I5 are fetched. Starting with time-step 3, nine instructions are processed concurrently. However, data dependencies among instructions in a program prevent full utilization of available hardware.

Studies of various benchmark programs reported indicate that there are a limited number of independent instructions that can be executed at the same time and in parallel [9]. This, therefore, sets a limit on how many transistors can be used to execute a single program efficiently, fully utilizing the available processing hardware during the execution of a program. Thus, while the transistor count was increasing, designers were not able to utilize the excess transistors to further increase the efficiency of a processing core. Sometimes, a core is designed to execute multiple (e.g., two) programs concurrently, called **multithreading**, in order to increase its efficiency. In addition, a common trend has been to utilize the increasing number of transistors and implement multiple identical cores within a single chip, thus creating a multicore processor. ILP and multithreading architectures are further discussed in Chap. 8.

Multicore Processors

Figure 1.13 illustrates a quad-core processor. Each of the cores can execute one or a small number of programs, each called a **thread**, thereby allowing a multicore processor to perform multiple tasks at the same time. Because at any given instance each processing core potentially executes a different instruction and operates on different data items, a multicore processor is said to use multiple instructions and multiple data (MIMD) architecture [5]. Recall that a single SIMD instruction operates on multiple data items at the same time. Likewise, single instruction and single data (SISD) defines the architecture of a single core when it is executing non-SIMD instructions; however, ILP may be used to accelerate SISD execution.

The architecture in Fig. 1.13 is also a representation of a **shared memory** system. In this case, to achieve faster program execution, programmers need to create multiple program threads, known as **multithreaded programming**, by

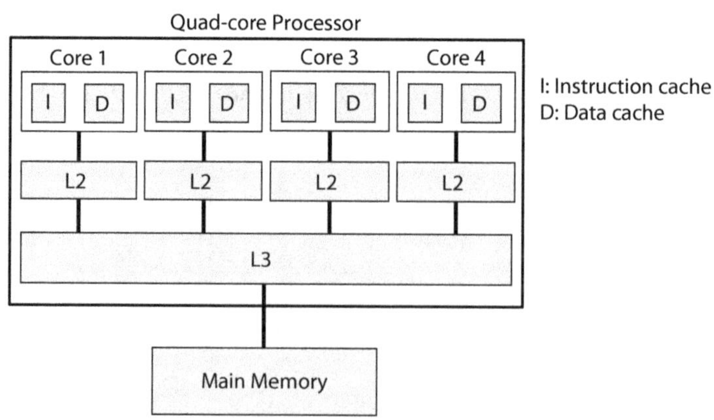

Figure 1.13 A quad-core processor with three levels of cache memories.

partitioning a program's data structures among several threads, which would be executing in parallel and/or concurrently on a multicore processor or, in general, on a shared-memory multiprocessor system. Each thread would operate on subsets of the program data and would need to synchronize and communicate with other threads. While each thread may access its own local variables, all the threads in a multithreaded program can share and operate on globally declared variables. Furthermore, the cores may execute threads from different programs.

Other types of multicore processors include heterogeneous cores that consist of cores with varying complexities to support the processing needs of different applications [10]. For instance, an ILP core could be used for sequential computation, while large SIMD cores could be used to operate on many data items in parallel, suitable for applications that also require data-parallel computations. It is expected that applications requiring both sequential and data-parallel computations will execute faster in a heterogeneous-core processor than in a homogenous-core processor. Typically, a processor uses multiple levels of cache memories (e.g., Fig. 1.2) to facilitate quick access to instructions and data inside the chip, as well as to share data among different cores.

Figure 1.14 illustrates the anatomy of the fourth-generation Intel i7 processor with four processing cores, each with two levels of cache memories (not shown), an L3 shared cache memory, and a graphic processor.

However, as the number of transistors and their switching speed increase, the amount of sustained power consumption and heat dissipation increase. This creates a limitation on how many cores can be implemented in a single processor. The power and cooling requirements can be formulated into a metric known as **thermal design power**, where several mechanisms, such as dynamically reducing the transistor switching speed (i.e., clock frequency), can be used to better balance the power consumption and cooling requirements. While these mechanisms help

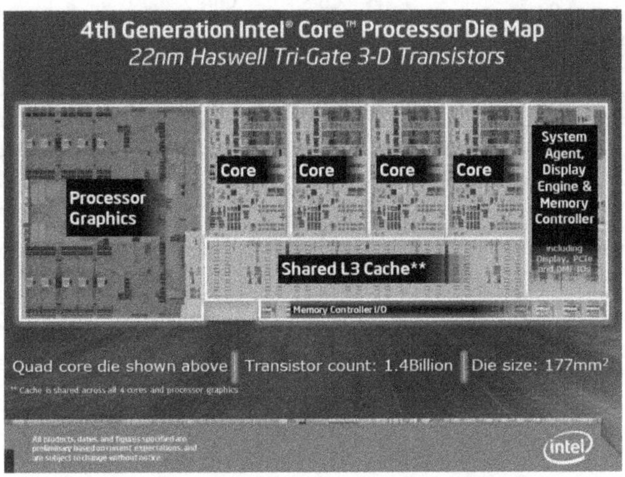

Figure 1.14 The anatomy of the fourth-generation Intel Core i7 processor. (*Source:* With permission of Intel Corporation.)

to increase the performance of a processor at times, further increase in performance requires using multiple processors (each potentially a multicore) to create a multiprocessor system.

Multiprocessor Systems

A multiprocessor system may be designed as either shared memory or **message passing**. In a shared-memory multiprocessor system, the operating system may need to be "thread aware" so that each thread is able to quickly access its local variables. In a message-passing system, however, threads must communicate by sending and receiving messages; there are no shared variables in a message-passing system that threads can access.

A multiprocessor system can execute many threads in parallel and concurrently. Hence, it increases the system's **throughput**, defined as the number of tasks the system can perform per unit time. The number of floating-point operations per second (FLOPS) that is necessary to run a scientific application, such as simulating ocean waves, or the number of Google searches the system can handle at the same time within, say, a couple of seconds, are examples of system throughputs. These systems also use a large amount of memory that may be partitioned among either individual processors or groups of processors that are interconnected.

Figure 1.15 illustrates a three-node, shared-memory multiprocessor system. In this case, each node consists of a two-core processor, memory, and an internode communication interface. Each core can access any of the M0, M1, and M2 memories. Shared-memory multiprocessor systems are used in the design of various servers.

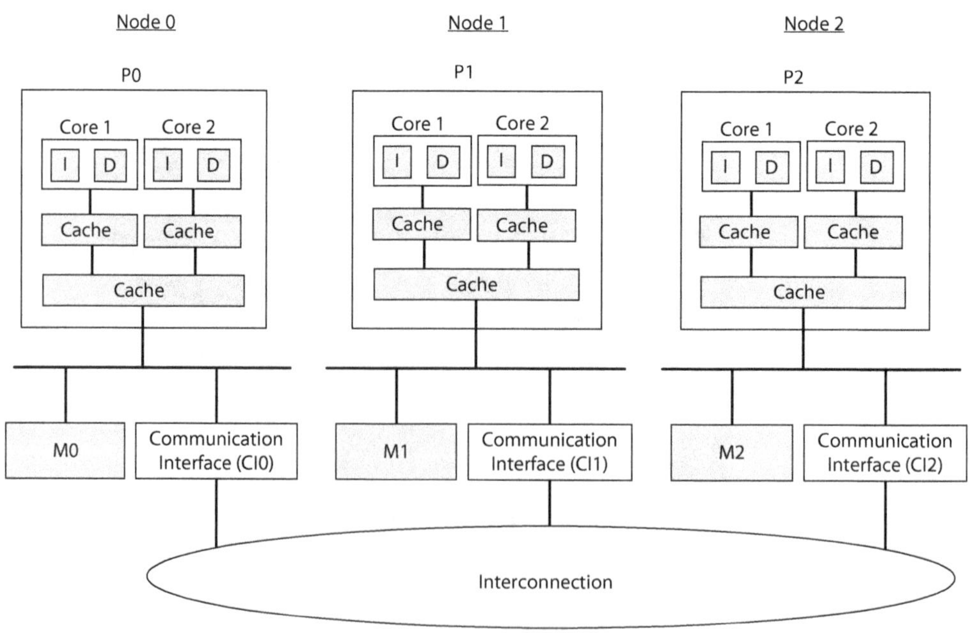

Figure 1.15 The architecture of a three-node, shared-memory multiprocessor system.

On the other hand, networked systems, each one a single processor system or server node, create a message-passing multiprocessor system. The nodes form a cluster and communicate by sending and receiving messages via the network. Finally, warehouse-scale computers are clusters that are designed from thousands of servers. Some warehouse-scale computers are designed as modern supercomputer systems used for scientific computations, requiring extensive FLOPS.

Clusters and warehouse-scale computers provide availability (if one server crashes, others continue to operate), interactive applications (e.g., online shopping, Google, Facebook, banking, etc.), and large-scale storage and computing (e.g., cloud computing). Power distribution and cooling problems are among the challenges facing the designers of very large-scale computing systems. Multicore and shared-memory multiprocessor systems are further discussed in Chap. 10.

1.5 Computer Security

As computer and network security have become more important, new computer architecture concepts are required to build computer systems that are able to detect malicious software (malware) attacks, such as from viruses or spyware. Individuals, government, and business organizations all have digital assets (programs and data) that need protection from attacks, and in many cases, also protection from unauthorized access by employees. Portable devices are additionally subject to physical attacks for the purpose of changing their functions or performing reverse-engineering tasks.

Because ICs are more secure than the hard disk and flash drives and memory, computer architects have become interested in special and general-purpose processors designed and securely built for computer security purposes, such as performing secure data storage, secure communication, secure e-commerce, and secure program execution. Chapter 11 introduces computer architecture for security.

References

1. J. Hennessy and D. Patterson, *Computer Architecture: A Quantitative Approach*, 5th edition, Morgan Kaufmann, 2012.
2. Jennifer Burg, Jason Romney, and Eric Schwarts, "Digital, Sound, and Music: Concepts, Application, and Science," http://csweb.cs.wfu.edu/~burg/CCLI/Documents/Chapter5.pdf.
3. Xilinx FPGA, http://www.xilinx.com/
4. NI Multisim, National Instruments, http://www.ni.com/multisim/.
5. Michael Flynn, Some computer organizations and their effectiveness, *IEEE Transactions on Computers*, Vol., No. 9, Sep. 192, pp. 948-960.
6. Intel SSE4 Programming Reference, white paper http://home.ustc.edu.cn/~shengjie/REFERENCE/sse4_instruction_set.pdf
7. AMD 3DNow, https://refspecs.linuxbase.org/AMD-3Dnow.pdf
8. Ramakrishna Rau and Fisher Josheph, Instruction-level parallel processing: History, overview, and perspective, *Journal of Supercomputing*, 7, 9-50, 1993.
9. David Culler, Jaswinder Pal Singh, and Anoop Gupta, *Parallel Computer Architecture: Hardware/Software Approach*, Morgan Kaufmann, 1999.
10. Mark D. Hill, Amdahl's law in the multicore era, *IEEE Computer*, July 2008, 33-38.

Exercises

1.1. Represent the following numbers as directed:

a. 12 as a 4-bit unsigned number
b. 12 as a 5-bit unsigned number
c. +1 as a 4-bit 2's complement number
d. −1 as a 4-bit 2's complement number
e. −1 as a 5-bit 2's complement number
f. +1 as a 4-bit signed-magnitude number
g. −1 as a 4-bit signed-magnitude number

1.2. Create a table similar to Table 1.1 for 4-bit unsigned, 2's complement, and signed-magnitude numbers.

1.3. What is the 16-bit FP number representation of −5.375 in hex with 1-bit sign, 4-bit biased exponent, and 11-bit fraction, where bias = 7?

1.4. What is the real number equivalent to FP number 0x3400 with 1-bit sign, 4-bit biased exponent, 11-bit fraction, and bias = 7?

1.5. What is the real number equivalent to FP number 0x3400 with 1-bit sign, 4-bit biased exponent, 11-bit fraction, and bias = 8?

1.6. What is the biggest positive FP number that can be represented in 16-bit format using 1-bit sign, 4-bit biased exponent, and 11-bit fraction, where bias is 7?

1.7. What is the biggest positive FP number that can be represented in 16-bit format using 1-bit sign, 4-bit biased exponent, and 11-bit fraction, where bias is 8?

1.8. Do the following assuming 16-bit FP numbers with 4-bit bias exponent, bias = 7, and 11-bit fraction:

a. What real number does an FP number with sign = 0, bias exponent = 1, and fraction = 0 represent?
b. What real number does an FP number with sign = 1, bias exponent = 14, and fraction = $(11111111111)_2$ represent?

1.9. Represent the following real numbers as 16-bit FP numbers with 4-bit biased exponent, bias = 7, and 11-bit fraction:

a. 1.0
b. 0.5
c. 0.25

1.10. Do the following assuming 16-bit FP numbers with 4-bit bias exponent, bias = 8, and 11-bit fraction:

a. What real number does an FP number with sign bit = 0, bias exponent = 1, and fraction = 0 represent?
b. What real number does an FP number with sign bit = 1, bias exponent = 14, and fraction = $(11111111111)_2$ represent?

1.11. Represent the following real numbers as 16-bit FP numbers with 4-bit biased exponent, bias = 8, and 11-bit fraction:

a. 1.0
b. 0.5
c. 0.25

1.12. Draw a data path similar to the one shown in Fig. 1.1 that would be used to generate the result for variable *A* described in the high-level language program statement "`A = A + B;`", where the values of *A* and *B* are brought in from an external memory and are stored in registers before use. Use only two registers. Also, the initial value of *A* and the final result A + B share the same register. Label all the logic modules in the data path and indicate the functions the controller would need to perform. Keep the final computed value $A + B$ in the register.

1.13. CPUs can perform addition, subtraction, multiplication, and division operations. Assuming that a separate module is used for each of the four math functions, draw a data path that can be used to generate the result for variable *A* described in the high-level language statement "`A=A+B*C;`", or "`A=A+B/C;`", where the values of variables *A*, *B*, and *C* are read from an external memory and stored in registers before use. Use no more than three registers. Your data path should be able to generate the result for $A + B * C$ or $A + B/C$. The final value should be kept in a register.

1.14. What is a Von Neumann architecture bottleneck?

1.15. Draw a transistor-level schematic of a three-input CMOS NAND gate and determine its truth table in terms of transistor ON and OFF positions.

1.16. Draw a transistor-level schematic of a three-input CMOS NOR gate and determine its truth table in terms of transistor ON and OFF positions.

1.17. What does the "C" in CMOS stand for, and why is that important?

1.18. What is the difference between pipelining and parallelism architectures? Identify their application areas.

1.19. Explain in which ways the increases in transistor count have influenced computer architecture.

1.20. What is an efficient processing core?

1.21. Explain why a further increase in performance comes from parallel processing.

1.22. Draw an SIMD data path to accelerate the execution of the following for-loop statement:

```
for(i = 0; i< 4; i++)
   sum = sum + array[i];
```

1.23. Draw an SIMD data path to accelerate the execution of the following for-loop statement:

```
for(i = 0; i< 4; i++)
   sum = sum + a[i] * b[i];
```

1.24. There is a limit to ILP. What is the source for this limitation, and how were processor designers able to increase performance beyond ILP?

1.25. Explain the reasons for multiprocessor systems.

1.26. Computer security (understanding security): Selective problems from Exercises 11.1 to 11.11. Also refer to Sec. 11.1. Students may be asked to read this section on their own.

CHAPTER 2

Combinational Circuits: *Small Designs*

2.1 Introduction

Combinational circuits and their application in digital systems were briefly discussed in Chap. 1. In this chapter, we will cover small combinational circuit design techniques that are different from the ones used to design large combinational circuits. Furthermore, here we will limit the number of the input signals to four when minimizing by hand, and we will use a minimization software to minimize designs that require more than four but still a small number (e.g., five or six) of inputs. In contrast, a large combinational circuit requires many more inputs and is implemented using smaller circuit modules. The design methodology of large combinational circuits is discussed in Chap. 3.

The relationships between an output and inputs of a small combinational circuit are defined by the truth table constructed from the description of the design problem. For example, consider the design of a 2-bit unsigned multiplier that multiplies a 2-bit unsigned multiplicant $A = a_1a_0$ by a 2-bit unsigned multiplier $B = b_1b_0$ and produces a 4-bit unsigned product $P = p_3p_2p_1p_0$, as illustrated in Fig. 2.1 using its block diagram. In the figure, uppercase and lowercase letters are used to indicate inputs and outputs that are multiple bits and single bit, respectively. In addition, an arrow with the line crossed indicates multibit inputs or outputs (Fig. 2.1(*a*)). Alternatively, multibit inputs or outputs may be shown with arrows that are bolded (Fig. 2.1(*b*)).

Table 2.1 shows the truth table of the unsigned multiplier. For example, the product of $A = 3 = (11)_2$ and $B = 2 = (10)_2$ is $P = 6 = (0110)_2$, as shown in the truth table. Each of the output bits p_3 to p_0 identifies a logic function in terms of the four inputs a_1, a_0, b_1, and b_0.

A truth table contains all the input logic conditions for which an output bit (e.g., p_0) is 0, as well as all the input conditions for which an output bit is 1. If an

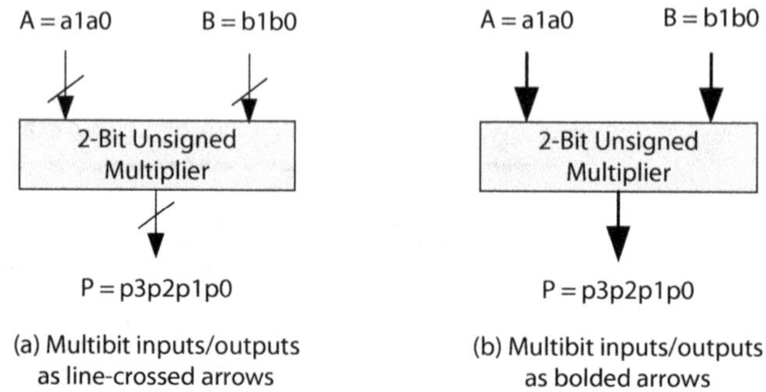

FIGURE 2.1 Block diagram for 2-bit unsigned multiplier circuit; two options to show multibit inputs/outputs.

Inputs				**Outputs**			
a_1	a_0	b_1	b_0	p_3	p_2	p_1	p_0
0	0	0	0	0	0	0	0
0	0	0	1	0	0	0	0
0	0	1	0	0	0	0	0
0	0	1	1	0	0	0	0
0	1	0	0	0	0	0	0
0	1	0	1	0	0	0	1
0	1	1	0	0	0	1	0
0	1	1	1	0	0	1	1
1	0	0	0	0	0	0	0
1	0	0	1	0	0	1	0
1	0	1	0	0	1	0	0
1	0	1	1	0	1	1	0
1	1	0	0	0	0	0	0
1	1	0	1	0	0	1	1
1	1	1	0	0	1	1	0
1	1	1	1	1	0	0	1

TABLE 2.1 Truth Table for 2-Bit Unsigned Multiplication Module

output bit is always 0 or always 1, then the bit is not a function of the inputs and should be deleted from the table. The number of rows in a truth table depends on the number of inputs to the circuit. With three inputs (each 1 or 0), there would be eight possible combinations, or eight rows in the table, and with four inputs, 16 possible rows, as in Table 2.1. In general, with n inputs, there would be 2^n rows in the truth table. There are two ways a truth table can be implemented in hardware:

- The entire truth table can be stored as a look-up-table (LUT). For example, Table 2.1 can be stored in a 16-entry, 4-bit-wide memory.
- A minimal logic circuit is determined for each of the outputs in terms of the inputs.

An LUT has the advantage of not requiring further design steps; the truth table is stored as-is in a memory module inside an integrated chip (IC). The disadvantage of an LUT, however, is twofold:

- Both 0 and 1 output values must be stored, which would require more hardware.
- An LUT is typically slower, as it requires a longer time to read its content.

In contrast, a minimal logic circuit implements either the input logic conditions for which the output is 1 or the input logic conditions for which the output is 0, thus requiring less hardware in terms of fewer gates, gates with fewer inputs, and fewer wire connections.

On the other hand, truth tables stored as LUTs have applications in configurable ICs, such as field programmable gate arrays (FPGAs). Each of the LUT modules in an FPGA chip can be updated with a different truth table to implement a different combinational logic.

In the rest of the chapter we will cover how to convert a truth table to its equivalent logic expressions for NAND or NOR circuit implementation. Manual and algorithmic logic minimization techniques, as well as the use of minimization software, Verilog hardware description language (HDL) circuit descriptions, and computer-aided design (CAD) tools for circuit designs, are discussed with examples. The chapter also presents circuit timing and potential timing hazards. Other gates, such as standard and tristate buffers, are also discussed. These have many applications, including design of modules used in the interconnection architectures. The chapter also includes design examples of some standard small combinational circuit modules.

2.1.1 Signal Naming Standards

Recall that a signal refers to a circuit's input or output as 1 or 0. Each signal name also has a polarity indicator that defines how a signal value, 0 or 1, is interpreted by the circuit. Signal polarities are defined as follows:

- **Active-high signal polarity**—A signal is called active-high if logic 1 represents the active, asserted, or enabled state of a logic condition and 0 represents the inactive, not asserted (deasserted), or not enabled (disabled) state of a logic condition. Typically, a signal name without a prefix or postfix symbol identifies an active-high signal (e.g., x).

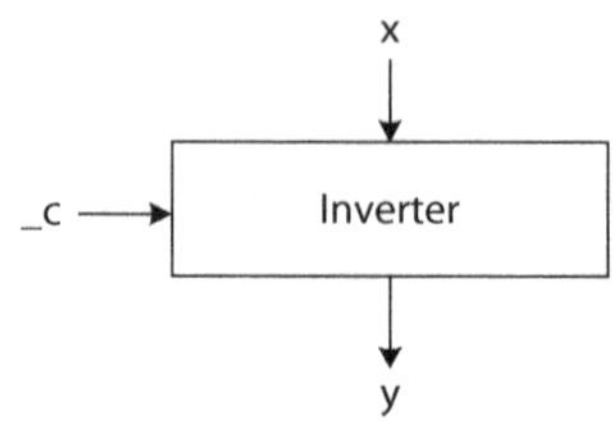

FIGURE 2.2 Block diagram of a 1-bit inverter circuit with an active-low control signal _c.

_c	x	y
0	0	1
0	1	0
1	0	0
1	1	1

TABLE 2.2 Truth Table for 1-Bit Inverter with Active-Low Control Signal _c

- **Active-low signal polarity**—A signal is called active-low if 0 represents the active, asserted, or enabled state of a logic condition and 1 represents the inactive, not asserted, or disabled state of a logic condition. Typically, it has a signal name with a prefix or postfix symbol. For example, *_x*, *x′*, */x*, or *x#* may be used to name an active-low signal.

Unless otherwise stated, we will use an underscore (_) prefix, such as in *_x* or *_X*, to identify an active-low signal or multiple active-low signals, respectively.

Example 2.1. Draw the block diagram of a 1-bit inverter circuit and appropriately label its input and output signals. The circuit inputs 1-bit data and a control signal labeled active-low as *_c*. The circuit outputs the 1-bit data input unchanged when *_c* is not active (deasserted, disabled) and inverts and outputs the data bit when the control signal is active (asserted, enabled). Since the input data is not interpreted by the inverter circuit, both the data and the output, for convenience, are labeled active-high as *x* and *y*, respectively.
Solution: Figure 2.2 illustrates the inverter's block diagram with 1-bit data *x*, an output *y*, and an active-low control signal *_c*. Table 2.2 shows the truth table for the 1-bit inverter. Since *_c* is an active-low signal, $y = \bar{x}$ when $\_c = 0$ (active), and $y = x$ when $\_c = 1$ (not active).

2.2 Logic Expressions

Figure 2.3 displays the symbols and the truth tables of gates used to implement a logic circuit. The AND, OR, NAND, NOR, and NOT gates were discussed in Chap. 1. XOR and XNOR are two-input gates. An XOR gate outputs 0 whenever its inputs are the same and 1 otherwise. The XNOR (XOR-NOT) gate, on the other hand, outputs 1 whenever its inputs are different and 0 otherwise. The XOR and XNOR gates may each be viewed as a 1-bit comparator. The NAND and NOR gates are universal gates because they can be used to implement any logic

NOT

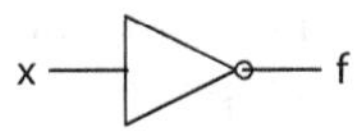

x	f
0	1
1	0

AND

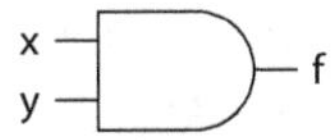

x	y	f
0	0	0
0	1	0
1	0	0
1	1	1

OR

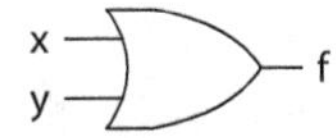

x	y	f
0	0	0
0	1	1
1	0	1
1	1	1

NAND

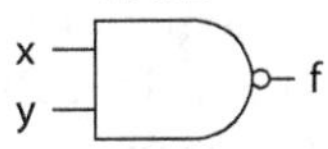

x	y	f
0	0	1
0	1	1
1	0	1
1	1	0

NOR

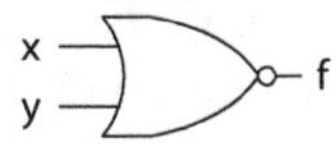

x	y	f
0	0	1
0	1	0
1	0	0
1	1	0

XOR

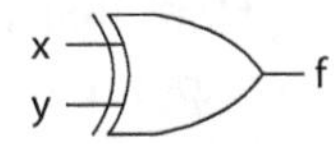

x	y	f
0	0	0
0	1	1
1	0	1
1	1	0

XNOR

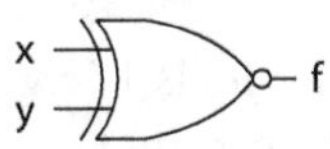

x	y	f
0	0	1
0	1	0
1	0	0
1	1	1

FIGURE 2.3 Primitive logic gate symbols and their truth tables.

expression. In addition, they require fewer transistors to build. Internally, all ICs are implemented with NAND or NOR gates.

Table 2.3 illustrates a truth table of a two-variable function *f* defined to be 1 when $x = 0$ AND $y = 0$ OR when $x = 1$ AND $y = 1$. The function is 0 for other values of *x* and *y*. Equation (2.1) is one way to express the truth table as a Boolean expression where the dot ("·") indicates an AND operator, "+" indicates an OR operator, and bar indicates a NOT operator.

$$f = \overline{x} \cdot \overline{y} + x \cdot y \tag{2.1}$$

x	y	f
0	0	1
0	1	0
1	0	0
1	1	1

TABLE 2.3 A Two-Variable Function

A logic expression is said to be equivalent to its truth table if it produces the exact same truth table when evaluated for all possible input values. For example, Eq. (2.1) generates the same output values in Table 2.3, as illustrated here for all values of x and y:

$$
\begin{array}{ll}
x=0 \text{ and } y=0: & f=\overline{0}.\overline{0}+0.0=1.1+0.0=1+0=1 \\
x=0 \text{ and } y=1: & f=\overline{0}.\overline{1}+0.1=1.0+0.0=0+0=0 \\
x=1 \text{ and } y=0: & f=\overline{1}.\overline{0}+0.1=0.1+0.1=0+0=0 \\
x=1 \text{ and } y=1: & f=1.\overline{1}+1.1=0.0+1.1=0+1=1
\end{array}
$$

Note that Eq. (2.1) contains only the two logic terms $\overline{x}\cdot\overline{y}$ and $x\cdot y$ that correspond to the input logic conditions for which f is 1. Often, an AND operator is implicitly indicated by a null operator (without the "·"), as illustrated in Eq. (2.2).

$$f=\overline{x}\,\overline{y}+x\,y \tag{2.2}$$

2.2.1 Sum of Product Expression

A Boolean expression is called sum of product (SOP) if it defines an output signal f in terms of its input conditions for which f is 1 (e.g., Eq. (2.2)). Each term in the expression is written by ANDing (product) of each distinct input condition, thus forming a **product term**. Output f is 1 if one or more of the product terms is 1; thus, f is the result of ORing (summing) the product terms. In Eq. (2.2), f is 1 when both x and y are zero; thus, $\overline{x}\,\overline{y}=1$, or when both x and y are 1, and thus $xy=1$.

An SOP expression can be translated into its equivalent logic circuit by using NOT, AND, and OR gates. The circuit for Eq. (2.2) is shown as a two-level AND-OR schematic in Fig. 2.4, not counting the NOT gates. The circuit contains seven signals; inputs x and y; output f; and $t=\overline{x}$, $u=\overline{y}$, $g=\overline{t}\,\overline{u}$, and $h=xy$ as intermediate signals. The two AND gates generate the intermediate signals g and h that are fed as input to the single OR gate to produce f.

Using NOT, AND, and OR gates to implement an expression is more intuitive than using the universal NAND or NOR gate.

Implementation of SOP Expressions with NAND Gates

Replacing all NOT, AND, and OR gates of an AND-OR circuit with NAND gates produces an equivalent but NAND-only circuit. This is based on the following DeMorgan's theorems shown in Eq. (2.3). A NAND operator is shown as an AND

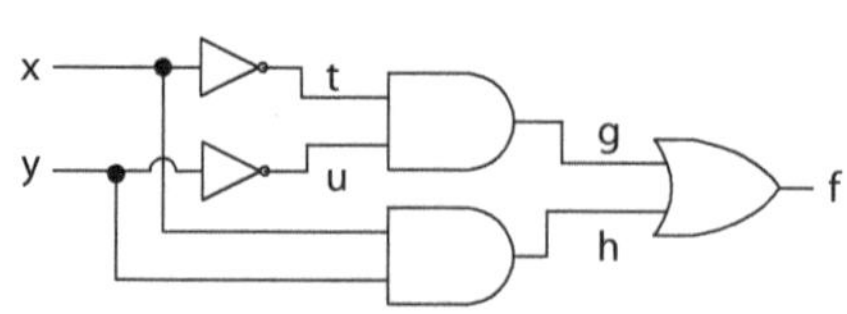

Figure 2.4 The AND-OR gate-level schematic of function $f=\overline{x}\,\overline{y}+x\,y$.

followed by a NOT (e.g., $\overline{xy}$), and a NOR is shown as an OR followed by a NOT (e.g., $\overline{x+y}$):

$$\text{Theorem 1:}\quad \overline{xy} = \overline{x} + \overline{y} \tag{2.3}$$

$$\text{Theorem 2:}\quad \overline{x+y} = \overline{x}\,\overline{y}$$

The two expressions $\overline{xy}$ and $\overline{x}+\overline{y}$ in Theorem 1 would produce the same truth table and thus are said to be equivalent. Likewise, the two expressions $\overline{x+y}$ and $\overline{x}\,\overline{y}$ in Theorem 2 are equivalent. These theorems apply to any number of variables. Theorem 1 states that an OR gate with inverted inputs is logically equivalent to a NAND gate without inverted inputs. Theorem 2 states that an AND operator with inverted inputs is logically equivalent to a NOR operator without inverted inputs.

Consider the AND-OR circuit in Fig. 2.4. The circuit can be changed to all NAND gates using the following procedure:

1. Replace each NOT gate (if any) with its equivalent NAND gate by connecting the inputs of a two-input NAND gate to the single input of the NOT gate, as shown. That is, $y = \overline{x \cdot x} = \overline{x}$.

2. Place two NOT gates on each end of the intermediate signals g and h as shown in Fig. 2.5. This will not change the circuit behavior since the two NOT gates will not alter the value of the original signal (e.g., $\overline{\overline{g}} = g$).
3. Replace each AND-NOT pair with a NAND gate as shown in Fig. 2.6; a NAND gate is equivalent to an AND gate followed by a NOT gate. The inner NOT gates in Fig. 2.5 are shown with small bubbles.

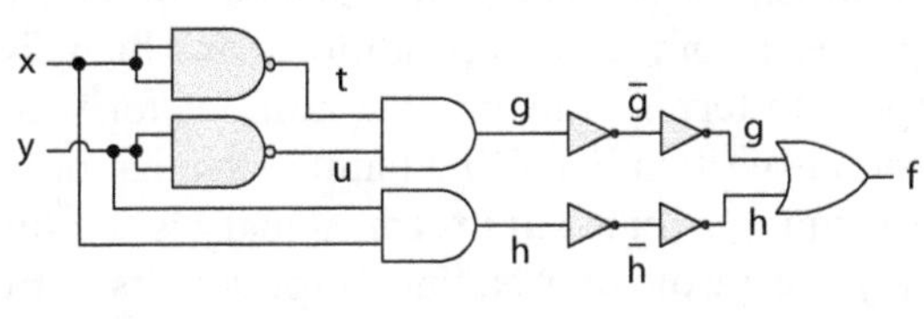

FIGURE 2.5 An AND-OR circuit with two NOT gates added at both ends of the wires used for intermediate signals *g* and *h*.

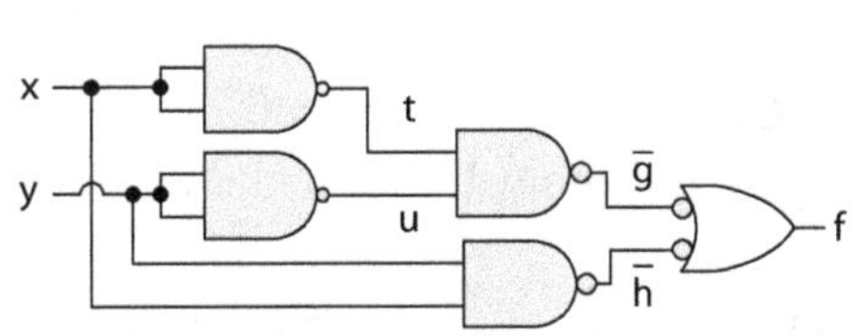

FIGURE 2.6 An AND-OR circuit with added bubbles.

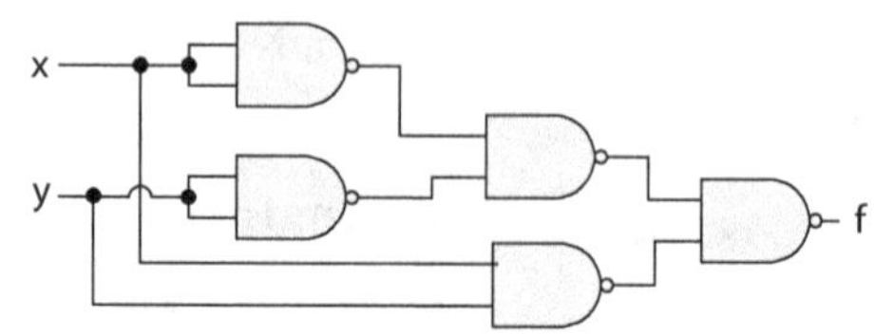

FIGURE 2.7 NAND-only implementation of the SOP expression $f = \overline{x}\,\overline{y} + xy$.

The only gate not yet converted to a NAND gate is the OR gate with inverted inputs. The gate is logically equivalent to a NAND gate according to DeMorgan's Theorem 1: $\overline{gh} = \overline{g} + \overline{h}$.

4. Replace the OR gate with inverted inputs with a NAND gate as shown in Fig. 2.7.

The circuit schematic with all NAND gates in Fig. 2.7 is equivalent to the AND-OR circuit of Fig. 2.4. Alternatively, the following Boolean algebra expresses an SOP expression with only NAND operators:

Recall that $f = \overline{\overline{f}}$;

substitute the SOP expression $\overline{x}\,\overline{y} + xy$ for f to yield:

$$f = \overline{\overline{\overline{x}\,\overline{y} + xy}};$$

apply DeMorgan's Theorem 2 to express $\overline{\overline{x}\,\overline{y} + xy}$ as $(\overline{\overline{x}\,\overline{y}})(\overline{xy})$. The two NAND terms $\overline{\overline{x}\,\overline{y}}$ and $\overline{xy}$ are again NANDed to define f as follows, requiring only NAND operators:

$$f = \overline{(\overline{\overline{x}\,\overline{y}})(\overline{xy})}$$

2.2.2 Product of Sum Expression

A Boolean expression is called a product of sum (POS) if it defines an output signal f in terms of the input conditions for which f is 0. Recall that an SOP defines an output f in terms of its input conditions for which f is 1. Each term in a POS expression is written by ORing (sum) each distinct input condition for which f is 0, thus forming a logic **sum term**; signal f is 0 if one of the sum terms is 0, thus making f the result of ANDing (product) its sum terms. Both SOP and POS expressions for an output signal are equivalent, and they would generate the same truth table. Only one expression (SOP or POS) is needed to describe an output signal. However, POS expressions are not as intuitive to understand as are SOP expressions.

SOP Versus POS

The following rules hold between the SOP and POS expressions of an output signal f:

Rule 1: POS expression of f = Complement of the SOP expression of $\overline{f}$

Rule 2: SOP expression of f = Complement of the POS expression of $\overline{f}$

Using Rule 1, the POS expression for a function *f* is derived by complementing the SOP expression of $\overline{f}$. Everywhere that *f* is 0 in its truth table, $\overline{f}$ is 1, and everywhere that *f* is 1, $\overline{f}$ is 0. Since the SOP of $\overline{f}$ defines $\overline{f}$ in terms of the input conditions for which $\overline{f}$ is 1, complementing the SOP expression of $\overline{f}$ results in a POS expression that defines *f* in terms of the input conditions for which *f* is 0. This is illustrated here:

x	y	f	$\overline{f}$
0	0	1	0
0	1	0	1
1	0	0	1
1	1	1	0

$$\text{SOP of } \overline{f} = \overline{x}\,y + x\,\overline{y};$$

thus,

$$f = \overline{\text{SOP of } \overline{f}} = \overline{\overline{x}\,y + x\,\overline{y}};$$

applying DeMorgan's Theorem 2, $\overline{x+y} = \overline{x}\,\overline{y}$, yields

$$f = (\overline{\overline{x}\,y})(\overline{x\,\overline{y}});$$

applying DeMorgan's Theorem 1, $\overline{xy} = \overline{x} + \overline{y}$, on each of the logic terms yields

$$f = (\overline{\overline{x}} + \overline{y})(\overline{x} + \overline{\overline{y}})$$

or

$$\text{POS of } f = (x + \overline{y})(\overline{x} + y) \tag{2.4}$$

Each of the logic terms $(x + \overline{y})$ and $(\overline{x} + y)$ is a sum (i.e., OR) term, and the sum terms are multiplied (ANDed) to create a POS expression. Rule 2 would be used to determine an SOP expression for *f* from the POS expression of $\overline{f}$.

Alternatively, a POS expression of *f* can be obtained from the dual principle (defined next) applied to an SOP expression of $\overline{f}$:

> **Dual Principle**—The dual of an expression $\overline{x}\,y + x\,\overline{y}$ is equal to $(\overline{x} + y)(x + \overline{y})$ where AND and OR operators are interchanged; ANDs are converted to ORs and ORs are converted to ANDs, but the variable names remain the same in their complemented or uncomplemented form.

In general, the dual of a Boolean algebraic rule, such as $x(y + z) = xy + xz$, results in $x + yz = (x + y)(x + z)$, which is another valid Boolean algebraic rule. If an algebraic rule contains a 1 or 0, such as in $x + 0 = x$, its dual expression requires that 0 be replaced with a 1 and 1 be replaced with a 0, for example, to produce $x.1 = x$ (dual of $x + 0 = x$) as another Boolean algebraic rule.

A simpler way to obtain the POS expression $f = (x + \overline{y})(\overline{x} + y)$ from its truth table is to first obtain the dual expression for $\overline{f} = \overline{x}\,y + x\,\overline{y}$, which would be $(\overline{x} + y)(x + \overline{y})$, and then complement each variable in the dual expression to obtain the POS of $f = (x + \overline{y})(\overline{x} + y)$.

Example 2.2. Find the POS expression of g given that an SOP expression of $\bar{g} = \bar{x}yz + xy\bar{z} + x\bar{y}z$.
Solution: First, obtain the dual expression of $\bar{g}$ as:

$$\text{Dual of} \quad \bar{g} = (\bar{x} + y + z)(x + y + \bar{z})(x + \bar{y} + z)$$

then complement (NOT) each variable in the dual expression to obtain the POS expression of g:

$$\text{POS of} \quad g = (x + \bar{y} + \bar{z})(\bar{x} + \bar{y} + z)(\bar{x} + y + \bar{z})$$

In summary, two different methods to determine a POS expression of a function f from an SOP expression of $\bar{f}$ were presented, as follows:

Method I: Use $f = \overline{SOP\ of\ \bar{f}}$ and then apply DeMorgan's theorems.

Method II: Find the dual expression for the SOP of $\bar{f}$ and then complement each variable.

Figure 2.8 presents the OR-AND circuit for the POS in Eq. (2.4). An OR-AND circuit, similar to an AND-OR circuit, is a two-level circuit not counting the initial NOT gates (if any). The intermediate outputs of two concurrently operating OR gates are then fed to the single AND gate to produce f.

Implementation of POS Expressions with NOR Gates

Replacing all the NOT, OR, and AND gates of an OR-AND circuit with NOR gates produces an equivalent NOR-only circuit. This is based on DeMorgan's Theorem 2 that states $\bar{x}\,\bar{y} = \overline{x + y}$. Consider the OR-AND circuit in Fig. 2.8. The circuit can be designed using all NOR gates with the following procedure:

1. Replace each NOT gate (if any) with its equivalent NOR gate by connecting the inputs of a NOR gate to the single input of the NOT gate as shown. That is, $y = \overline{x + x} = \bar{x}$.

2. Place two NOT gates on each end of the intermediate signals m and n as shown in Fig. 2.9. This will not change the circuit behavior since the two NOT gates will not alter the output of the original signal (e.g., $\bar{\bar{m}} = m$).
3. Replace each OR-NOT pair with a NOR gate as shown in Fig. 2.10; a NOR gate is equivalent to an OR gate followed by a NOT gate.
   The only gate symbol not yet converted to a NOR gate is the AND gate with inverted inputs (shown with bubbles). The gate symbol is logically equivalent to NOR according to DeMorgan's Theorem 2, $\overline{m + n} = \bar{m}\,\bar{n}$.

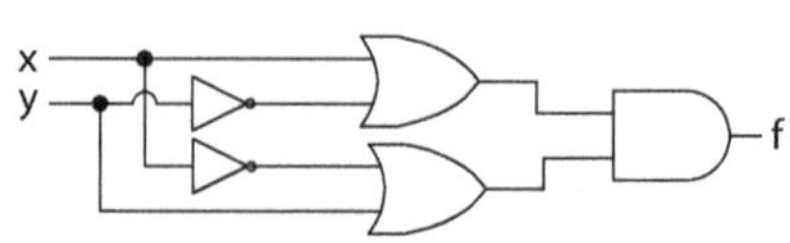

FIGURE 2.8 The AND-OR gate-level schematic of POS expression $f = (x + \bar{y})(\bar{x} + y)$.

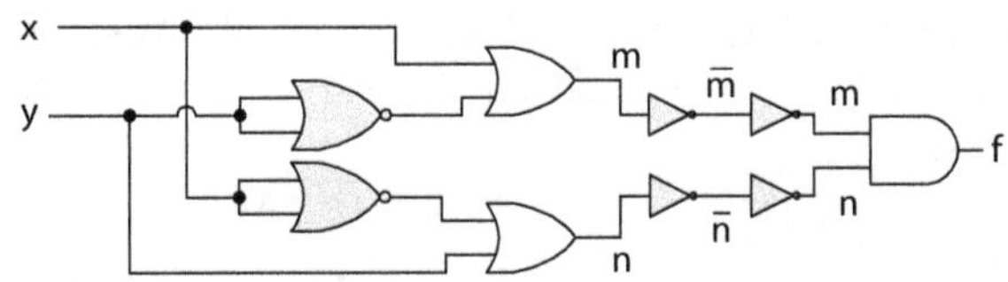

FIGURE 2.9 An OR-AND circuit with two NOT gates added at both ends of the wires used for intermediate signals *m* and *n*.

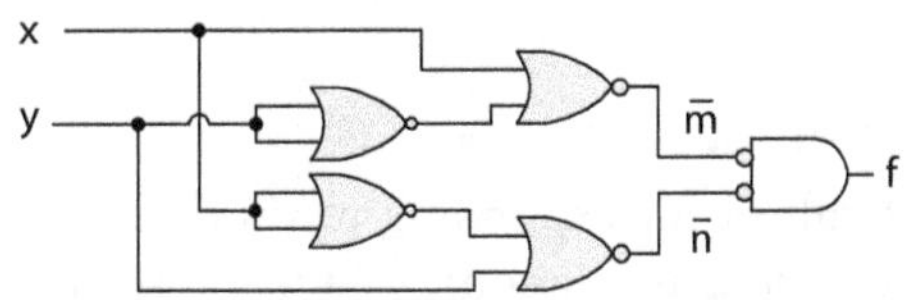

FIGURE 2.10 POS circuit with added bubbles.

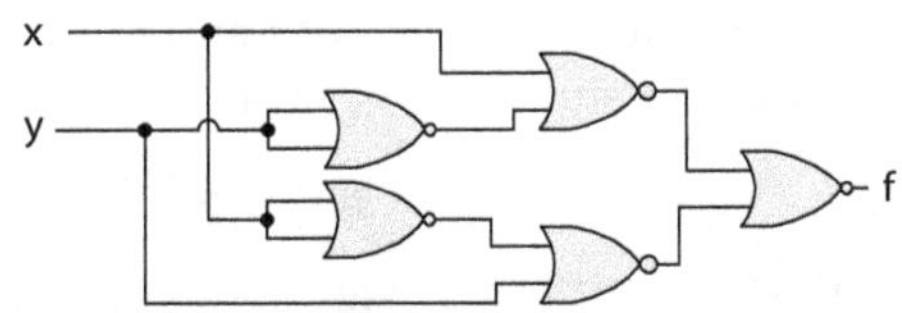

FIGURE 2.11 NOR-only implementation of the POS expression $f = (x + \overline{y})(\overline{x} + y)$.

4. Replace the AND gate with inverted inputs with a NOR gate as shown in Fig. 2.11.

The following Boolean algebra alternatively expresses a POS expression with NOR-only operators:

$$\begin{aligned} f &= (x+\overline{y})(\overline{x}+y) \\ &= \overline{\overline{(x+\overline{y})(\overline{x}+y)}} \\ &= \overline{\overline{(x+\overline{y})}+\overline{(\overline{x}+y)}} \end{aligned} \tag{2.5}$$

The logic terms $\overline{(x+\overline{y})}$ and $\overline{(\overline{x}+y)}$ represent NOR terms, which are NORed to produce *f*. In summary, the SOP and POS expressions of a function are equivalent, and each would produce the same truth table. Therefore, use SOP expressions for NAND-only and POS expressions for NOR-only circuits.

2.3 Canonical Expression

An expression is called canonical if each logic term contains all of the input variables either in their complemented or uncomplemented forms. For example, a two-variable $f = \overline{x}\,\overline{y} + x\,y$ is a canonical SOP expression. Both of the product terms in the expression include both variables x and y either in their complemented or uncomplemented forms. Similarly, a two-variable $f = (x+\overline{y})(\overline{x}+y)$ is an example of a canonical POS expression. A noncanonical expression contains one or more logic terms that do not include all the variables. For example, a three-variable SOP expression

$$g = x\overline{y} + \overline{x}z + xyz$$

is a noncanonical expression since the logic term $x\overline{y}$ is missing variables z and $\overline{z}$ and $\overline{x}z$ is missing variables y and $\overline{y}$. A given noncanonical SOP or POS expression may or may not be minimal; however, it can be first converted to its equivalent canonical expression and then minimized using the minimization techniques presented below.

2.3.1 Min-Terms

The input values corresponding to product terms are called the min-terms. For example, consider the canonical SOP expression $f = \overline{x}\,\overline{y} + x\,y$ with two product terms $\overline{x}\,\overline{y}$ and $x\,y$. The two product terms of f correspond to input values $x = 0$ and $y = 0$ or $(00)_2 = 0$ if x and y are concatenated, and $x = 1$ and $y = 1$ or $(11)_2 = 3$. The 2 and 3 are called the min-terms of f and are written as follows using the Greek symbol Σ (sigma) for sum to indicate an SOP:

$$f(x, y) = \Sigma(2, 3)$$

Using min-terms, it is a straightforward process to write the canonical SOP expression for an output. If the min-terms are given in decimal numbers, they are first converted to binary and then the product terms are determined from the binary numbers, as illustrated here for an arbitrary function g:

$$g(x, y, z) = \Sigma(0, 1, 6, 7)$$
$$g(x, y, z) = \Sigma((000)_2, (001)_2, (110)_2, (111)_2)$$
$$g = \overline{x}\,\overline{y}\,\overline{z} + \overline{x}\,\overline{y}\,z + x\,y\,\overline{z} + xyz$$

Note that for an output variable, its truth table, its list of min-terms, and its canonical SOP expression are three equivalent representations.

2.3.2 Max-Terms

Likewise, the input values corresponding to sum-terms are called max-terms. Max-terms are also written as integer numbers and identified by the Greek symbol Π (pi) for product to indicate a POS. Each max-term corresponds to a sum-term in the canonical POS expression. For example, the expression

$$f(x, y) = \Pi(0, 1)$$

describes f in terms of its max-terms; f is 0 when its two inputs x and y concatenated equal $(00)_2 = 0$ or $(01)_2 = 1$.

The max-terms of a function f are the min-terms of its complement function $\overline{f}$ and vice versa. The following steps 1 to 3 illustrate how to determine the canonical POS expression of an arbitrary function h from its list of max-terms. Steps i and ii are listed for convenience, and they are used to determine the SOP expression of $\overline{h}$ from its min-terms.

1. $h(x, y, z) = \prod(0, 1, 6, 7)$

   $h(x, y, z) = \prod(000, 001, 110, 111)$

2. $h(x, y, z) = \overline{\text{SOP of } \bar{h}}$

   $h(x,y,z) = \overline{\bar{x}\bar{y}\bar{z} + \bar{x}\bar{y}z + xy\bar{z} + xyz}$

3. $h(x,y,z) = (\overline{\bar{x}\bar{y}\bar{z}})(\overline{\bar{x}\bar{y}z})(\overline{xy\bar{z}})(\overline{xyz})$

   POS of $h(x,y,z) = (x + y + z)(x + y + \bar{z})$
   $(\bar{x} + \bar{y} + z)(\bar{x} + \bar{y} + \bar{z})$

i. $\bar{h}(x, y, z) = \Sigma(0,1,6,7)$

   $\bar{h}(x, y, z) = \Sigma(000, 001, 110, 111)$

ii. $\bar{h}(x, y, z) = \bar{x}\bar{y}\bar{z} + \bar{x}\bar{y}z + xy\bar{z} + xyz$

Steps 2 and 3, however, may be replaced using the dual principle (i.e., Method II), as follows:

> Apply the principle of duality to the SOP expression of $\bar{h}$ to determine its dual expression $(\bar{x} + \bar{y} + \bar{z})(\bar{x} + \bar{y} + z)(x + y + \bar{z})(x + y + z)$; then complement each of the variables in the dual expression to obtain the POS expression of h.

Again, note that for an output variable, its truth table, its list of max-terms, and its canonical POS expression are three equivalent representations.

2.4 Logic Minimization

As was discussed in Chap. 1, it is important that a minimum number of gates and each gate with a minimum number of inputs are used to generate an output signal. A minimal SOP or POS expression contains the least number of terms, and each term has the least minimum number of variables. Figure 2.12 illustrates the advantage of logic minimization. In both cases, circuit implementations of the canonical expressions require more gates and, therefore, more transistors and wires.

In Fig. 2.12, the circuit for the SOP canonical expression would require four 3-input AND gates, three NOT gates, and one 4-input OR gate, or a total of eight NAND gates, including the NOTs, each with 4 or fewer inputs. On the other hand, its equivalent minimal SOP expression would require only two 2-input AND gates, one NOT gate, and one 2-input OR gate, or a total of four NAND gates, each with fewer inputs. Likewise, the circuit for the canonical POS expression of f would require a total of eight NOR gates versus five NOR gates, each with fewer inputs.

Implement with NAND gates

Canonical SOP: $f = \bar{x}\bar{y}z + \bar{x}yz + xy\bar{z} + xyz$

Minimal SOP: $f = \bar{x}z + xy$

Implement with NOR gates

Canonical POS: $f = (x + y + z)\,(x + \bar{y} + z)\,(\bar{x} + y + z)\,(\bar{x} + y + \bar{z})$

Minimal POS: $f = (x + z)\,(\bar{x} + y)$

x	y	z	f
0	0	0	0
0	0	1	1
0	1	0	0
0	1	1	1
1	0	0	0
1	0	1	0
1	1	0	1
1	1	1	1

FIGURE 2.12 SOP and POS minimal vs. canonical expressions.

As was discussed earlier, canonical SOP or POS expressions are easily derived from min-terms or max-terms, respectively. The min-/max-terms are read directly from truth tables. It is also equally straightforward to create a truth table from a given canonical SOP or POS expression; however, this is not recommended and not necessary.

Determining min-terms or max-terms from a given noncanonical expression is not a straightforward process. One option, which is not recommended, is to evaluate the expression for every input signal condition. For example, to construct the truth table for $f = y(\overline{x} + xz)$, one has to evaluate the function for all the possible 3-bit values that make up the bit values of inputs x, y, and z. For example, $x = 0$, $y = 1$, and $z = 0$ yields $f = 1$; $x = 1$, $y = 1$, and $z = 0$ yields $f = 0$; and so on to construct the entire truth table. The list of min-terms or max-terms of f is then read from the table.

An alternative option, which is recommended, is to use Boolean algebra and convert a noncanonical expression to its canonical expression and then directly convert the canonical expression to its list of min-terms, if the expression is an SOP, or its list of max-terms, if the expression is a POS. This is done by reversing the minimization steps, as illustrated here.

$f = y(\overline{x} + xz)$	Noncanonical
$f = \overline{x}y + xyz$	Distribute y
$f = \overline{x}y(z + \overline{z}) + xyz$	Insert the missing z and $\overline{z}$ in the first term
$f = \overline{x}yz + \overline{x}y\overline{z} + xyz$	Canonical SOP

Use the following steps to convert a noncanonical POS of a function f to its corresponding canonical expression:

1. Convert the noncanonical POS of f to its noncanonical SOP of $\overline{f}$ using SOP $\overline{f} = \overline{POS\ f}$.
2. Determine the canonical SOP of $\overline{f}$ from its noncanonical expression.
3. Convert the canonical SOP of $\overline{f}$ to its corresponding canonical POS of f using POS $f = \overline{SOP\ \overline{f}}$.

2.4.1 Karnaugh Map

A Karnaugh map (K-map) is a graphical technique used to identify and eliminate redundancies in canonical expressions and obtain one or more equivalent minimal expressions. Two min-terms or max-terms with binary representations that differ in only one bit can be simplified to one less variable. For example, with three variables x, y, and z, two min-terms $2 = (010)_2$ and $3 = (011)_2$ differ in the bit corresponding to variable z. The terms minimizes to $\overline{x}y$ as illustrated here:

$$\overline{x}y\overline{z} + \overline{x}yz = \overline{x}y(\overline{z} + z)$$
$$= \overline{x}y$$
$$\text{since} \quad \overline{z} + z = 1 \quad \text{and} \quad \overline{x}y.1 = \overline{x}y$$

Likewise, two max-terms with binary representations that differ in only one bit simplify to a logic term with one less variable. Consider the following canonical POS expression consisting of two max-terms $2 = (010)_2$ and $3 = (011)_2$ that differ in only one bit:

$$
\begin{aligned}
&(x+\overline{y}+z)(x+\overline{y}+\overline{z}) & \\
&=((x+\overline{y})+z)((x+\overline{y})+\overline{z}) & \text{Regroup smaller terms and distribute}\\
&=(x+\overline{y})(x+\overline{y})+(x+\overline{y})\overline{z}+z(x+\overline{y})+z\overline{z} & \text{Simplify and factor out}\\
&=(x+\overline{y})+(x+\overline{y})(\overline{z}+z)+0 & \text{Simplify}\\
&=(x+\overline{y})+(x+\overline{y}) & \text{Simplify}\\
&=(x+\overline{y}) &
\end{aligned}
$$

A K-map organizes the min-/max-terms in such a way that any two terms with binary representations that differ in only one bit become physically adjacent in the map, making it easier to identify such terms. For example, with three variables *x*, *y*, and *z*, the eight possible min-/max-terms are organized as either a 2 × 4 or 4 × 2 map, as shown in Fig. 2.13(*a*) and Fig. 2.13(*b*), respectively. Each cell in a K-map represents a min-/max-term, and is identified from the row and column labels in binary. For example, row 0 and column 00 identify the cell for the min-/max-term 0; row 0 and column 01 identify the min-/max-term 1; etc.

Note that in the figure, the adjacent rows and columns are labeled with binary numbers that differ in only one bit. The two columns with binary labels 00 and 10 in Fig. 2.13(*a*) are also considered adjacent; so are rows with labels 00 and 10 in Fig. 2.13(*b*). This makes it easier to visually identify two min-/max-terms with binary representations that differ in only one bit. Each term is adjacent to its physically east, west, north, or south term. For example, in Fig. 2.13(*a*), term $0 = (000)_2$ is adjacent to its west term $2 = (010)_2$, to its east term $1 = (001)_2$, and to

yz:	00	01	11	10
x: 0	0	1	3	2
1	4	5	7	6

(a) A 2 × 4 organization

z:	0	1
xy: 00	0	1
01	2	3
11	6	7
10	4	5

(b) A 4 × 2 organization

Figure 2.13 A three-variable K-map organization; each cell represents a min- or max-term; choose either organization (*a*) or (*b*).

	yz: 00	01	11	10
wx: 00	0	1	3	2
01	4	5	7	6
11	12	13	15	14
10	8	9	11	10

FIGURE 2.14 A four-variable K-map; cells are numbered by their corresponding min-/max-terms.

its south term $4 = (100)_2$. Term $(000)_2$ has 1-bit difference with $(001)_2$, $(010)_2$, and $(100)_2$. Figure 2.14 illustrates a four-variable K-map. Term 0, for example, is adjacent to terms 1 (east), 2 (west), 4 (south), and 8 (north).

The K-map for function $g(x, y, z) = \Sigma(2, 6, 7)$ is shown here with 1 in each corresponding min-term cell. The remaining cells correspond to max-terms and are left blank.

yz:	00	01	11	10
x: 0				1
1			1	1

In the K-map, min-term $2 = (010)_2$ is adjacent to min-term $6 = (110)_2$ and min-term 6 is adjacent to min-term $7 = (111)_2$. Min-term 2, however, is not adjacent to min-term 7 since binary $(010)_2$ and $(111)_2$ differ in two bits. A K-map for POS $g(x, y, z) = \Pi(0, 1, 3, 4, 5)$ is shown here:

yz:	00	01	11	10
x: 0	0	0	0	
1	0	0		

2.4.2 K-Map Minimization

A K-map always produces a minimal expression. A function may have more than one minimal expression, yet they are equivalent. Since any pair of adjacent terms reduces to a simplified expression with one less variable, multiple adjacent terms can result in an even simpler logic term with fewer variables. This is illustrated in Fig. 2.15 using a K-map with four min-terms: 2, 3, 6, and 7.

The min-term $(010)_2$ is adjacent to min-term $(011)_2$ and $(110)_2$, the term $(011)_2$ is adjacent to $(010)_2$ and $(111)_2$, and the min-term $(110)_2$ is adjacent to

yz:	00	01	11	10
x: 0			1	1
1			1	1

FIGURE 2.15 A K-map with four min-terms: 2, 3, 6, and 7.

$(010)_2$ and $(111)_2$. Using Boolean algebra, these min-terms simplify to y as follows:

$$
\begin{aligned}
\Sigma(2,3,6,7) &= \bar{x}y\bar{z} + \bar{x}yz + xy\bar{z} + xyz && (2.6)\\
&= \bar{x}y(\bar{z}+z) + xy(\bar{z}+z) && \text{Factor out smaller terms and simplify}\\
&= \bar{x}y + xy && \text{Factor out } y \text{ and simplify}\\
&= y(\bar{x}+x) && \text{Simplify}\\
&= y
\end{aligned}
$$

The following steps use Eq. (2.6) and Fig. 2.15 as an example to illustrate the K-map minimization technique:

1. Compare the two column labels 11 and 10 that are associated with the group of min-terms $(010)_2$, $(011)_2$, $(110)_2$, and $(111)_2$. The bit label associated with variable z changes, while the bit label associated with variable y remains unchanged; thus, z should be eliminated. (Algebraically, two smaller terms $\bar{x}y$ and xy are factored out to eliminate z.)
2. Compare the two row labels. Signal x changes, while the bit label associated with y (the remaining variable) remains unchanged; thus, x should be eliminated. (Algebraically, y is factored out to eliminate x.)
3. Write only the signals that were not deleted (only y in this case) as the result of the simplification in steps 1 and 2, but write the signal either as y if its bit label is 1, or as complement $\bar{y}$ if its bit label is 0. This produces y as the minimal expression for this group of min-terms, which was also illustrated in Eq. (2.6) using Boolean algebra.

Using K-maps eliminates the need to apply Boolean algebra to reduce a canonical to its equivalent minimal expression. Consider the following K-map with five min-terms: 1, 2, 3, 6, and 7.

yz	00	01	11	10
x: 0		1	1	1
1			1	1

It was just illustrated that the group of adjacent min-terms 2, 3, 6, and 7 reduces to y. For the min-term 1 = $(001)_2$, the only adjacent min-term is 3 = $(011)_2$. By comparing the columns 01 and 11, y changes and thus is eliminated from the resultant logic term $\bar{x}z$, which corresponds to $x = 0$ and $z = 1$. Therefore, the final minimal expression for min-terms 1, 2, 3, 6, and 7 is

$$y + \bar{x}z \tag{2.7}$$

Note that min-term 3 was used twice, once in the grouping of min-terms 2, 3, 6, and 7, and again in the grouping of min-terms 1 and 3. This redundant use of a term in the grouping of min-terms is based on the Boolean algebra rule $x = x + x$

(or $x.x = x$ for max-terms). The rules state that repeated terms in a logic expression do not change its truth table, but they do help create larger groups of adjacent terms that eliminate more variables in the final simplified expression. The following Boolean algebra illustrates this point:

Given a canonical SOP expression $\bar{x}\,\bar{y}z + \bar{x}y\bar{z} + \bar{x}yz + xy\bar{z} + xyz$,

repeat the logic term $\bar{x}yz$ associated with min-term 3 once to yield

$$(\bar{x}\,\bar{y}z + \bar{x}y\bar{z} + \bar{x}yz + xy\bar{z} + xyz) + \bar{x}yz;$$

regroup the terms using the adjacency property:

$$(\bar{x}y\bar{z} + \bar{x}yz + xy\bar{z} + xyz) + (\bar{x}\,\bar{y}z + \bar{x}yz);$$

minimize each group to yield the final minimal SOP expression $y + \bar{x}z$.

Without a K-map, it is difficult to know which terms must be repeated and how many times. A minimal expression is derived by making sure that each grouping of the terms in the K-map contains a maximum number of adjacent terms. The K-map minimization rules are summarized in the following section.

K-Map Minimization Rules

1. Min-/max-terms that differ in only one bit are adjacent, and they are said to form an **implicant**. A K-map is assumed to wrap around on both sides; for example, column 00 is also adjacent to column 10.
2. A set of implicants may be combined to form a large group called a **prime implicant**. The number of terms in each group must be powers of 2; that is, a group with one term, two terms, four terms, or eight terms.
3. Each prime implicant must contain at least a single term that doesn't belong to any other prime implicant (i.e., no redundant groups). A prime implicant that satisfies this rule is called an **essential prime implicant** (EPI). The final minimal expression must include the logic term for all the EPIs.
4. All terms must be grouped.

Example 2.2. Minimize $f(x, y, z) = \Sigma(1, 3, 6, 7)$.
Solution: A K-map of f:

yz:	00	01	11	10
x: 0		1	1	
1			1	1

The four min-terms 1, 3, 6, and 7 form not one but three prime implicants as shown in the K-map. This is because min-term $1 = (001)_2$ and $6 = (110)_2$ are adjacent to only one other term in the K-map. The prime implicant associated with min-terms $3 = (011)_2$ and $7 = (111)_2$ is not an EPI. For the prime implicant associated with min-terms 1 and $3 = (011)_2$, the column labels for variable y changes, resulting in logic term $\bar{x}z$. For the prime implicant associated with min-terms 6 and 7, the column labels change for variable *z*, thus resulting in logic term *xy*. The final minimal SOP expression is:

$$f = \bar{x}z + xy$$

Example 2.3. Minimize $f(x, y, z) = \Sigma\ (0, 1, 2, 3, 4, 6)$.
Solution:

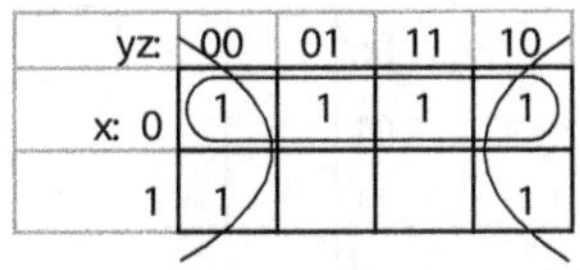

Both the four corner min-terms 0, 2, 4, and 6 and the four min-terms 0, 1, 2, and 3 on the first row form two EPIs. For the prime implicant associated with the min-terms 0, 1, 2, and 3, labels of both variables *y* and *z* change; thus, the term reduces to $\bar{x}$. Likewise, for the prime implicant associated with the four corner min-terms, labels of variables *x* and *y* change, resulting in $\bar{z}$. The final minimal SOP is:

$$f(x,y,z) = \bar{x} + \bar{z}$$

Example 2.4. Minimize $f(w, x, y, z) = \Sigma\ (0, 2, 3, 4, 5, 6, 7, 8, 10, 12, 13)$.
Solution:

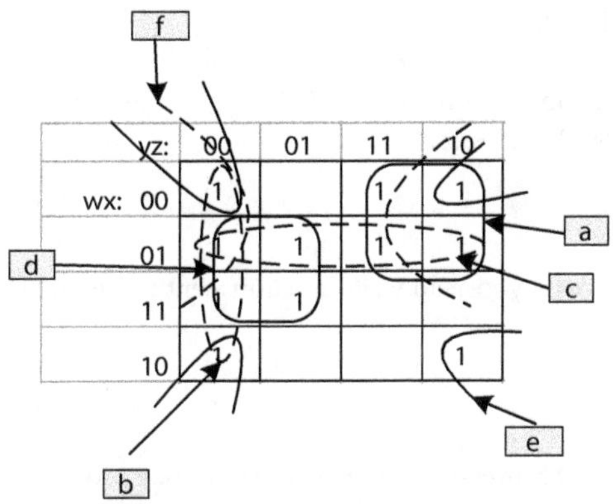

There are five prime implicants listed here with their corresponding list of min-terms. Only the prime implicants labeled *a*, *d*, and *e* are EPIs. This is because the prime implicants *a*, *d*, and *e* include the min-terms that are part of prime implicants *b*, *c*, and *f*.

a: Σ(2, 3, 6, 7)
b: Σ(0, 4, 8, 12), Redundant
c: Σ(4, 5, 6, 7), Redundant
d: Σ(4, 5, 12, 13),
e: Σ(0, 2, 8, 10),
f: Σ(0, 2, 4, 6) Redundant

For prime implicant *a*, the labels for variables *x* and *z* change, resulting in the minimal term $\bar{w}y$. For prime implicant *d*, the labels for variables *w* and *z* change, resulting in $x\bar{y}$, and for *e*, the labels for variables *w* and *y* change, resulting in $\bar{x}\bar{z}$. The final minimal SOP expression is:

$$f(w,x,y,z) = \bar{w}y + x\bar{y} + \bar{x}\,\bar{z}$$

Example 2.5. Minimize $f(w, x, y, z) = \Sigma\,(1, 9, 11, 14, 15)$.
Solution:

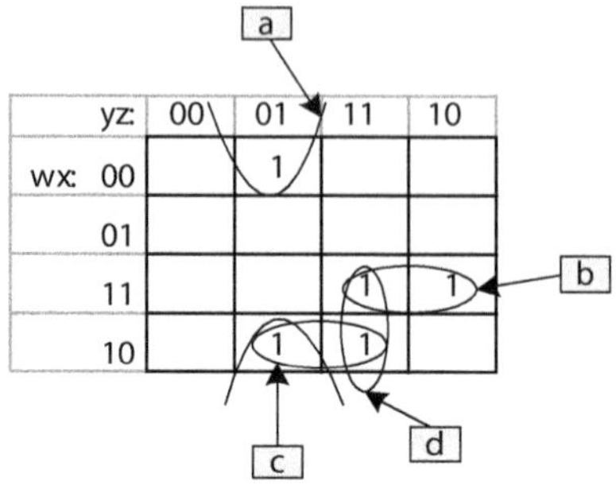

The list of prime implicants is:

a: $\Sigma(1, 9)$

b: $\Sigma(14, 15)$

c: $\Sigma(9, 11)$

d: $\Sigma(11, 15)$.

The prime implicants *a* and *b* and either *c* or *d* are EPIs. The EPIs *a*, *b*, and *c* yield the following minimal SOP expression:

$$f(w,x,y,z) = \bar{x}\,\bar{y}z + wxy + w\bar{x}z$$

Likewise, we may instead write an equivalent minimal SOP expression using the EPIs *a*, *b*, and *d* as follows:

$$f(w,x,y,z) = \bar{x}\,\bar{y}z + wxy + wyz$$

Two minimal expressions are equivalent if they produce the same truth table; that is, the same list of min-/max-terms.

Example 2.6. Minimize $g(w, x, y, z) = \Pi\,(1, 9, 11, 14, 15)$.
Solution:

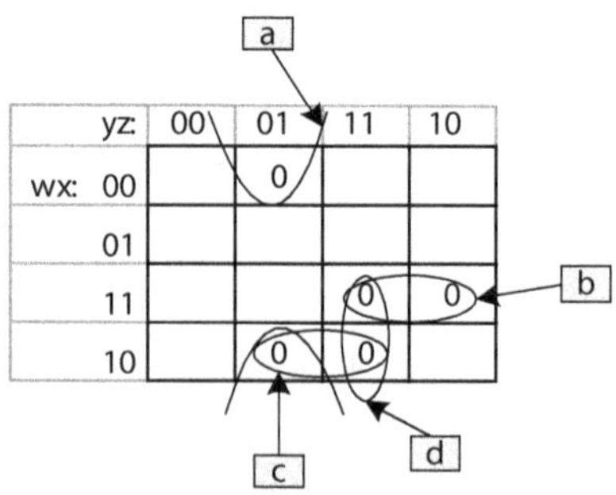

The prime implicants of a POS expression are determined exactly the same way as the prime implicants of an SOP expression. However, in this case, 0's of *g* in the map, which are equivalent to the 1's of $\bar{g}$, are grouped to determine the POS expression of g using the technique ($POS\ f = SOP\ \bar{f}$) discussed earlier. In this example, similar to Example 2.5, the list of SOP prime implicants of $\bar{g}$ are $\Sigma(1, 9) = \bar{x}\,\bar{y}z$, $\Sigma(14, 15) = wxy$, $\Sigma(9, 11) = w\bar{x}z$, and $\Sigma(11, 15) = wyz$. Thus, the list of POS prime implicants of g are:

a: $\Pi\ (1, 9) = \overline{\bar{x}\,\bar{y}z} = (x + y + \bar{z})$

b: $\Pi\ (14, 15) = \overline{wxy} = (\bar{w} + \bar{x} + \bar{y})$

c: $\Pi\ (9, 11) = \overline{w\bar{x}z} = (\bar{w} + x + \bar{z})$

d: $\Pi\ (11, 15) = \overline{wyz} = (\bar{w} + \bar{y} + \bar{z})$.

Again, there exist two minimal POS expressions for g: (1) using the EPIs a, b, and c; or (2) using the EPIs a, b, and d. Option 2 yields the following POS expression:

$$g = (x + y + \bar{z})(\bar{w} + \bar{x} + \bar{y})(\bar{w} + \bar{y} + \bar{z})$$

Example 2.7. Minimize $f(w, x, y, z) = \Pi\ (0, 1, 2, 5, 8, 9, 14)$.
Solution:

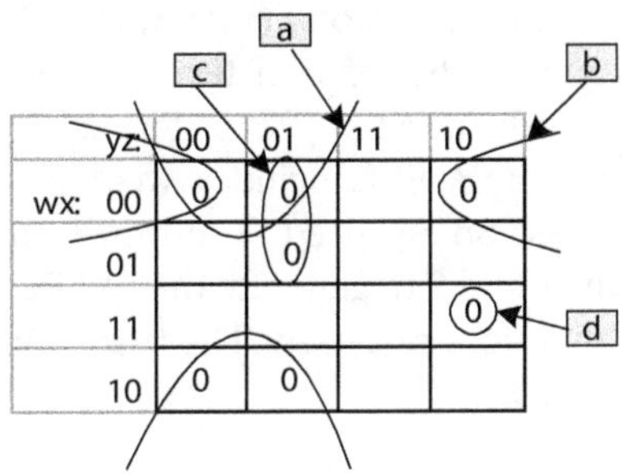

The prime implicants are:

a: $\Pi\ (0, 1, 8, 9)$

b: $\Pi\ (0, 2)$

c: $\Pi\ (1, 5)$

d: $\Pi\ (14)$.

These prime implicants are all essential, and the corresponding minimal POS expression is determined as follows:

a: Labels for variables w and z change, resulting in $(x + y)$.

b: Labels for variable y change, resulting in $(w + x + z)$.

c: Labels for variable x change, resulting in $(w + y + \bar{z})$.

d: The implicant consists of only one term, resulting in $(\bar{w} + \bar{x} + \bar{y} + z)$.

The final minimal expression is:

$$f(w,x,y,z) = (x + y)(w + x + z)(w + y + \bar{z})(\bar{w} + \bar{x} + \bar{y} + z)$$

Don't-Cares

Occasionally, an output may be defined only for a subset of input conditions, and for the remaining input conditions, the input is undefined. For example, consider a 7-segment display unit (7SDU) and its converter module in Fig. 2.16. Suppose the converter is designed to display binary coded decimal (BCD) numbers 0 to 9. Given a 4-bit input between $0 = (0000)_2$ and $9 = (1001)_2$, the converter generates seven signals *fa* through *fg*: one signal to turn on each of the seven segments *a* to *g* and display a corresponding decimal digits 0 to 9.

For example, to display 0, all seven segments except *g* must be turned on; thus, all signals except *fg* need to be 1 (assuming active-high outputs). To display 9, all segments but *ff* must be turned on; thus, all but *ff* must be 1. A BCD-to-7SDU converter is expected to generate correct values for signals *fa* to *fg* when its 4-bit input is between 0 and 9. For inputs 10 to 15, the outputs are undefined and can be considered *don't-cares*, which would be marked as "d" in the truth table.

A don't-care min-/max-term is interpreted as a "wild card"—either as 0 or 1 in the K-map as needed. Thus, it helps eliminate variables and simplify the final expression. Consider an expression $f(w, x, y, z) = \Sigma\,(1, 9, 14) + \Sigma_d\,(3, 7, 11)$ where Σ_d is used to indicate the list of min-terms for which f is don't-care. Likewise, symbol Π_d is used to indicate a list of max-terms for which a function is don't-care. As illustrated next, one of the EPIs includes two don't-cares. Only the don't-cares that help reduce the expression are used. The min-terms 3 and 11 as don't-cares are used with min-terms 1 and 9 to generate the EPI $\bar{x}z$. Min-term 7, also a don't-care, is not needed and therefore not used. The final minimal SOP expression reduces to $f(w,x,y,z) = \bar{x}z + wxy\bar{z}$.

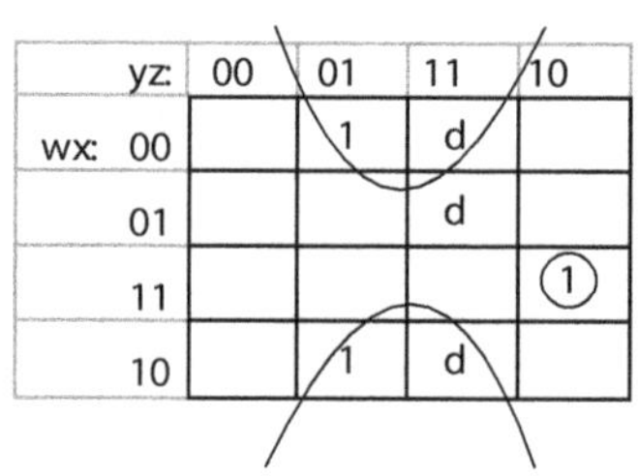

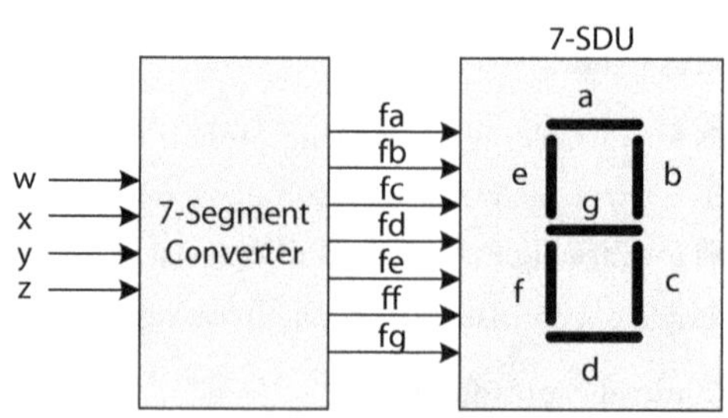

FIGURE 2.16 A 7-segment display unit and converter.

2.5 Logic Minimization Algorithm

A K-map, being a graphical method, is only suitable for a small number of variables, such as four. For more than four variables, an algorithmic method that was developed in the mid-1950s and known as the Quine-McCluskey algorithm is more appropriate. The algorithm finds a minimal logic expression using steps similar to those used with K-maps. The min-terms are grouped into different sets, where each set contains only the min-terms that have a specific number of 1's in their binary representation. Consider the expression $f(w, x, y, z) = \Sigma\,(0, 2, 3, 4, 5, 6, 7, 8, 10, 12, 13)$ from Example 2.4. In binary, the min-terms are 0000, 0010, 0011, 0100, 0101, 0110, 0111, 1000, 1010, 1100, and 1101. They can be grouped into four sets as follows:

Set 1:	0000	Min-terms with no 1's
Set 2:	0010	Min-terms with one 1
	0100	
	1000	
Set 3:	0011	Min-terms with two 1's
	0101	
	0110	
	1010	
	1100	
Set 4:	0111	Min-terms with three 1's
	1101	

One pair at a time, one min-term from Set 1 with one term from Set 2, are compared. The single bit change in any pair indicates an implicant. The changing bit is replaced by a dash (–) indicating the omission of the corresponding variable.

For example, min-term 0000 from Set 1 is compared with 0010 from Set 2 to generate the implicant 00–0 with signal y omitted. Min-term 0000 is again compared with min-term 0100 to generate the implicant 0–00 with variable x omitted, and so on. The same process (one pair at a time) is applied to the min-terms from Set 2 with those of Set 3 and Set 3 with those of Set 4 to generate the first set of implicants for the output. The initial Sets 1 to 4 are shown as Sets I.1 to I.4 under column I in Table 2.4.

The list of implicants generated using the terms in Sets I.1 to I.4 is shown as Sets II.1 to II.3 under column II in Table 2.4. Next to each min-term in Sets I.1 to I.4, an "x" is placed if the min-term contributes to an implicant in column II; otherwise, the term is marked by an asterisk (*), identifying a prime implicant (none exists in this case). Once the implicants of column II are generated, the process repeats; one pair at a time, one implicant from Set II.1 with one implicant from Set II.2 is compared. In this case, any dashes between each pair of the implicants must line up.

For example, implicant 00–0 from Set II.1 is compared with implicant 10–0 from Set II.2 to generate the implicant –0–0. The generated implicants are

	Column I						Column II						Column III				
	w	x	y	z			w	x	y	z			w	x	y	z	
Set I. 1:	0	0	0	0	x	Set II. 1:	0	0	–	0	x	Set III. 1:	–	0	–	0	*
							0	–	0	0	x		0	–	–	0	*
Set I. 2:	0	0	1	0	x		–	0	0	0	x		–	–	0	0	*
	0	1	0	0	x												
	1	0	0	0	x	Set II. 2	0	0	1	–	x	Set III. 2:	0	–	1	–	*
							0	–	1	0	x		0	1	–	–	*
Set I. 3:	0	0	1	1	x		–	0	1	0	x		–	1	0	–	*
	0	1	0	1	x		0	1	0	–	x						
	0	1	1	0	x		0	1	–	0	x						
	1	0	1	0	x		–	1	0	0	x	x: Not Prime Implicant					
	1	1	0	0	x		1	0	–	0	x	*: Prime Implicant					
							1	–	0	0	x						
	0	1	1	1	x												
Set I. 4:	1	1	0	1	x	Set II. 3:	0	–	1	1	x						
							0	1	–	1	x						
							–	1	0	1	x						
							0	1	1	–	x						
							1	1	0	–	x						

TABLE 2.4 List of All Prime Implicants (Marked *) Generated from Min-terms 0, 2, 3, 4, 5, 6, 7, 8, 10, 12, and 13

shown in the table as Sets III.1 and III.2 in column III. Again, an implicant pair is marked by an "x" in column II if it contributes to the list of implicants for the next cycle (column III); otherwise, it is marked by an asterisk (∗)—again, none exist in this case. The process repeats using the implicants listed in column III, but this time, no further processing is possible because none of the implicants from Set III.1 when compared with those of Set III.2 generates new implicants; thus, all the implicants in Sets III.2 and III.3 are marked with asterisks and are listed in Fig. 2.17 along with their corresponding logic terms.

The next step in the process is to use a minimum-set algorithm to select EPIs from the list of prime implicants *a* through *f*. Figure 2.17 is an organization of the prime implicants and their corresponding min-terms. An "x" is placed in each cell in the table when a prime implicant covers a min-term. For example, the prime implicant 0––0 with two dashes covers min-terms $0 = (0000)_2$, $2 = (0010)_2$, $4 = (0100)_2$, and $6 = (0110)_2$, and thus these cells are marked with an "x" in row 1, as illustrated in the table.

a. 0–1-: $\bar{w}y = \Sigma(2, 3, 6, 7)$
b. – –00: $\bar{y}\bar{z} = \Sigma(0, 4, 8, 12)$
c. 01--: $\bar{w}x = \Sigma(4, 5, 6, 7)$
d. –10-: $x\bar{y} = \Sigma(4, 5, 12, 13)$
e. –0–0: $\bar{x}\bar{z} = \Sigma(0, 2, 8, 10)$
f. 0– –0: $\bar{w}\bar{z} = \Sigma(0, 2, 4, 6)$

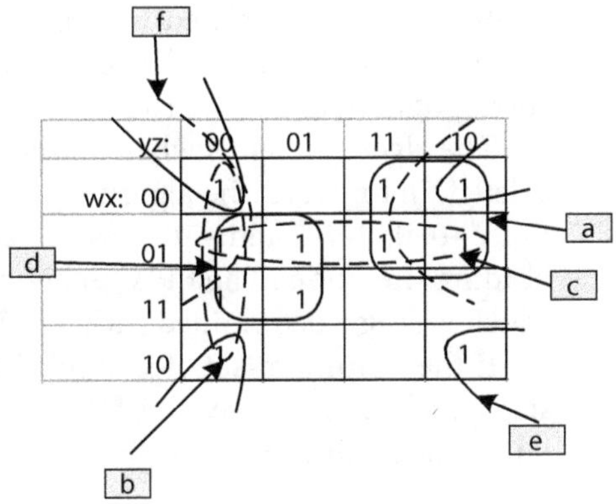

FIGURE 2.17 The list of prime implicants obtained from Table 2.4.

The minimum-set algorithm is also an iterative process and starts by first selecting a prime implicant with only a single "x" in any one column, which always identifies an EPI. In this case, the columns associated with min-terms 3, 10, and 13 each contain only a single "x"; these are bolded and underlined in the table.

Suppose the min-terms are processed from left to right in Table 2.5. During iteration 1, the column associated with min-term 3 contains only a single "x," and thus the corresponding prime implicant 0–1– is selected as an EPI because it is the only one covering min-term 3. It also covers min-terms 2, 3, 6, and 7, and therefore, the associated columns and the row are marked deleted (D), as illustrated in the table. This effectively reduces the size of the table for the next iteration. In iteration 2, the prime implicant –0–0 that corresponds to a single "x" in the column associated with min-term 10 is selected as the next EPI. Thus, the columns 0, 8, and 10 and row 2 are marked D. Finally, in iteration 3, the prime

	Prime Implicant				Min-terms											Iteration		
	w	x	y	z	0	2	3	4	5	6	7	8	10	12	13	1:	2:	3:
	0	–	–	0	x	x		x		x								
EPI	–	0	–	0	x	x						x	**x**				D	
	–	–	0	0	x			x				x		x				
	0	1	–	–				x	x	x	x							
EPI	0	–	1	–		x	**x**			x	x					D		
EPI	–	1	0	–				x	x					x	**x**			D
Iteration 1:						D	D			D	D							
Iteration 2:					D							D	D					
Iteration 3:								D	D					D	D			

TABLE 2.5 Illustrating the Minimum-Set Algorithms Using the Prime Implicants Obtained from Table 2.4

implicant –10– associated with min-term 13 is selected as the next EPI, which results in deleting all of the remaining columns in the table, as well as the row corresponding to EPI = –10–.

The algorithm ends when all the columns associated with the min-terms are deleted. In this case, the algorithm stops after three iterations and produces three EPIs, –0–0, 0–1–, and –10– associated with the three rows marked D under the heading "Iteration". The EPIs generate the minimal SOP expression $\bar{x}\bar{z} + \bar{w}y + x\bar{y}$, which was also determined earlier in Example 2.5 using a K-map.

If the minimum-set algorithm determines that there are no columns with a single "x" in the table, the following rules are used to select the next prime implicant candidate:

1. Identify the columns with the least number of "x" markings and then select the corresponding prime implicants as candidates.
2. From the list of prime implicants in step 1, select those prime implicants that cover the largest number of the remaining min-terms (excluding columns that are marked D).
3. If multiple prime implicants are obtained in step 2, choose the one that has the largest number of dashes; the corresponding logic term would have fewer variables.
4. If there is more than one prime implicant satisfying step 3, then there are two or more equivalent minimal expressions.

The algorithm for obtaining a minimum POS expression is similar, except that max-terms are used in Table 2.4 and Table 2.5 instead of min-terms.

Typically, if a circuit has multiple outputs, a minimal expression for each output is not determined independently. Instead, the minimization goal would be to select those prime implicants that are common among the different expressions in order to minimize the total number of gates needed to implement all the expressions as a single circuit. Output signal from some gates may be shared and connected as inputs to more than one gate. **Espresso** minimization software [1] does exactly that when there are two or more expressions to minimize at the same time. CAD tools for logic design typically include this and other minimization software.

2.5.1 Minimization Software

Example 2.9 illustrates the format of Espresso's input and output files using the min-terms of $f(w, x, y, z) = \Sigma\ (0, 2, 3, 4, 5, 6, 7, 8, 10, 12, 13)$. A dot (·) in the first column indicates a parameter. For example, ".i 4" indicates the number of inputs—four in this case—and ".o 1" indicates the number of output bits—1 in this case. The label ".ilb w x y z" lists input variables—four in this case; ".ob f" lists output variables—one in this case; and ".e" indicates the end of the input file. The symbol "#"indicates a comment line. In the output file, ".p" indicates the number of EPIs.

Example 2.8. Use Espresso to minimize the function $f(w, x, y, z) = \Sigma\,(0, 2, 3, 4, 5, 6, 7, 8, 10, 12, 13)$.
Solution:

(a) Input file

```
#Inputs: 4, Outputs: 1
.i 4
.o 1
#Input labels
.ilb w x y z
#output bit label
.ob f
#list of min-terms separated by space and a single output
bit separated by a tab
0 0 0 0	1
0 0 1 0	1
0 0 1 1	1
0 1 0 0	1
0 1 0 1	1
0 1 1 0	1
0 1 1 1	1
1 0 0 0	1
1 0 1 0	1
1 1 0 0	1
1 1 0 1	1
#end of list
.e
```

(b) Output file; all comment lines are printed first

```
#Inputs: 4, Outputs: 1
#Input signal labels
#output bit label
#list of min-terms and output
#end of list
.i 4
.o 1
.ilb w x y z
.ob f
.p 3
-10- 1
-0-0 1
0-1- 1
.e
```

The output file lists three EPIs: –10–, –0–0, and 0–1–. These EPIs are the same as those obtained earlier manually, as illustrated using Table 2.4 and Table 2.5.

The following Espresso output displays the EPIs for two output signals *f* and *g*. The "11," "11," and "10" printed next to the EPIs indicate that all the three EPIs belong to *f*, and the first two (-10- and -0-0) are shared and also belong to *g*.

```
.o 2
.ob f g
.p 3
-10- 11
-0-0 11
0-1- 10
.e
```

Table 2.6 shows the results of the minimization algorithm applied to expression $f(w, x, y, z) = \Sigma\,(1, 9, 14) + \Sigma_d\,(3, 7, 11)$, which was also minimized earlier in Sec. 2.4.2. In column I, each of the don't-care terms is marked with "d." The minimization algorithm discussed earlier is the same, except that only one min-term in each pair can be a don't-care; two don't-care terms are never compared. For example, min-term $(0011)_2$ in Set I.2 and min-term $(0111)_2$ in Set I.3, both don't-cares, are never paired to generate the unnecessary prime implicant 0–11.

The algorithm produces three prime implicants (marked with asterisks), one in each column. Table 2.7 is the organization of the prime implicants obtained in Table 2.6 for the minimum-set algorithm. The prime implicant $(1110)_2$ in Set I.3 is an EPI. From the two remaining prime implicants -0-1 and -001 that cover both min-terms 1 and 9, –0–1 is selected because it has more dashes than -001. These EPIs were also produced using Espresso in Example 2.9, where a dash (–) in the input file indicates a don't-care.

	I						II						III				
	w	x	y	z			w	x	y	z			w	x	y	z	
Set I.1:	0	0	0	1	x	Set II.1:	0	0	–	1	x	Set III.1:	–	0	–	1	*
							–	0	0	1	*						
Set I.2:	0	0	1	1	d												
	1	0	0	1	x	Set II.2	1	0	–	1	x	Set III.2:					
	0	1	1	1	d		x: Not Prime Implicant										
Set I.3:	1	0	1	1	d		*: Prime Implicant										
	1	1	1	0	*												

TABLE 2.6 Prime Implicants of $f(w, x, y, z) = \Sigma\,(1, 9, 14) + \Sigma_d\,(3, 7, 11)$

	w	x	y	z	1	9	14	1:	2:
EPI	1	1	1	0			X	D	
	–	0	0	1	X	X			
EPI	–	0	–	1	X	X			D
1:							D		
2:					D	D			

TABLE 2.7 Illustrating the Minimum-Set Algorithms Using the Prime Implicants Obtained from Table 2.6

Example 2.9. Use Espresso to minimize $f(w, x, y, z) = \Sigma\,(1, 9, 14) + \Sigma_d\,(3, 7, 11)$.

(a) Input file

```
.i 4
.o 1
.ilb w x y z
.ob f
0 0 0 1  1
0 0 1 1  -
0 1 1 1  -
1 0 0 1  1
1 0 1 1  -
1 1 1 0  1
.e
```

(b) Output file

```
.i 4
.o 1
.ilb w x y z
.ob f
.p 2
1110     1
-0-1     1
.e
```

2.6 Circuit Timing Diagram

We will start with the timing diagram of a NOT gate before we discuss a circuit's timing diagram. There is a delay associated with each gate. It is the time required for the gate output to change from logic 0 to logic 1 or vice versa from the time that one or more of its inputs change. Figure 2.18 illustrates the timing of a NOT gate with 0.1 ns gate delay. As shown in Fig. 2.18(a), the output z transitions from 1 to 0 in 0.1 ns from the time that input x transitions from 0 to 1. Likewise,

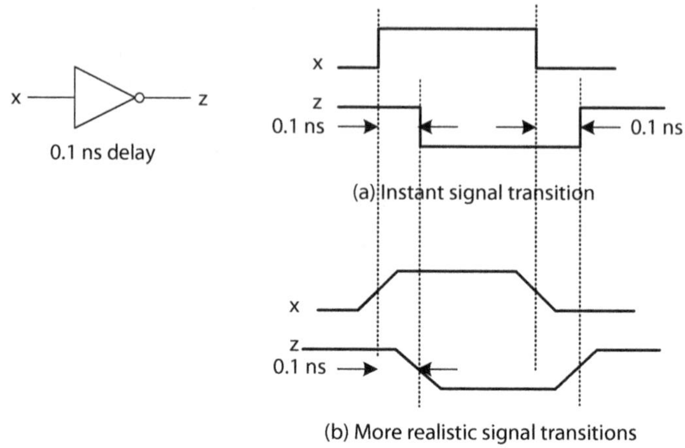

FIGURE 2.18 A NOT gate timing diagram with 0.1-ns delay: (*a*) a simplified timing diagram; (*b*) a more realistic timing diagram.

z transitions from 0 to 1 in 0.1 ns from the time *x* transitions from 1 to 0. In Fig. 2.18(*a*), the signal transitions are shown to happen instantly. However, in reality, signals do not change instantly.

A signal's **rise time** is the amount of time that the output voltage associated with logic 0 rises (increases) to that associated with logic 1. Similarly, a signal's **fall time** is the amount of time required for the output voltage associated with logic 1 to fall (decrease) to that associated with logic 0. The rise and fall times of a gate may not be the same. Figure 2.18(*b*) illustrates a more realistic timing diagram of a NOT gate with the signal rise and fall times. The midpoints of the rise and fall times are often used to indicate instant signal transition as shown in the figure.

With a +5.0 voltage source, any value between 0 and 0.8 V at the input is interpreted as logic 0, and between 2.0 and 5.0 V as logic 1. An input voltage greater than 0.8 and less than 2.0 V is considered undefined. For logic 1 output, the voltage range is between 2.4 and 5.0 V, and for logic 0, it is between 0 and 0.4 V. A lower voltage source (e.g., 1.8 or 1.2 V) is typically used in battery-powered systems.

Recall that a minimal Boolean expression defines the logic relationship between a circuit's inputs and its output without considering gate or wire delays. A circuit's timing diagram is an illustration of the actual changes that happen at the output of each gate in the circuit due to gate and wire delays. A timing diagram provides a more realistic view of a circuit behavior when changes take place at its inputs. For example, consider the expression $f(a, b, c, d) = \Sigma(1, 3, 5, 7, 10, 11, 14, 15)$ that has the minimal SOP expression $f = \bar{a}d + ac$; note that *f* does not depend on *b*. Figure 2.19 shows its equivalent NAND-only circuit with the intermediate signal $\bar{a}$, *x*, and *y*; the gates are also labeled G1 through G4 for reference.

Figure 2.20 illustrates the timing diagram of the circuit when its input changes from concatenated *acd* = 111 to *acd* = 011; that is, *a* changes from 1 to 0. The

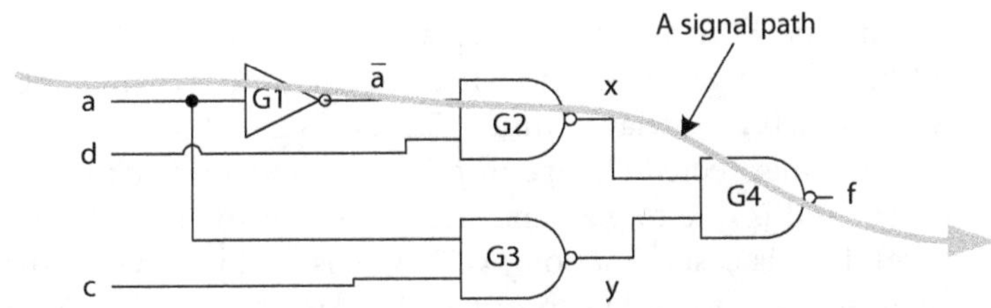

Figure 2.19 Circuit for $f = \bar{a}d + ac$ with intermediate signal names; *f* does not depend on *b*.

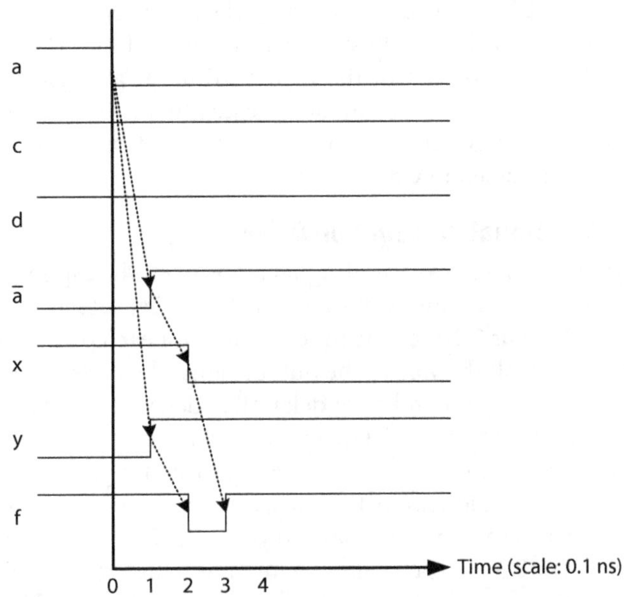

Figure 2.20 A timing diagram for the circuit in Fig. 2.19 when input *a* changes from 1 to 0.

delay for all the gates is assumed to be 0.1 ns, and wire delays are ignored. (Wire delays are outside the scope of this book.) Note that $f = \bar{a}d + ac$ produces $f = 1$ when input is $acd = 111$ or $acd = 011$. However, due to gate delays, *f* does not remain at logic 1, as illustrated in the timing diagram, between time = 0.2 ns and 0.3 ns.

The change in signal *a* causes $\bar{a}$ to change to 1 after a 0.1 ns delay at gate G1 from its initial value of 0 to 1 at time-step 1. This change is shown by an arrow from the time signal *a* changes to 0 to the time when signal $\bar{a}$ changes to 1. An arrow indicates signal dependency. At the same time, signal *y* also changes from 0 to 1 at time-step 1 after a 0.1 ns delay at gate G3. The change in $\bar{a}$ at time = 0.1 ns causes signal *x* to change from its initial value of 1 to 0 at time-step 2 after a 0.1 ns delay at gate G2. This change is also shown by an arrow from the time $\bar{a}$ changes to 1 and *x* changes to 0.

At time = 0.1 ns when both *x* and *y* are at logic 1, *f* changes from 1 to 0 after a 0.1 ns delay at gate G4. And when *x* changes to 0 at time = 0.2 ns, *f* changes back to 1 after a 0.1 ns delay at time = 0.3 ns. Signal *f* then remains at 1 thereafter.

The unexpected change in *f* from 1 to 0 and back to 1 is called a **hazard** or **glitch**, and is due to gate and wire delays in the circuit (wire delays are ignored here). In this case, function $f = \bar{a}d + ac$ is said to have a 1-hazard when its input changes from *acd* = 111 to *acd* = 011. An OR-AND (or NOR-only) circuit has a 0-hazard when its timing diagram shows an unexpected change from 0 to 1 and back to 0 when its input changes.

Hazards violate the expected behavior of a combinational circuit and must be prevented from affecting the state (i.e., register contents) of a digital system. In Chap. 4, we will introduce a clock, an alternating 1 0 1 0 1 0 1 0 . . ., signal with a fixed period to control the loading time of a register. In Fig. 2.20, output *f* is valid only after 0.3 ns from the time that there is a change at the circuit's input. The clock period is determined from the signal propagation delay as well as other delays discussed in Chaps. 4 and 5.

2.6.1 Signal Propagation Delay

Typically, there are several signal paths from the inputs to one or more outputs of a circuit. For example, output signal *f* in Fig. 2.19 is determined by signal paths G1-G2-G4 or G3-G4. The time required for each path to propagate a signal change at its input all the way to the output signal depends on the number and size of the gates on the path and wire delay. The delay of the longest path is known as the circuit's propagation delay. Ignoring wire delay, a circuit's propagation delay is proportional to the number of gates in its longest path. The path G1-G2-G4 in Fig. 2.19 is the longest path. Thus, the circuit's propagation delay is proportional to three gate delays, or 0.3 ns, as shown in Fig. 2.20; for simplicity, assume a 0.1 ns gate delay. The 0.3 ns delay is also the minimum time required for the 1-hazard at output *f* to disappear and for the circuit to output *f* = 1, as defined by its logic expression when the input to the circuit changes from concatenated *acd* = 111 to *acd* = 011.

In general, the propagation delay of a circuit module with multiple outputs is determined by the longest path from its inputs to its many outputs. In this case, some of the individual outputs may have shorter propagation delays; however, at least one of the output signals would have the longest path that would determine the propagation delay for the entire module.

The circuits that implement SOP and POS expressions always have signal paths consisting of either two levels of gates or three levels of gates, including the initial NOT gate. This results in a propagation delay proportional to only two or three gate delays. Thus, the SOP or POS expressions are always preferred to speed up outputs and thus increase performance.

However, an expression that contains logic terms with many variables may not be implemented as a two- or three-level circuit due to fan-in limitations of the gates (discussed next). In this case, an expression must be partitioned into smaller expressions, each implemented with a smaller circuit requiring gates with fewer fan-in. The smaller circuits are then connected to create the final multilevel circuit

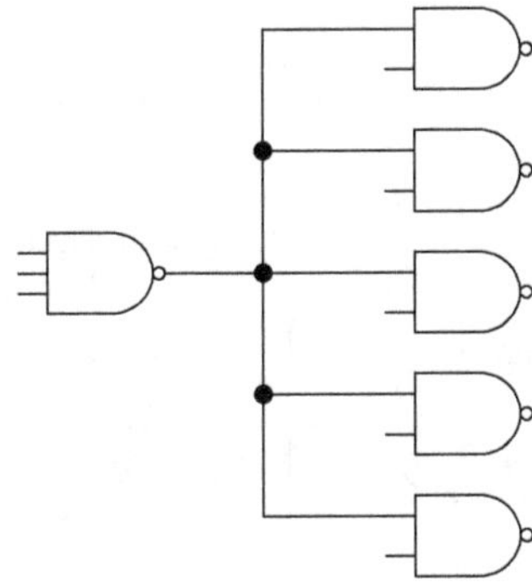

Figure 2.21 A NAND gate with fan-in of 3 and fan-out of 5.

with more than two levels, not counting the initial NOT gates. For instance, FPGAs, being a programmable chip, typically have restrictive resources and therefore cannot implement any arbitrarily sized SOP or POS expression. FPGAs are typically slower than the custom chips built for high performance.

2.6.2 Fan-In and Fan-Out

Fan-in is the number of inputs a gate can have, and the fan-out is the number of connections to which a gate's output can connect. For example, a NAND gate with a fan-in of 3 and a fan-out of 5 is shown in Fig. 2.21. The fan-in of a NOT gate is always 1. The fan-in of an XOR and XNOR gate is typically 2. The fan-in of AND, OR, NAND, and NOR gates could vary; however, each gate has a maximum fan-in and fan-out limit (e.g., 8) in order to operate normally.

2.7 Other Gates

In addition to the standard gates discussed earlier, there are other useful gates necessary to design a digital system. These gates are known as a buffer, an open collector (OC) buffer, and a tri-state buffer.

2.7.1 Buffer

The symbol and truth table of a buffer are shown in Fig. 2.22(*a*). A buffer does not alter its input signal; it simply amplifies it. Assuming that the fan-out of an AND gate is 5, a buffer can increase the fan-out of the AND gate from 5 to, for example, 9, as illustrated in Fig. 2.22(*b*).

2.7.2 Open Collector Buffer

An OC buffer is similar to a buffer gate, except that when its input is logic 1, its output becomes high impedance, shown as Z (Fig. 2.23(*a*)). A high-impedance signal is neither driven to logic 0 nor logic 1, and is shown electrically isolated, as if the wire is "floating" and not connected. Figure 2.23(*b*) illustrates a circuit with

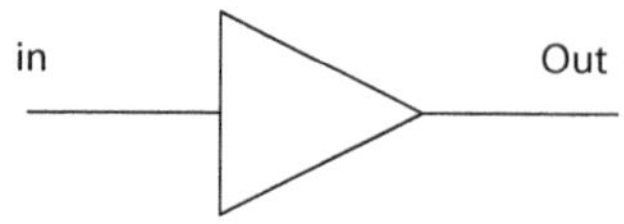

In	Out
0	0
1	1

(a) Buffer gate and its truth table

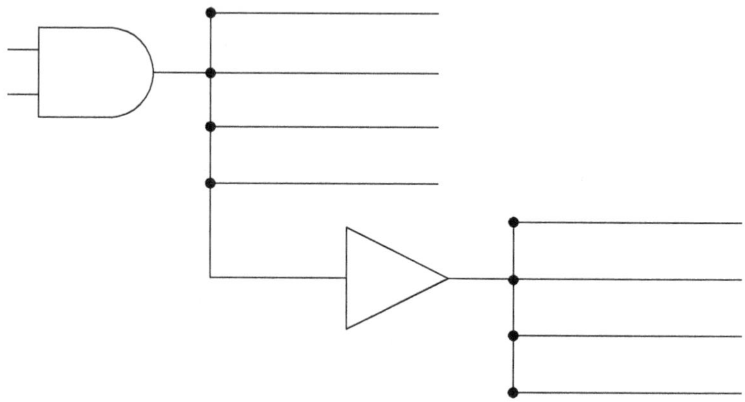

(b) Increasing fan-out

Figure 2.22 A buffer gate: (*a*) buffer symbol and truth table; (*b*) buffer used to increase the fan-out.

two OC buffers. The output of each gate is either 0 or Z, and thus the outputs can be connected together to generate a single output *f*. A Z-output can be changed to logic 1 or logic 0 using a "pull-up" or "pull-down" resistor connected to either the power source (e.g., 5.0 V) or ground (0.0 V), respectively. The Z-output in Fig. 2.23(*b*) is pulled up.

Figure 2.24 illustrates the behavior of a high-impedance output using the circuit in Fig. 2.23(*b*) with two inputs *a* and *b*. When $a = 0$ and $b = 0$, the OC buffers B1 and B2 output 0, thus connecting *f* to ground, logic 0 (Fig. 2.24(*a*)). When $a = 1$ and $b = 1$, the outputs of both OC buffers B1 and B2 become Z (floating); this leaves *f* connected to power, logic 1 (Fig. 2.24(*d*)). When $a = 0$ and $b = 1$ or $a = 1$ and $b = 0$, one of the buffers outputs 0 while the output of the other becomes Z, thus connecting *f* to logic 0, as illustrated in Fig. 2.24(*b*) and (*c*). The four cases are summarized as a truth table in Fig. 2.23(*b*). The truth table illustrates an AND logic, and the circuit in this case is called a **wired-AND logic**. A wired-logic circuit can have a large fan-in.

Wired-AND and wired-OR are two commonly used wired-logic circuits. For example, a wired-logic circuit is used in the design of computer systems with expansion slots. In this case, one can add a new functionality to a computer system by inserting an expansion card, a device controller interface (DCI), in one of the computer's expansion slots. An *n*-input wired-OR circuit, for example, can be used to OR *n* signals, one from each device, and generate an output to inform the CPU when a device needs service, as illustrated in Fig. 2.23(*c*). The device interfacing will be discussed in more details in Chap. 9.

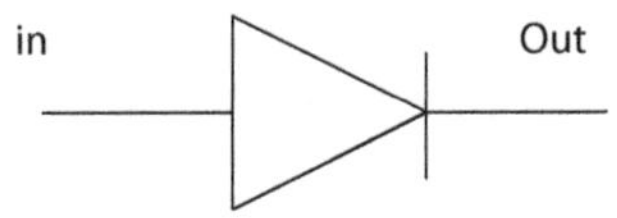

In	Out
0	0
1	Z

(a) Open-collector buffer

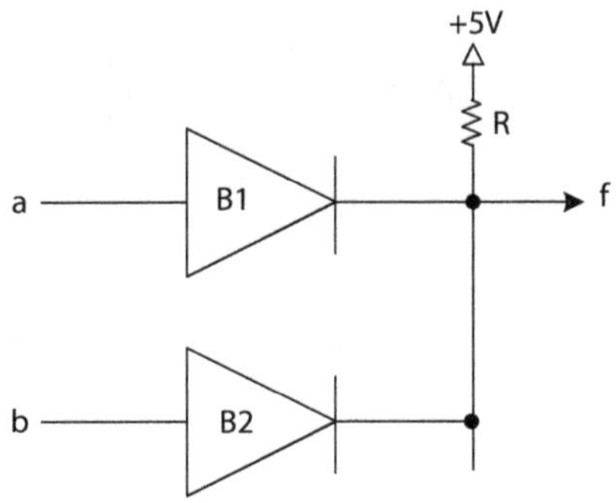

a	b	f
0	0	0
0	1	0
1	0	0
1	1	1

(b) A two-input wired-AND logic

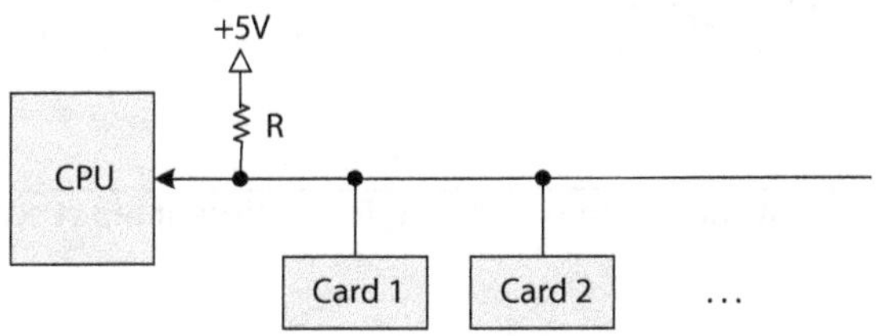

(c) An application of wired-logic

Figure 2.23 Open collector buffer and application: (*a*) symbol and truth table; (*b*) a two-input wired-AND logic and truth table; (*c*) the design of expansion slots.

2.7.3 Tri-State Buffer

Figure 2.25 shows a tri-state buffer and its truth table. It is a combination of a buffer and an OC buffer. It operates like a buffer when enabled ($e = 1$), but its output becomes Z when disabled ($e = 0$). Tri-state buffers are useful when two or more signals need to share a common wire known as a **bus line** (or simply a bus). Typically, a bus has many lines. Figure 2.26 illustrates the connection of three tri-state buffers to a 1-bit bus. One at a time, one of the enabling signals e1, e2, or e3 may be asserted to place the corresponding signal *a*, *b*, or *c* on the bus. The other disabled tri-state buffers make their outputs Z ("floating"), and thus become isolated from the bus.

A bus connection can be bidirectional if a module outputs to the bus and inputs from the bus. Figure 2.27(*a*) illustrates an example of bidirectional bus connections, using a buffer to input from the bus and a tri-state buffer to output to the bus. A data item that is transmitted over a bidirectional bus has a source module and a destination module. The source module places a data item using a

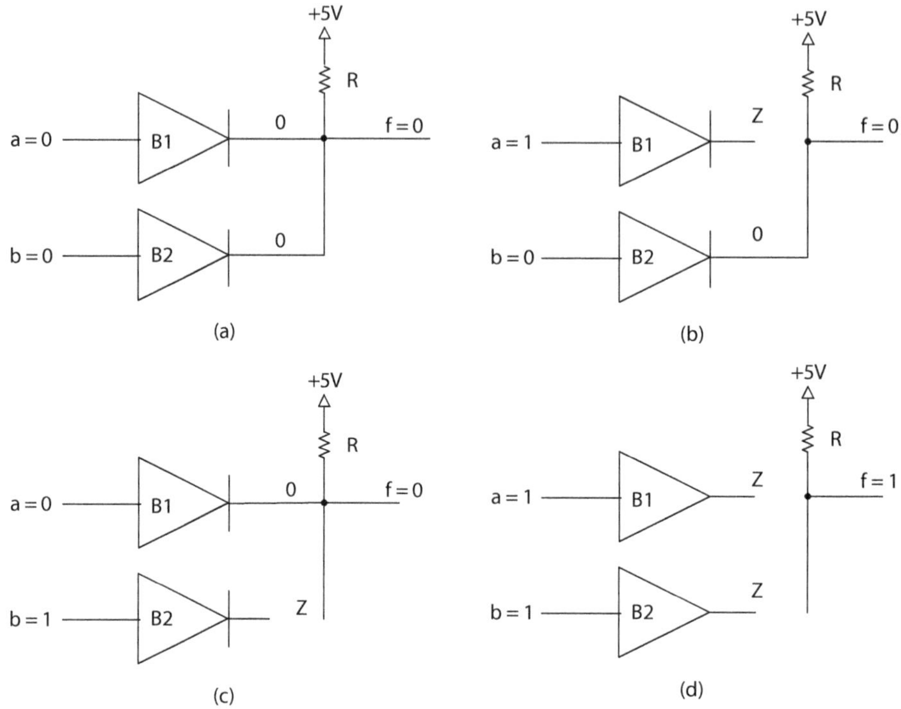

Figure 2.24 A two-input wired-AND logic with four different input scenarios.

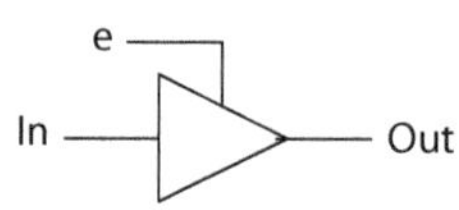

e	In	Out
0	0	Z
0	1	Z
1	0	0
1	1	1

Figure 2.25 A tri-state buffer and its truth table.

tri-state buffer on the bus, and a destination module inputs the data using a buffer. A buffer protects a system from fan-out violation at the source module if the bus fan-out at the destination module is greater than 1; that is, a bus signal is connected to two or more gates inside a destination module, as illustrated in Fig. 2.27(*a*) for destination module C.

A bus reduces interconnection overhead. It can replace many one-to-one connections among several modules (Fig. 2.27(*b*)) as long as the rate (how often) at which data (number of bits) is transmitted over the bus is sufficiently large enough to handle the load. The data rate of a bus is called **bandwidth**. For example, in Fig. 2.27(*a*), 1-bit data can be transmitted over the 1-bit bus every 10 ns if the enabling signals e1, e2, and e3 are asserted one at a time every 10 ns, or if

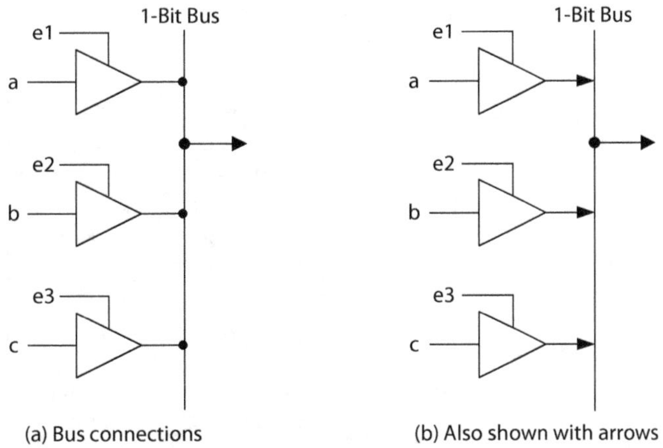

Figure 2.26 Three tri-state buffers sharing a 1-bit bus: (*a*) actual connections; (*b*) connections typically are shown with arrows.

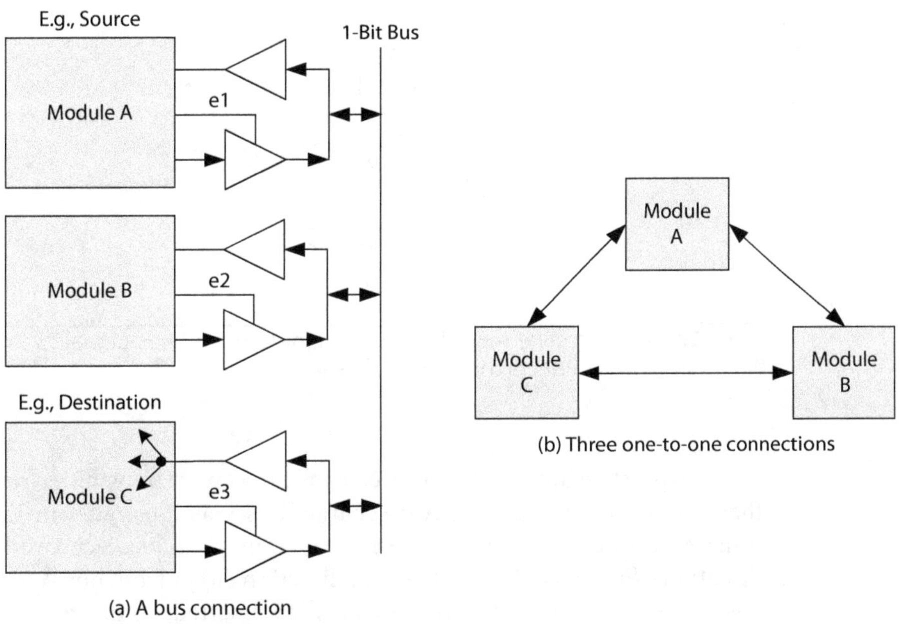

Figure 2.27 Module interconnection: (*a*) three modules interconnected using a bus; (*b*) three modules with one-to-one connections.

one of the enable signals remains asserted for multiples of 10 ns durations. A 1-bit bus that is capable of transferring a 1-bit data every 10 ns has the same bandwidth as a 10-bit bus capable of transferring a 10-bit data every 100 ns; in 100 ns both buses would transfer 10 bits. Therefore, the width of a bus (number of lines) and the speed of the bus (how often) determine the bandwidth of the bus.

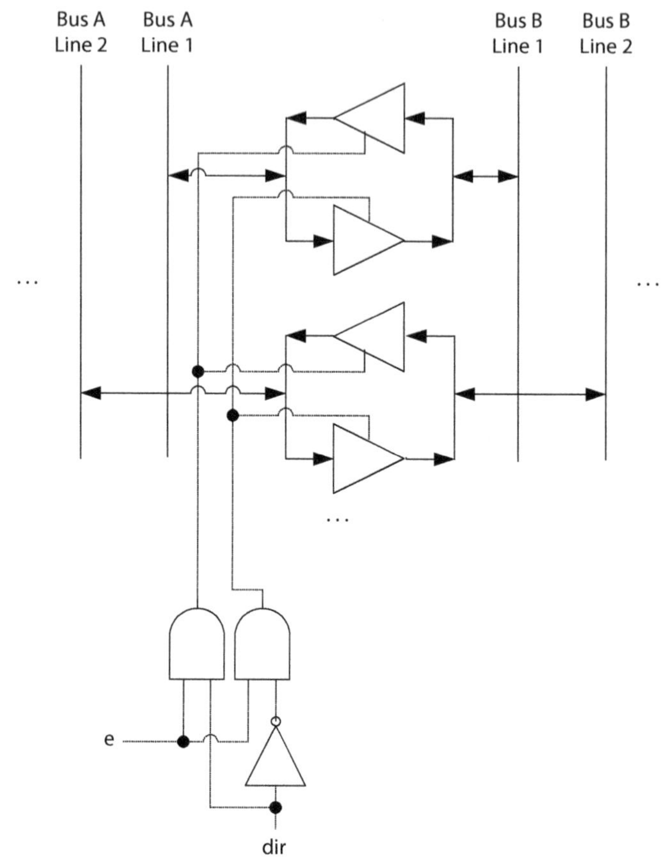

Figure 2.28 A transceiver module [2] shown for two bus lines; signal *dir* indicates data direction and *e* connects the two bus lines.

Two tri-state buffers may be used as a transceiver (transmitter/receiver) circuit that connects, for example, two separate buses as illustrated in Fig. 2.28. Each transceiver circuit creates a bidirectional connection between two bus lines. The direction (*dir*) signal decides the data direction either from bus A to bus B or from bus B to bus A. The enable (*e*) signal, when asserted, connects the two bus lines, keeping them connected while active.

2.8 Design Examples

As discussed in Chap. 1, a data path includes many circuit modules. This section covers a few commonly used but small combinational circuit modules. The "selector" module in Fig. 1.1 (Chap. 1) is known as a multiplexer. Other examples discussed here are a simple adder, decoder, and encoder modules.

A 1-bit adder, known as a full adder (FA), generates the sum of two 1-bit inputs plus an incoming carry-in bit as 0 or 1. A decoder module converts a number *A* (0, 1, 2, etc.) to a corresponding output signal (e.g., f_0, f_1, f_2, etc.). Only one of the outputs f_0, f_1, etc. can be active at any time. An encoder, on the other hand, performs the reverse operation and generates a number associated with an active input signal. Specifically, the following examples are discussed:

- The design of an FA with active-high signals
- The design of a 1-bit, 2-to-1 multiplexer and 1-bit, 4-to-1 multiplexer
- The design of a 1-to-2 decoder with active-low output signals
- The design of a 3-to-2 encoder with active-low input signals

2.8.1 Full Adder

An FA has three 1-bit inputs, one of which is the carry-in (c_{in}) and outputs a 1-bit sum (s) and a 1-bit carry-out (c_{out}), as illustrated in Fig. 2.29. Table 2.8 shows its truth table. In each case in the table, s and c_{out} are determined as the sum of three-bits a, b, and c_{in}. Multiple FA modules, as will be illustrated in the next chapter, can be used to design a large adder.

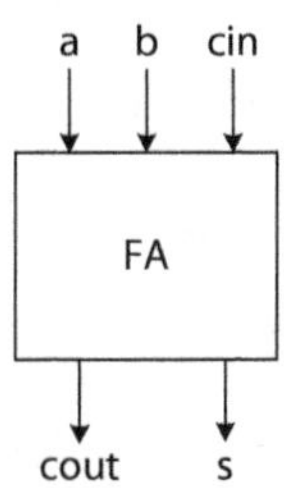

FIGURE 2.29 The block diagram of an FA.

a	b	c_{in}	c_{out}	s
0	0	0	0	0
0	0	1	0	1
0	1	0	0	1
0	1	1	1	0
1	0	0	0	1
1	0	1	1	0
1	1	0	1	0
1	1	1	1	1

TABLE 2.8 FA Truth Table

Minimal SOP expressions for s and c_{out} are determines as follows:

$$s(a, b, c_{in}) = \Sigma\,(1, 2, 4, 7)$$
$$c_{out}(a, b, c_{in}) = \Sigma\,(3, 5, 6, 7)$$

b cin:	00	01	11	10
a: 0		1		1
1	1		1	

b cin:	00	01	11	10
a: 0			1	
1		1	1	1

$$\begin{aligned} s &= \bar{a}\bar{b}c_{in} + \bar{a}b\overline{c_{in}} + a\bar{b}\,\overline{c_{in}} + abc_{in} \\ c_{out} &= ab + ac_{in} + bc_{in} \end{aligned} \tag{2.8}$$

Alternatively, the expressions of s and c_{out} can be written as illustrated in Eq. (2.9) also using XOR gates, thus simplifying the gate-level schematic of the circuit, as shown in Fig. 2.30. However, this solution would result in a longer propagation delay as compared to the circuit designed from the SOP expressions in Eq. (2.8).

$$\begin{aligned} s &= \bar{a}\bar{b}c_{in} + \bar{a}b\overline{c_{in}} + a\bar{b}\,\overline{c_{in}} + abc_{in} \qquad \text{(canonical SOP)} \\ &= c_{in}(\bar{a}\bar{b} + ab) + \overline{c_{in}}(\bar{a}b + a\bar{b}) \\ &= c_{in}(\overline{a \oplus b}) + \overline{c_{in}}(a \oplus b) \\ &= a \oplus b \oplus c_{in} \\ c_{out} &= \bar{a}bc_{in} + a\bar{b}c_{in} + ab\overline{c_{in}} + abc_{in} \qquad \text{(canonical SOP)} \\ &= (\bar{a}b + a\bar{b})c_{in} + ab(\overline{c_{in}} + c_{in}) \\ &= (a \oplus b)c_{in} + ab \end{aligned} \tag{2.9}$$

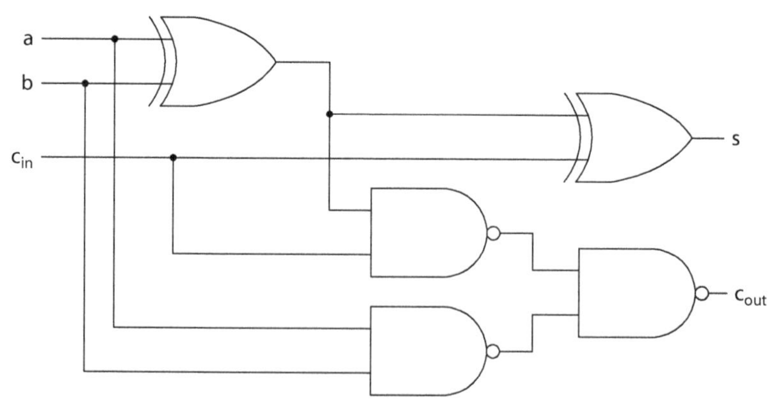

FIGURE 2.30 An alternative circuit for an FA.

Propagation Delay Estimation

Assuming that a NAND gate has 0.1 ns delay, Eq. (2.10) shows the estimated propagation delay, denoted by symbol Δ, for signals s and c_{out} in Eq. (2.8). The SOP expressions for s and c_{out} have three and two levels of gates, respectively. Wire delays are ignored in the calculation of the estimated delays for s and c_{out}.

$$\begin{aligned} \Delta s &= \Delta not + \Delta nand + \Delta nand \\ &= 0.1\text{ ns} + 0.1\text{ ns} + 0.1\text{ ns} \\ &= 0.3\text{ ns} \\ \Delta c_{out} &= \Delta nand + \Delta nand \\ &= 0.1\text{ ns} + 0.1\text{ ns} \\ &= 0.2\text{ ns} \end{aligned} \tag{2.10}$$

Equation (2.11) shows the estimated delays for signals s and c_{out} in Eq. (2.9) where an XOR designed from its SOP expression has 0.3 ns delay:

$$\begin{aligned} \Delta s &= 2 * \Delta xor \\ &= 2 * 0.3\text{ ns} \\ &= 0.6\text{ ns} \\ \Delta c_{out} &= \Delta xor + \Delta nand + \Delta nand \\ &= 0.3\text{ ns} + 0.1\text{ ns} + 0.1\text{ ns} \\ &= 0.5\text{ ns} \end{aligned} \tag{2.11}$$

2.8.2 Multiplexer

A 1-bit, 2-to-1 multiplexer, or MUX for short, is a simple combinational circuit as shown in Fig. 2.31. The inputs x and y are each 1-bit data, and s (a selector signal) causes the MUX to output either x or y. As illustrated in the block diagram, labels 1 and 0 are arbitrarily assigned to inputs x and y, respectively, and are reflected in the MUX's truth table (Table 2.9). The MUX outputs y when $s = 0$ or x when $s = 1$. Its minimal SOP expression is determined as follows:

$$r(s, x, y) = \Sigma(1, 3, 6, 7)$$

$$r = \bar{s}y + sx \tag{2.12}$$

xy:	00	01	11	10
s: 0		1	1	
1			1	1

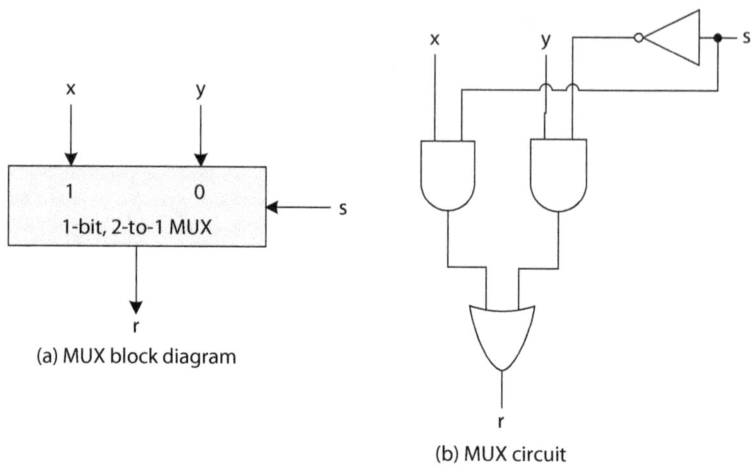

FIGURE 2.31 The block diagram and circuit of a 1-bit, 2-to-1 MUX.

For example, when $s=0$, Eq. (2.12) evaluates, as expected, to $r=y$, as illustrated here:

$$\begin{aligned} r &= \bar{0}y + 0x \\ &= 1y + 0 \\ &= y + 0 \\ &= y \end{aligned}$$

Likewise, when $s=1$, the MUX outputs (i.e., selects) input x. Figure 2.32 shows the block diagram of a 1-bit, 4-to-1 MUX with four data bits, w, x, y, and z, which are labeled as input numbers 3 to 0, respectively. The MUX requires two select signals, labeled s_1 and s_0. Table 2.10 shows its simplified truth table. Its expanded truth table would have six inputs, larger than the limit of four we have assumed for using K-maps. There are two ways to determine the minimal SOP expression of the 4-to-1 MUX without using a K-map: (1) use Espresso software, or (2) extrapolate

s	x	y	r
0	0	0	0
0	0	1	1
0	1	0	0
0	1	1	1
1	0	0	0
1	0	1	0
1	1	0	1
1	1	1	1

TABLE 2.9 Truth Table of the 1-bit, 2-to-1 MUX

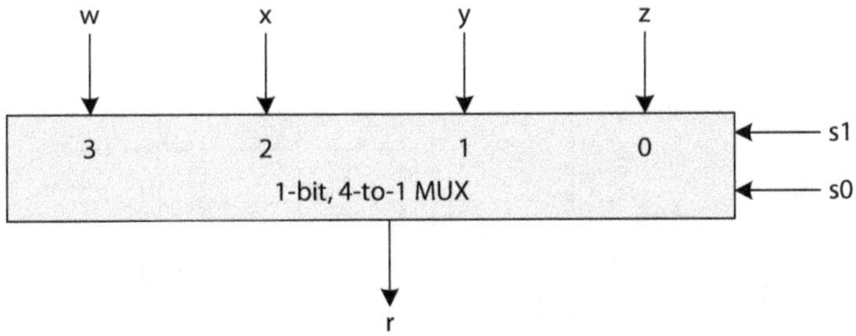

FIGURE 2.32 The block diagram of a 1-bit, 4-to-1 MUX.

Eq. (2.12) to four inputs and two selection signals. That is, when $s_1s_0 = 0 = (00)_2$, the MUX should output *z*, and for $s_1s_0 = 1 = (01)_2$, it outputs *y*; for $s_1s_0 = 2 = (10)_2$, it outputs *x*, and for $s_1s_0 = 3 = (11)_2$, it outputs *w*. Its minimal SOP expression, therefore, is:

$$r = \overline{s_1}\,\overline{s_0}\,z + \overline{s_1}\,s_0\,y + s_1\,\overline{s_0}\,x + s_1\,s_0\,w \tag{2.13}$$

For example, when $s_1s_0 = 2 = (10)_2$, Eq. (2.13) evaluates, as expected, to $r = x$, illustrated here:

$$\begin{aligned} r &= \overline{1}.\overline{0}.z + \overline{1}.0.y + 1.\overline{0}.x + 1.0.w \\ &= 0.1.z + 0.0.y + 1.1.x + 1.0.w \\ &= 0 + 0 + x + 0 \\ &= x \end{aligned}$$

The circuit for the 4-to-1 MUX is given in Fig. 2.33. As the size of an MUX increases, so do its fan-in and fan-out requirements. Consider the circuits for the aforementioned 2-to-1 and 4-to-1 MUXs. Their respective maximum fan-in and fan-out requirements are 2 and 2 and 4 and 3. Large MUXs, if designed using the methods discussed here, will lead to fan-in and fan-out problems. In Chap. 3, we will discuss design methodologies for large combinational circuits by first partitioning a problem into smaller design problems, and then for each smaller problem, a circuit is designed using the techniques learned in this chapter. The smaller circuits are then assembled to create a large combinational circuit that would be free of any fan-in and fan-out problems.

s_1	s_0	r
0	0	z
0	1	y
1	0	x
1	1	w

TABLE 2.10 Simplified Truth Table of the 1-bit, 4-to-1 MUX

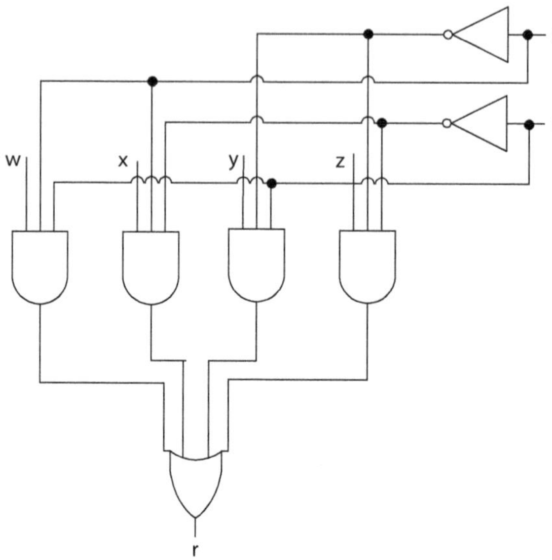

Figure 2.33 Circuit of a 1-bit, 4-to-1 MUX; maximum fan-in = 4; maximum fan-out = 2.

2.8.3 Decoder

The block diagram and the circuit of a 1-to-2 decoder circuit with active-low outputs are shown in Fig. 2.34. Active-low labeled pins are typically shown with bubbles at the base of each pin, as illustrated in Fig. 2.34(*b*). However, the bubbles in Fig. 2.34(*b*) do not necessarily imply that they are NOT gates. Table 2.11 shows the truth table of the decoder. Only one or none of the outputs $\_f_1$ and $\_f_0$ is asserted, depending on the values of signals v and e. When $e = 1$ and $v = 0$, $\_f_0 = 0$ (asserted). When $e = 1$ and $v = 1$, $\_f_1 = 0$ (asserted). Otherwise, when $e = 0$ (not active), both $\_f_1$ and $\_f_0$ are 1 (deasserted).

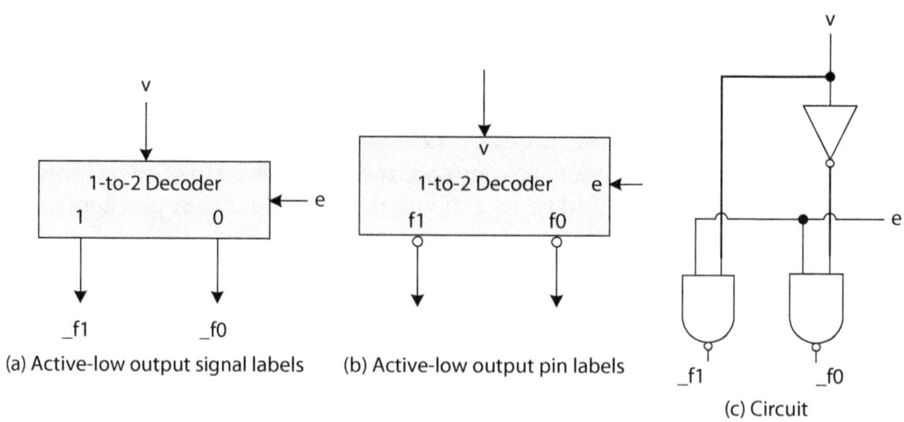

Figure 2.34 Block diagram and circuit of a 1-to-2 decoder: (*a*) block diagram with signal names; (*b*) block diagram with pin labels; (*c*) decoder circuit.

e	v	$\_f_1$	$\_f_0$
0	0	1	1
0	1	1	1
1	0	1	0
1	1	0	1

TABLE 2.11 A 1-to-2 Decoder Truth Table with Active-Low Outputs

The expressions of signals $\_f_1$ and $\_f_0$ may be expressed as SOP or POS. However, in this case, both SOP and POS expressions for these signals are the same as determined here. The NAND-only decoder circuit has one gate delay, not counting the NOT gate (Fig. 2.34(*c*)).

$$\_f_1(e, v) = \Sigma(0, 1, 2) = \Pi(3)$$

$$\_f_1 = \overline{e} + \overline{v} = \overline{ev}$$

$$\_f_0(e, v) = \Sigma(0, 1, 3) = \Pi(2)$$

$$\_f_0 = \overline{e} + v = \overline{e\overline{v}}$$

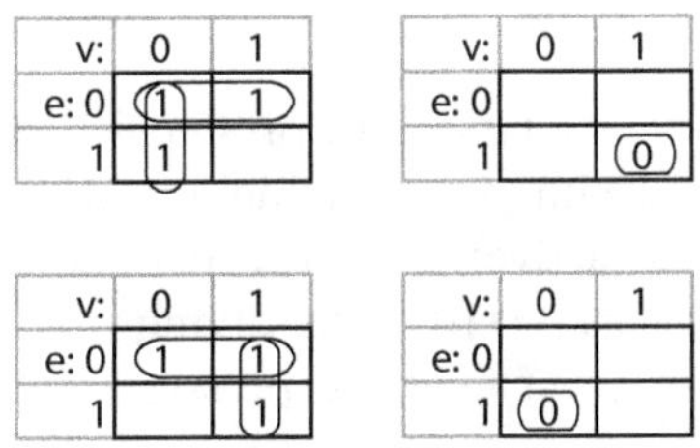

Large decoders are also designed using a different methodology to avoid fan-in and fan-out problems. Decoders have many applications and are used in the design of, for example, memory and CPU data paths. Decoders are used to decode a memory address so that the content of the address can be read or written. Decoders are also used to decode a register number when writing a register file (Chap. 1).

2.8.4 Encoder

The block diagram and circuit of a 3-to-2 encoder with active-low inputs are given in Fig. 2.35(*a*), and the active-low labeled pins are shown with bubbles in Fig. 2.35(*b*). Table 2.12 shows its truth table. The input signals are arbitrarily numbered 0 through 2 as illustrated in the figure. An encoder circuit outputs the number assigned to an active input signal. For example, when $\_z = 0$ (active), $\_y = 1$ (not active), and $\_x = 1$ (not active), the encoder outputs $r_1r_0 = (00)_2 = 0$, correctly identifying the asserted signal $\_z$ as the input number 0. However, when none of the encoder's inputs are active, another output signal named a (input-active) is necessary. When $a = 1$ (asserted), it indicates that one or more of the signals $\_x$, $\_y$, and $\_z$ are active, and thus the 2-bit result r_1r_0 identifies the active signal. On the other hand, when $a = 0$ (not asserted), the output $r_1r_0 = (00)_2$ is ignored.

It is also possible that two or more of the encoder inputs become active at the same time. For example, when $\_x = 0$, $\_y = 0$, and $\_z = 1$, the encoder must output either the number assigned to active signal $\_x$ or active signal $\_y$ based on some signal priority. Such an encoder is called a **priority encoder**.

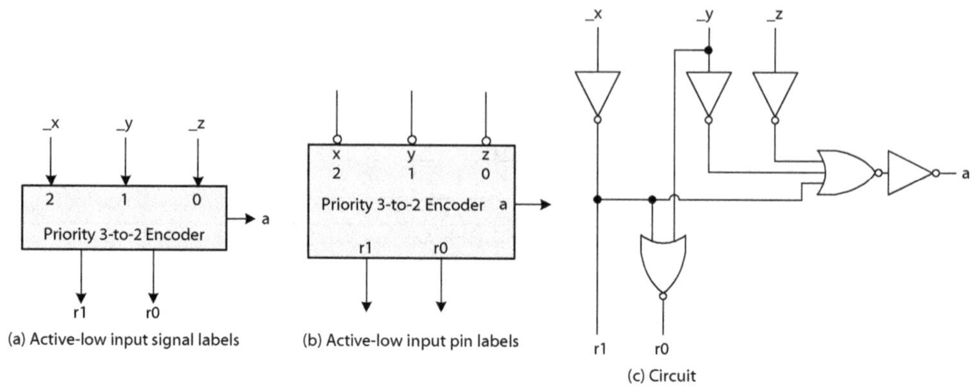

FIGURE 2.35 Block diagram and circuit of a 3-to-2 encoder: (*a*) block diagram with signal names; (*b*) block diagram with pin labels; (*c*) encoder circuit.

Table 2.12 presents the truth table of the 3-to-2 priority encoder (Fig. 2.35) with $\_x$ as the highest-priority input and $\_z$ the lowest. Therefore, when $\_x = 0$ (active), $\_y = 0$ (active), and $\_z = 1$ (not active), the encoder outputs $a = 1$ and $r_1 r_0 = (10)_2$, identifying $\_x$ as the higher-priority input signal.

The POS expressions for output signals a, r_1, and r_0 are determined as follows:

$$r_1(x, y, z) = \Pi(4, 5, 6) + \Pi_d(7)$$

$$r_1(x, y, z) = \overline{\_x}$$

_y_z:	00	01	11	10
_x: 0				
1	0	0	d	0

$$r_0(x, y, z) = \Pi(0, 1, 2, 3, 6) + \Pi_d(7)$$

$$r_0(x, y, z) = (\_x)(\_\overline{y}) = \overline{\overline{\_x} + \_y}$$

_y_z:	00	01	11	10
_x: 0	0	0	0	0
1			d	0

$$a(x, y, z) = \Pi(7)$$

$$a(x, y, z) = \overline{\_x} + \overline{\_y} + \overline{\_z} = \overline{\overline{\overline{\_x} + \overline{\_y} + \overline{\_z}}}$$

_y_z:	00	01	11	10
_x: 0				
1			0	

_x	_y	_z	a	r_1	r_0
0	0	0	1	1	0
0	0	1	1	1	0
0	1	0	1	1	0
0	1	1	1	1	0
1	0	0	1	0	1
1	0	1	1	0	1
1	1	0	1	0	0
1	1	1	0	d	d

TABLE 2.12 Truth Table of a 3-to-2 Encoder with Active-Low Inputs

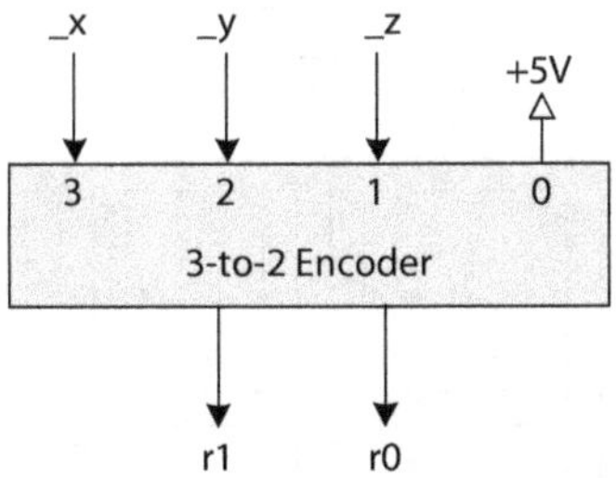

Figure 2.36 A 3-to-2 encoder block diagram without an input-active output signal.

Encoders are also designed without the output signal *a*, such as the one shown in Fig. 2.36. It is designed as a 4-to-2 encoder without the signal *a*, but the input number 0 is not used and is tied to power (or ground for active-high inputs), and thus effectively changes the circuit to a 3-to-2 encoder. When none of the inputs *_x*, *_y*, and *_z* are asserted, the encoder outputs $r_1r_0 = (00)2$, indicating that inputs are not active. When one or more of the input signals become active, the encoder outputs 3, 2, or 1, and respectively identifies input *_x*, *_y*, or *_z* as an active signal. This eliminates the logic required to generate *a*, and reduces one less signal when compared to the design shown in Fig. 2.35. The input that is tied to power can be implemented internally. Again, a different methodology is used to avoid fan-in and fan-out problems when designing a large encoder.

Encoders also have many applications, especially in the design of a motherboard. An encoder, for example, can be used to quickly inform the CPU when an external signal becomes active. The active external signal may be generated by an input/output (I/O) device or by a module on the board requesting a service from the CPU.

2.9 Implementation

Modern digital circuit designers rely on CAD tools to translate a design into implementation data. A digital design CAD tool synthesizes (translates) a description of a digital circuit into an optimized and technology-dependent gate-level description called a netlist. The application-specific integrated chip (ASIC) and FPGA are examples of noncustom IC technologies. A processor chip would typically be a custom IC. A circuit may be described schematically or using an HDL or a combination of both. However, modern CAD tools require circuits described in HDL.

2.9.1 Programmable Logic Devices

Programmable logic devices (PLDs) are prefabricated, off-the-shelf devices that contain no manufacturing faults. They can be programmed (i.e., configured) to implement a netlist, instantly, and sometimes dynamically on demand. Simple PLDs (SPLDs) are the simplest among all PLDs. An SPLD uses wired-logic to implement logic expressions, and is suitable for implementing small digital circuits. Complex PLDs (CPLDs) are the next generation of PLDs that contain configurable wiring channels within the chip to implement a more complex digital circuit.

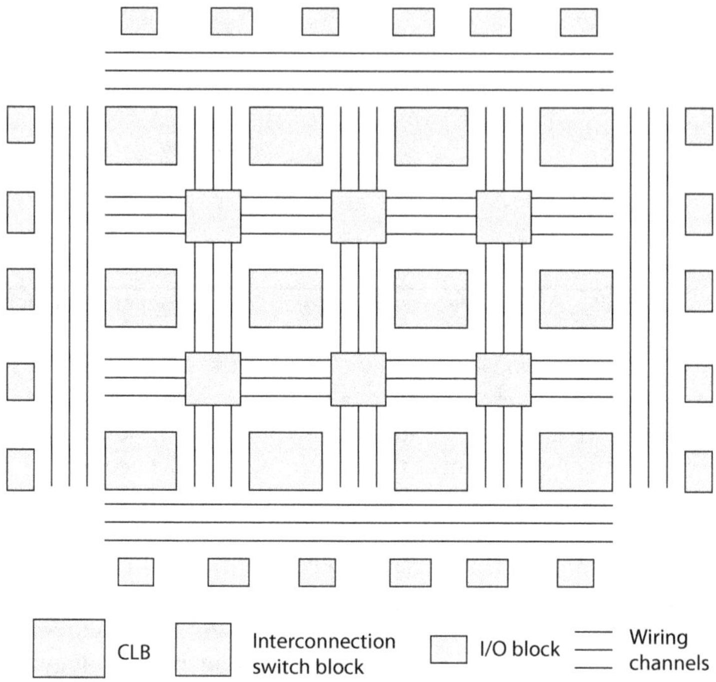

FIGURE 2.37 A simple FPGA block diagram.

FPGAs, which briefly were discussed in Chap. 1, are the modern version of PLDs that contain many configurable logic blocks (CLBs), configurable wiring channels, and configurable IO blocks that interface with the chip's (I/O) pins. An FPGA may be viewed as the modern-day equivalent of both the TTL 7400 chip series and circuit boards. The 7400 series were the first family of ICs that were designed for general use. The series included the standard logic gates, as well as larger combinational logic modules, such as MUX, decoder, and adder, and modules used to design sequential circuits. They were used to build the mini and mainframe computers during the 1960s and 1970s. Today, the 7400 series chips are sometimes used in education, especially in some introductory logic design courses.

An FPGA requires programming data to configure and interconnect CLBs and I/O blocks as indicated by a netlist. Some I/O blocks are configured as input pins and some as output pins. Figure 2.37 illustrates the internal organization of a simple FPGA with nine CLBs, each capable of implementing one or two simple logic functions. The wiring channels and the switch blocks are used to interconnect the inputs and outputs of each CLB to other CLBs and via the I/O blocks to I/O pins.

Some FPGAs contain a configuration memory for implementing a different netlist on demand. Modern FPGAs typically contain thousands of CLBs, and some

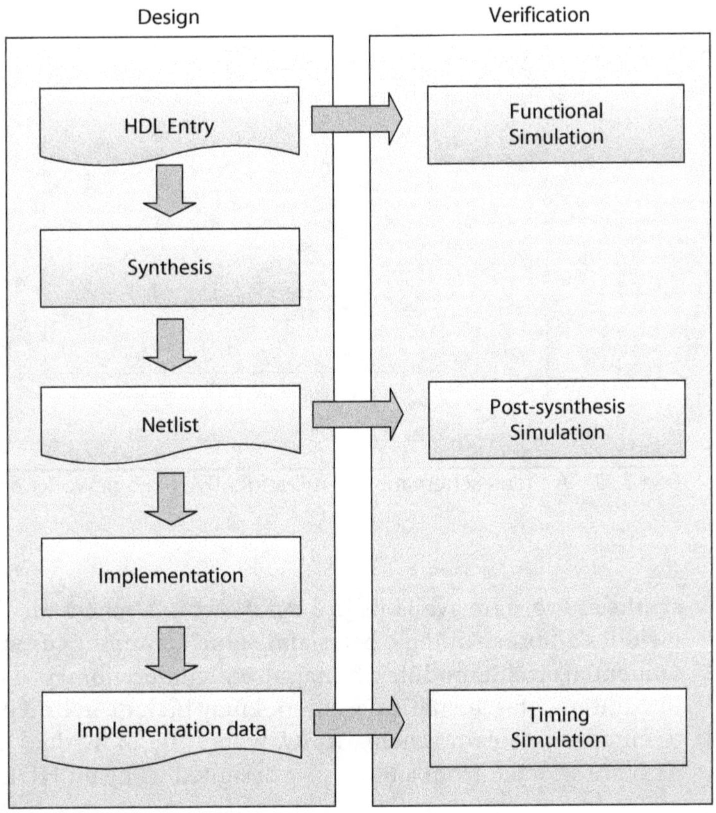

Figure 2.38 Digital circuit design flow [6].

also contain memory blocks. There are system-on-chip (SoC) FPGA chips [3–4] that also contain complex modules such as CPU and digital signal processor (DSP). With these chips, it is easier to design custom and complex digital circuits without requiring fabrication. Both Altera and Xilinx provide FPGA design kits with a universal serial bus (USB) interface [3–5].

2.9.2 Design Flow

Figure 2.38 illustrates a typical digital circuit design flow. It includes design entry, synthesis, and implementation phases. Each step in the design flow produces a different description of a target circuit, where each description is verified for design errors.

Design Entry

During this phase of the design, a target digital circuit is described manually using a schematic design tool, an HDL, or a combination of both. The strict uses of schematic design tools in the industry have diminished over the years in favor of HDLs. The Verilog HDL is discussed in Sec. 2.10. Figure 2.39 shows the design of an FA using a schematic design and simulation tool called LogicWorks [7]. No

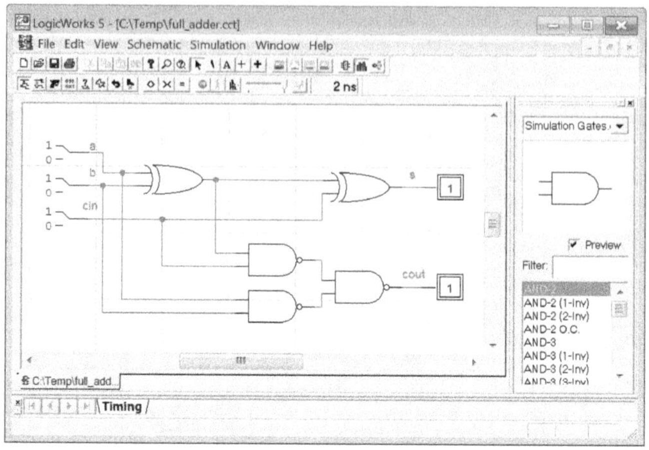

FIGURE 2.39 A circuit schematic of a full-adder (FA) in LogicWorks, a schematic design tool.

synthesis tools are available in LogicWorks. A schematic design tool typically includes a library of logic gates and some commonly used combinational and sequential circuit modules. It may also include a library of the 7400 chip series. In addition, an schematic design tool may include hybrid design entry features to enter a data path schematically, where the individual modules in the data path are selected from a library or designed using an HDL.

Functional Simulation

A schematically and/or HDL designed circuit must be verified to make sure it operates as expected. For example, for a given input, does the FA shown in Fig. 2.39 output the same results as indicated in its truth table? Because this verification process can take a long time, especially for large circuits, the process is typically divided into functional, post-synthesis, and timing simulations. A functional simulation is used only to verify the correctness of a design without concern for its implementation issues. It is the first step to make sure a design is correct. If the functional simulation of a circuit is error free, then synthesis is started as the first phase of implementation; however, the design may still have synthesis and timing errors.

Post-Synthesis Simulation

During the synthesis phase, a design is translated into its corresponding netlist based on the available resources in a given technology, such as the logic resources available in each CLB in a specific FPGA chip. Each CLB can only implement a few and simple logic expressions—for example, two 4-variable logic expressions. A design's logic expressions may need to be divided into simpler expressions—for example, no more than four variables. The simpler expressions are saved as the circuit's netlist.

A post-synthesis simulation phase may be necessary to make sure a design has been translated correctly and that the generated netlist accurately describes the target circuit. A post-synthesis simulation requires less processing time than does a timing simulation. In addition, some delay information—for example, the CLB's signal propagation delay—may be available for post-synthesis simulation.

Timing Simulation

A timing simulation is performed after a netlist is mapped to the available resources of a target device that is virtually modeled in the computer. For example, using the virtual model of an FPGA chip, a netlist is used to configure the CLBs, the I/O blocks, and the wiring channels through a process called **placement-and-route**.

During the placement phase, the minimized expressions in the netlist are assigned to the CLBs and the circuit's primary input and output signals are assigned to the I/O pins via the I/O blocks. Some designs may also require complex modules—for example, CPU, DSP, and memory, that already exist inside the chip.

During the routing phase, the signal dependency information in the netlist is used to wire (interconnect) the signals among the different CLBs and the I/O blocks using the available on-chip wiring channels and switch blocks. However, the placement and the routing tasks are not typically done independently; the assignments of logic expressions to the CLBs and the primary I/O signals to I/O blocks may be changed in order to (1) maximize the use of the available resources in the chip or (2) minimize propagation delays. A timing simulation is used to make sure the timing requirement of a design is met.

2.10 Hardware Description Languages

Verilog and VHDL (VHSIC, or very-high-speed integrated circuit) are two industry-standard HDLs used to describe digital circuits. An HDL is used to formally describe a digital circuit and a **test-bench** is used to generate tests (i.e., test vectors) for the circuit.

An HDL description is called structural if a circuit is described in terms of a set of interconnected modules. The modules can be small, like AND, OR, NAND, etc., logic gates, or large circuits, like a decoder, multiplexer, adder, etc. Some commonly used large modules, such as an adder, may be predefined and used during synthesis.

An HDL description is called behavior if the HDL code describes the relationships between a module's inputs and its outputs using high-level language statements such as "if-else" or "case" (i.e., switch).

2.10.1 Structural Model

In this section, we briefly introduce the Verilog HDL. Other examples are included elsewhere in the book. However, the description is not complete and additional references may be necessary. Example 2.11 illustrates a structural model of an FA with two XOR gates and three NAND gates, as shown schematically in Fig. 2.39. A

Verilog model starts with the keyword "module" and includes a name (e.g., full_adder) and a list of input and output ports (e.g., *a*, *b*, *cin*, *s*, and *cout*). A module description ends with the keyword "endmodule." The port listing can appear in any order, but must be specifically declared as "input" or "output." Signal names that are not declared as input and output ports are considered local and should be declared as a wire when the design is structural. For instance, the three signals *out1*, *out2*, and *out3* in the example are all local.

Example 2.11. A Verilog structural model for the FA shown in Fig. 2.39:

```
module full_adder
(
  input a, b, cin,
  output s, cout
); //defines a module's name and its interface signals

wire out1, out2, out3; //defines local signal names

xor   x1(out1, a, b);
xor   x2(s, out1, cin);
nand  n1(out2, out1, cin);
nand  n2(out3, a, b);
nand  n3(cout, out2, out3);

endmodule
```

The standard gates are called **primitive gates** and are known to the Verilog complier and need not be described. The x1, x2, n1, n2, and n3 in the example are optional and are names given to two instantiated XOR and three NAND primitive gates. The leftmost argument in each instantiated primitive gate is output, and the others are inputs. For instance, the signals *out1*, *s*, *out2*, *out3*, and *cout* are all outputs and thus are listed as the leftmost argument. A primitive gate can be instantiated with one or more input arguments, depending on its type. For example, a three-input primitive NAND gate would have one output (the leftmost) and three input arguments.

The modules can be instantiated in any order, similar to the way they are instantiated on the screen when using a schematic design tool (e.g., Fig. 2.39). The interconnections of the modules are determined from the list of their ports. For instance, the first instantiated primitive XOR gate has *out1* as its output port, and the second XOR uses *out1* as an input port. This implies that there is a wire that connects the two *out1* ports. Likewise, the *out2* and *out3* signals are each connected by a wire. These signals are declared as "wire."

A test-bench module, such as the one given in Example 2.12, is also described in HDL and is used to test a circuit model, such as the FA model in Example 2.11. A test-bench module has no input or output ports. The `include directive may be needed (depending on the design tool) to import a nonprimitive but already created HDL model in another model—for example, a test-bench. In general, each imported module may be instantiated one or more times as needed to create the target circuit model before testing. In Example 2.12, the "full_adder" HDL model is imported and is instantiated once for testing.

An initial block is used to list the test vectors in the "full_adder" module. All the statements inside an initial block are processed in sequence. All the variables to the left of an assignment operator (e.g., =) within an initial block or an always block (discussed later) must be declared as type `reg`. When describing a combinational circuit, the type `reg` has no specific significance. The type `reg`, however, becomes important when designing sequential circuits.

Example 2.12. A test-bench for testing the FA structural model in Example 2.10:

```
'include "full_adder.v" //input FA's description from local folder
module tester();
reg a, b, cin;

wire s, cout;

full_adder fa1(a, b, cin, s, cout); //instantiate FA

initial begin //start the test
$display("Time a b cin cout s"); //header for the output
$monitor ("%4d %b  %b", $time, cout, s);
a = 0; b = 0; cin = 0; $display("%4d %b %b %b", $time, a, b, cin);
//test 1
#1 //simulate
a = 1; b = 1; cin = 0; $display("%4d %b %b %b", $time, a, b, cin);
//test 2
#1 //simulate
a = 1; b = 1; cin = 1; $display("%4d %b %b %b", $time, a, b, cin);
//test 3
#1 //simulate
$finish; //stops simulation
end
endmodule
```

When using a design tool without a debugger, a `$display` statement is used to output the value of one or more input signals. A `$monitor` statement, on the other hand, is entered only once to track the values of one or more signals (input or output) during a simulation run. Each time that there is a change in the value of one or more signals listed in the `$monitor` statement, the statement is executed to output the signal values. The syntax for both the `$display` and `$monitor` are the same and are similar to that of a "`printf`" statement in the C programming language. The output format "`%d`", "`%h`", "`%o`" and "`%b`" can be used to display values in decimal, hexadecimal, octal, and binary, respectively. Additional display formats, such as "`%s`" and "`%f`" are used to display string and floating-point numbers.

A simulation time-step is specified using the symbol "#" followed by the length of simulation time as an integer number. If no delays are assigned to modules, a functional simulation will assume each module has zero propagation delay and one simulation time (#1) is sufficient to generate outputs.

Using the Synopsys design tool, the simulation output to test the FA model in Example 2.12 is shown here. It can be seen that when, for example, $a = 1$, $b = 1$, $cin = 1$, the `$monitor` outputs $s = 1$ and $cout = 1$ at simulation time = 2.

```
Chronologic VCS simulator
Contains Synopsys proprietary information.
Compiler version D-2009.12; Runtime version D-2009.12;

Time   a    b   cin cout  s
 0     0    0    0
 0                    0    0
 1     1    1    0
 1                    1    0
 2     1    1    1
 2                    1    1
$finish called from file "tester2.v", line 18.
$finish at simulation time   3
```

VCS Simulation Report

```
Time: 3
CPU Time:  0.460 seconds;    Data structure size:  0.0Mb
```

The "=" symbol is called a blocking assignment. All the blocking assignment statements within an initial or an always block are evaluated sequentially one at a time, much like in a programming language. On the other hand, a nonblocking assignment (discussed later) is indicated by the symbols "<=" and is evaluated simultaneously with the other nonblocking statements in an initial or an always block. The HDL statements within an initial block are evaluated only once, whereas those of an always block are evaluated as long as the circuit is being simulated, much like a real circuit that operates as long as it is powered.

Verilog also supports "for-loop," "case" (switch), "forever," and other control statements. However, not all Verilog statements are synthesizable. A test-bench, such as the one shown in Example 2.13, uses a for-loop to fully test the FA model. In the example, signals *a*, *b*, and *cin* are declared as a 1-bit `reg`, and variable *k*, used in the for-loop, is declared as a 4-bit `reg` in Little Endian bit order. The "`reg [0:3] k;`" would define *k* in Big Endian bit order. A multibit variable can be referenced as a group or individually bit by bit. For example, in the example, `k[2]` refers to the third bit of the multibit variable *k*, and `k[0]` refers to the least significant bit (LSB) in *k*.

Example 2.13. A test-bench model to fully test the FA structure model given in Example 2.10:

```
'include "full_adder.v"
module tester();
reg a, b, cin;

reg [3:0] k;
wire s, cout;
```

```
full_adder  fa1(a, b, cin, s, cout);

initial begin
$display("Time a b cin cout s");
$monitor ("%4d  %b %b", $time, cout, s);

for (k = 0; k <= 7; k = k+1) begin
   #1  a = k[2]; b = k[1]; cin = k[0]; $display("%4d %b %b %b",
$time, a, b, cin);
end
#10 $finish; //stops simulation
end
endmodule
```

In the test-bench in Example 2.13, a new value is assigned to each of the inputs *a*, *b*, and *cin* during each simulation step. The inputs are also displayed before each simulation step. After each simulation step, the `$monitor` statement automatically displays the values of the output signals, provided that one or more signal values change. The output of the simulation run is shown next. Note that the `$monitor` does not display outputs when signals *s* and *cout* do not change between the test vector $a = 0$, $b = 0$, and $cin = 1$ and the test vector $a = 0$, $b = 1$, and $cin = 0$, and again between the test vector $a = 1$, $b = 0$, and $cin = 1$ and the test vector $a = 1$, $b = 1$, and $cin = 0$.

```
Chronologic VCS simulator
Contains Synopsys proprietary information.
Compiler version D-2009.12; Runtime version D-2009.12;

Time  a    b   cin cout  s
  0                 x    x
  1   0    0    0
  1                 0    0
  2   0    0    1
  2                 0    1
  3   0    1    0
  4   0    1    1
  4                 1    0
  5   1    0    0
  5                 0    1
  6   1    0    1
  6                 1    0
  7   1    1    0
  8   1    1    1
  8                 1    1

$finish called from file "tester.v", line 16.
$finish at simulation time   18
```

VCS Simulation Report

```
Time: 18

CPU Time:  0.390 seconds;    Data structure size:  0.0Mb
```

2.10.2 Propagation Delay Simulation

It is also possible to include an optional propagation delay for each of the primitive gates at the time of their instantiations. This provides a more realistic functional simulation. Example 2.14 is a description of the FA with a 1 ns delay assigned to each of the primitive NAND gates and a 3 ns delay to each of the primitive XOR gates. Therefore, $\Delta s = 6$ ns and $\Delta cout = 5$ ns.

Example 2.14. Structural modeling of an FA using primitive gates with delays:

```
'timescale 1ns/100ps
module full_adder
(
  input a, b, cin,
  output s, cout
);

wire out1, out2, out3;

xor    #3 x1(out1, a, b);
xor    #3 x2(s, out1, cin);
nand   #1 n1(out2, out1, cin);
nand   #1 n2(out3, a, b);
nand   #1 n3(cout, out2, out3);

endmodule
```

The compiler directive `'timescale` indicates the timing scale applied during the simulation. The `'timescale` in Example 2.14 defines the scale as 1 ns with 100 ps (picoseconds) increments. However, the timing scale is an estimation and the simulation results do not provide any real timing data.

The test-bench in Example 2.15 contains two test vectors entered at simulation times 0 and 10. As illustrated in the following simulation output, initially, the values of both signals *cout* and *s* are unknown ("x"). For the test vector $a = 0$, $b = 0$, and $cin = 1$, which is applied at simulation time 0, the `$monitor` outputs, as expected, $cout = 0$ at time = 5 and $s = 1$ at time = 6. For the test vector $a = 0$, $b = 1$, and $cin = 1$ at time = 10, the outputs, as expected, are $cout = 1$ at time = 15 and $s = 0$ at time = 16. The two test vectors were selected to expose the worst-case delay scenarios.

Example 2.15. A test-bench for the FA model with delays:

```
'include "full_adder.v"
'timescale 1ns/100ps
module tester();
reg a, b, cin;

wire s, cout;

full_adder  fa1(a, b, cin, s, cout);
initial begin
```

```
$display("Time  a  b  cin  cout  s");
$monitor ("%4d  %b  %b", $time, cout, s);
a = 0; b = 0; ci = 1;  $display("%4d  %b  %b  %b", $time, a, b, ci);
#10
a = 0; b = 1; ci = 1;  $display("%4d  %b  %b  %b", $time, a, b, ci);
#10 $finish;
end
endmodule
```

```
Chronologic VCS simulator copyright 1991-2005
Contains Synopsys proprietary information.

Time  a    b    c   cout   s
  0   0    0    1
  0                 x      x
  5                 0      x
  6                 0      1
 10   0    1    1
 15                 1      1
 16                 1      0
$finish at simulation time   200
```

VCS Simulation Report

```
Time: 20000 ps
CPU Time:  0.040 seconds;  Data structure size:  0.0Mb
```

Nonprimitive modules cannot be instantiated with delay information. For these modules, the delay information is determined from the delay values specified within the module. However, it is possible to use a parameterized delay to overwrite the delay information of any module during instantiation.

2.10.3 Behavioral Modeling

The basic behavioral description in Verilog is the assign statement. It is used to directly enter a Boolean expression using the symbols "~", "&", "|", and "^" to express bit-wise NOT, AND, OR, and XOR operators, respectively. The NAND, NOR, and XNOR operators are also expressed using combined symbols "&~", "|~", and "~^" or "^~", respectively. Table 2.13 is a summary of the operators used in the Verilog HDL. Example 2.16 illustrates a behavior description of a 1-bit 2-to-1 MUX using an assign statement.

Example 2.16. A behavior model of a 1-bit 2-to-1 MUX using an assign statement:

```
module mux1bit

 (
   input s,
   input x, y,
   output r
 );
 assign r = ~s & x | s & y;

 endmodule
```

Precedence	Operator Type	Symbol	Example
Highest	Unary	+, −, !, ~	+a, −a (negate a), !a (logical not), ~a (bidwise not)
	Exponential	**	a ** 3 (a cubed)
	Arithmetic 1	*, /, %	a * b (multiply), a / b (divide), a % b (mod)
	Arithmetic 2	+, −	a + b (add), a − b (subtract)
	Shift:		
	Logical	<<, >>	a << 2 (shift left twice)
	Arithmetic	<<<, >>>	a >>> 3 (shift right 3 times extending the sign bit)
	Relational	<, <=, >, >=	a >= b (a greater or equal to b)
	Equality		
	Logical	==, !=	a == b if a is identical to b excluding x and z
	Case	===, !==	a === b is identical to b including x and z
	Bit-wise		
	Basics	&, |, ~, ^	a & b (and), a | b (or), ~a (not), a ^ b (xor)
	Combined	&~, |~, ~^, ^~	a &~ b (nand), a |~ b (nor), a ~ ^b (xnor), a ^~ b (xnor)
	Logical	&&, ||, !	a &&b (and), a || b (or), !a (not)
Lowest	Conditional	?:	(a >= b) ? a − b : b − a

Table 2.13 A Summary of the Verilog HDL Operators

The most commonly used statement to describe the behavior of a circuit is an `always` block. It is declared using the keyword "always" followed by the symbol @ and a set of signal names as its **sensitivity list**, and like an `initial` block, contains a `begin-end` block. Example 2.17 illustrates an alternative behavior model of a 1-bit 2-to-1 MUX using an "if-else" statement. Because the *r* signal depends on signals *s*, *x*, and *y*, these signals are included in the sensitivity list of the `always` block.

Example 2.17. A behavior modeling of a 1-bit, 2-to-1 MUX using an "if-else" statement:

```
module mux1bits
(
  input s,
  input x, y,
  output reg r
);

always@(s or x or y)
begin
```

```
    if (s == 1'b0)
        r = x;
    else
        r = y;
end

endmodule
```

Like an `initial` block, all the variables to left of an assignment (=) symbol within an `always` block must additionally be declared as type `reg`. A proper way to declare an output variable both as type `output` and type `reg` is the combined syntax `output reg`, as illustrated in Example 2.17.

The syntax `1'b0` is used to enter a 1-bit binary number. Other examples to enter numbers are `5'b11111`, `8'hFF`, and `9'o777` to enter a 5-bit binary number $(11111)_2$, an 8-bit hex number 0xFF, and a 9-bit octal number $(777)_8$, respectively, where "8" is used here to indicate an octal number.

In addition, depending on the compiler version, alternative syntaxes are available to describe a sensitivity list. For example, "`always@(s or x or y)`", "`always@(s, x, y)`", "`always@(*)`", or "`always@*`" are all acceptable syntaxes to enter a sensitivity list for a combinational circuit model. Furthermore, using a * as the sensitivity list in a combinational `always` block is the most preferred syntax, allowing the compiler to determine the list of sensitivity variables. A missing variable in a sensitivity list can result in an incorrect combinational circuit behavior.

Example 2.18 illustrates the behavior model of an FA using a `case` statement to enter its truth table. In the example, the curly brackets ({}) indicate concatenation. A "`default`" case is also normally required to handle missing cases. The acceptable signal values are 0, 1, *x* (unknown), and *z* (high-impedance, Z). Alternatively, "`casex`" ignores the *x* and *z* signal values and treats them as don't-cares.

Example 2.18. A behavior model of a FA using a "`case`" statement:

```
module full_adder
(
  input a, b, cin,
  output reg s, cout
);

always@(a or b or cin)
begin
  case ({a, b, cin})
  3'b000:  begin s = 0; cout = 0; end
  3'b001:  begin s = 1; cout = 0; end
  3'b010:  begin s = 1; cout = 0; end
  3'b011:  begin s = 0; cout = 1; end
  3'b100:  begin s = 1; cout = 0; end
  3'b101:  begin s = 0; cout = 1; end
  3'b110:  begin s = 0; cout = 1; end
```

```
    3'b111:  begin s = 1; cout = 1; end
    default: begin s = 0; cout = 0; end
    endcase
  end

  endmodule
```

Example 2.19 illustrates a behavior description of a 4-bit tri-state buffer with an active-low enable signal, where `4'bz` indicates a 4-bit high-impedance Z value.

Example 2.19. A behavior model of a 4-bit tri-state buffer:

```
module tristate4bits
(
  input _e,   //declared active-low
  input [3:0] x,
  output reg [3:0] r
);
always@(*)
begin
  if (_e == 1'b0)
         r = x;
  else
         r = 4'bz;   //or r = 4'bzzzz;
end
endmodule
```

2.10.4 Synthesis and Simulation

The FA HDL model in Example 2.18 was synthesized using the Altera Quartus II design and synthesis tool [4]. For synthesis, EP4CGX15BF14A7 1.2V, one of the Altera Cyclone IV GX family of programmable chips, was used. The synthesized circuit was then simulated with the Altera ModelSim simulation tool. Figure 2.40 illustrates the synthesized circuit with two CLBs used to implement the expressions for the sum and carry-out bits. In this case, the inputs *a*, *b*, and *cin* and the outputs *s* and *cout* are also buffered to prevent possible fan-out violations. The simulation of the synthesized circuit with eight test vectors using a waveform is shown in Fig. 2.41.

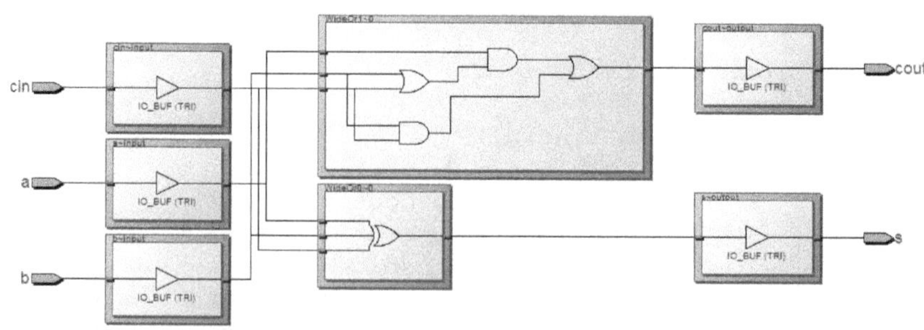

Figure 2.40 The synthesized circuit of the FA behavioral model given in Example 2.18.

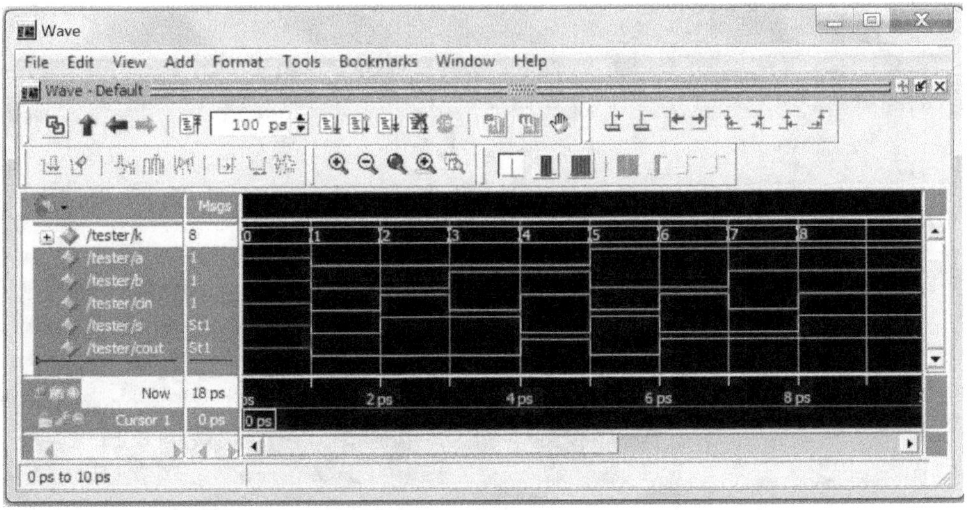

Figure 2.41 A gate-level simulation of the synthesized FA in Fig. 2.40.

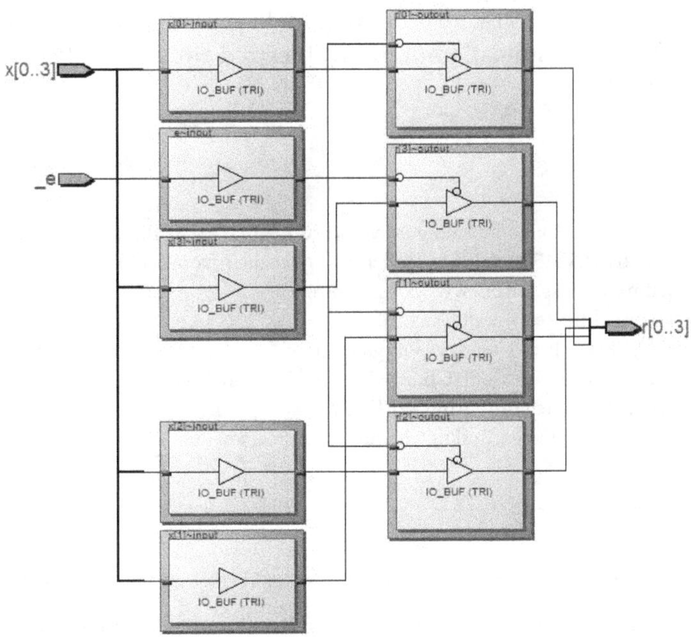

Figure 2.42 A synthesized 4-bit tri-state buffer using the Altera Quartus II design tool.

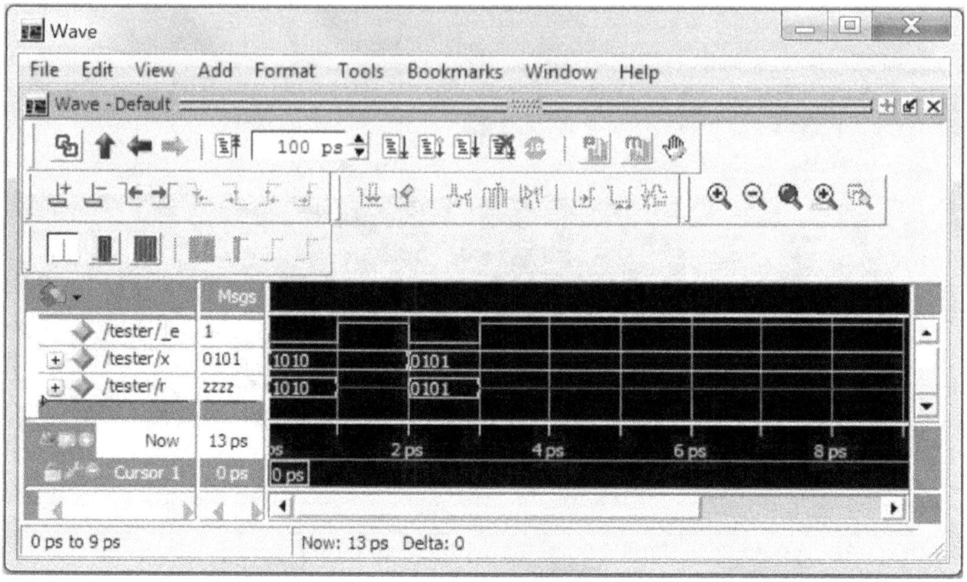

FIGURE 2.43 The simulation output of the 4-bit tri-state buffer in Fig. 2.42.

Likewise, the 4-bit tri-state buffer model in Example 2.19 was synthesized and simulated using the Altera design, synthesis, and simulation tools. The synthesized circuit is shown in Fig. 2.42. Because the enable signal *_e* is active-low, an extra NOT gate is not required to operate the active-low enabled tri-state buffers. The simulation of the synthesized circuit with two test vectors is shown in Fig. 2.43.

References

1. Espresso, http://diamond.gem.valpo.edu/~dhart/ece110/espresso/tutorial.html.
2. SN54/74LS245 Octal Bus Transmitter/Receiver from 7400 chip series.
3. Xilinx FPGAs, http://www.xilinx.com/.
4. Altera , http://www.altera.com.
5. EasyFPGA, http://www.easyfpga.com/.
6. Ducan Buel, Tarek El-Ghazawi, Kris Gaj, and Voldymyr Kindratenko, "High-performace reconfigurable computing," *IEEE Computer*, March 2007, pp. 23–27.
7. LogicWorks, Digital design schematic tool, Pearson Publishing, http://www.pearsonhighered.com/.

Exercises

2.1 Evaluate $f = x\bar{y} + yz$ for $x = 1, y = 0$, and $z = 1$ and for $x = 1, y = 1$, and $z = 0$.

2.2 Evaluate $y = \overline{\_c\bar{x}} + \_cx$ for $\_c = 0$ and $x = 1$ and for $\_c = 1$ and $x = 1$ where $\_c$ is an active-low signal.

2.3 Proof Demorgan's theorem $\overline{xy} = \bar{x} + \bar{y}$ by creating truth tables for $f = \overline{xy}$ and $g = \bar{x} + \bar{y}$. Are the two truth tables identical?

2.4 Proof Demorgan's theorem $\overline{x+y} = \bar{x}\,\bar{y}$ by creating truth tables for $f = \overline{x+y}$ and $g = \bar{x}\,\bar{y}$. Are the two truth tables identical?

2.5 Draw the circuit schematic for $f = x\bar{y} + yz$ and then convert the schematic to NAND gates using the steps illustrated in the textbook.

2.6 Evaluate $f = (x + y)(\bar{y} + z)$ for $x = 1, y = 0$, and $z = 1$ and for $x = 1, y = 1$, and $z = 0$.

2.7 Draw the circuit schematic for $f = (x + y)(\bar{y} + z)$ and then convert the schematic to NOR gates using the steps illustrated in the textbook.

2.8 Given $f = x\bar{y} + yz$ (an SOP expression) determine its equivalent POS expression. Hint: First find the SOP of $\bar{f}$ and then use the rule "POS expression of *f* = Complement of the SOP expression of $\bar{f}$ ".

2.9 Obtain the POS expression of *f* by applying the Dual Principle to the SOP of $\bar{f}$ where $f = x\bar{y} + yz$.

2.10 Suppose we would like to build function Y = 2X + 3 where X denotes a 3-bit unsigned value $(x2x1x0)_2$ and Y = *y*4..*y*0 is a 5-bit value in hardware. Construct its truth table where input bits are *x*2, *x*1, and *x*0 and output bits are *y*4 through *y*0. Then do the following for output *y*2 (you may repeat this for the other outputs):

a. Determine the canonical SOP expression for output bit *y*2.

b. Write the min-terms for *y*2.

c. Use K-map and find a minimal SOP expression for *y*2.

d. Draw a minimal NAND circuit for *y*2.

e. Compare the number of transistors required to implement the canonical and the minimal SOP expressions.

2.11 Repeat Problem 2.10 but this time use POS expressions for *y*2.

2.12 Repeat Problem 2.10 *b* through *d* but this time use 3-bit 2's complement values for X and output *y*4.

2.13 Repeat Problem 2.12 *b* through *d* but this time use POS expressions for *y*4.

2.14 Find a minimal SOP expression for each of the following functions using K-maps:

a. $f(w, x, y, z) = \Sigma(0, 2, 8, 10) + \Sigma_d(12, 14)$

b. $g(a, b, c, d) = \Sigma(5, 7, 13, 15) + \Sigma_d(6, 14)$

c. $h(w, x, y, z) = \Pi(0, 2, 8, 10) + \Pi_d(12, 14)$

d. $t(a, b, c, d) = \Pi(5, 7, 13, 15) + \Pi_d(6, 14)$

2.15 Find minimal POS expressions for each of the functions given in Problem 2.14.

2.16 Use Espresso software to generate the minimal SOP expressions for all the output bits of function Y = 2X + 3 where X is a 4-bit unsigned value less than 10. The X values 10 to 15 are ignored and are treated as don't cares.

2.17 Repeat Exercise 2.16 but use 4-bit 2's complement values for $-5 \le X \le 5$ and don't care for $X \le -5$ and $X \ge 5$.

2.18 Given the function Y = X mod 7 where X = $x3x2x1x0$ is a 4-bit unsigned input and Y = $y2y1y0$ is a 3-bit unsigned result, create a truth table for Y and determine SOP and POS expressions for $y2$, $y1$, and $y0$.

2.19 Use the logic minimization algorithm to determine a minimal SOP expression for the output bit $y0$ of function Y = X − 3 where X = $x3..x0$ and Y = $y3..y0$ are 4-bit 2's complement numbers.

2.20 Use the logic minimization algorithm to determine a minimal SOP expression for $y = \Sigma(2, 3, 6, 9, 10, 13)$.

2.21 The prime implicants for $f(a, b, c, d) = \Sigma(1, 3, 5, 7, 10, 11, 14, 15)$ are $\bar{a}d + ac$ and cd. The timing diagram for its minimal expression $f = \bar{a}d + ac$ is shown in Fig. 2.25. Draw the circuit for the non-minimal $f = \bar{a}d + ac + cd$, which includes all its prime implicants, and label its internal signals. Draw a timing diagram for the new circuit when its input change from $acd = 111$ to $acd = 011$. Does the circuit produce a glitch?

2.22 The minimal POS expression for $f(a, b, c, d) = \Pi(0, 2, 4, 6, 8, 9, 12, 13)$ has two essential prime implicants $(a + d)$ and $(\bar{a} + c)$ and a non-essential prime implicant $(c + d)$.

a. Draw a timing for minimal $f = (a + d)(\bar{a} + c)$ when its inputs change from $acd = 000$ to $acd = 100$. Does the circuit produce a glitch?

b. Draw a timing for the non-minimal $f = (a + d)(\bar{a} + c)(c + d)$ when its inputs change from $acd = 000$ to $acd = 100$. Note f includes all its prime implicants. Is there a 1-hazard?

2.23 Design a two-input wired-OR gate (Hint: Use Demorgan's theorems.).

2.24 Determine a POS expression for the 2-to-1 MUX in Fig. 2.31.

2.25 Design an arbitrary function $f(w, x) = \Sigma(0, 2)$ using a 4-to-1 MUX.

2.26 Use a 2-to-4 decoder to connect four modules each outputting one bit to a one-bit bus. Only one module at a time can place data on the bus. At times no module may be allowed to place data on the bus. Show details.

2.27 Design a circuit for the 3-to-2 encoder in Fig. 2.35 using NAND gates.

2.28 Suppose, the decoder in Problem 26 is able to activate each output signal every 10 ns in a round-robin fashion and allow each module to output 1 bit once every 10 ns. What is the peak rate of transfer for each module in bytes? Also, what is the peak bus bandwidth? Hint: Rate of transfer and bandwidth are measured in bytes per second. Peak rate of transfers is the maximum number of bytes (in KB, MB, etc.) a module can send per second. Peak bus bandwidth is the maximum number of bytes a bus can transfer per second.

2.29 Create and simulate for all values of x, y, and z a Verilog model for $f = x\bar{y} + yz$ using:

a. A structural description with NOT, AND, and OR gates

b. A structural description with NOT and NAND gates

c. A structural description with delays using 1 ns delay for NOT and NAND and 2 ns delay for AND and OR gates

d. A behavioral description using an "assign" statement

e. A behavioral description using an "always" statement

2.30 Create and simulate for all values of x, y, and z a Verilog model for $f=(x+y)(\bar{y}+z)$ using:

a. A structural description with NOT, AND, and OR gates

b. A structural description with NOT and NOR gates

c. A structural description with delays using 1 ns delay for NOT and NOR and 2ns delay for AND and OR gates

d. A behavioral description using an "assign" statement

e. A behavioral description using an "always" statement

2.31 For the cases below create and simulate a Verilog behavioral description for a 1-to-4 MUX.

a. Use "if-else" statements

b. Use a "case" statement

2.32 For the cases below create and simulate a Verilog behavioral description for a 2-to-4 decoder. Label signal names with correct polarity (e.g., _x can be used to indicate an active-low signal and x an active-high signal).

a. Active-high inputs and active-high outputs using an "always" statement

b. Active-high inputs and active-low outputs using an "always" statement

2.33 For the cases below create and simulate a Verilog behavioral description for a 3-to-2 decoder. Label signal names with correct polarity (e.g., _x would indicate an active-low signal and x an active-high signal).

a. Active-high inputs and active-high outputs using an "always" statement

b. Active-low inputs and active-high outputs using an "always" statement

CHAPTER 3

Combinational Circuits: *Large Designs*

3.1 Introduction

The design techniques that were presented in the previous chapter apply only to combinational circuits with a small number of inputs. Combinational circuits that have many inputs must be designed differently. For example, consider a combinational circuit with $n = 32$ inputs. Its truth table would have more than four billion rows—too large to be designed using the techniques of Chap. 2. Moreover, large circuits must meet design fan-in and fan-out requirements. This requires a top-down methodology to repeatedly partition a large combinational circuit design problem into smaller problems until the final set of design problems is small enough to use the techniques learned in Chap. 2. The larger circuit is then created by assembling the smaller circuit modules.

Circuits that perform four elementary arithmetic operations—addition, subtraction, multiplication, and division—are examples of large combinational circuits used in modern processors. Like software solutions that may implement alternative algorithms with each requiring different processing time and memory usage, large combinational circuits may be implemented using a different amount of hardware (e.g., transistor count). A circuit solution with more transistors implies a higher number of logic operations will be performed in parallel, which typically means smaller circuit propagation delay, but also more power consumption.

In general, more hardware implies less processing time. A CPU that contains fast arithmetic modules is expected to execute programs faster; a multicore processor is expected to perform tasks faster than a single-core processor, etc.

In this chapter, we provide design examples of arithmetic circuits. In particular, we discuss the design of a commonly known fast adder and also present the design of a subtractor, a 2's complement adder, an arithmetic logic unit (ALU), a multiplier, and a divider. The chapter also presents IEEE floating point (FP) number standards and arithmetic.

3.1.1 Top-Down Design Methodology

A top-down methodology, also known as hierarchical, refers to a design flow, such as a tree, that consists of parent and leaf nodes. A large design problem at the root node is successively partitioned into a set of smaller design problems at the leaf nodes. The problem at the root is first partitioned into smaller design problems as children nodes. The design problem at each child node (now a parent node) is again divided into yet smaller design problems, if necessary. The process continues until each of the design problems at the leaf nodes is small and has fewer inputs.

For each design problem at the leaf nodes, the techniques of Chap. 2 are applied to design a circuit. These circuits are then successively combined to build the target large combinational circuit. A design may require one or more copies of each smaller circuit. A final circuit must be free from any fan-in and fan-out problems.

At each step in the design flow, various methods and algorithms are examined, design tradeoffs such as circuit delay and gate counts are analyzed, and the best solutions that meet the overall design requirements are selected. In general, a bit-parallel or bit-serial design partitioning technique (defined next) is applied at each parent node, starting at the root node.

A partitioning is called **bit-parallel** when a design problem is partitioned into smaller design problems, each a single function. For example, the design of the combined AND/ADD circuit module that generates the result of either an n-bit bitwise AND or an n-bit addition can be viewed as three separate design problems: an n-bit bitwise AND, an n-bit adder, and an n-bit 2-to-1 multiplexer (MUX), as illustrated for $n = 8$ in Fig. 3.1. The MUX would select either the result generated by the bitwise AND module or the sum generated by the adder. Even though the bitwise AND, the adder, and the MUX are now associated with only a single function, they are still too big to design using the techniques of Chap. 2. Each of these modules must be further partitioned into yet smaller design problems.

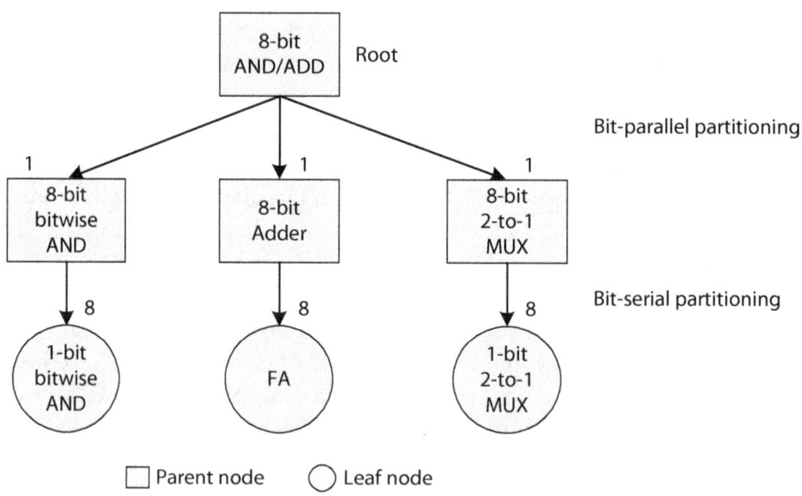

FIGURE 3.1 A top-down, bit-parallel and bit-serial design partitioning example.

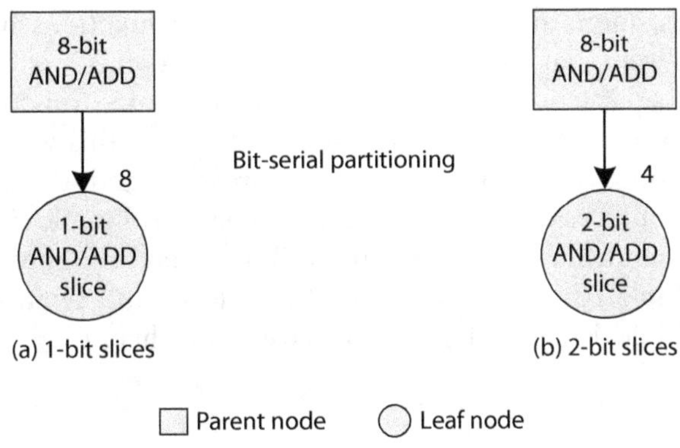

FIGURE 3.2 A top-down, bit-serial design partitioning example: (*a*) eight 1-bit slices; (*b*) four 2-bit slices.

The **bit-serial** methodology is used to partition a design problem into smaller problems, each requiring fewer inputs. For example, the design of a large circuit with n-bit inputs can be partitioned into a k-bit input design problem called a slice, with k (preferably) evenly dividing n. Each slice may perform one or more functions, but only operates on fewer bits. For instance, an 8-bit adder can be designed using 8 copies of a full-adder (FA), or the 8-bit, 2-to-1 MUX can be designed using eight copies of a 1-bit, 2-to-1 MUX. In the figure, the number next to each node indicates the number of copies. The constant k is selected in such a way that a k-bit slice would not violate the fan-in and fan-out limitations of the gates used.

It is also possible to use the bit-serial methodology to design an 8-bit AND/ADD module using eight copies of a 1-bit AND/ADD slice, four copies of a 2-bit AND/ADD slice, etc., as illustrated in Fig. 3.2 for 1-bit and 2-bit AND/ADD slices. In general, an n-bit module may be designed using n 1-bit, $n/2$ 2-bit, $n/4$ 4-bit slices, etc. Each slice may also need to output additional signals—for example, a carryout bit that would be used by an adjacent slice when the ADD function is selected. For a minimum propagation delay, a bit-serial slice is modeled with a truth table and either designed as a minimized sum of product (SOP) or product of sum (POS) circuit.

Which design methodology, bit-parallel or bit-serial, to use in each step of the design process will have an impact on the maximum propagation delay of the resultant circuit and the total number of required gates. This chapter presents the design of some commonly used large combinational circuits.

3.2 Arithmetic Functions

Adding, subtracting, multiplying, and dividing are four basic arithmetic functions. More complex functions such as square root, exponential, sine, etc., that operate on floating-point (FP) numbers use the basic four functions to produce outputs. As the number of transistors in an integrated chip (IC) has increased over the years, more and more of the arithmetic functions have been implemented in hardware. For example, the earlier microprocessors only implemented addition and subtraction functions, and the remaining functions were implemented in software. Many calculators are still designed this way, where a small number of functions are implemented in hardware and the rest in software. Modern microprocessors typically include integer arithmetic circuits called an integer unit (IU) and floating-point unit (FPU). Often, a modern microprocessor includes multiple IUs and FPUs as well as integer and floating-point single instruction multiple data (SIMD) units, discussed in Chap. 1.

3.3 Adder

An n-bit integer adder inputs two n-bit numbers and an optional carry-in bit (c_{in}) to generate an n-bit sum and a final carryout bit (c_{out}). As illustrated in Fig. 3.3, starting from the right (i.e., the least significant bits first), two bits and the previous carry-bit are added to generate a sum-bit and a carry-bit (0 or 1) to be used with the next two bits. In the figure, the initial c_{in} is assumed to be 0. The algorithm repeats until the last two bits are added and the final sum bit and the final carryout bit, c_{out}, are generated.

In the figure, the first two bits (both 1) and $c_{in} = 0$ are added to generate 0 as the sum bit and 1 as the carry bit. The next two bits, again both 1, and the previous carry-bit (1) are added to generate 1 for both the next sum-bit and the next carry-bit. Other bits are similarly added. This simple addition algorithm can be viewed as multiple 1-bit addition steps, each an FA, as illustrated in Fig. 3.4 for the ith step. During each step of the algorithm, two 1-bit inputs a_i and b_i and the previous carry-bit c_{i-1} are added to generate the sum-bit s_i and the next carry-bit c_i. For n bits, the steps can be repeated n times for $i = 0, 1, \ldots, n - 1$ with $c_{-1} = c_{in}$ and $c_{out} = c_{n-1}$.

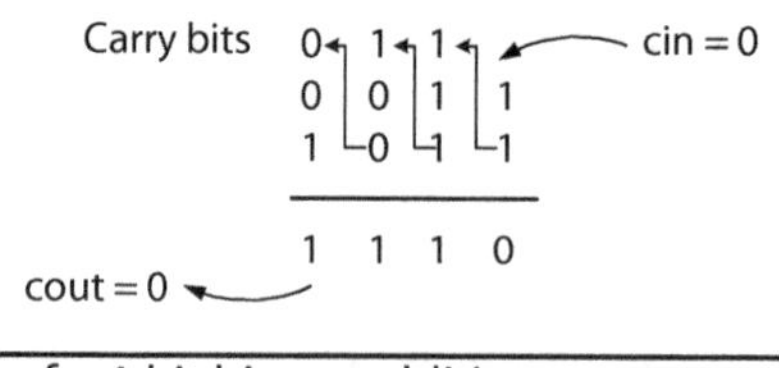

Figure 3.3 The illustration of a 4-bit binary addition.

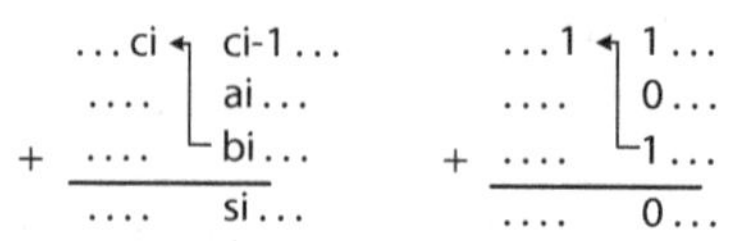

Figure 3.4 Viewing each binary addition step as an FA.

3.3.1 Carry Propagate Adder

An n-bit adder that implements the simple addition algorithm illustrated in Fig. 3.4 is designed by connecting n FA slices in series, as shown in Fig. 3.5. The carry-bits c_0 through c_{n-1} are generated one at a time, starting from the least significant carry-bit, similar to the way addition is performed by hand. The adder is called a carry propagate adder (CPA) because the carry-bits propagate from one FA slice to the next. The adder is also called a ripple carry adder (RCA) because the carry-bits ripple through the circuit as one carry-bit is fed to the next FA slice in a chain.

The CPA is the simplest circuit for an adder and has the longest propagation delay that is proportional to the number of the carry-bits. Each carry signal depends on the preceding carry signal; c_0 depends on c_1, c_1 depends on c_0, c_2 depends on c_1, etc. Equation (3.1) is used to estimate the propagation delay of an n-bit CPA, where $\Delta\text{CPA}(n)$ represents the propagation delay of an n-bit CPA and the ΔFAc and ΔFAs, respectively, stand for the carry and sum signal propagation delays of an FA.

$$\Delta\text{CPA}(n) = (n-1) * \Delta\text{FAc} + \Delta\text{FAs} \tag{3.1}$$

Equation (3.2) presents the propagation delay calculation for an n-bit CPA using ΔNAND = 0.1 ns and for an FA, ΔFAc = 0.2 ns and ΔFAs = 0.3 ns, as given in Eq. (2.8) in Chap. 2 as SOP expressions. For $n = 8$, $\Delta\text{CPA}(8) = 1.7$ ns.

$$\begin{aligned} \Delta\text{CPA}(n) &= (n-1)\,(0.2\text{ ns}) + 0.3\text{ ns} \\ &= (0.2n + 0.1)\text{ ns} \end{aligned} \tag{3.2}$$

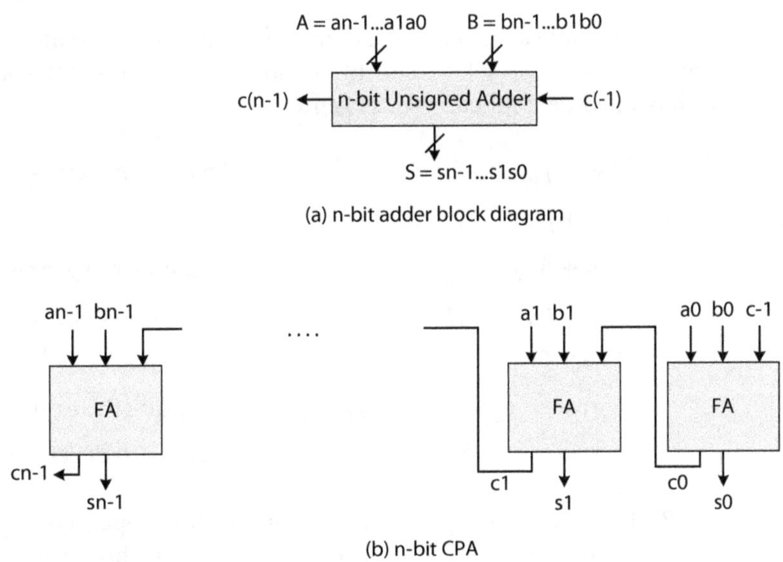

FIGURE 3.5 An n-bit adder: (a) block diagram; (b) an n-bit CPA.

3.3.2 Carry Look-Ahead Adder

For large values of n, the propagation delay of a CPA would be prohibitively long. The circuit would consist of a long carry-generate chain. However, it is possible to use more gates and generate the carry-bits independently and concurrently in less time. To illustrate this, two signals, named *propagate* (p) and *generate* (g), are defined as follows for the bit position i, for $i = 0, 1, 2, \ldots, n - 1$.

$$p_i = a_i \oplus b_i \qquad (3.3)$$
$$g_i = a_i b_i$$

Note that all the p and g bits can be generated at the same time and in parallel using n XOR and n AND gates, respectively. Recall from Eq. (2.9) (Chap. 2),

$$s_i = a_i \oplus b_i \oplus c_{i-1} \qquad (3.4)$$
$$c_i = (a_i \oplus b_i)c_{i-1} + a_i b_i$$

The substitution of Eq. (3.3) into Eq. (3.4) yields:

$$s_i = p_i \oplus c_{i-1} \qquad (3.5)$$
$$c_i = g_i + p_i c_{i-1}$$

Equation (3.6) lists the logic expressions for the first three carry-bits in terms of the p, g, and preceding carry-bits. The expressions clearly illustrate the recursive dependency of the carry-bits.

$$c_0 = g_0 + p_0 c_{-1} \qquad \text{(depends on carry-in } c_{-1}\text{)} \qquad (3.6)$$
$$c_1 = g_1 + p_1 c_0 \qquad \text{(depends on carry-in } c_0\text{)}$$
$$c_2 = g_2 + p_2 c_1 \qquad \text{(depends on carry-in } c_1\text{)}$$

It is possible to generate the carry-bits c_1, c_2, etc. in parallel if the expression of each successive carry-bit is substituted in the expression of the next carry-bit. This is illustrated next for carry-bits c_1 and c_2.

$$c_0 = g_0 + p_0 c_{-1} \qquad \text{depends on carry-in bit } c_{-1} \qquad (3.7)$$
$$c_1 = g_1 + p_1 c_0$$
$$= g_1 + p_1(g_0 + p_0 c_{-1}) \qquad \text{substitute the expression for } c_0$$
$$= g_1 + p_1 g_0 + p_1 p_0 c_{-1} \qquad \text{now } c_1 \text{ also depends on carry-in bit } c_{-1}$$
$$c_2 = g_2 + p_2 c_1$$
$$= g_2 + p_2(g_1 + p_1 g_0 + p_1 p_0 c_{-1}) \qquad \text{substitute the expression for } c_1$$
$$= g_2 + p_2 g_1 + p_2 p_1 g_0 + p_2 p_1 p_0 c_{-1} \qquad \text{now } c_2 \text{ also depends on carry-in bit } c_{-1}$$

Both of the final expressions of c_1 and c_2 now depend on c_{-1} and thus can be generated in parallel. Each expanded expression, however, will require more hardware, including gates, with larger fan-in and fan-out requirements, as compared to those required for implementing Eq. (3.6).

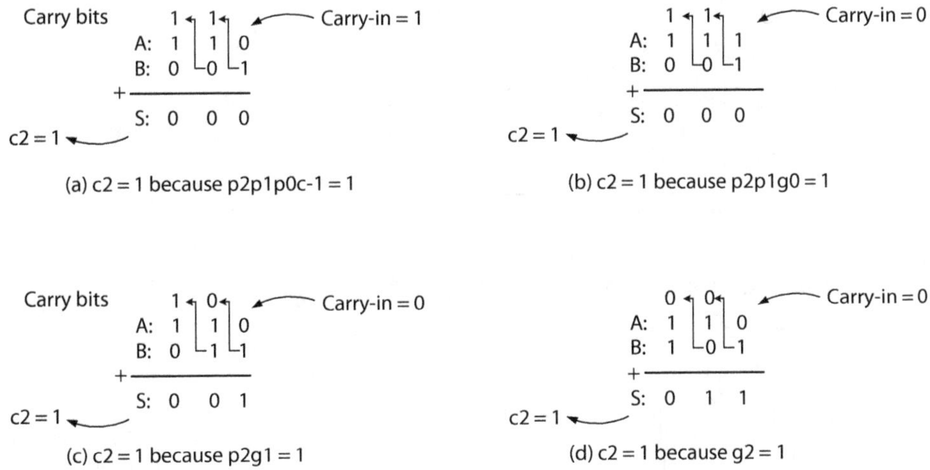

FIGURE 3.6 Examples illustrating the concept of carry look-ahead adder.

The p and g bits are used to identify certain bit patterns that can be used to quickly determine a carry-bit for a block of inputs. The examples in Fig. 3.6 illustrate this point using 3-bit inputs $A = a_2a_1a_0$ and $B = b_2b_1b_0$. In example (*a*), p_2, p_1, and p_0 are all 1; therefore, 1 as the carry in ($c_{-1} = 1$) will propagate out as $c_2 = 1$ because $p_2p_1p_0c_{-1} = 1$ in Eq. (3.7). In example (*b*), $p_2 = 1$, $p_1 = 1$, and $g_0 = 1$; therefore, a carry-bit generated at bit position 0 (i.e., $a_0 = 1$ and $b_0 = 1$, and thus $g_0 = 1$) will propagate out as $c_2 = 1$ because $p_2p_1g_0 = 1$. The remaining examples illustrate other bit patterns that cause c_2 to become 1. In example (*c*), $p_2g_1 = 1$, thus $c_2 = 1$, and in example (*d*), $g_2 = 1$, thus $c_2 = 1$.

The sum-bits s_0, s_1, and s_2 are also generated in parallel, as determined by their respective equations shown next, once the carry-bits become available.

$$s_0 = p_0 \oplus c_{-1}$$
$$s_1 = p_1 \oplus c_0$$
$$s_2 = p_2 \oplus c_1$$

Figure 3.7 illustrates the circuit of a 3-bit carry look-ahead (CLA) adder. The p and g bits are generated in parallel using three XOR and three AND gates within 0.3 ns, the maximum of ΔXOR = 0.3 ns and ΔAND = 0.2 ns, assuming 0.1 ns delay for a NAND gate. The circuit is called a PG-unit (PGU). The p and g bits and the initial carry-in c_{-1} are fed into a carry generate unit (CGU). The three carry circuits in the CGU are independent and would in parallel generate all the carry-bits c_0, c_1, and c_2 within 0.2 ns, assuming the circuits are implemented with NAND gates. Using another set of XOR gates, the final sum-bits are also generated in parallel using the p and the carry-bits as inputs. These XOR gates are combined into a module called a sum-unit (SU).

An 8-bit CLA with carry-bit c_0 through c_7 is still small enough that the design will not violate the fan-in and fan-out limitations of the gates. In this case, the circuit that implements the carry-bit c_i uses gates with fan-in $\leq i + 2$. The delay of the CLA(8) would be 0.8 ns, as determined by Eq. (3.8), using 0.1 ns delay for a

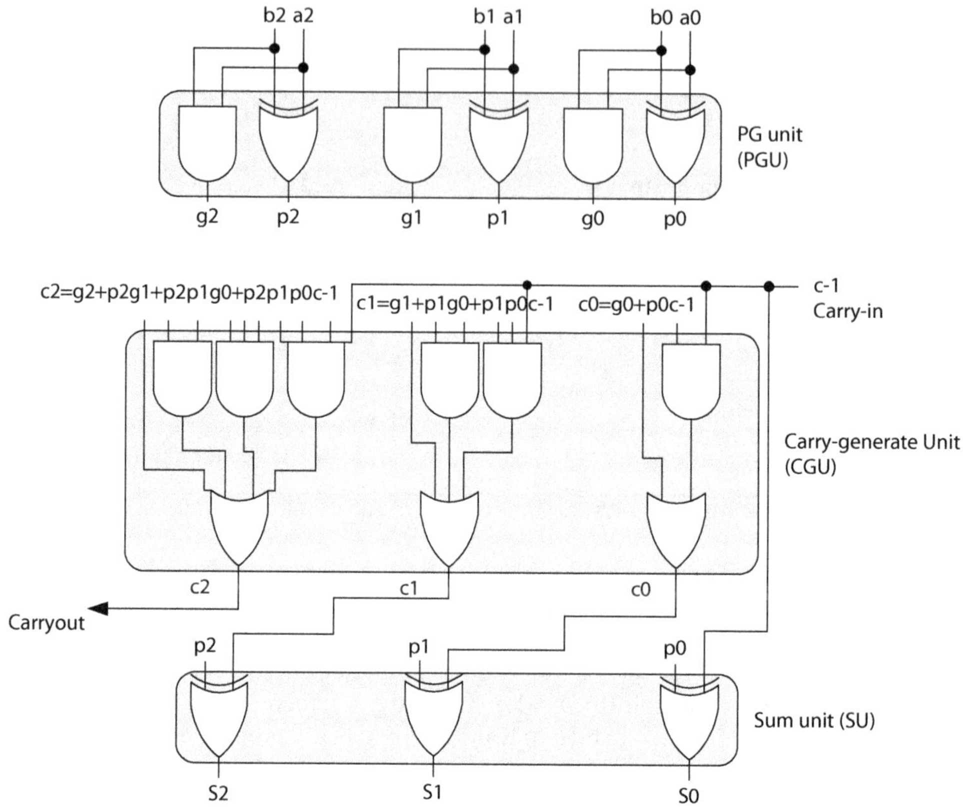

Figure 3.7 A 3-bit CLA with carry-bits c_0, c_1, and c_2 generated in parallel.

NAND gate. The CLA(8) is more than two times faster than the CPA(8); however, its implementation would require more hardware.

$$\begin{aligned}\Delta\text{CLA}(8) &= \Delta\text{PGU} + \Delta\text{CGU} + \Delta\text{SU} \\ &= 0.3\text{ ns} + 0.2\text{ ns} + 0.3\text{ ns} \\ &= 0.8\text{ ns}\end{aligned} \tag{3.8}$$

An alternative CLA circuit uses AND gates for g bits, OR gates for p bits, and FAs for the sum bits [1].

Large CLA Adder

For large values of n (e.g., $n > 8$), the successive substitution of the carry expressions, as it was illustrated in Eq. (3.7) for the carry-bits c_1 and c_2, would eventually result in long carry expressions, and thus the corresponding circuits would require gates with large fan-in numbers. In addition, some of the gates, such as the XOR gates that generate the p signals, require large fan-out values, as the p's are the inputs to multiple carry-generate circuits in the CGU, as well as the sum circuits in the SU. Therefore, when n is a large number (e.g., 32 or 64 bits), a different method is used that limits the fan-in and fan-out requirements of each circuit module.

In this case, the carry expressions are grouped into different sets so that the maximum fan-in and fan-out of the gates in the corresponding circuits are within an acceptable range. For simplicity, this is illustrated for $n=8$. The carry expressions of a CLA(8) are grouped into three sets, as illustrated in Eq. (3.9). An expression in each of the sets has fewer than three logic terms and fewer than three variables. The corresponding circuits would have fan-in ≤3 and fan-out ≤2. For example, for c_1 in set 1, the circuit would have a maximum fan-in = 3 and a maximum fan-out = 2 due to signal p_0. The carry expressions in each set are data independent; thus, the carry-bits can be generated in parallel as soon as the inputs, including the carry-in signals, are available. For the expressions in set 1, c_{-1} is the carry-in bit; in set 2, carry in is c_2; and in set 3, carry in is c_5.

Set 1: (3.9)

$$c_0 = g_0 + p_0 c_{-1} \qquad \text{(requires } c_{-1} \text{ as carry-in)}$$
$$c_1 = g_1 + p_1 g_0 + p_1 p_0 c_{-1} \qquad \text{(requires } c_{-1} \text{ as carry-in)}$$

Set 2:

$$c_3 = g_3 + p_3 c_2 \qquad \text{(requires } c_2 \text{ as carry-in)}$$
$$c_4 = g_4 + p_4 g_3 + p_4 p_3 c_2 \qquad \text{(requires } c_2 \text{ as carry-in)}$$

Set 3:

$$c_6 = g_6 + p_6 c_5 \qquad \text{(requires } c_5 \text{ as carry-in)}$$
$$c_7 = g_7 + p_7 g_6 + p_7 p_6 c_5 \qquad \text{(requires } c_5 \text{ as carry-in)}$$

Among the carry-in signals c_{-1}, c_2, and c_5, the carry-in bit c_{-1} is a primary input, while the other two must be generated. They are defined in Eq. (3.10) and grouped as set 4.

$$c_2 = (g_2 + p_2 g_1 + p_2 p_1 g_0) + (p_2 p_1 p_0) c_{-1} \qquad (3.10)$$
$$c_5 = (g_5 + p_5 g_4 + p_5 p_4 g_3) + (p_5 p_4 p_3) c_2$$

Let

$$g^*_0 = g_2 + p_2 g_1 + p_2 p_1 g_0$$
$$p^*_0 = p_2 p_1 p_0$$
$$g^*_1 = g_5 + p_5 g_4 + p_5 p_4 g_3$$
$$p^*_1 = p_5 p_4 p_3$$

Thus, Set 4:

$$c_2 = g^*_0 + p^*_0 c_{-1} \qquad \text{(requires carry-in } c_{-1})$$
$$c_5 = g^*_1 + p^*_1 c_2$$
$$= g^*_1 + p^*_1 (g^*_0 + p^*_0 c_{-1})$$
$$= g^*_1 + p^*_1 g^*_0 + p^*_1 p^*_0 c_{-1} \qquad \text{(now also requires carry-in } c_{-1})$$

Note that the expressions of c_2 and c_5 are data independent when they are written in terms of the $p*$ and $g*$ signals. The $p*$ and $g*$ signals are also data independent and are defined in terms of p and g signals. Also, the final expressions of c_2 and c_5 are similar to those in sets 1 to 3, except that these expressions require $p*$ and $g*$ signals as inputs. The * in a $p*$ implies a carry propagation within a block of bits. Likewise, the * in a $g*$ implies a carry generation and propagation within a block of bits. When a $p* = 1$, it implies that a 1 as the carry bit entering the block will be propagated as 1 for the carryout. Likewise, when a $g* = 1$, it implies a carry of 1 generated within the block will be propagated out as the carryout.

The carry bits of set 1 depend on the carry bit c_{-1} and the p and g signals; thus, they can all be generated in parallel as soon as the p and g bits are available. The carry-bits of set 4 also depend on c_{-1} and can be generated in parallel as soon as the $p*$ and $g*$ bits are available. Note that since all the $p*$ and $g*$ bits can be generated at the same time and when c_0 and c_1 are generated, they are moved into sets 1 to 3, as shown in Eq. (3.11).

Set 1: (3.11)

$$c_0 = g_0 + p_0c_{-1} \qquad \text{(requires carry-in } c_{-1}\text{)}$$

$$c_1 = g_1 + p_1g_0 + p_1p_0c_{-1} \qquad \text{(requires carry-in } c_{-1}\text{)}$$

$$g*_0 = g_2 + p_2g_1 + p_2p_1g_0$$

$$p*_0 = p_2p_1p_0$$

Set 2:

$$c_3 = g_3 + p_3c_2 \qquad \text{(requires } c_2 \text{ as carry-in)}$$

$$c_4 = g_4 + p_4g_3 + p_4p_3c_2 \qquad \text{(requires } c_2 \text{ as carry-in)}$$

$$g*_1 = g_5 + p_5g_4 + p_5p_4g_3$$

$$p*_1 = p_5p_4p_3$$

Set 3:

$$c_6 = g_6 + p_6c_5 \qquad \text{(requires } c_5 \text{ as carry-in)}$$

$$c_7 = g_7 + p_7g_6 + p_7p_6c_5 \qquad \text{(requires } c_5 \text{ as carry-in)}$$

$$g*_2 = g_8 + p_8g_7 + p_8p_7g_6 \qquad \text{(needed for } n > 8\text{; for } n = 8,\ g_8 = 0,\ p_8 = 0\text{)}$$

$$p*_2 = p_8p_7p_6 \qquad \text{(needed for } n > 8\text{; for } n = 8,\ p_8 = 0\text{)}$$

A detailed block diagram of the 8-bit CLA is illustrated in Fig. 3.8. All the eight carry-bits c_0 through c_7 are generated in three steps, as follows:

1. The carry-bits c_0 and c_1 in set 1 and all the $p*$ and $g*$ bits in sets 1, 2, and 3 are generated first.
2. Next, the carry-bits c_2 and c_5 of set 4 are generated.
3. Finally, the carry-bits c_3, c_4, c_6, and c_7 of sets 2 and 3 are generated.

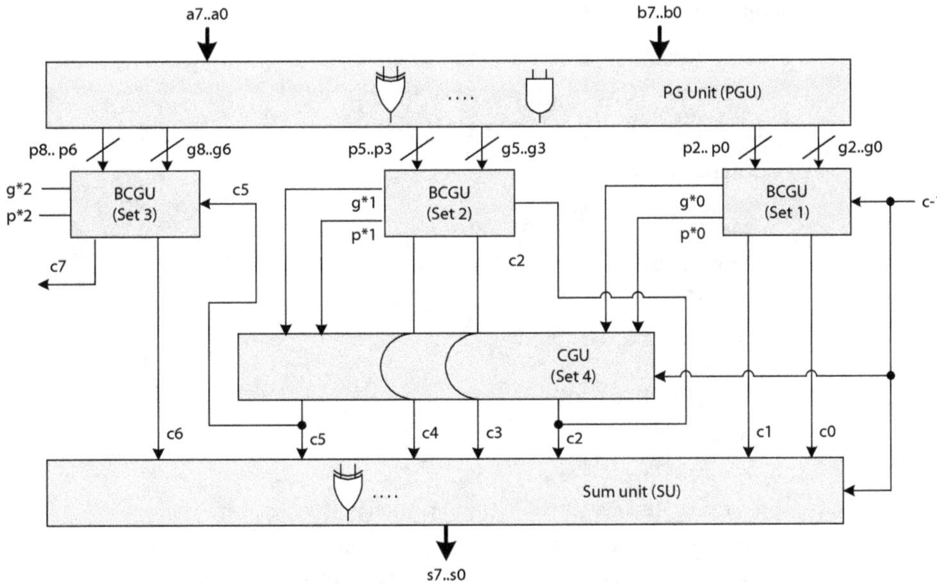

Figure 3.8 An 8-bit CLA using BCGUs. Signals p_8 and g_8 are set to 0.

The circuit that implements the expressions in each of the sets 1 to 3 is called a block carry generate unit (BCGU). The circuit that implements the expressions in set 4 is still a CGU. Once all the carry-bits are generated, c_{-1} to c_6 and p_0 to p_7 are fed to the SU to generate all the sum-bits s_0 to s_7 in parallel.

For a large n (e.g., $n = 32$ or 64), the circuit for a CLA(n) is the same as the one shown in Fig. 3.8 for CLA(8). Except that when n is a large number, the carry-bits must be sliced into multiple sets such that there are no BCGU or CGU fan-in and fan-out violations. Based on the design shown in Fig. 3.8, Eq. (3.12) estimates the propagation delay of a CLA(n), assuming 0.1 ns delay for a NAND gate.

$$\begin{aligned}\Delta\text{CLA}(n) &= \Delta\text{PGU} + \Delta\text{BCGU} + \Delta\text{CGU} + \Delta\text{BCGU} + \Delta\text{SU} \\ &= 0.3\text{ ns} + 0.2\text{ ns} + 0.2\text{ ns} + 0.2\text{ ns} + 0.3\text{ ns} \\ &= 1.2\text{ ns}\end{aligned} \tag{3.12}$$

A large CLA adder is a lot faster than an equivalent CPA, but requires more hardware. Alternatively, one may design a hybrid adder that is partly a CLA adder and partly a CPA. A hybrid adder would be faster than a CPA but slower than a CLA adder. For example, a 16-bit hybrid adder may be designed using two CLA(8) slices, where the carry-bit c_7 from the first slice is fed as carry-in to the second slice. The resultant 16-bit adder would be faster than a CPA(16) but slower than a CLA(16) and would require fewer hardware to implement than a CLA(16).

HDL Model

Examples 3.1 to 3.6 present a Verilog description of an 8-bit CLA using a PGU, three BCGUs, a CGU, and an SU. Both the behavioral and structural models are used to design the adder.

Example 3.1. A structural description of an 8-bit CLA adder:

```
`include "pg_unit_8bits.v"
`include "carry_generate_8bits.v"
`include "sum_unit_8bits.v"
module adder_8bits
(
  input [7:0] a, b,
  input ci,
  output [7:0] s,
  output c6, c7
);
wire [7:0] c, p, g;
assign c6 = c[6];
assign c7 = c[7];
pg_unit_8bits           pgu1(a, b, p, g);
carry_generate_8bits         cgu1(p, g, ci, c);
sum_unit_8bits      su1(p, {c[6:0], ci}, s);
endmodule
```

Example 3.2. A structure model of an 8-bit carry-generate module using BCGUs and a CGU:

```
`include "block_cla_carry_generate_3bits.v"
`include "cla_carry_generate_2bits.v"
module carry_generate_8bits (p, g, ci, c);
input [7:0] p, g;
input ci;
output [7:0] c;
wire [2:0] ps, gs;
block_cla_carry_generate_2bits  bccg1(p[2:0], g[2:0], ci, c[1:0], gs[0], ps[0]);
block_cla_carry_generate_2bits  bccg2(p[5:3], g[5:3], c[2], c[4:3], gs[1], ps[1]);
block_cla_carry_generate_2bits  bccg3({1'b0,p[7:6]}, {1'b0,g[7:6]}, c[5], c[7:6], gs[2], ps[2]);
cla_carry_generate_2bits        ccg1(ps[1:0], gs[1:0], ci, c[2], c[5]);
endmodule
```

Example 3.3. A behavior description of a 3-bit BCGU:

```
module block_cla_carry_generate_2bits (p, g, ci, c, gs, ps);
input [2:0] p, g;
input ci;
output [1:0] c;
output gs, ps;
```

```
assign    c[0] = g[0] | p[0] & ci;
assign    c[1] = g[1] | p[1] & g[0] | p[1] & p[0] & ci;
assign    gs = g[2] | p[2] & g[1] | p[2] & p[1] & g[0];
assign    ps = p[2] & p[1] & p[0];
endmodule
```

Example 3.4. A behavior description of a 2-bit CGU:

```
module cla_carry_generate_2bits (ps, gs, ci, c2, c5);
input [1:0] ps, gs;
input ci;
output c2, c5;

assign c2 = gs[0] | ps[0] & ci;
assign c5 = gs[1] | ps[1] & gs[0] | ps[1] & ps[0] & ci;
endmodule
```

Example 3.5. A behavior description of an 8-bit PGU:

```
module pg_unit_8bits (a, b, p, g);
input [7:0] a, b;
output [7:0] p, g;
assign p = a ^ b;
assign g = a & b;
endmodule
```

Example 3.6. A behavior description of an *n*-bit SU:

```
module sum_unit_8bits (p, c, s);
input [7:0] p, c;
output [7:0] s;
assign s = p ^ c;
endmodule
```

3.4 Subtractor

A subtractor generates the difference D of its two inputs X and Y using one of the two methods illustrated in Fig. 3.9 for 4-bits [2]. In method (a), each time that $x_i < y_i$, a 1 as a borrow is subtracted from x_{i+1} (the next higher bit) if $x_{i+1} > 0$ and a 2 is added to x_i. If $x_{i+1} = 0$, a borrow comes from the next higher bit x_{i+2} if $x_{i+2} > 0$ and a 2 is added to x_{i+1}. Again, if $x_{i+2} = 0$, the borrow comes from x_{i+3} and a 2 is added to x_{i+2}, and so on. This process recursively continues until $x_i \geq y_i$, and $d_i = x_i - y_i$ result in a 0 or 1. In the figure, $x_0 = 0$ is less than $y_0 = 1$; thus, a borrow from x_1 is needed. However, since $x_1 = 0$, a 1 is borrowed from x_2. The remaining bits are

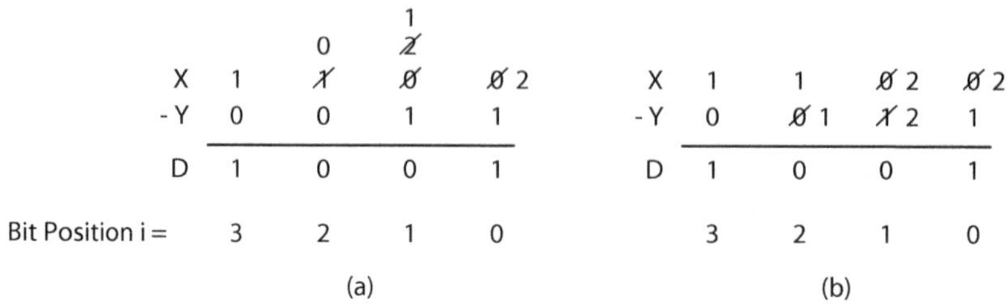

FIGURE 3.9 Illustrating unsigned subtraction: (*a*) borrow method; (*b*) credit method.

handled in the same way. This example does not produce a 1 as borrow out since $X = 12 = (1100)_2$ is greater than $Y = 3 = (0011)_2$ and thus $D = 12 - 3 = 9 = (1001)_2$.

In method (*b*), each time that $x_i < y_i$, a 1 as a credit is added to y_{i+1} (the next higher bit) and a 2 is added to x_i to produce $d_i = x_i - y_i$ as a 0 or 1. In the example, $x_0 = 0$ is less than $y_0 = 1$; thus, a 1 as a credit is added to y_1, making $y_1 = 2$, and a 2 is added to x_0, also making it 2. At this point, $d_0 = x_0 - y_0$ is 1 $(2 - 1 = 1)$. Next, $x_1 = 0$ is less than $y_1 = 2$; thus, again, a 1 as credit is added to y_2, making it 1, and a 2 is added to x_1, also making it 2. This results in $d_1 = 2 - 2 = 0$. The remaining bits are handled in the same way to produce the final $D = 9 = (1001)_2$.

An *n*-bit borrow propagate subtractor (BPS), similar to an *n*-bit CPA, is designed using *n* copies of a 1-bit subtractor slice. Each of the slices inputs one bit from *X*, one bit from *Y*, and a borrow/credit bit (0 or 1) and outputs one bit difference and the next borrow/credit bit, as illustrated for the *i*th bit. An *n*-bit BSP is illustrated in Fig. 3.10.

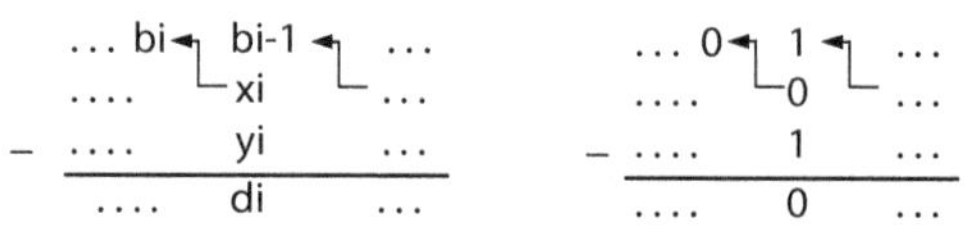

Borrow and credit subtraction algorithms are described next for the *i*th bit. Note that the equation for difference bit $d[i]$ is the same in both algorithms, except that parentheses are used to enforce precedence and thus illustrate the concept of borrow in method (*a*) and credit in method (*b*).

Subtraction Algorithms

Method (*a*): Borrow (parentheses illustrate how borrow is used)

if $x[i] > y[i]$
    $d[i] = (x[i] - b[i{-}1]) - y[i]$;   //borrow made to previous step is subtracted //from *x*

    $b[i] = 0$;
else if $x[i] < y[i]$
    $d[i] = (x[i] + 2 - b[i{-}1]) - y[i]$;   //*x* borrows a 2 and borrow made to previous //step is subtracted from *x*

    $b[i] = 1$;   //indicates the borrow from the next *x* bit
else if $b[i{-}1] == 0$ //$x[i] = y[i]$

```
    d[i] = x[i] − y[i];
    b[i] = 0;
else //b[i−1] = 1 and x[i] = y[i]
    d[i] = (x[i] + 2 − b[i−1]) − y[i];      // x borrows a 2 and borrow made to
                                            //previous step is subtracted from x
    b[i] = 1;                               // indicates the borrow from the next x bit
```

Method (*b*): Credit (parentheses illustrate how a credit is used)

```
    If x[i] > y[i]
    d[i] = x[i] − (b[i−1] + y[i]);          //the credit from the previous step is
                                            //added to y
    b[i] = 0;
else if x[i] < y[i]
    d[i] = (x[i] + 2) − (b[i−1] + y[i]);    //x borrows a 2 and the credit from the
                                            //previous step is added to y
    b[i] = 1;                               //indicates a credit given to the next y bit
else if b[i−1] == 0 //x[i] = y[i]
    d[i] = x[i] − y[i];
    b[i] = 0;
else //b[i−1] = 1 and x[i] = y[i]
    d[i] = (x[i] + 2) − (b[i−1] + y[i]);    // x borrows a 2 and the credit from the
                                            //previous step is added to y
    b[i] = 1;                               // indicates a credit given to the next y bit
```

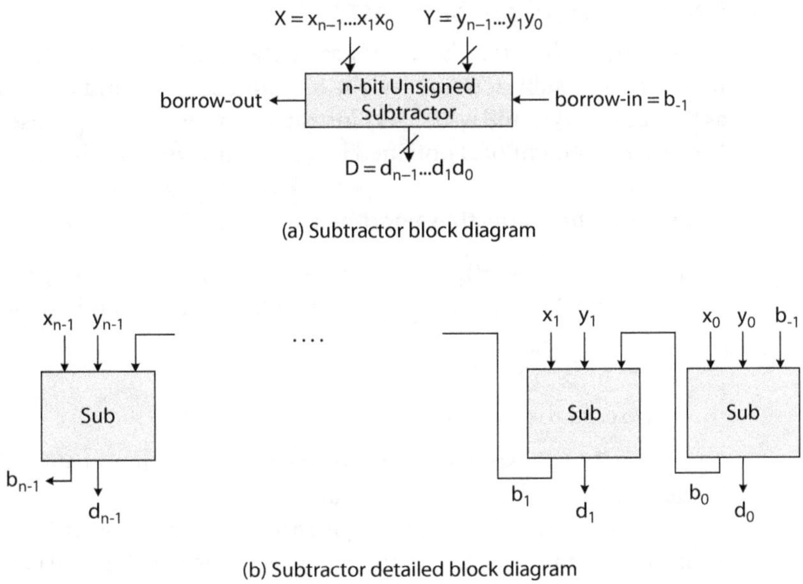

Figure 3.10 An *n*-bit BPS: (*a*) its block diagram; (*b*) its detailed block diagram.

xi	yi	bi-1	bi	di
0	0	0	0	0
0	0	1	1	1
0	1	0	1	1
0	1	1	1	0
1	0	0	0	1
1	0	1	0	0
1	1	0	0	0
1	1	1	1	1

TABLE 3.1 The Truth Table for a 1-Bit Subtrator (Sub)

Table 3.1 and Eq. (3.13) show the truth table and logic expressions for a 1-bit subtractor slice. The truth table is easily determined using one of the two subtraction algorithms discussed earlier. Note that the difference and borrow (credit) expressions are similar to the sum- and carry-bit expressions of an FA. Alternatively, a borrow-look-ahead (BLA) subtractor can be created using the techniques to design a CLA.

$$d_i = x_i \oplus y_i \oplus b_{i-1} \tag{3.13}$$

$$b_i = (\overline{x_1 \oplus y_1})b_{i-1} + \overline{x}_1 y_i$$

3.5 2's Complement Adder/Subtractor

For signed arithmetic, the values are either positive or negative 2's complement numbers. The subtraction of two n-bit numbers $(A)_{2s}$ and $(B)_{2s}$ can be interpreted as the sum of $(A)_{2s}$ and $(-B)_{2s}$, as illustrated next, where $(B)_{1s}$ is used here to indicate the 1's complement of B obtained by simply inverting each bit.

2's complement subtraction algorithm:

1: $(S)_{2s} = (A)_{2s} - (B)_{2s}$
$= (A)_{2s} + [-(B)_{2s}]_{2s}$ // + indicates a sum and not an OR function
$= (A)_{2s} + [(B)_{1s} + 1]_{2s}$
$= [(A)_{2s} + (B)_{1s} + 1]_{2s}$

2: Discard the carryout bit.

Figure 3.11 illustrates the 2's complement subtraction algorithm. The "+ 1" is implemented by using a 1 for the carry in.

Likewise, the following algorithm describes an n-bit 2's complement addition. In this case, the input B is used as-is (unchanged) and the carry in is set to 0.

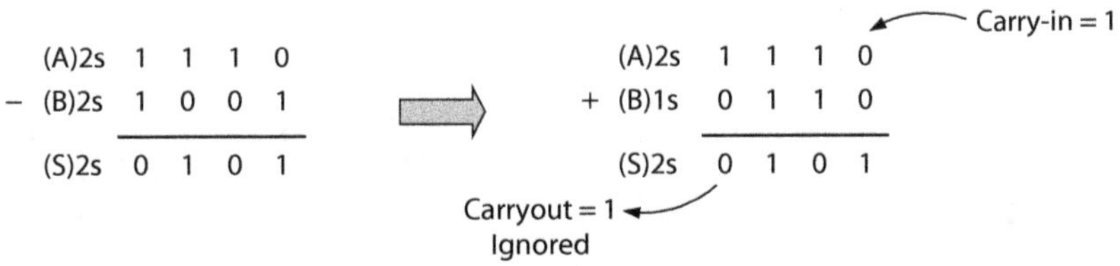

FIGURE 3.11 Illustrating 2's complement subtraction.

2's complement addition algorithm:

1: $(S)_{2s} = (A)_{2s} + (B)_{2s}$
$\quad = [(A)_{2s} + (B)_{2s} + 0]_{2s}$

2: Discard the carryout bit.

A 2's complement arithmetic generates an n-bit 2's complement output, except that the carryout is ignored and is not counted as part of the final result. However, the result may overflow. For instance, consider subtracting a 1 from the smallest n-bit 2's complement negative number, or adding a 1 to the largest n-bit 2's complement positive number. In both cases, the resultant difference and the sum will exceed the ranges defined for the n-bit negative and positive 2's complement numbers.

When both A and B are 2's complement positive numbers, their sign-bits are 0. In this case, in order for the quantity $A + B$ not to overflow, the carry-bit that is added to the sign-bits must be 0. This will also produce 0 as the carryout; otherwise, there will be an overflow. For example, consider examples (a) and (c) in Fig. 3.12. In example (a), $A = (0111)_{2s} = 7$ and $B = (0001)_{2s} = 1$ are both positive 2's complement numbers, and in addition, A is the largest 4-bit positive number. As shown in the figure, when $B = 1$ is added to $A = 7$, the result is $(1000)_{2s} = -8$, a negative number, which indicates an overflow. Note that, in this case, $c_3 = 1$ and when it is added to the sign-bits, both 0, this makes the sign of the resultant sum 1 (negative) and carryout $c_4 = 0 \neq c_3 = 1$. For the overflow not to occur, both c_3 and c_4 must be 0 when adding two positive 2's complements numbers.

In example (c), when two negative numbers $A = -1$ and $B = -2$, both with sign-bits equal to 1, are added, the quantity $A + B$ will not overflow if $c_3 = 1$ generates $c_4 =$ 1. Otherwise, the result will overflow, resulting in a positive number for the result. Therefore, the conclusions from examples (a) and (c) is that when $c_3 = c_4$, the resultant sum will not overflow, regardless if both A and B are positive or negative 2's complement numbers.

Examples (b) and (d) illustrate subtraction. In this case, the quantity $A - B$ is computed as $A + (B)_{1s} + 1$. In example (b), when $B = 1$ (a positive number) is subtracted from $A = -8$ (the smallest 4-bit negative 2's complement number), c_3, which is 0, when added to the sign bits (both 1), generates 7 (an incorrect value). Note that $c_4 = 1 \neq c_3 = 0$. In example (d), both $A = -1$ and $B = -2$ are negative 2's

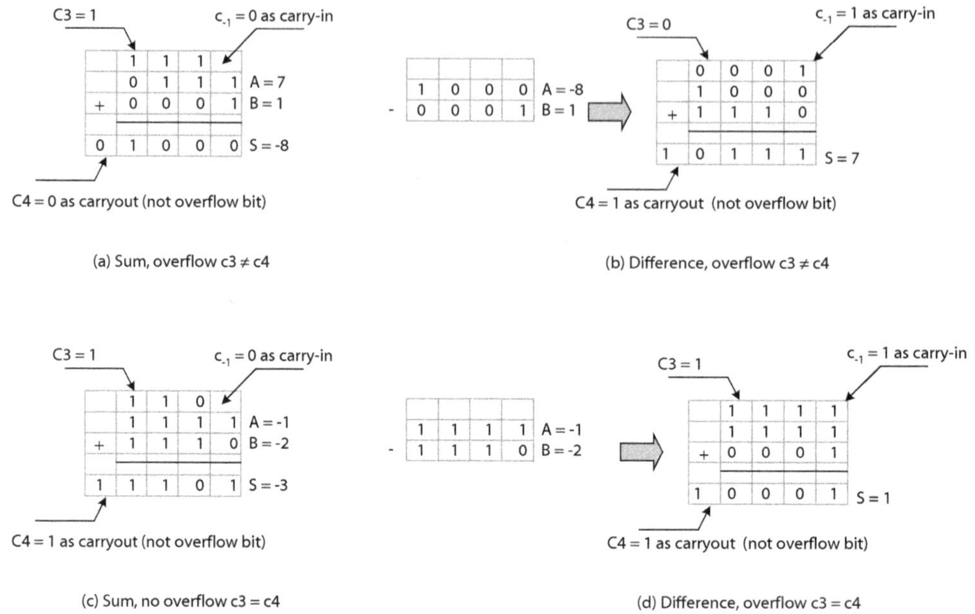

FIGURE 3.12 Examples of 2's complement arithmetic.

complement numbers and when subtracted, the result should never overflow. Note that in this case, $c_3 = 1$ is the same as $c_4 = 1$ (i.e., again, $c_3 = c_4$). The subtraction of two positive numbers also should never result in an overflow.

Table 3.2 presents the truth table for the overflow signal *ovf* for an n-bit 2's complement adder/subtractor. The carry-bit c_{n-2} is added to the sign bits to generate the final carryout c_{n-1}. Equation (3.14) defines the overflow signal.

$$ovf = c_{n-1} \oplus c_{n-2} \tag{3.14}$$

Figure 3.13 shows two block diagrams: (*a*) a 2's complement adder, and (*b*) a 2's complement subtractor. The two block diagrams differ only in the way the input $(B)_{2s}$ and the initial carry-in bit are handled. In Fig. 3.13(*a*), the input B is added to A as-is, unchanged. In Fig. 3.13(*b*), the input B is first bitwise NOTed before it is added to A. Therefore, it is possible to combine the two block diagrams into a single adder/subtractor module using only one adder module. An inverter module is used to output either B when adding, or bitwise NOT of B, denoted as

c(n-1)	c(n-2)	ovf
0	0	0
0	1	1
1	0	1
1	1	0

TABLE 3.2 Truth Table for Arithmetic Overflow Signal

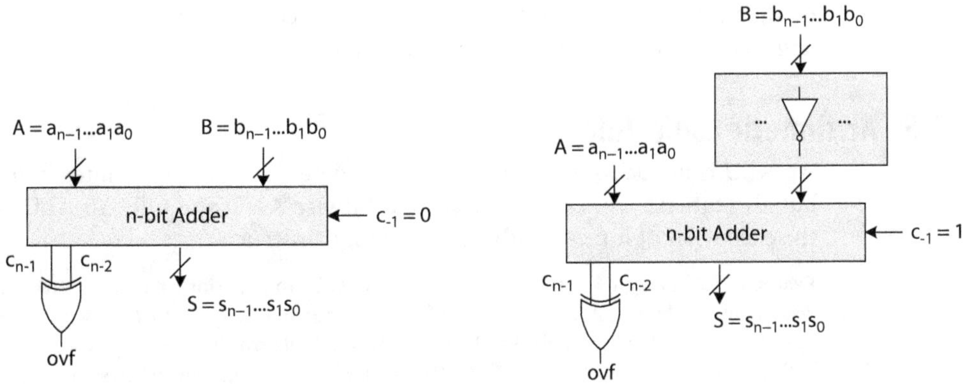

(a) 2's complement adder (b) 2's complement subtractor

Figure 3.13 2's complement adder and subtractor block diagrams: (*a*) adder block diagram; (*b*) subtractor block diagram that also includes an adder module.

	m	bi	ei
Add	0	0	0
Add	0	1	1
Subtract	1	0	1
Subtract	1	1	0

Table 3.3 Truth Table of a 1-Bit Inverter

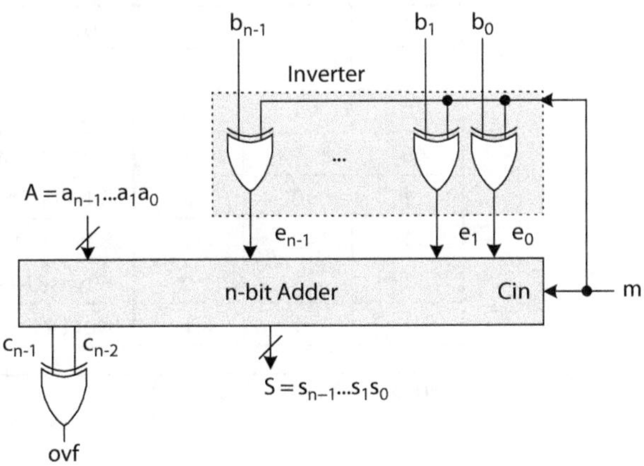

Figure 3.14 2's complement adder/subtractor data path.

$E = e_{n-1} \cdots e_1e_0$ equals to the 1s complement of B or $(B)_{1s}$, when subtracting. A mode signal m is used to perform either addition when $m = 0$ (carry in is 0) or subtraction when $m = 1$ (carry in is 1).

Table 3.3 presents the truth table of a 1-bit inverter slice. The final combined design is illustrated in Fig. 3.14, where m also serves as the initial carry-in value. Note that this module with m connected to the carry-in input cannot be used to design a bit-serial larger adder/subtractor circuit. For that, you must use a separate input for carry-in.

3.6 Arithmetic Logic Unit

An ALU is included in every processor and performs not only integer arithmetic, but also bitwise logic functions such as bitwise AND and OR. An ALU is used in the execution of integer arithmetic and logic instructions.

Example 3.7. Design an n-bit, seven-function ALU that performs add, subtract, increment, decrement, and bitwise AND, OR, and NOT. A 3-bit function code $F=f_2f_1f_0$ is used to select an ALU operation as specified in Table 3.4. The ALU also outputs overflow flag *ovf* when it performs an arithmetic operation ($F=0, 1, 2,$ or 3). $F=7$ is not used in this design, and when selected, the ALU may perform an unknown operation. The details of bit-parallel and bit-serial designs are discussed next.

Figure 3.15 illustrates the top-level block diagram of the ALU. Each of the input values A and B and result value R is interpreted as a 2's complement number when F specifies an arithmetic operation. Signal *ovf* indicates an arithmetic overflow when asserted. Step-by-step bit-parallel and bit-serial ALU designs are discussed in the following sections.

3.6.1 Design Partitioning: Bit-Parallel

The ALU is considered a large combinational circuit when n is large. Using the top-down bit-parallel design methodology discussed earlier, the functions of the

f2	f1	f0	Function
0	0	0	Add
0	0	1	Sub
0	1	0	Increment
0	1	1	Decrement
1	0	0	Bitwise AND
1	0	1	Bitwise OR
1	1	0	Bitwise NOT
1	1	1	Not Defined

Table 3.4 A List of ALU Functions, $F=f_2f_1f_0$

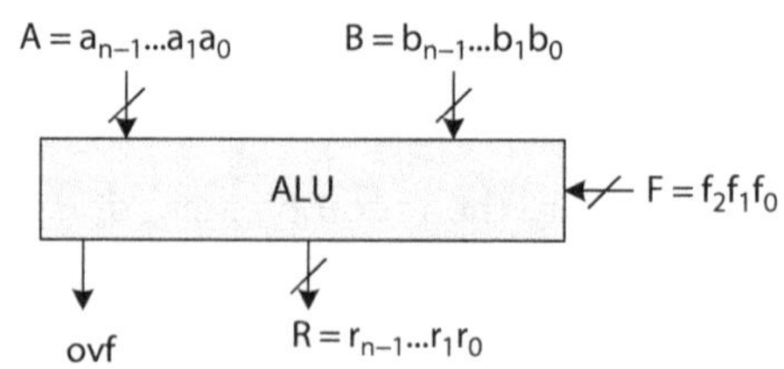

Figure 3.15 An ALU example.

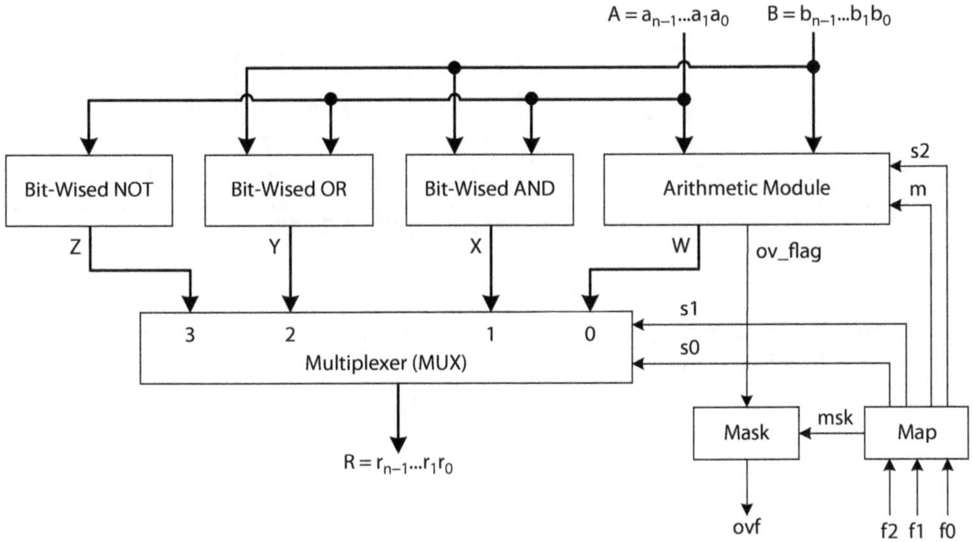

Figure 3.16 A data path for the ALU of Example 3.1.

ALU are first partitioned into arithmetic and logic operations. The circuits for the four arithmetic operations add, subtract, increment, and decrement are combined into a single arithmetic module in the ALU data path shown in Fig. 3.16. The three bitwise operations are implemented using AND, OR, and NOT gates. An 8-bit, 4-to-1 MUX selects one of the outputs *W*, *X*, *Y*, and *Z* as the ALU output.

In the ALU data path, the *A* and *B* inputs are both connected to the arithmetic and the three bitwise modules. These modules simultaneously operate on inputs *A* and *B*, and each generates a result, but only one must be selected as the final output of the ALU. Therefore, when *F* specifies a logic operation, it is still possible for the arithmetic module to assert *ov_flag* depending on the values of *A* and *B* at the time. However, in such cases, the overflow signal should be masked and not output as asserted *ovf* by the ALU. This is done using the Mask module.

The data path also includes a Map module that translates a given ALU function code *F* to internal signals s_0, s_1, s_2, *m*, and *msk* in the ALU data path. The design of the ALU is complete when all its modules are designed and interconnected.

The 4-to-1 MUX in the data path selects and outputs one of the four *n*-bit results generated by the arithmetic and the three bitwise modules. The Mask module sets the *ovf* to 0 (not active) if *F* indicates a logic operation ($F = 4$ to 6); otherwise, *ovf* is set to *ov_flag*, an output of the arithmetic module.

If necessary, the top-down design methodology is successively applied to partition all the larger modules in the data path to smaller circuit modules until each of the lowest-level circuit modules is small and requires fewer inputs. The design techniques presented in Chap. 2 are then used to design each small circuit module. The bitwise logic modules are designed using *n* 2-input AND, *n* 2-input OR, and *n* NOT gates. These modules generate *n*-bit values $X = x_{n-1} \dots x_0$, $Y = y_{n-1} \dots y_0$,

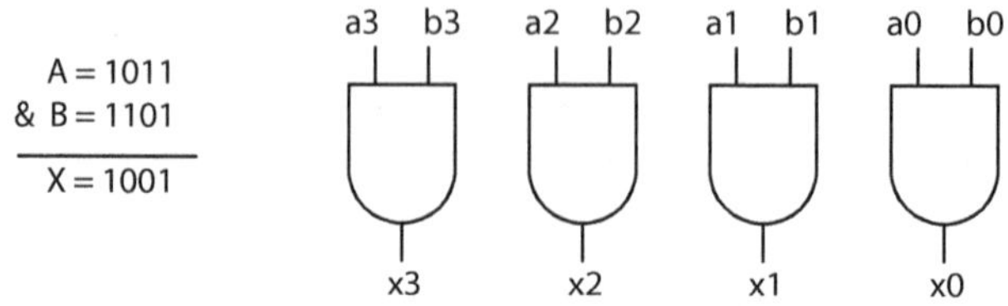

FIGURE 3.17 A 4-bit bitwise AND and its corresponding logic circuit.

and $Z = z_{n-1}...z_0$, respectively, as shown in the figure, where the ith bit in each case is defined as follows:

$$x_i = a_i b_i$$
$$y_i = a_i + b_i$$
$$z_i = \overline{a_i}$$

As an example, Fig. 3.17 illustrates the circuit of a 4-bit bitwise AND. When $A = a_3a_2a_1a_0 = (1011)_2$ and $B = b_3b_2b_1b_0 = (1101)_2$, $X = x_3x_2x_1x_0 = (1001)_2$ is determined as $x_0 = a_0.b_0 = 1.1 = 1$, $x_1 = a_1.b_1 = 0.1 = 0$, etc. Other bitwise logic modules are similarly designed.

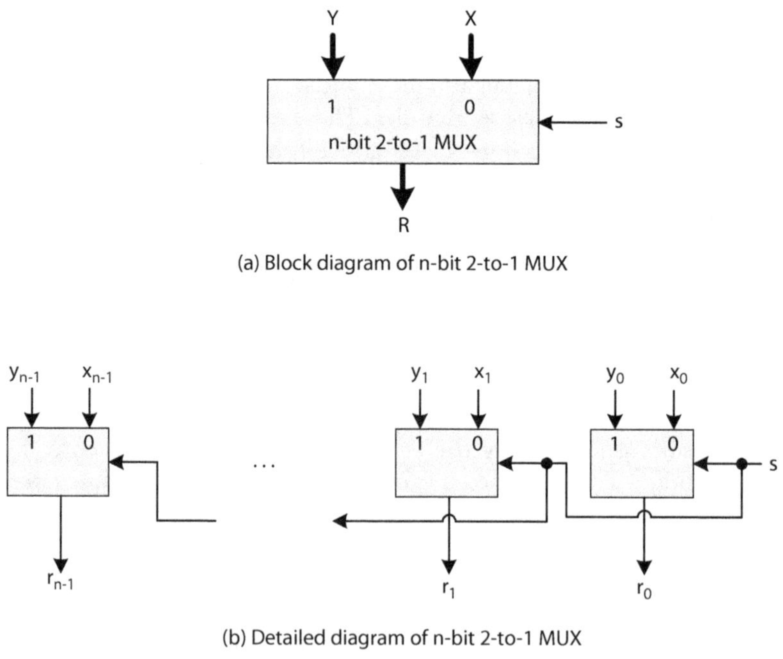

FIGURE 3.18 An *n*-bit, 2-to-1 MUX: (*a*) block diagram; (*b*) designed using 1-bit, 2-to-1 MUX slices.

s1	s0	R
0	0	W
0	1	X
1	0	Y
1	1	Z

TABLE 3.5 Simplified Truth Table of an *n*-Bit, 4-to-1 MUX

The design of a small multiplexer circuit was discussed in Chap. 2. The ALU requires an *n*-bit, 4-to-1 MUX that can be designed in one of two ways as follows:

1. Use *n* copies of a 1-bit, 4-to-1 MUX.

or

2. Use *n*-bit, 2-to-1 MUXs. An *n*-bit, 2-to-1 MUX is designed using *n* copies of a 1-bit, 2-to-1 MUX, as illustrated in Fig. 3.18.

Option 2 has the advantage of extending the design to any *n*-bit, *k*-to-1 MUX without running into any fan-in and fan-out problems. Table 3.5 shows a minimized truth table of an *n*-bit, 4-to-1 MUX. The s_1 signal selects either *W* or *X* if it is 0, or *Y* or *Z* if it is 1. On the other hand, the s_0 signal selects either *W* or *Y* if it is 0, or *X* or *Z* if it is 1. Thus, it is possible to apply the process of elimination and use three 2-to-1 MUXs to design a 4-to-1 MUX, as illustrated in Fig. 3.19 for *n* bits. For example, when s_1s_0 as a 2-bit number is 2 (i.e., $s_1 = 1$ and $s_0 = 0$), the top two 2-to-1 MUXs correctly choose the two possible input candidates *W* and *Y*, and the bottom 2-to-1 MUX correctly chooses the input *Y* as the final output. This is illustrated in the figure.

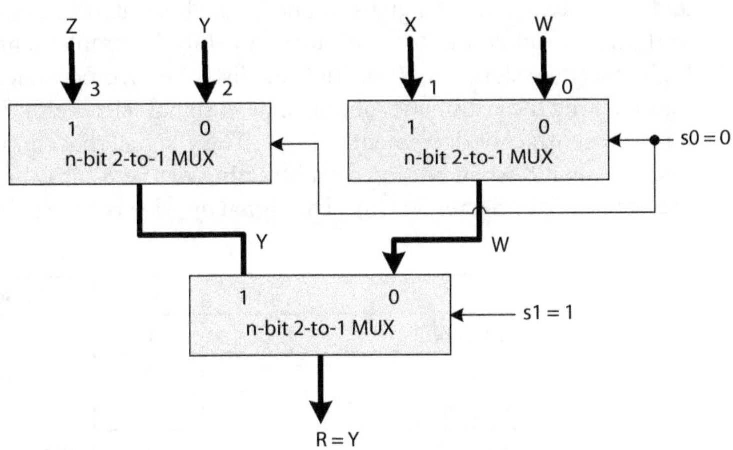

FIGURE 3.19 An *n*-bit, 4-to-1 MUX designed using *n*-bit, 2-to-1 MUXs.

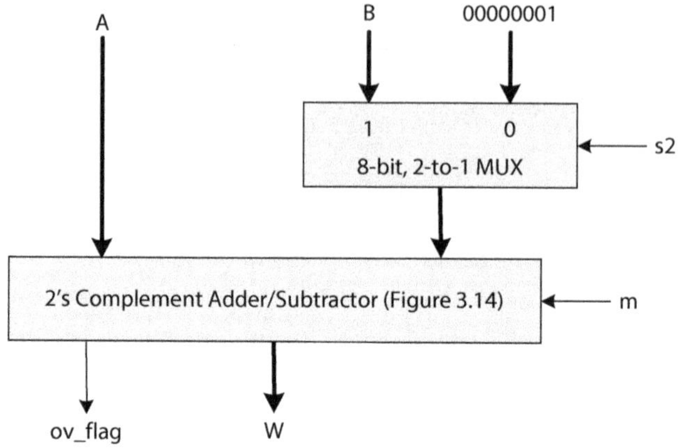

FIGURE 3.20 The detailed block diagram of the arithmetic module; the result is $A + B$, $A - B$, $A + 1$, or $A - 1$.

In general, the previous approach requires $\log_2 k$ levels of 2-to-1 MUXs. Equation (3.15) is used to estimate the propagation delay of a k-to-1 MUX designed using 2-to-1 MUXs.

$$\Delta_{k\text{-to-1 MUX}} = \log_2 k * \Delta_{2\text{-to-1 MUX}} \tag{3.15}$$

For example, an 8-to-1 MUX would require three levels of 2-to-1 MUXs, and a 64-to-1 MUX would require six levels of 2-to-1 MUXs. The number of levels and thus the total delay could be reduced if a combination of different MUXs is used. For instance, an 8-to-1 MUX can also be designed using a combination of 2-to-1 and 4-to-1 MUXs. In this case, a 4-to-1 MUX would be designed as an AND-OR (SOP) or OR-AND (POS) circuit to minimize its propagation delay.

The inputs to the arithmetic module are two n-bit 2's complement numbers A and B and two control signals m and s_2, as illustrated in Fig. 3.20 for $n = 8$. The arithmetic module has two outputs: an n-bit 2's complement W and an (active-high) overflow signal *ov_flag*. Table 3.6 lists the specific values of signals m and s_2 for operating the arithmetic module. The m signal selects either add or increment, if it is 0, or subtract or decrement, if it is 1. The s_2 signal that controls the 2-to-1 MUX selects either B when adding or subtracting or the 8-bit value $(00000001)_2$ when incrementing or decrementing. The signal *ov_flag* is asserted if W overflows.

	m	s2	Output
Increment	0	0	$W = A + 1$
Add	0	1	$W = A + B$
Decrement	1	0	$W = A - 1$
Subtract	1	1	$W = A - B$

TABLE 3.6 Operating Signal Values for the ALU's Arithmetic Module

msk	ov_flag	ovf
0	0	0
0	1	1
1	0	0
1	1	0

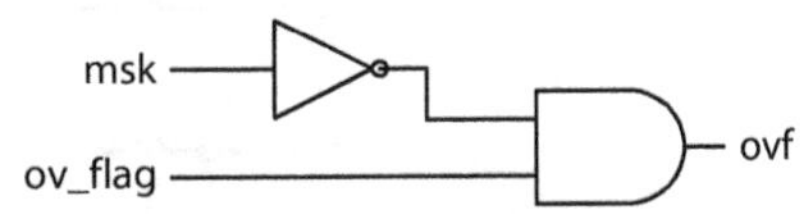

Figure 3.21 The overflow signal-masking module: its truth table and logic circuit.

The truth table and the circuit for the Mask module are given in Fig. 3.21. The module outputs *ovf* = 0 when *msk* = 1, masking the *ov_flag* generated by the arithmetic module; it outputs *ovf* = *ov_flag* when *msk* = 0.

Table 3.7 presents the truth table for the Map module. Some of the control signals in the table, except the *msk*, can be don't-cares (*d*). For example, when *F* specifies a logic operation (i.e., *F* = 4 to 6), output *W* from the arithmetic module will never be selected as the ALU output. Therefore, the control signals *m* and s_2 can both be don't-cares to minimize the circuit size. However, signal *ov_flag* must be masked so it won't be output as *ovf* from the ALU. The circuit for Map is shown in Fig. 3.22.

In addition, because no operation is assigned to *F* = 7, all the control signals, except the *msk*, can be considered as don't-cares when *F* = 7. However, because the Map circuit in Fig. 3.22 generates $s_2 = 0$, $s_1 = 1$, $s_0 = 0$, *m* = 1, and *msk* = 1 when *F* = 7 and these signal values correspond to bitwise OR, the ALU is said to perform a bitwise OR when *F* = 5 or 7.

3.6.2 Design Partitioning: Bit-Serial

A bit-serial design, as opposed to a bit-parallel design, requires the input data bits divided into slices. For a bit-serial ALU design, each ALU slice would operate on a fraction of the data bits but perform all the ALU functions. For example, Fig. 3.23 illustrates a bit-serial ALU designed using *n* copies of a 1-bit ALU slice. Table 3.8 presents the truth table of the 1-bit ALU slice. For inputs not shown in the table, the *r* and *co* signals are considered zero. In addition, in order for the increment and decrement operations to work correctly, the carry-in signal c_{-1} must be 1. This requires that the Inc/Dec module in the figure must generate $c_{-1} = 1$ when the ALU function code *F* is either 2 (increment) or 3 (decrement); that is, $c_{-1} = \overline{f_2} f_1$.

	f2	f1	f0	s2	s1	s0	m	msk
Add	0	0	0	1	0	0	0	0
Subtract	0	0	1	1	0	0	1	0
Increment	0	1	0	0	0	0	0	0
Decrement	0	1	1	0	0	0	1	0
Bit-wised AND	1	0	0	d	0	1	d	1
Bit-wised OR	1	0	1	d	1	0	d	1
Bit-wised NOT	1	1	0	d	1	1	d	1
Not Defined	1	1	1	d	d	d	d	1

Table 3.7 The Truth Table for the ALU Map Module

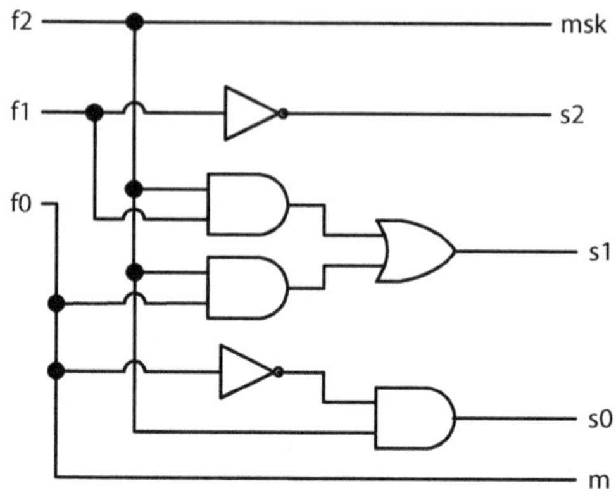

FIGURE 3.22 The circuit for the ALU map function.

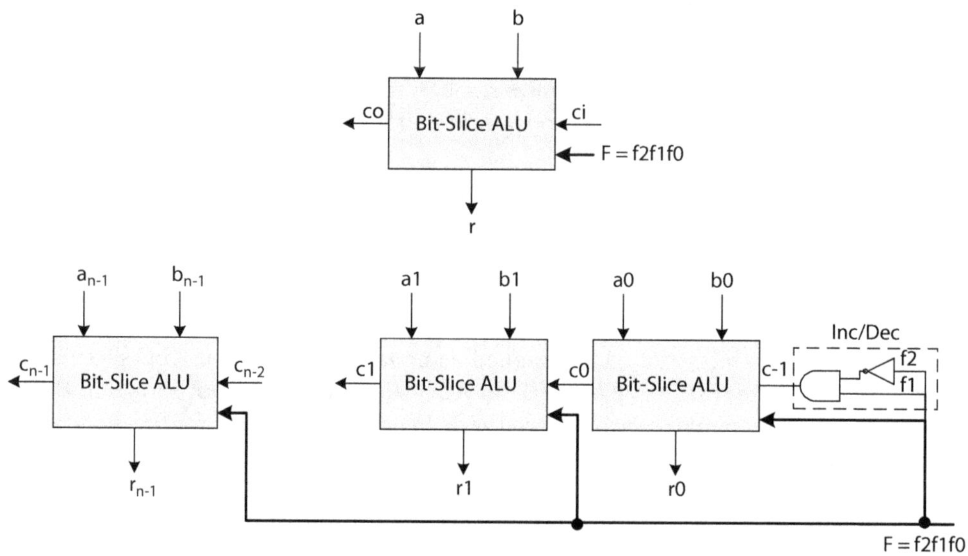

FIGURE 3.23 One-bit ALU slice and *n*-bit ALU designed from the 1-bit slices.

Note that because the ALU performs no operations when $F = 7$, the corresponding table entries are set to don't-cares. The truth table is too big to be minimized manually. The Espresso-generated essential prime implicants are listed next. Although no specific function was defined for $F = 7$, the 1-bit ALU slice outputs 1 when $F = 7$. Therefore, an 8-bit bit-serial ALU would output eight 1s or $(11111111)_2 = 0xFF$ when $F = 7$. Note that the output 0xFF may also be interpreted as 8-bit 2's complement number for −1. Table 3.9 presents the final list of the bit-serial ALU operations where ALU outputs −1 when $F = 7$.

f2	f1	f0	a	b	ci	co	r	Function
0	0	0	0	0	0	0	0	Add
			0	0	1	0	1	
			0	1	0	0	1	
			0	1	1	1	0	
			1	0	0	0	1	
			1	0	1	1	0	
			1	1	0	1	0	
			1	1	1	1	1	
0	0	1	0	0	0	0	0	Sub
			0	0	1	1	1	
			0	1	0	1	1	
			0	1	1	1	0	
			1	0	0	0	1	
			1	0	1	0	0	
			1	1	0	0	0	
			1	1	1	1	1	
0	1	0	0	d	1	0	1	Increment
			1	d	0	0	1	
			1	d	1	1	0	
0	1	1	0	d	1	1	1	Decrement
			1	d	0	0	1	
1	0	0	1	1	d	d	1	Bit-wised AND
1	0	1	0	1	d	d	1	Bit-wised OR
			1	0	d	d	1	
			1	1	d	d	1	
1	1	0	0	d	d	d	1	Bit-wised NOT
			1	d	d	d	0	
1	1	1	d	d	d	d	d	Not Defined

TABLE 3.8 The Truth Table of a 1-Bit ALU Slice

The essential prime implicants of the 1-bit ALU slice:

```
#1-bit ALU-slice: Add, Sub, Inc, Dec, And, Or, and Not
#Input signal labels
#output bit label
#list of min-terms
.i 6
.o 2
.ilb f2 f1 f0 a b ci
```

```
.ob co r
.p 15
00-010 01
0-01-1 10
-0011- 10
0-10-1 10
-0101- 10
0--100 01
01-1-0 01
00--11 10
0--001 01
10-11- 01
-0-111 01
11-0-- 01
-1-0-1 01
1-1-1- 01
1-11-- 01
.e
```

In general, a bit-serial design may be advantageous if the word sizes are nonstandard (e.g., 256- or 1024-bit operands used by a bit-serial encryption hardware), or the equivalent bit-parallel design would require prohibitively more hardware or creates less concurrency compared to bit-serial.

3.7 Design Examples

In addition to combinational circuits that perform integer addition and subtraction operations discussed earlier, the following subsections present the design of combinational integer multiplier and divider circuits. The basic arithmetic operation for a multiplier is addition and for a divider is subtraction. However, some multiplier and divider algorithms use both addition and subtraction operations. A 2's complement multiplier using addition and subtraction operations is discussed in Chap. 6.

f2	f1	f0	Function
0	0	0	Add
0	0	1	Sub
0	1	0	Increment
0	1	1	Decrement
1	0	0	Bit-wised AND
1	0	1	Bit-wised OR
1	1	0	Bit-wised NOT
1	1	1	−1

Table 3.9 The Final List of ALU Functions Performed by an *n*-Bit Bit-Serial ALU Using 1-Bit ALU Slices

```
          1001    A
      *   1011    B
          ----
          1001
         1001
        0000
    +  1001
       -------
       1100011    P=B*A
```

Figure 3.24 A 4-bit unsigned binary multiplication example.

3.7.1 Multiplier

Figure 3.24 illustrates the multiplication of the 4-bit unsigned multiplier $B = b_3b_2b_1b_0$ and the 4-bit unsigned multiplicand $A = a_3a_2a_1a_0$. Each multiplication step generates an addend. In the figure, the values $(1001)_2$, $(1001)_2$, $(0000)_2$, and $(1001)_2$ are four addends generated, respectively, by ANDing b_0, b_1, b_2, and b_3 with the bits of A. Each new addend is shifted left $k - 1$ times, where k is the multiplication step. As illustrated in the figure, the four addends are shifted left, in order, by 0-, 1-, 2-, and 3-bits before they are added to produce the final product P.

One way to design a combinational n-bit multiplier circuit, such as the one illustrated in Fig. 3.25 for $n = 4$, is to use $n - 1$ n-bit adders and n n-bit bitwise AND modules. The design is straight forward and is based on the steps illustrated in Fig. 3.24. The design, however, has a long propagation delay because, except for

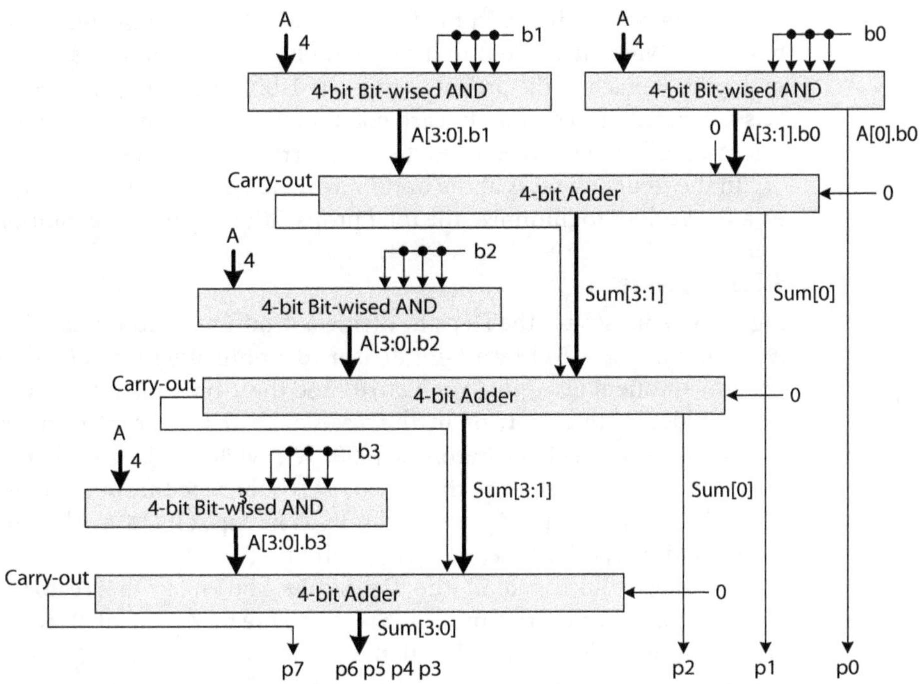

Figure 3.25 A 4-bit unsigned multiplier using n-bit adder modules.

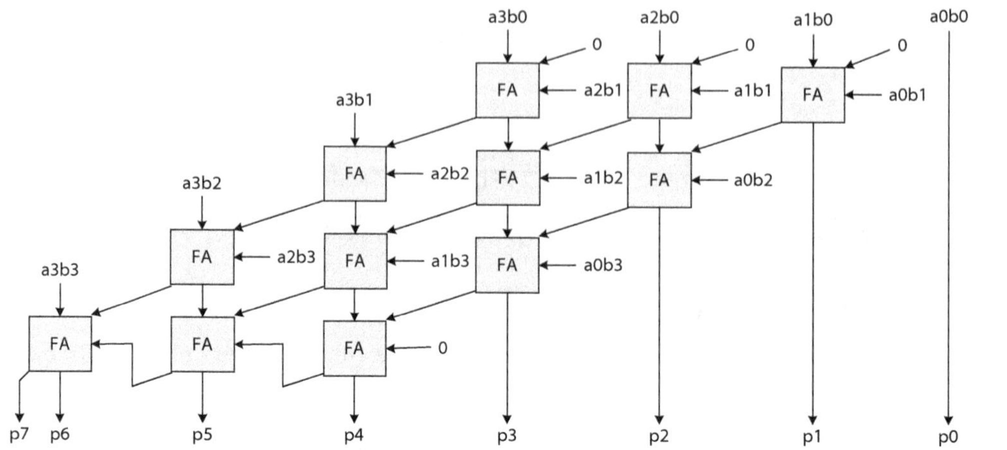

Figure 3.26 A 4-bit array multiplier using an array of FAs.

the first two addends, the remaining addends are added one at a time, creating long signal paths from the input to the final output signals.

Alternatively, one can use FA slices to add the addends, one column at a time. When connected, the FA slices create a two-dimensional structure called an **array multiplier**. Figure 3.26 illustrates a 4-bit array multiplier using six columns of FAs. The sum of the addends is determined one column at a time, similar to how one adds several numbers by hand. In the figure, an addend is represented by its individual bits as a_ib_j where a_i is the ith bit of the multiplicand A and b_j is the jth bit of the multiplier B. Each of the product bits is the 1-bit final sum generated by a chain of FAs organized in a column. In each column, the unused inputs are connected to 0. The last product bit p_7 is equal to the final carryout bit from the last FA column.

In the figure, the FAs at the bottom row create a CPA, which may be replaced by a CLA adder to minimize the total propagation delay of the multiplier.

3.7.2 Divider

Figure 3.27 illustrates the steps to divide a 4-bit unsigned dividend (numerator) $N = n_3n_2n_1n_0 = 4\text{'}b1011$ by a 4-bit divisor (denominator) $D = 4\text{'}b0010$ to generate the 4-bit quotient $Q = q_3q_2q_1q_0 = 4\text{'}b0101$ and the 4-bit remainder $R = 4\text{'}b0001$. The N is padded with $n - 1$, or in this case, with three zeroes from the left; that is, the starting dividend becomes {000, N}, where {} is used here to indicate concatenation. In each step, the divisor D is subtracted from the higher n bits of the dividend, denoted as A_k, during the division step k. If $D \leq A_k$, the corresponding quotient bit is 1; otherwise, the quotient bit is 0.

The steps illustrated in the figure are known as the **restoring division** algorithm because every time that the $A_k < D$ (e.g., $A_3 = 4\text{'}b0001 < D = 4\text{'}b0010$) and thus the difference $A_k - D < 0$, the A_k, not the remainder $R_k = A_k - D$, is used to start the next division step—thus, the name "restoring." In this case, the lower $n - 1$ bits from A_k is concatenated with the next bit in N to make up the next n-bit

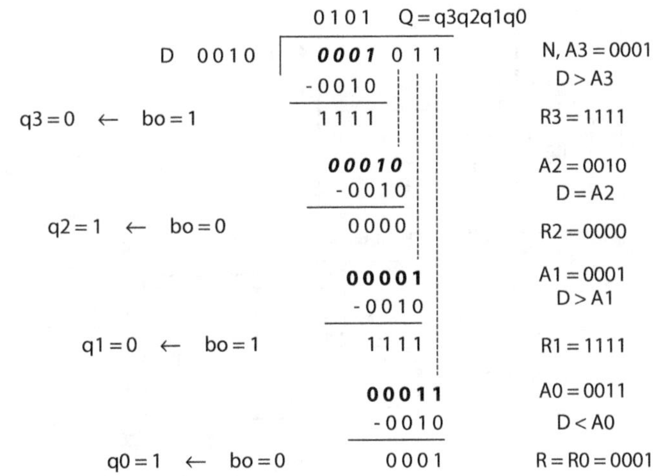

FIGURE 3.27 A restoring division example.

dividend A_{k-1}. Specifically, for $n = 4$, the steps to divide {000, N} by D and generate the 4-bits Q and 4-bits R, as illustrated in the figure, are as follows:

1. $R_3 = A_3 - D$ (4'*b*0001 – 4'*b*0010 = 4'1111) results in $R_3 = -1$ and thus borrow-out = 1 (i.e., $bo_3 = 1$) and $q_3 = 0$.
2. $R_2 = A_2 - D$ (4'*b*0010 – 4'*b*0010) results in $R_2 = 0$ and thus $bo_2 = 0$ and $q_2 = 1$.
3. $R_1 = A_1 - D$ (4'*b*0001 – 4'*b*0010) results in $R_1 = -1$ and thus $bo_1 = 1$ and $q_1 = 0$.
4. $R_0 = A_0 - D$ (4'*b*0011 – 4'*b*0010) results in $R_0 = 1$ and thus $bo_0 = 0$ and $q_0 = 1$. The final remainder is R_0 = 4'*b*0001.

Figure 3.28 illustrates the data path of a 4-bit bit-parallel restoring divider. A subtractor, a NOT gate, and a MUX are needed to implement each division step. The subtractor computes $R_k = A_k - D$ and generates the bo_k. The NOT gate generates the quotient bit as $q_k = \overline{bo_k}$. The MUX is used to select either A_k or R_k using the quotient bit q_k. Equation (3.16) estimates the propagation delay for each division step.

$$\Delta_{\text{division-step}} = \Delta_{\text{subtractor}} + \Delta_{\text{MUX}} \tag{3.16}$$

An array divider, similar to an array multiplier, can be designed using an array made of 1-bit divide slices. Each slice would perform a combined subtractor-MUX function. The combined function can be translated into a truth table for minimal SOP or POS expressions (refer to the Exercise section for more details).

For a large n, an arithmetic function designed as a combinational circuit may require a prohibitively large number of gates. This is especially true when the algorithm, such as that of the multiplier and divider, is repetitive and can alternatively be implemented iteratively. For example, instead of using four subtractors, four MUXs, and four NOT gates to implement a 4-bit restoring divider, as shown in Fig. 3.28, one can use only one subtractor, one MUX, one

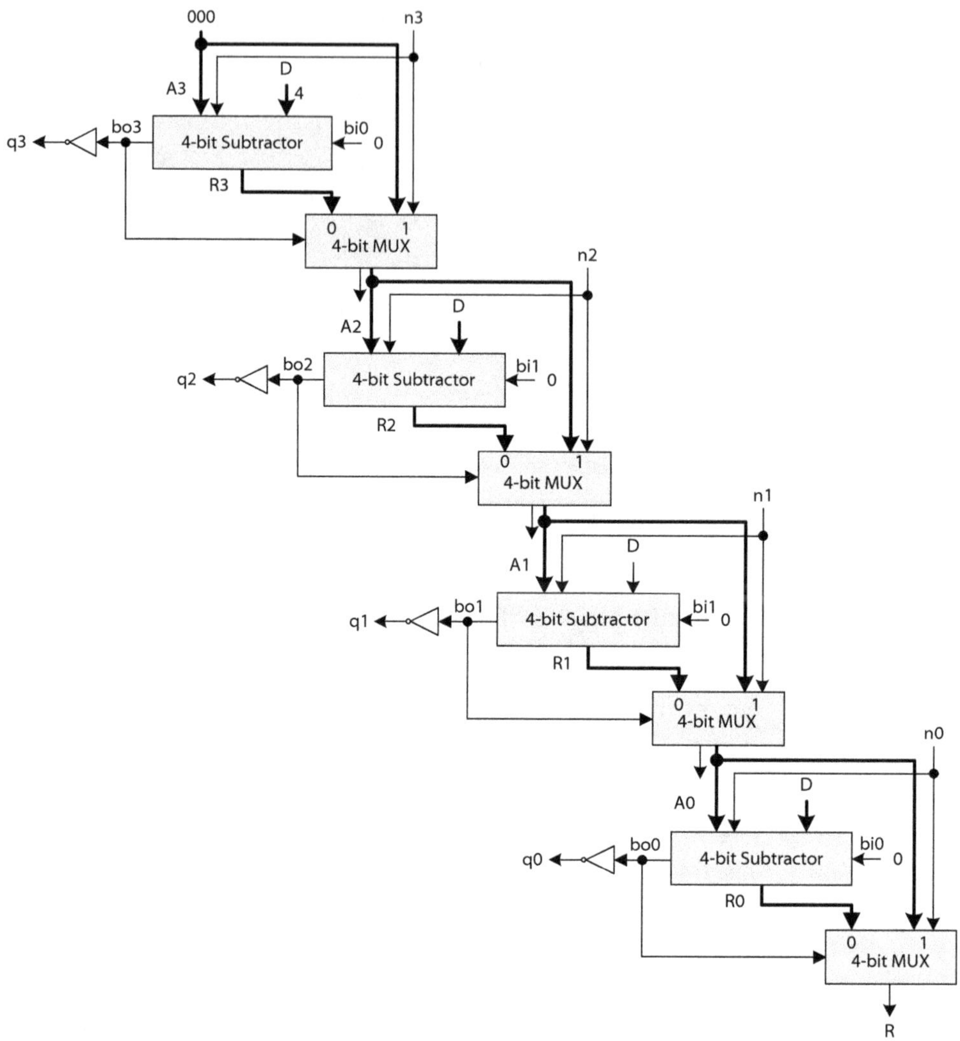

FIGURE 3.28 A 4-bit "restoring" divider data path.

NOT gate, and a set of registers to generate the four quotient bits in four steps. The results of each step would be saved in the registers. However, a repeated use of a hardware module requires some extra hardware to control the timing of each step, and would slightly increase the total time required to produce the final result. The design of such circuits is discussed in Chap. 6.

3.8 Real Number Arithmetic

The representation of real numbers as FP numbers was briefly discussed in Chap. 1. Table 3.10 presents, as an example, three different representations of 3-bit exponent values as 2's complement signed numbers, as biased numbers with bias = 3, and as biased

3-Bit Exponents	Exponent Range as 2's Complement Numbers	Exponent Range as Biased Exponents, Bias = 3	Exponent Range as Biased Exponents, Bias = 4
000	0	–3 = 0 – 3	–4 = 0 – 4
001	1	–2 = 1 – 3	–3 = 1 – 4
010	2	–1 = 2 – 3	–2 = 2 – 4
011	3	0 = 3 – 3	–1 = 3 – 4
100	–4	1 = 4 – 3	0 = 4 – 4
101	–3	2 = 5 – 3	1 = 5 – 4
110	–2	3 = 6 – 3	2 = 6 – 4
111	–1	4 = 7 – 3	3 = 7 – 4

Table 3.10 The List of 3-Bit Signed vs. Biased Exponent Values

numbers with bias = 4. The exponent range for each of the three representations, respectively, are –4 to +3, –3 to +4 when the bias = 3, and –4 to +3 when the bias = 4.

In general, with biased exponents, the designers have more freedom when deciding which set of real numbers to represent in a computer system. Note that, in this case, the highest positive exponent is 4 when bias = 3 vs. 3 when bias = 4. Likewise, the smallest negative exponent is –3 when bias = 3 and –4 when bias = 4. This implies that with bias = 3, there would be more real numbers > |1| (absolute values) that can be represented as FP numbers, but more real numbers < |1| (again absolute values) that would be represented as FP numbers when bias = 4.

3.8.1 Floating-Point Standards

The IEEE 754 standards [3] include three types of FP representations known as single, double, and double extended. Table 3.11 lists the exponent and fraction sizes of each type. Both the single and double FP numbers have in-memory, or external, representation with 23 and 52 fraction bits, respectively, and also in-register, or internal, representation inside the FPU with 24 and 53 fraction bits, respectively. The double extended representation has a 64-bit fraction and is used to increase the accuracy of FP arithmetic. The fractions are represented as signed magnitude numbers, using a separate bit for the sign bit. There is no external (in-memory) representation for the double extended data type.

Format	Fraction Sign	Biased Exponent	External Fraction	Internal Fraction	Total Size (Internal)
Single	1 bit	8 bits	23 bits	24 bits	32 bits
Double	1 bit	11 bits	52 bits	53 bits	64 bits
Double-Extended*	1 bit	15 bits	—	64 bits	80 bits
* As implemented in Pentium Processor family					

Table 3.11 IEEE 754 FP Standards

In addition, the IEEE standards group FP numbers into five data classes, listed here for the single data type:

- Zero
- Denormal with 23-bit fraction
- Normal with 24-bit fraction (only 23 bits are stored in memory)
- Infinity
- NaN (not-a-number), which indicates an invalid FP number or operation

Equation (3.17) presents the relationship between the unbiased exponent e and the biased exponent E. The format of a representable single or double normal FP number is $1.F \times 2^E$ with an explicit 1 before the decimal point; however, the 1 is not stored in memory. The F is the external (in-memory) fraction, and $1.F$ is the internal (in-register) fraction. A denormal FP number is defined as $0.F \times 2^E$ with an explicit 0 before the decimal point.

$$E = e + \text{bias} \qquad (3.17)$$
$$e = E - \text{bias}$$

Table 3.12 presents the ranges for each of the data classes. In the table, the quantities e_{min} and e_{max} indicate the actual (not biased) exponent range for the normal FP numbers.

For example, assuming 3-bit biased exponents with bias = 3, Eq. (3.18) defines the E and F values for each of the data classes.

$$\begin{aligned} &\text{Zero:} && E = 0,\ F = 0 \\ &\text{Denormals:} && E = 0,\ F > 0 \\ &\text{Normals:} && 1 \le E \le 6,\ F \ge 0 \\ &\text{Infinity:} && E = 7,\ F = 0 \\ &\text{NaNs:} && E = 7,\ F > 0 \end{aligned} \qquad (3.18)$$

Unbiased Exponent	External Fraction	Data Class Range	In Memory Representation {Sign, E, F}
$e = e_{min} - 1$	$F = 0$	+/– 0 (zero)	{0/1, 0, 0}
$e = e_{min} - 1$	$F > 0$	+/– $0.F \times 2^e$(Denormal)	{0/1, 0, F}
$e_{min} \le e \le e_{max}$	$F \ge 0$	+/– $1.F \times 2^e$ (Normal)	{0/1, E, F}
$e = e_{max} + 1$	$F = 0$	+/– ∞ (Infinity)	{0/1, E, 0}
$e = e_{max} + 1$	$F > 0$	+/– NaN (Not-a-Number)	{0/1, E, F}
{} Indicates concatenation.			

Table 3.12 IEEE FP Data Classes

3.8.2 Floating-Point Data Space

A FP data space refers to all the real numbers that can be represented as FP numbers in a computer system. Figure 3.29 displays an FP data space using the positive real axis. The dotted horizontal line between 0 and $2^{e_{min}}$ (not including 0 or $2^{e_{min}}$) indicates the denormal data space. The thin and bold vertical lines indicate the specific real numbers that can be represented as FP numbers when fractions are only 2-bits each. The more fraction bits there are in the representation, the more real numbers can be represented as FP numbers.

For instance, in the single precision representation, the 8-bit unbiased exponent rage is from $e_{min} = -126$ to $e_{max} = 127$ with bias = 127. Each fraction is 23-bits. Its data space would have $2^{22} - 1$ thin lines between each pairs of the bold lines. The data space of the double precision representation would have $2^{52} - 1$ thin lines between each pair of bold lines.

Figure 3.30 displays the list of positive real numbers that can be represented as 6-bit FP numbers using 1-bit for sign, 2-bits for a fraction, and 3-bits for an exponent with bias = 3. As shown in the figure, 17 real numbers ranging between 0.25 and 14.0 (inclusive) can be represented with the 6-bit FP format. The FP numbers between 0 and 0.25 (exclusive) would be considered denormals.

Example 3.2. Determine the external (in-memory) single precision FP representation of real number +10.75.

Solution: First, the number is converted into binary. Then the representation is converted into its scientific format in binary.

$$\begin{aligned} +10.75 &= (1010.11)_2 \\ &= (1010.11)_2 * 2^0 \\ &= (1.01011)_2 * 2^3 && \text{(unbiased scientific format as } 1.F \times 2^e) \\ &= (1.01011)_2 * 2^{3+127} && \text{(biased scientific format, bias = 127)} \\ &= (1.01011)_2 * 2^{130} && \text{(IEEE format as } 1.F \times 2^E) \end{aligned}$$

The result is a 32-bit number created from concatenating the 1-bit sign = 0, the 8-bit biased exponent $E = 130 = (10000010)_2$, and the 23-bit fraction $F = (010110....0)_2$ as follows:

0	10000010	01011000000000000000000

Or, 0x412C0000 in hex. The 1 in the 1.*F* format is not stored in memory.

Other than the sign bit, the representation of negative and positive FP numbers is the same. Negative FP numbers have a 1 as the sign bit. As an example, the single precision FP in-memory representation of –10.75 is 0xC12C0000.

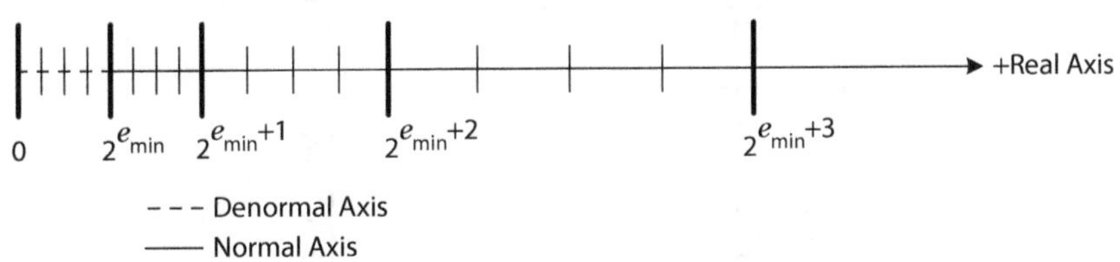

FIGURE 3.29 An FP data space with 2-bit fractions.

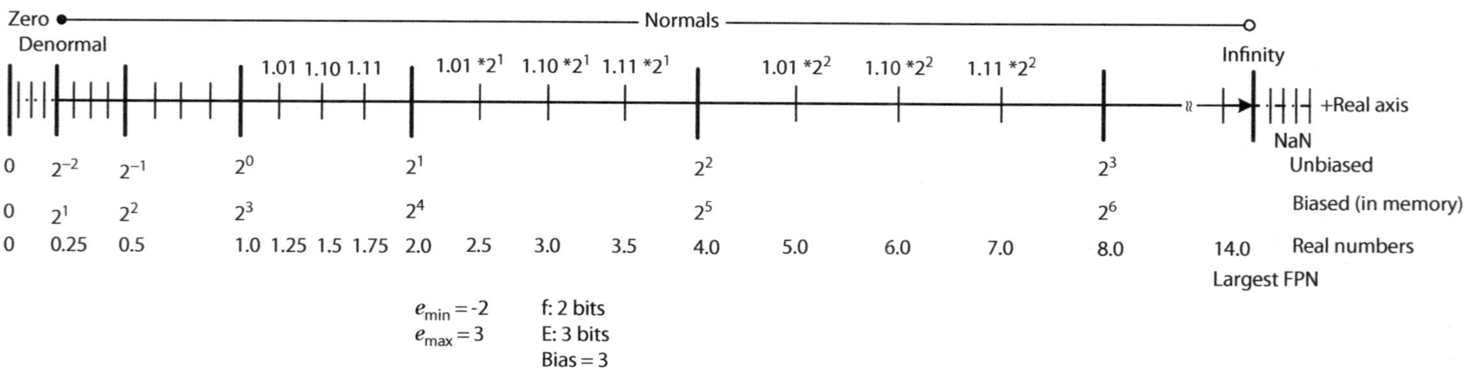

Figure 3.30 Real numbers represented as 6-bit FP numbers with 2-bit fractions and 3-bit exponents with bias = 3.

Two-Dimensional Display

Alternatively, another way to display an FP data space is to use rectangular regions [4], as illustrated in Fig. 3.31 for the single precision. A separate two-dimensional display is shown for each positive and negative FP data spaces. In the figure, the *x*-axis is arbitrarily selected to represent the exponent values and the *y*-axis to represent fraction values. A different scale is used in each of the axes.

When compared to the one-dimensional display in Fig. 3.29, it is easier with the two-dimensional display to identify the location of a special FP number or the domain and/or the range for an FP function, such as cosine. For instance, the range for the cosine function is all the representable FP numbers between 0 and 1.0, inclusive. A function's domain and range may be used, for example, to generate test vectors for an FPU [4]. Figure 3.32 illustrates the location of few sample FP numbers in the two-dimensional FP data space.

The smallest positive normal FP number is marked as item 1 in the figure and is located at the bottom-left corner of the positive normal region with $E = 1$ and $F = 0$. The smallest (i.e., largest magnitude) negative normal FP number is marked as item 2 and is located at the upper-right corner of the negative normal region with exponent $E = 254$ and F all 1's. The +1.0 is marked as item 3 and is located in the bottom center of the +normal region with $E = 127$ and $F = 0$. The ±zero is shown with $E = 0$ and $F = 0$. The ±infinity is shown with $E = 255$ and $F = 0$. Also, as implemented in the Intel Pentium processors, there are other special FP numbers, such as, the quite NaN (QNaN) that indicates an invalid operand—for example, −1 in the computation of $\sqrt{-1}$ —or the signaling NaN (SNAN) that indicates an invalid operation [5].

The two-dimensional display of the double precision FP data space is similar, with the exception that the number of bits used to represent each exponent and fraction are, respectively, 11- and 52-bits.

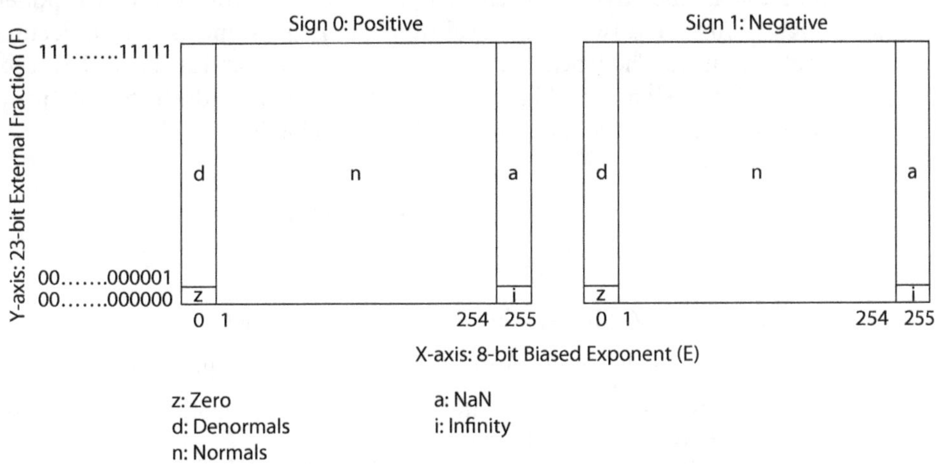

FIGURE 3.31 Two-dimensional display of the single precision FP data space.

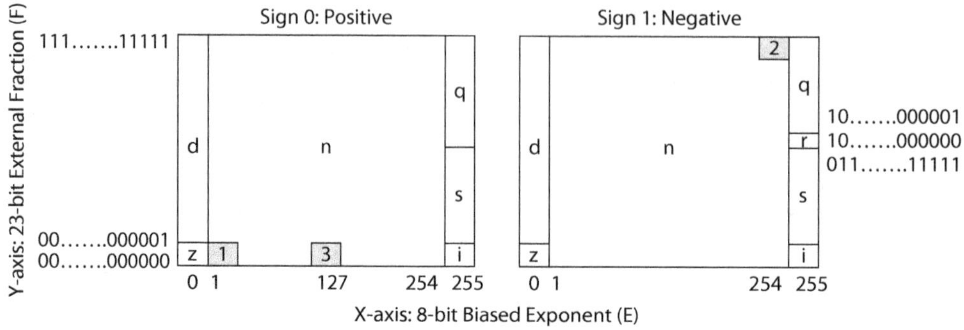

[1] Smallest Normal positive FPN: 0, 00000001, 00000000000000000000000 =~ 1.175*10^{-38}
[2] Smallest Normal negative FPN: 1, 11111110, 11111111111111111111111 =~ −3.40*10^{+38}
[3] +1.0: 0, 01111111, 00000000000000000000000
[z] + Zero: 0, 00000000, 10000000000000000000000
[i] - Infinity: 1, 11111111, 00000000000000000000000
[q] Quiet Nan (QNaN): Denotes indeterminate operation, such as $\sqrt{-1}$
[s] Signalling NaN (SNaN): Denotes invalid operation
[r] Real-Indefinite (a special QNaN): 1, 11111111, 10000000000000000000000

FIGURE 3.32 Examples of the single precision FPNs.

3.8.3 Floating-Point Arithmetic

Both the exponent and the fraction of an FPN are integer numbers and are operated on separately when FP arithmetic is performed. For example, in order to add the two real numbers 0.1075 (10.75×10^{-2}) and 72.5 (0.725×10^{2}) in decimal, the smaller fraction is shifted right to line up the decimal points before the two fractions are added to produce the sum 72.5075 (0.725075×10^{2}). The number of bits the smaller fraction is shifted is determined from the two exponents. The algorithm to add two binary real numbers is the same as that in decimal and is outlined next. The labels *A.s* and *B.s*, *A.E* and *B.E*, and *A.F* and *B.F*, respectively, refer to the sign bits, biased exponent values, and the external (in-memory) fraction values of two FP numbers *A* and *B*. That is,

$$A = \{A.s,\ A.E,\ A.F\}$$
$$B = \{B.s,\ B.E,\ B.F\}$$

where {} is used to indicate concatenation.

Floating-Point Addition Algorithm: Normal Data Class

1. *Initialize the inputs*: *A* must be bigger than or equal to *B*.
   i. Let $A.F = \{1,\ A.F\}$ and $B.F = \{1,\ B.F\}$ be the internal (in-register) representations of the fractions.
   ii. Make sure $|A| \geq |B|$. If $|A| < |B|$, then switch *B* with *A*; $|X|$ indicates the absolute value of *X*.

2. *Line up decimal points*: Shift *B.F* right by *D* bits as determined next.

   i. Let $D = A.E - B.E$.
   ii. Shift *B.F* right *D* number of times, entering 0 from left.

3. *Generate the result* (*R*): Compute $R.F = A.F \pm B.F$.

   i. Generate the *R.F* as the sum or the difference of *A.F* and *B.F* depending on the values of *A.s* and *B.s* as follows:

A.s	B.s	Computation
0	0	$A.F + B.F$
0	1	$A.F - B.F$
1	0	$A.F - B.F$
1	1	$A.F + B.F$

   ii. Let $R.s = A.s$ and $R.E = A.E$.

4. *Normalize the result*: Convert the *R.F* to 1.*F* format, if it is not already.

   i. If the format of the *R.F* is 1*x.F* where *x* is either 0 or 1 in step 3.*i*, then add a 1 to *R.E* and shift *R.F* 1-bit right to change its format to 1.*F*; the LSB (least significant bit) of 1*x.F* will be lost.
   ii. Or, if *R.F* in step 3.*i* has leading zeroes, then subtract a 1 from *R.E* each time that *R.F* is shifted left in order to remove the leading zeros necessary to represent *R.F* as 1.*F*. For example, if *R.F* = 0.01xxxxxx....x, then it would be shifted left twice to 1.xxx...x (1.*F* format). In this case, *R.E* must be reduced by two.

5. *Round the result*:

   The final output, *S.F* = {1, *S.F*}, is selected from the upper bits of *R.F* depending on the size of the target fraction: 24-bits for single precision and 54-bits for the double precision internal representations. The unused lower bits of *R.F* will be lost. However, the lost bits can be used to round up the result by adding a 1 to the LSB of *S.F*. If the format of the rounded *S.F* becomes 1*x.F* again, another normalization step would be necessary to change its format back to 1.*F*. A more complete discussion on rounding involving the guard (*G*), round (*R*), and sticky (*S*) bits, as outlined in the IEEE FP standards, is beyond the scope of this book.

6. *Final output*:

   The *R.s*, *R.E*, and *S.F* are concatenated to create either a 32-bit number, if single precision, or a 64-bit number, if double precision, before storing it in memory. That is,

$$S = \{R.s,\ R.E,\ S.F\}$$

The size of *R.F* is 64-bits, and all the integer arithmetic and shift functions are performed in 64-bit fractions to minimize rounding errors when billions of computations are performed.

Example 3.3. Determine $S = A + B$ given that $A = 17.875$ and $B = 15.75$. Assume *A*, *B*, and *S* are 16-bit FP numbers with 1-bit sign, 7-bit *e* exponents with bias = 63, and 8-bit fractions.

Solution: The five steps to determine $S = A + B$ are as follows:

1. Convert each of the real numbers A and B into their binary representations.
2. Convert the binary representation to their equivalent unbiased scientific format.
3. Convert the unbiased scientific formats to their equivalent biased format.
4. Convert each of the biased scientific formats to a 16-bit in-memory representation.
5. Follow the FP addition algorithm to add the two FP numbers.

Step 1: Convert to binary

$A = 10001.111 \times 2^0$ $\qquad$ $B = 1111.11 \times 2^0$

Step 2: Scientific format (unbiased exponent)

$A = 1.0001111 \times 2^4$ $\qquad$ $B = 1.11111 \times 2^3$

Step 3: Scientific format (biased exponent), IEEE format as $1.F \times 2^E$

$A = 1.0001111 \times 2^{4+63}$ $\qquad$ $B = 1.11111 \times 2^{3+63}$
$A = 1.0001111 \times 2^{67}$ $\qquad$ $B = 1.11111 \times 2^{66}$

Step 4: In-memory representations:

$A = 0x431E$ (0, 1000011, 00011110) $\qquad$ $B = 0x42F8$ (0, 1000010, 11111000)

Step 5: Apply the FP addition algorithm discussed earlier.

1. *Initialize the inputs:*
   Because $|A| \geq |B|$, there is no need to switch A with B.

$$A.s = 0,\ A.E = 1000011\ (67),\ A.F = 1.00011110$$
$$B.s = 0,\ B.E = 1000010\ (66),\ B.F = 1.11111000$$

2. *Line up the decimal points:*

$$\begin{aligned} D &= A.E - B.E \\ &= (1000011)_2 - (1000010)_2 \quad \text{or} \quad 67 - 66 = 1 \\ &= 0000001 \end{aligned}$$

   Shift $B.F$ right once ($D = 1$) to produce $B.F = 0.111111000$

3. *Generate the sum:*

$$\begin{aligned} R.F &= A.F + B.F \\ &= 1.00011100 + 0.111111000 \\ &= 10.00011010 \text{ (not in the 1.}F\text{ format)} \\ R.s &= A.s = 0 \\ R.E &= A.E = 67 \end{aligned}$$

4. *Normalize the result:*

   i. Add 1 to the $R.E$ to yield $R.E = 68$ $(67 + 1)$.
   ii. Right-shift the $R.F$ by 1 to yield $R.F = 1.000011010$ (now in the $1.F$ format).

5. *Round the result:*

   Select the upper 9-bits from the new $R.F$ as $S.F = 1.00001101$. The lower only remaining bit of $R.F$ is 0 and thus would be ignored. The result is:

$$S.F = 1.00001101 \text{ with } S.F = 00001101$$

(However, if the new *R.F* was instead 1.000011011 with 1 as its LSB, we would have had the option to round up the *S.F* to 1.00001110 (i.e., 1.00001101 + 0.00000001).

6. *Final sum:*

$$S = \{R.s, R.E, S.F\}$$
$$S = \{0, 1000100, 00001101\} \text{ (or, 0x440}D\text{ in memory)}$$

The *S* is converted to a decimal number as follows:

$$\begin{aligned} S &= 1.00001101 \times 2^{68} && \text{(biased scientific format)}\\ S &= 1.00001101 \times 2^{68-63} && \text{(make it unbiased)}\\ S &= 1.00001101 \times 2^{5} && \text{(unbiased scientific format)}\\ S &= 100001.101 && \text{(in binary)}\\ S &= 33.625 && \text{(in decimal)}\\ &= 17.875 + 15.75 \end{aligned}$$

FP subtraction, multiplication, and division are similarly performed in several steps. For subtraction, the fractions are first lined up, as in the case of the FP add, and then they are subtracted if $A.s = B.s$ or added if $A.s \neq B.s$. For multiplication, the fractions are multiplied, the exponents are added, and the sign bits are XORed. Finally, for division, the fractions are divided (integer division), the exponents are subtracted, and the sign-bits are XORed. The rounding and normalization steps are the same as those discussed for the FP addition.

However, because for FP division the most significant bits of the numerator internal fraction *N.F* and denominator internal fraction *D.F* are both 1, *N.F* is not padded with 0's from left as it was done in Fig. 3.28 for integer division (refer to Exercise 3.29).

3.8.4 Floating-Point Unit

Figure 3.33 illustrates the data path of a floating-point adder. The data path includes combinational circuit modules that implement the tasks outlined in the algorithm. In the data path, the Switch module is used to test if $|A| < |B|$. In order for $|A|$ to be less than $|B|$, either *A.E* must be less than *B.E*, or in case of $A.E = B.E$, *A.F* must be less than *B.F*. Two subtracting modules inside the Switch (not shown) generate the difference $A.E - B.E$ and $A.F - B.F$. The borrow-out signals from these modules would indicate whether or not $|A| \geq |B|$.

If $|A| < |B|$, then *A* and *B* must be switched so that the larger number is used as *A*, the left input of the data path. Two 2-to-1 MUXs (not shown) would be needed to switch *A* with *B* and *B* with *A* when necessary. The two inputs *A* and *B* are switched if $A.E < B.E$ or if $A.F < B.F$ if $A.E = B.E$.

The Right Shifter module is used to align the decimal points when computing $A.F \pm B.F$. A Shift module is also used within each of the normalization and rounding modules. A combinational shifter is described next.

Combinational Shifter

A combinational shifter is designed using $\log_2(k)$ 2-to-1 MUXs organized in $\log_2(k)$ levels, where *k* indicates the range for the number of shifts. For example, for $k = 8$, a combinational shifter can shift its input 0 to $k - 1$ or 7 bits. Figure 3.34

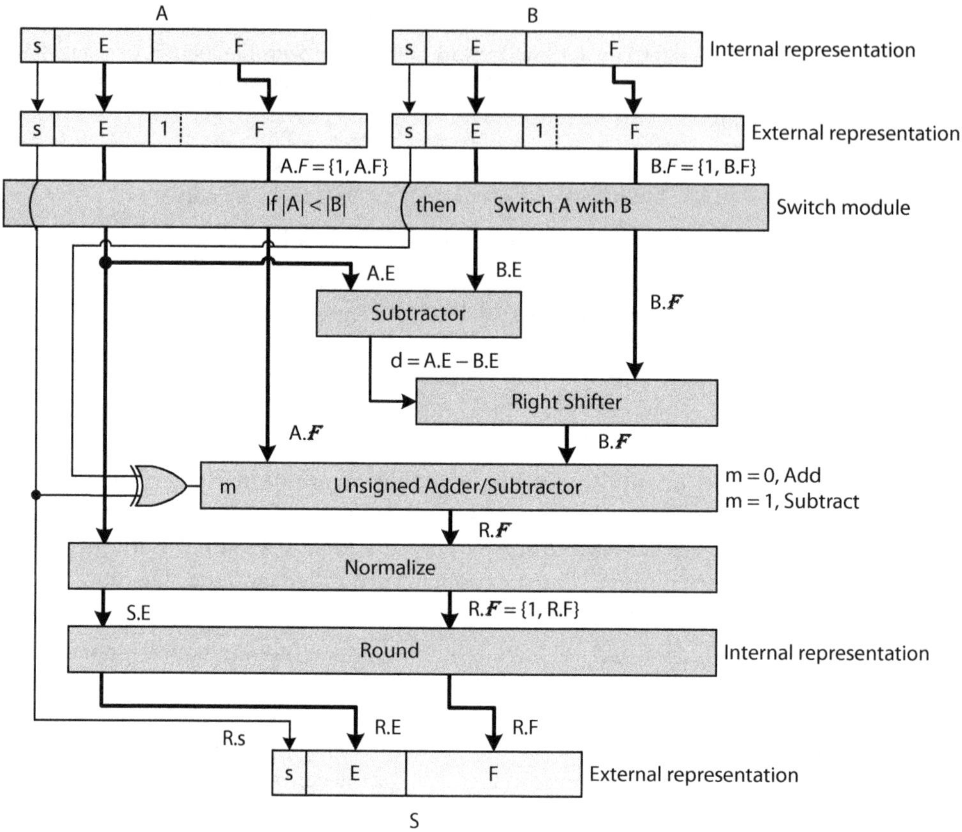

Figure 3.33 A data path for FP addition.

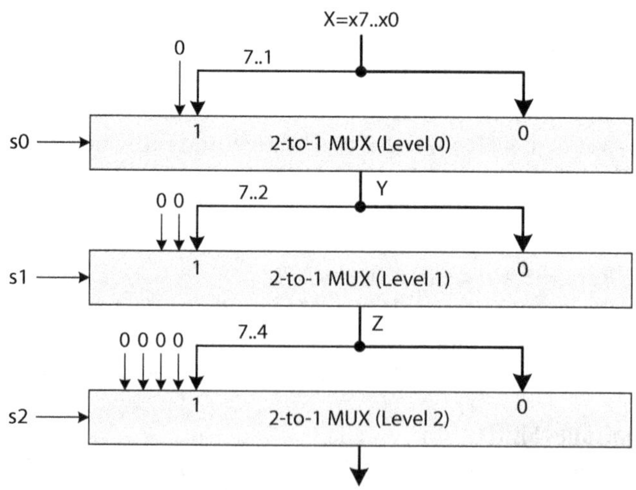

Figure 3.34 An 8-bit combinational right shifter with zero fill.

illustrates an 8-bit combinational right shifter with $k = 8$. The 3-bit $S = s_2s_1s_0$ specifies the shift size between 0 and 7 bits. The bits s_2, s_1, and s_0 are used as the select inputs to the three MUXs.

As illustrated in the figure, the top (Level 0) MUX selects either the input X if $s_0 = 0$ or the X shifted right by 1-bit if $s_0 = 1$. The next MUX selects either the Y if $s_1 = 0$ or the Y shifted right by 2-bits if $s_1 = 1$. Finally, the bottom (Level 2) MUX selects either the Z if $s_2 = 0$ or the Z shifted right by 4 bits if $s_2 = 1$.

For example, when $S = s_2s_1s_0 = (011)_2$, $s_0 = 1$ causes the Level 0 MUX to shift input X right by 1-bit. The signal $s_1 = 1$ causes the Level 1 MUX to shift Y by 2 bits. Finally, the signal $s_2 = 0$ causes the Level 2 MUX to pass Z as-is to its output. As a result, the shifter shifts its input X 3 bits to the right. When $S = (101)_2$, X would be shifted right five times. The maximum number of times that X can be shifted is seven when $S = (111)_2$.

References

1. Vincent P. Heuring and Harry F. Jordan, *Computer Systems Design and Architecture*, Pearson Education, Inc., 2004.
2. Steve Wilson, "Alternative Subtraction Algorithms," http://www.sonoma.edu/users/w/wilsonst/courses/math_300/groupwork/altsub/default.html.
3. W. Kahan, "Lecture Notes on the Status of IEEE Standard 754 for Binary Floating-Point Arithmetic," http://www.cs.berkeley.edu/~wkahan/ieee754status/.
4. Nikrouz Faroughi, "A floating-point data space model: domain and range of a function," WORLDCOMP'03, The 2003 World Congress in Computer Science, Computer Engineering, and Applied Computing, June 25-28, 2007, Las Vegas, USA.
5. Intel 64 and IA-32 Architecture Software Developer's Manual; Volume 1: Basic Architecture.

Exercises

3.1 Consider a CPA(8), an 8-bit CPA, and do the following:

a. Design the adder using NOT and NAND gates and determine the total number of each gate used. Use SOP expressions for the FA from Chap. 2.
b. Determine the total number of transistors.
c. How many transistors would be needed to design a CPA(32)?

3.2 Calculate the following 2's complement sum and difference for the following values. Indicate if any will result in an overflow.

```
a)                   b)
A: 11001100          A: 11111000
B: 00110100          B: 00001000
+                    +
c)                   d)
A: 00000001          A: 11111000
B: 00000010          B: 00001000
-                    -
e)                   f)
A: 10000010          A: 01111101
B: 00000011          B: 11111010
-                    -
```

3.3 Determine the POS expressions of a 1-bit PGU and 1-bit CGU and then draw their circuits using NOT and NOR gates, but combine and use the minimum number of gates.

3.4 Suppose an 8-bit adder is designed using two 4-bit CPAs (labeled CPA1 and CPA2), where carry-out c_3 is fed as carry-in into CPA2. CPA1 inputs the initial carry-in c_{-1}. In order to speed up the adder, c_3 is generated as $c_3 = g_3 + p_3g_2 + p_3p_2g_1 + p_3p_2p_1g_0 + p_3p_2p_1p_0c_{-1}$, where p's and g's are generated in 0.3 ns. Determine how much faster the new adder will be if ΔFAc = 0.5 ns.

3.5 Design and estimate the propagation delay of a 16-bit hybrid adder using

a. Two CLA(8) modules
b. Four CLA(4) modules
c. Eight CLA(2) modules

3.6 Compare ΔCPA(8) with ΔCLA(8) by calculating the speedup as the ratio of ΔCPA(8) over ΔCLA(8). Use SOP expressions for an FA.

3.7 Design a 15-bit CLA using 4-bit BCGUs, where each BCGU outputs three carry-bits and p^* and g^* signals.

3.8 Design a 1-bit, 8-to-1 MUX using 1-bit, 2-to-1 MUXs. Also estimate its propagation delay assuming that NAND gates have 0.1 ns delay.

3.9 Design a 1-bit, 8-to-1 MUX using a combination of 1-bit, 2-to-1 and 1-bit, 4-to-1 MUXs. Also estimate its propagation delay assuming that NAND gates have 0.1 ns delay and a 4-to-1 MUX implements SOP expressions.

3.10 Estimate the delay of the 8-bit ALU in Fig. 3.16 assuming 0.1 ns delay for NAND gates, the adder is a CLA(8), and there are only 2-to-1 MUXs available.

3.11 Verify that the circuit in Fig. 3.22 implements the ALU Map module.

3.12 Consider an 8-bit bit-parallel ALU with only four functions: add, subtract, and bitwised AND and XOR. Assume an adder/subtractor module uses a hybrid adder designed from CLA(2) modules. In addition, the overflow flag must be masked out when performing bitwise computation. Assume only 8-bit, 2-to-1 MUXs are available. Do the following:

a. Draw the data path and estimate its propagation delay assuming 0.1 ns delay for each NOT and NAND gate.
b. Construct the truth table for its Map module.

3.13 Design a 4-to-2 encoder using 2-to-1 encoder modules. Hint: You also need a 1-bit, 2-to-1 MUX and a two-input OR gate.

3.14 Design an 8-to-3 encoder using 4-to-2 encoder modules. Hint: You also need a 2-bit, 2-to-1 MUX and a two-input OR gate.

3.15 Design the 8-bit bit-serial ALU shown in Fig. 3.23 and use the truth table given in Table 3.8 for the 1-bit ALU slice.

3.16 Design a 2-bit 2's complement comparator, and then use four of the comparator modules to design an 8-bit comparator. Hint: A 2-bit 2's complement comparator inputs two 2-bit inputs *A* and *B* and also *gti* (greater than), *eqi* (equal), and *lti* (less than) signals, where "*i*" stands for input from the previous module and then outputs three signals, *gto*, *eqo*, and *lto*, where "*o*" stands for output. First design a 2-bit 2's complement with only inputs *A* and

B to generate three outputs *gtt* if $A > B$, *eqt* if $A = B$, and *ltt* if $A < B$, where "*t*" stands for temporary. Then combine *gtt*, *eqt*, and *ltt* with the *gti*, *eqi*, and *lti* to generate the *gto*, *eqo*, and *lto*. For example, $gto = 1$ if ($gti = 1$ and $gtt = 1$) or ($gti = 1$ and $eqt = 1$) or ($eqi = 1$ and $gtt = 1$) or ($lti = 1$ and $gtt = 1$).

3.17 Consider the 4-bit array multiplier given in Fig. 3.26. Do the following:

a. Estimate its propagation delay in terms of propagation delays of the carry- and sum-bits of an FA; that is, in terms of ΔFAc and ΔFAs.
b. Give a generalized equation for the propagation of an *n*-bit array multiplier.

3.18 Consider an 8-bit array multiplier where a CLA(8) is used to replace the CPA(8) in the last row. Assuming that ΔFAc = 0.2 ns, ΔFAs = 0.3 ns, and CLA(8) = 0.8 ns, how much faster will the new multiplier will be compared to the original?

3.19 Use the restoring division algorithm discussed to divide $N = 10101101$ by $D = 1110$. Note you can use binary calculation on your calculator to double-check your results.

3.20 Design an array divider by first designing a 1-bit combined subtract-MUX bit slice with minimum delay (i.e., SOP or POS expressions); then use it to design a 4-bit bit-serial subtract-MUX module; then replace each of the 4-bit subtract and 4-bit MUX pairs in the divider in Fig. 3.28. with a 4-bit bit-serial subtract-MUX module.

3.21 Write a program in the language of your choice (or use Excel) to implement the following reciprocal division algorithm and make an observation when *D* becomes 1.0. Then verify that the reciprocal division algorithm, which computes Q, is equal to *N*/*D* computed by using the divide (/) operator. The reciprocal division algorithm is as follows:

```
#define ITERATIONS 15
Procedure ()
{
  float D, N, Q;
  float R;
input D, N;  //e.g., D = 1.99 and N = 2.4
For j = 1 to INTERATIONS do
    R = 2 - D;
    D = D * R;
    N = N * R;
    print j, D, N, R
Endfor

}
```

Note that no division operator is used in the calculation of Q; only multiplication and subtraction operations are used to compute the result of *N* divided by D. This algorithm was implemented in hardware as the FP divide instruction in an Intel x486 processor. Run the program twice, once with inputs $D = 1.99$ and $N = 2.4$, and again with $D = 1.56$ and $N = 2.4$. Note the fraction of *D* is always less than 2 (e.g., maximum fraction of *D* in binary is $1.1111111\ldots1 < 2$ with 23 1s after the decimal point for the "float" data type and 52 1s for the "double" data type).

Compare the N value when D becomes 1.0 with the Q value computed as the original N (N_0) divided (/) by the original D (D_0) (e.g., $Q = 2.4/1.99$). Compare N_i with $Q = N/D$ when D_i becomes 1.0 for some i. Is $Q = N_i$ when $D_i = 1.0$? What do you notice between the two runs?

3.22 Determine the IEEE floating point representations of 6.725 for:

3.23 Determine the IEEE floating point Single precision representation of 0.35.

3.24 0x41DD0000 is an IEEE single precision FP number. What number does it represent in decimal?

3.25 Show the steps to compute the sum (S) of two single precision FP numbers A = 0xC18D0000 = −17.625 and B = 0x41080000 = 8.5.

3.26 Design an 8-bit combinational arithmetic right shifter. Each time a number is arithmetic right shifted, the sign bit is repeated. Also illustrate −80 right shifted three times (i.e., −80 >>> 3).

3.27 Design and simulate a Verilog behavior and structural model for an 8-bit 2's complement adder/subtractor using a CPA(8). Use a "case" statement to describe an FA; then use a Verilog structural model to design the CPA(8) with eight copies of the FA modules. Use an "assign" statement to enter the expression for the overflow flag (*ovf*) and an "if-else" statement to describe the inverter module. Create a Verilog tester module and test your design using test vectors 0x80 − 0x01 and 0x7F + 0x01. Is the overflow flag asserted in both cases?

3.28 Design an 8-bit restoring divider in Verilog. Use a behavior Verilog model to create a 1-bit subtractor similar to an FA. Also create an 8-bit BPS similar to an 8-bit CPA. Create a behavior description of 8-bit, 2-to-1 MUX. Combine several BPS and MUX modules to design the divider. Create a tester module to test your design.

3.29 Consider the FP numerator with external fraction N.*F* = {1, N.F} and denominator external fraction D.*F* = {1, D.F} where N.F and D.F are 4-bit numbers. Use four 4-bit bit-serial subtractor-MUX modules from Exercise 3.20 to design an integer divider used in an FPU. Hint: N.F will be padded with 0's from the right; in this case, Ak in each step is always a 5-bit number and Ak[4], the most significant bit (MSB), is not used to determine the next remainder (i.e., Ak[3:0] – D.F); Ak[4] is used instead with the conjunction of borrow-out (bo) in each step to determine the next quotient bit as $q_k = A_k[4] + (\overline{bo_k})$.

3.30 Computer security (hardware Trojans): Exercise 11.12 to understand how a computational malicious circuit is designed. Do not implement the triggering mechanism; instead, directly operate the multiplexer to cause a computational attack (also see Sec. 11.2).

3.31 Computer security: (access control) Exercise 11.26 to design a hierarchical access control scheme suitable for hardware implementation (also see Sec. 11.1.4 and Sec. 11.1.5).

CHAPTER 4
Sequential Circuits: *Core Modules*

4.1 Introduction

While combinational circuits are necessary and are an important part of a digital system, their outputs depend only on the inputs currently applied. Any change in the inputs is expected to change the outputs. On the other hand, the outputs of a sequential circuit depend not only on the current inputs, but also on the inputs previously entered. For example, as illustrated in Fig. 4.1, if a single adder is used to generate the sum of several numbers, then a partial sum, initially zero, must be saved and added with each new number to generate the next partial sum. Therefore, the circuit is a sequential circuit because the current partial sum is the sum of all the previously entered numbers, and the next partial sum is determined by adding the next number, now the current input to the adder, to the saved partial sum.

In general, a complex sequential data path requires combinational circuits to generate outputs as results and storage modules (typically registers) to save those results. The data path also requires a control unit (a controller) that follows a specific set of steps (i.e., an algorithm) to compute a complex function using the data path. A control unit is a sequential circuit and uses storage modules to save its current state (e.g., a specific step in the algorithm) in order to determine its next state using the inputs it currently receives.

Sequential circuits also require a timing control signal called a *clock*. It is used to determine when a value—for example, the sum from the adder in Fig. 4.1—should be stored in the storage module. Note that the adder does not generate all the sum bits at the same time. Each sum bit is the output of a combinational circuit with a specific propagation delay. The circuits that have shorter propagation delays generate their outputs sooner than those that have longer propagation delays. Therefore, the sum is valid when the circuit

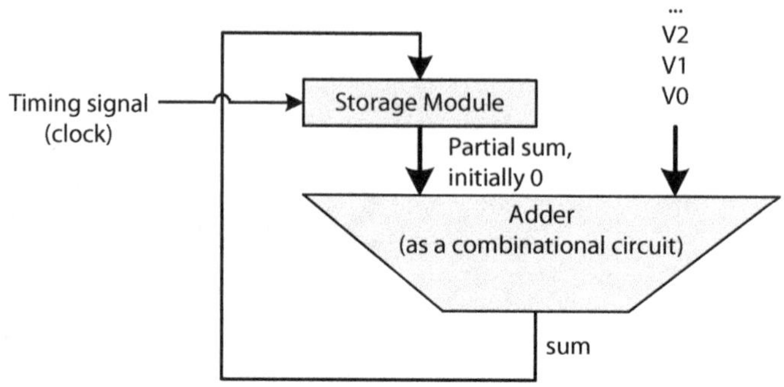

FIGURE 4.1 An illustration of a sequential circuit; it computes the sum of several numbers, V_0, V_1, etc.

with the longest propagation delay in the adder generates its output. Otherwise, one or more of the sum bits would be still changing, and thus the sum would be invalid. The clock signals the validity of the sum and allows the storage module to save it.

This chapter presents the core logic circuits required to design a storage module. In a core circuit, outputs are fed back as inputs so the circuit can retain an output value, thereby creating a storage module. However, these core circuits require certain operating constraints to remain stable and retain their stored values. They are used in the design of small and large sequential circuits. Small sequential circuit designs are covered in Chap. 5, large designs in Chap. 6, and CPU design in Chap. 8.

4.2 SR Latch

The circuit of an SR latch is illustrated in Fig. 4.2. The circuit has two inputs: s and r. They, respectively, stand for set q, meaning q becomes 1, and reset q, meaning q becomes 0. Signals p and q are interdependent and are defined as $q=\overline{r+p}$ and $p=\overline{s+q}$. In order to determine the value of q, one must know the value of p, or vice versa. Therefore, because the initial values of q and p are not known, the circuit activities are best understood by examining the following four cases.

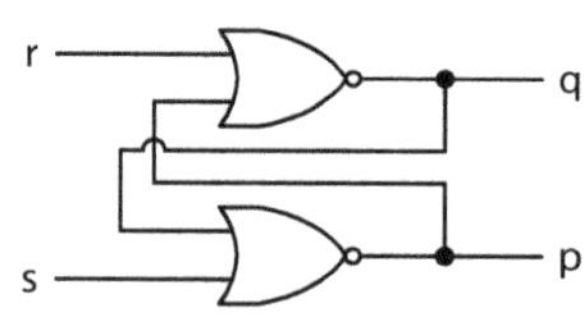

FIGURE 4.2 A basic SR latch.

Case 1: $s = 0$ and $r = 0$

a. Assuming that the current value of $q = 0$, let $s = 0$ and $r = 0$ and then determine the new values of p and q. In this case,

$$p^{new} = \overline{s + q}$$
$$= \overline{0 + 0}$$
$$= 1$$
$$q^{new} = \overline{r + p^{new}}$$
$$= \overline{0 + 1}$$
$$= 0$$

b. Assuming that the initial value of $q = 1$, let $s = 0$ and $r = 0$ and then determine the new values of p and q.

$$p^{new} = \overline{s + q}$$
$$= \overline{0 + 1}$$
$$= 0$$
$$q^{new} = \overline{r + p^{new}}$$
$$= \overline{0 + 0}$$
$$= 1$$

That is, when $s = 0$ and $r = 0$, $q^{new} = 0$ (no change) and $p^{new} = 1$.
That is, when $s = 0$ and $r = 0$, $q^{new} = 1$ (again no change) and $p^{new} = 0$.

Case 1 indicates that when both $s = 0$ and $r = 0$, the values of p and q remain unchanged and $p = \overline{q}$.

Case 2: $s = 0$ and $r = 1$

a. Assuming that $q = 0$, let $s = 0$ and $r = 1$. Thus,

$$p^{new} = \overline{s + q} = \overline{0 + 0} = 1$$
$$q^{new} = \overline{r + p^{new}} = \overline{0 + 1} = 0$$

That is, when $s = 0$ and $r = 1$, $q^{new} = 0$ (no change) and $p^{new} = 1$.

b. Assuming that $q = 1$, let $s = 0$ and $r = 1$.

$$p^{new} = \overline{s + q} = \overline{0 + 1} = 0$$
$$q^{new} = \overline{r + p^{new}} = \overline{1 + 0} = 0, \qquad q \text{ changes, thus}$$

$p^{new} = \overline{s + q^{new}} = \overline{0+0} = 1,$ p changes, thus

$q^{new} = \overline{r + p^{new}} = \overline{1+1} = 0$ no change in q^{new}, both p and q are now stable.

In this case, the value of q changes, but eventually stabilizes and remains at 0.

Case 2 indicates that no matter what the current value of q is, if $s = 0$ and $r = 1$, then $q^{new} = 0$ and $p^{new} = 1$. That is, q is reset to 0 (or the latch stores logic 0) and $p = \bar{q}$.

Case 3: $s = 1$ and $r = 0$

a. Assuming that $q = 0$, let $s = 1$ and $r = 0$.

$p^{new} = \overline{s+q} = \overline{1+0} = 0$

$q^{new} = \overline{r + p^{new}} = \overline{0+0} = 1$ q changes, thus

$p^{new} = \overline{s + q^{new}} = \overline{1+1} = 0$ both p and q are now stable.

That is, if $s = 1$ and $r = 0$, then $q^{new} = 1$ (changes from 0 to 1) and $p^{new} = 0$.

b. Assuming that $q = 1$, let $s = 1$ and $r = 0$.

$p^{new} = \overline{s+q} = \overline{1+1} = 0$

$q^{new} = \overline{r + p^{new}} = \overline{0+0} = 1$ remains the same

Case 3 indicates that no matter what the current value of q is, if $s = 1$ and $r = 0$, then $q^{new} = 1$ and $p^{new} = 0$. That is, q is set to 1 (or the latch stores logic 1) and $p = \bar{q}$.

Case 4: $s = 1$ and $r = 1$

a. Assuming that $q = 0$, let $s = 1$ and $r = 1$.

$p^{new} = \overline{s+q} = \overline{1+0} = 0$

$q^{new} = \overline{r + p^{new}} = \overline{1+0} = 0$

In this case, both q^{new} and p^{new} become 0.

b. Assuming that $q = 1$, let $s = 1$ and $r = 1$.

$p^{new} = \overline{s+q} = \overline{1+1} = 0$

$q^{new} = \overline{r + p^{new}} = \overline{1+0} = 0$ q changes, thus

$p^{new} = \overline{s + q^{new}} = \overline{1+0} = 0$ both p and q are now stable.

Again, in this case, both q^{new} and p^{new} become 0.

Case 4 is a special case. It indicates that no matter what the initial value of q is, if $s = 1$ and $r = 1$, then p^{new} and q^{new} both become 0, unlike the other three cases. This case produces inconsistent values for p and q as compared to the

s	r	q^{t+1}
0	0	q^t
0	1	0
1	0	1
1	1	–

(a)

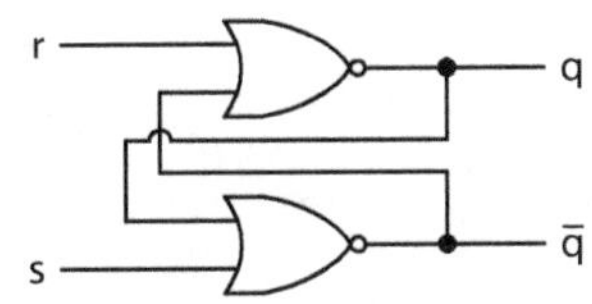

(b)

FIGURE 4.3 SR latch and its characteristic table: (*a*) SR latch's characteristic table; (*b*) a NOR-gate SR latch.

previous three cases, where p and q always have the opposite values. This case is also inconsistent because if both r and s are simultaneously set to 0 after both being 1, p and q signals will oscillate, both will become 1 and then both will become 0, and then it would repeat again. While the oscillation will continue forever (never stabilizing) during simulation, the oscillation would eventually stop in a real circuit. However, the final and stabilized value of q would be random, 0 or 1, and $p = \overline{q}$. Case 4 can alter the state of a system, randomly causing problems. For this reason, Case 4 should never happen. That is, r and s should never become 1 at the same time.

Cases 1 to 3, on the other hand, provide the necessary functions one would expect from a storage module: retain q (Case 1), reset q or store 0 (Case 2), or set q or store 1 (Case 3), with p always being equal to $\overline{q}$.

Figure 4.3 illustrates the final SR latch circuit, with p replaced with $\overline{q}$. Its characteristic table (also called a truth table) is shown in the figure with Case 4 ($s = 1$ and $r = 1$) marked unused. In the table, the current value of q is shown as q^t to mean the value of q at time t. The new (next) value of q is shown as q^{t+1} to indicate the stabilized values of both q and $\overline{q}$.

The SR latch circuit still lacks certain features to be a useful circuit:

- We must make sure that r and s signals both do not become active at the same time (i.e., Case 4 never happens).
- We must be able to initialize q to a known value, 0 or 1, during a system startup.

4.2.1 Clocked SR Latch

A clock is a signal generated by an electronic device called an oscillator that repeatedly outputs 1 and 0, each with a fixed duration. It is used to sample signals at specific times, and the sampling is done either when the clock is 0 or when the clock is 1. Figure 4.4 presents the circuit for a clocked SR latch. The clock signal is individually ANDed with r and s signals. It controls the timing when both r and s signals are allowed to change the state of the core circuit (Fig. 4.3(*b*)), defined by the q and $\overline{q}$ signal values.

When $clk = 0$ in Fig. 4.4, both G1 and G2 AND gates in the circuit would output 0, independent of the values of r and s. This, in turn, will cause the core SR latch to retain its current state (Case 1). However, when clk becomes 1, G1 will

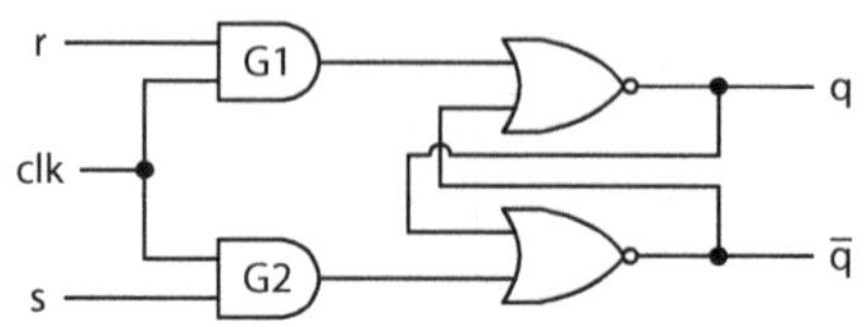

FIGURE 4.4 A positive-level SR latch.

output an incoming *r* value and G2 an incoming *s* value. At this time, the *r* and *s* values could alter the state of the core SR latch according to the characteristic table in Fig. 4.3(*a*).

The circuit in Fig. 4.4 is called a **positive-level** SR latch if *r* and *s* signal values "enter" the core circuit when $clk = 1$, and a **negative-level** SR latch if the signals "enter" the core circuit when $clk = 0$. Figure 4.5 illustrates a few timing scenarios for a positive-level SR latch, assuming that signal *s* has a shorter propagation delay than *r*.

During time slot 1, both *s* and *r* signals are changing, and since *s* has a shorter propagation delay than *r*, it changes to 1 while *r* is still 1. Thus, as illustrated in the figure, both *r* and *s* are 1 momentarily (i.e., Case 4), as illustrated by the shaded area in the timing diagram. However, because $clk = 0$ during this time slot, both G1 and G2 AND gates output 0 and prevent $s = 1$ and $r = 1$ from "entering" the core. The latch, therefore, retains its current state, unaffected.

During time slot 2, when $clk = 1$, $r = 0$, and $s = 1$, the AND gates would pass the $r = 0$ and $s = 1$ to the core circuit, changing *q* to 1. During this time slot, the latch is said to be sampling its *r* and *s* input values.

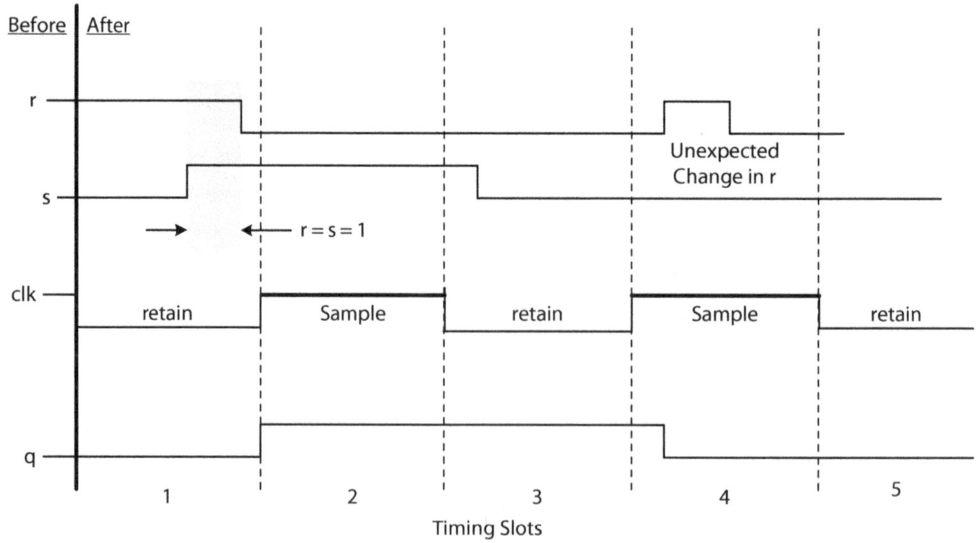

FIGURE 4.5 SR latch timing examples.

During time slot 3, when *clk* = 0 (retain), the latch retains $q = 1$, keeping it at 1. Finally, during time slot 4, when *clk* = 1 and the latch is sampling, a momentary change in the value of *r* (e.g., due to a glitch) unexpectedly changes *q* to 0. For this reason, both *r* and *s* signal values must stabilize prior to each sampling time. This is a disadvantage of an SR latch.

Another disadvantage of an SR latch is that the latch requires two inputs (*r* and *s*) to operate. This is especially true today with modern integrated chips (ICs), where wires also occupy chip real estate. A modern IC implements thousands of latches to build registers. If each latch requires two input signals to operate, this will double the space required to wire those signals.

A **D-latch**, designed from an SR latch, resolves both of these disadvantages. The D-latch will be discussed in the next section.

Because the *r* and *s* inputs affect the core circuit only during the sampling time, they are known as the synchronous reset and synchronous set signals, respectively. However, it is often necessary to initialize the latch to either $q = 0$ or $q = 1$ asynchronously, independent of the clock levels. Figure 4.6 illustrates a clocked SR latch with asynchronous reset and asynchronous preset signals. These signals are direct inputs to the NOR gates, and each can independently change both *q* and $\overline{q}$ when active. When *reset* = 1 and *preset* = 0, *q* becomes 0, independent of the values of *r*, *s*, and *clk* signals. Likewise, when *preset* = 1 and *reset* = 0, *q* becomes 1. The *reset* and *preset* signals both should not be active at the same time; only one needs to be active to either initialize *q* to 0 or 1.

NAND Version

The NOR-version SR latch in Fig. 4.6 is easier to understand because it uses active-high signals. An equivalent NAND version of the latch is shown in Fig. 4.7 with asynchronous active-low *_reset* (or *_r*) and *_preset* (or *_s*) signals. Note that, when compared to the NOR version, the position of the *r* and *s* signals have changed in the NAND version; the *s* is now lined up with the *q* signal and *r* with the $\overline{q}$ signal. The figure also shows the symbol for a clocked SR latch, where the input *c* indicates "clock." Unless otherwise stated, the term "SR latch" implies a clocked SR latch, either positive or negative level.

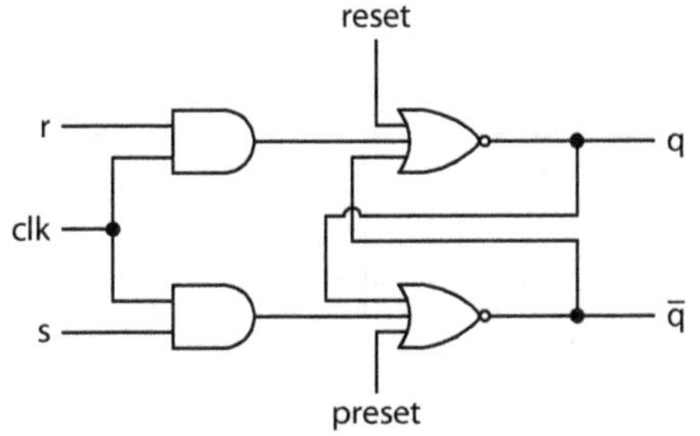

Figure 4.6 An SR latch with asynchronous reset and preset signals.

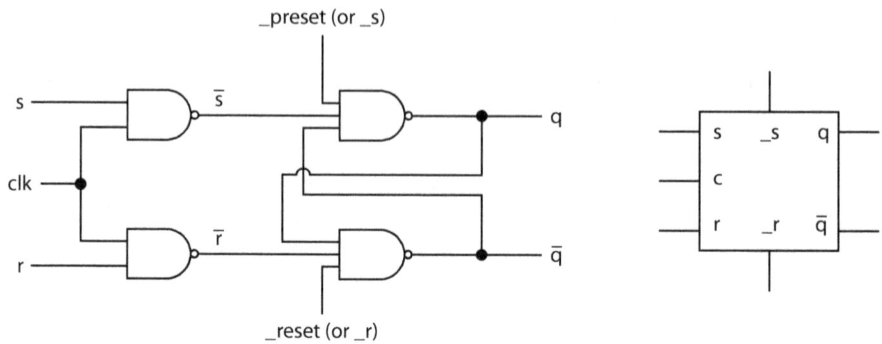

(a) SR latch, NAND version (b) SR latch symbol

FIGURE 4.7 A positive-level SR latch: (*a*) NANDs-only SR latch; (*b*) SR latch logic symbol.

4.3 D-Latch

An SR latch can be converted to a D-latch by eliminating the retain option (Case 1) when both s and r signals are 0. In synchronous mode, a D-latch requires a single input d in addition to the clock signal to operate it. This is done by connecting d to s and $\bar{d}$ to r, as illustrated in Fig. 4.8; the d stands for data. A D-latch operates in only two modes, setting q ($q = 1$) synchronously when $d = 1$ or resetting q ($q = 0$) synchronously when $d = 0$. The q signal always becomes d when the clock signal indicates sampling. However, a D-latch, just like an SR latch, does retain q when the clock is in the "off" and nonsampling level.

4.4 Disadvantage of Latches

Both SR and D-latches have a disadvantage that limits their application in the design of most sequential circuits. When two or more latches operate with the same clock signal, no data dependency can exist among their d signals. Consider two or more D-latches that operate with the same clock signal. In this case, none of the d inputs among the latches can be a function of any of the q's or $\bar{q}$'s; otherwise, a condition called **signal chasing** will prevent the circuit from operating correctly.

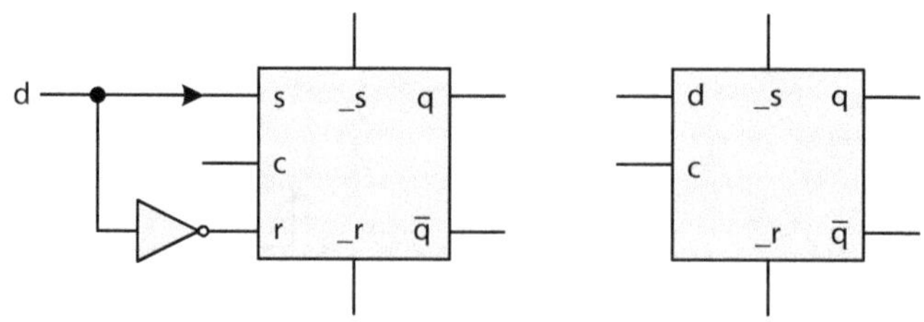

(a) D-Latch circuit (b) D-Latch symbol

FIGURE 4.8 The circuit and logic symbol of a D-latch.

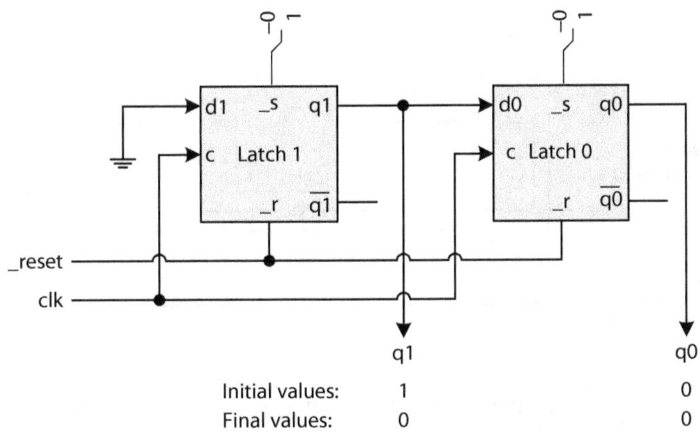

FIGURE 4.9 Illustrating an incorrectly designed 2-bit right-shift register; no latches should be used.

For instance, consider the two positive-level D-latches illustrated in Fig. 4.9. Both latches operate using the same clock signal *clk*. In this case, d_0 depends on q_1 (i.e., $d_0 = q_1$). The circuit is supposed to operate as a 2-bit right-shift register, but it does not. Each time that *clk* changes from 0 to 1, the current register content, indicated by q_1 and q_0 signals, should change to its new content as $q_1^{new} = 0$ and $q_0^{new} = q_1^{current}$.

To illustrate this, suppose that after using the active-low *_reset* signal to reset the latches, making their q bits 0, the q_1 is set to 1 using the switch connected to the Latch 1 active-low asynchronous set (*_s*) input. This makes the concatenated values of q_1 and q_0 or $q_1q_0 = (10)_2$, indicated in the figure as the initial values of q_1 and q_0. Now, when *clk* changes from 0 or 1, the next value of q_1q_0 should be $(01)_2$ indicating a right shift where $d_1 = 0$ is entered from the left. However, that will not happen with latches.

Specifically, when *clk* becomes 1, both the D-latches will simultaneously begin to sample their inputs, changing the current (in this case, initial) values of $q_1^{current} = 1$ and $q_0^{current} = 0$ to $q_1^{new} = d_1 = 0$ and $q_0^{new} = d_0 = q_1^{current} = 1$. However, if *clk* remains at 1, the D-latches will continue sampling their inputs, causing q_0^{new} to take the value of q_1^{new}, making the final $q_1q_0 = (00)_2$ instead of, as expected, $(01)_2$. In this case, signal $d_1 = 0$ chases and changes d_0 to 0 when *clk* stays 1 for a longer time.

In general, a set of interconnected latches with dependent inputs will fail to operate independently, and thus cannot be used in the design of many important sequential circuits, such as shift registers and control units. The circuit that prevents signal chasing is called a flip-flop.

4.5 D Flip-Flop

A flip-flop can be designed from two connecting latches, as illustrated in Fig. 4.10 for a D flip-flop. During each clock level, only one of the latches samples its input signal, while the other latch retains its current value. The two latches operate like a double-door entry system, much like the ones used in many buildings. Only one door at a

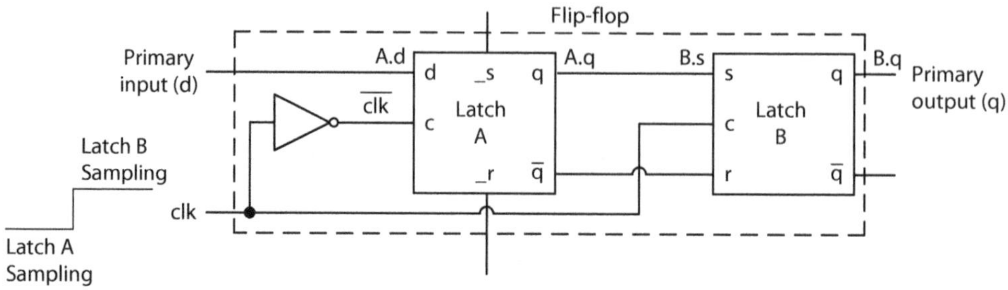

Figure 4.10 A D flip-flop.

time is kept open while the other door is closed. To enter the building, one must enter through the first door and then the second door. In the figure, the two latches are labeled A (door A) and B (door B). Their inputs and outputs are, respectively, referred to as A.*d*, A.*q*, B.*s*, etc. Both A and B are positive-level latches but operate with complementing clock levels. Latch A is a D-latch and Latch B is an SR latch.

When $clk = 1$ and thus $\overline{clk} = 0$, Latch A would not be sampling (i.e., door A is closed), and thus retains its *q*, A.*q*. During this time, a change in the primary input *d* (A.$d = d$) will not affect A.*q* and A.$\overline{q}$. These two signals are connected, respectively, to B.*s* and B.*r* inputs of Latch B, and thus if *d* changes, B.*q* will not change even though Latch B is still sampling when $clk = 1$ (i.e., door B is open).

However, when *clk* transitions from 1 to 0 and $\overline{clk}$ transitions from 0 to 1, the latches switch modes; Latch A starts sampling A.*d* (i.e., door A opens), and Latch B stops sampling A.*q* (i.e., door B closes); thus, Latch B retains B.*q*. At this time, any change in the primary input *d* will change A.*q* but not B.*q*. Therefore, the two latches that operate with opposite clock levels prevent signal chasing. The output of Latch A can only affect the output of Latch B if the *clk* transitions again from 0 to 1. Thus, the flip-flop requires a clock pulse to operate.

In the figure, when the *clk* signal changes from 1 to 0 and again back to 1 (a 1-0-1 transition), the flip-flop samples and stores its primary input *d* as B.*q*, its primary output.

4.5.1 Alternative Circuit

A different D flip-flop circuit with fewer total required gates is illustrated in Fig. 4.11. The D flip-flop operates like the one shown in Fig. 4.10. It samples the primary input *d* when the *clk* makes a 1-0-1 transition. Specifically, when *clk* transitions from 1 to 0, both signals $\overline{s}$ and $\overline{r}$ become 1, causing the flip-flop to retain *q*, independent of the current value of *d*. When *clk* transitions from 0 to 1, the flip-flop starts sampling *d*. If $d = 1$, then $\overline{s} = 0$ and $\overline{r} = 1$ will change *q* to 1; or, if $d = 0$, then $\overline{s} = 1$ and $\overline{r} = 0$ will change *q* to 0. The values of both $\overline{s}$ and $\overline{r}$ remain unchanged until the next time that *clk* makes a 1-0-1 transition.

4.5.2 Operating Conventions

Latch A and Latch B in Fig. 4.10 do not operate (i.e., sample) during the same clock level (0 or 1). One latch is always not sampling (one door is closed) while the

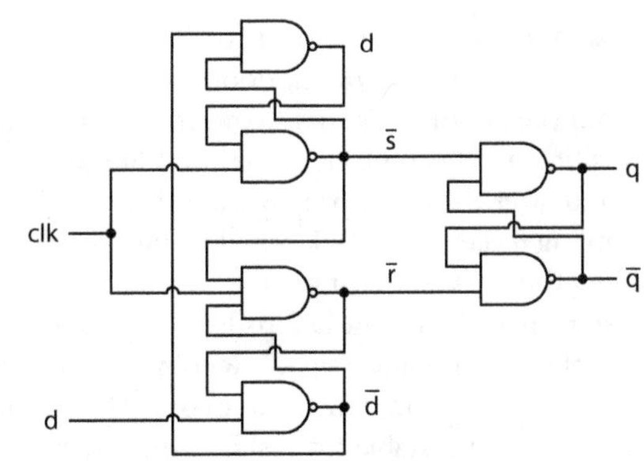

Figure 4.11 An alternative D flip-flop with fewer gates.

other is always sampling (one door is open). The only time that a sampled input *d* is passed from Latch A to Latch B (i.e., B.*q*) is when *clk* makes a 0-1 transition and causes Latch A, which was sampling, to stop sampling and Latch B to start sampling A.*q*. Note that A.*q* will not change even if *d* changes since Latch A is now not sampling. The next time that Latch A will start sampling is when *clk* makes a 1-0 transition, which makes Latch B stop sampling.

With D flip-flops, data moves from one latch to the next on the clock edge and not on the clock level; thus, a D flip-flop is called **edge triggered**. It is called a positive- or rising-edge triggered flip-flop if a 1-0-1 clock transition causes *d* to be loaded (A.*q* = *d* on 1-0 transition) and stored as *q* (B.*q* = A.*q* on 0-1, positive, up arrow transition). Likewise, a negative- or falling-edge triggered flip-flop would require a 0-1-0 clock transition to load *d* (A.*q* = *d* on 0-1 transition) and store it as *q* (B.*q* = A.*q* on 1-0, negative, down arrow transition). As illustrated in Fig. 4.12, the clock input of an edge triggered flip-flop is marked with the right angle (>) symbol, and also with a small bubble if the flip-flop is negative edge triggered.

4.5.3 Setup and Hold Times

A D flip-flop, as opposed to, say, an SR flip-flop, that is designed (but not used today) using positive-level and negative-level SR latches has the advantage of allowing its *d* input to change while Latch A (the first latch) is still sampling. The only requirement is that the *d* signal must stabilize and remain stable during a small period when the clock is transitioning.

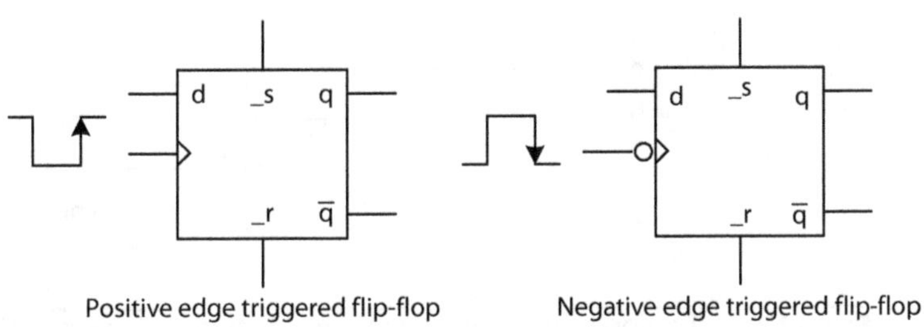

Figure 4.12 Positive- and negative-edge triggered flip-flop symbols.

For example, consider the positive edge-triggered flip-flop in Fig. 4.10. When $clk = 0$, Latch B retains B.q (i.e., door B is closed), and because $\overline{clk}$ is 1, Latch A samples A.d (i.e., door A opens). Now, the instant that the clk makes a rising-edge (0-1) transition and becomes 1, both the clk and the $\overline{clk}$ are still at logic 1 until the $\overline{clk}$ changes to 0 after the short propagation delay of the NOT gate and stops Latch A from sampling d (i.e., door A closes). During this short period, both the latches will be sampling their respective inputs.

Therefore, in order for the D flip-flop to operate correctly during this time, d must become stable—so the s and r signals within Latch A can become stable—a short time before clk transitions from 0 to 1, and in order to continue keeping the s and r signals stable, d must remain stable for a short time after clk has transitioned to 1. This is required because the signals clk and $\overline{clk}$ do not change simultaneously at the same time, thus causing Latch A to stop sampling ("closing") with a delay (see Fig. 4.10). The amount of time that d must remain stable before and after a 0-1 clk transition is called, respectively, setup time (τ_{st}) and hold time (τ_{ht}) of the flip-flop. For a negative-edge flip-flop, d must remain stable before and after a 1-0 clock transition.

The violation of setup and hold times destabilizes a D flip-flop—a condition known as **metastability**, as illustrated in the following two figures. In Fig. 4.13(a), signal d changes too close to clk transitioning from 0 to 1; thus, it violates the flip-flop's setup time and causes B.q and B.$\overline{q}$ to oscillate. On the other hand, in Fig. 4.13(b), because d stabilizes earlier before clk transitions to 1 by at least an amount $\geq \tau_{st}$, the flip-flop correctly loads d, making B.$q = d$ and B.$\overline{q} = \overline{d}$.

Likewise, in Fig. 4.14(a), when d changes too soon after clk transitions from 0 to 1, it violates the hold-time requirement of the flip-flop. This would also cause signal oscillations at B.q and B.$\overline{q}$. On the other hand, in Fig. 4.14(b), the D flip-flop operates

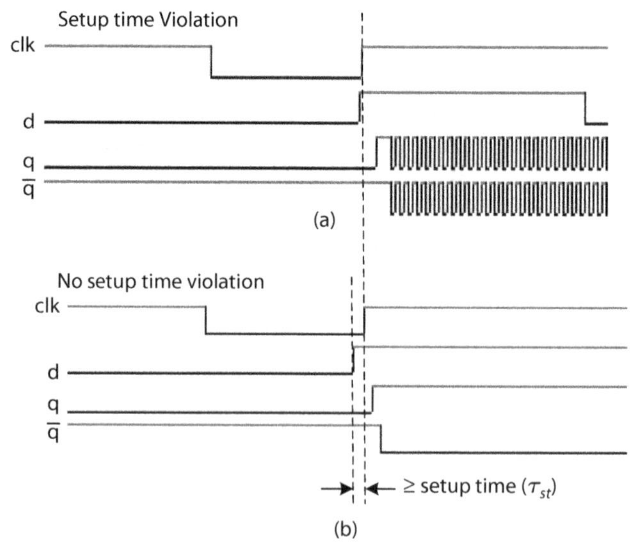

FIGURE 4.13 D flip-flop timing: (a) setup time violation; (b) no setup time violation.

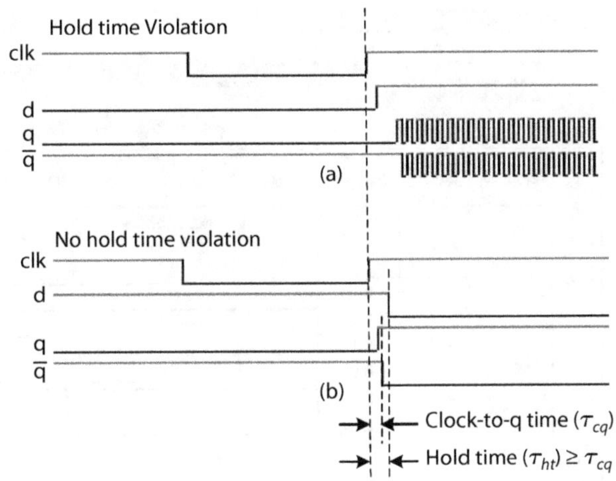

Figure 4.14 D flip-flop timing: (*a*) hold-time violation; (*b*) no hold-time violation.

normally and makes $B.q = d$ and $B.\overline{q} = \overline{d}$ when d continues to remain stable after *clk* transitions to 1. The hold time (τ_{ht}) is the amount of time that d must remain stable after *clk* transitions to 1. The τ_{ht} must be greater than or equal to the clock-to-q (τ_{cq}) delay, which is the time required for the flip-flop to stabilize and output the final values of $B.q$ and $B.\overline{q}$. The τ_{cq} is also known as clock-to-output (τ_{co}).

Negative Hold Time

The setup and hold times discussed earlier pertain to *d* and *clk* signal sources at the flip-flop and not at the boundary of an IC or a module within a chip. These setup and hold times may also be determined with respect to the data and clock sources at a chip's or a module's boundary. Figure 4.15 illustrates a flip-flop inside a chip with five interface signals *d*, *clock*, *q*, and active–low signals *_preset* and *_reset* at the chip's boundary. In this case, the input signals will affect the flip-flop after

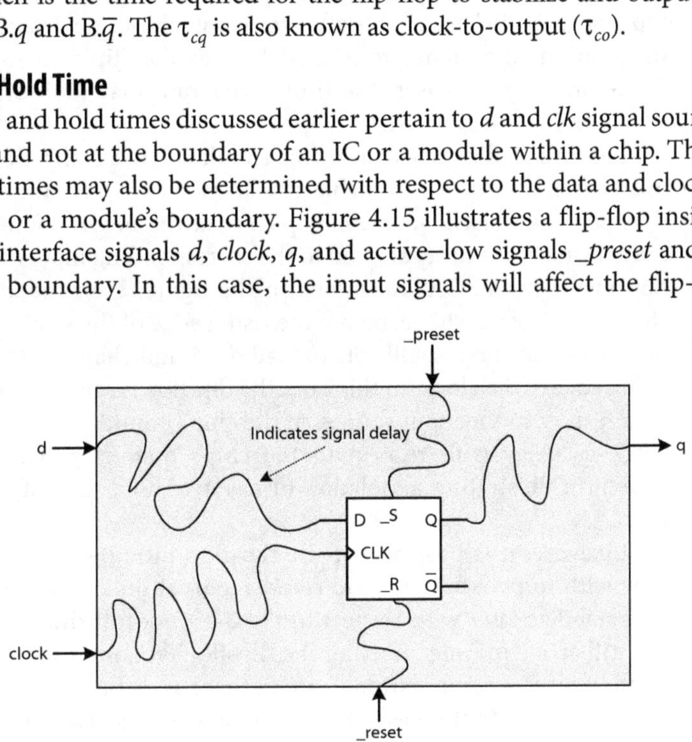

Figure 4.15 Illustrating signal delay within an integrated chip (IC) or a module.

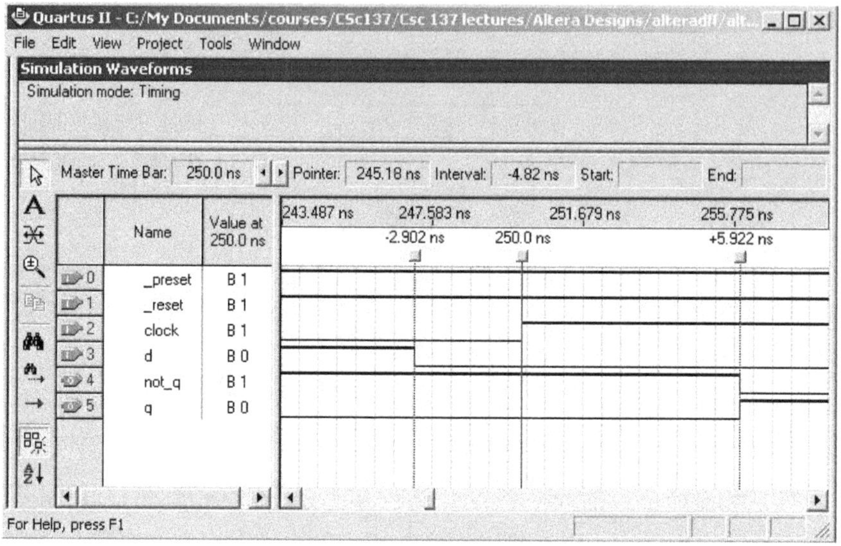

Figure 4.16 A D flip-flop timing diagram illustrating its normal operation with a negative hold time.

some signal routing delay, as illustrated in the figure. Likewise, the output of the flip-flop as *q* will be available after some signal routine delay at the chip's boundary. In addition, the signal routing delays to the flip-flop may not be the same. Therefore, in such cases, the timing diagram may look different from the ones shown in Figs. 4.13 and 4.14.

Figure 4.16 illustrates the timing simulation of a D flip-flop that was synthesized in an Altera field programmable gate array (FPGA). All the input/output signals—*d*, *reset*, *preset*, *clock*, *q*, and *not_q*— are at the FPGA's boundary, which are then routed with some delay and fed to the corresponding inputs and outputs of the flip-flop inside the FPGA. As shown in the figure, even though initially, $d = 1$ and changes to 0 at −2.902 ns before the rising edge of the *clock* at 250.0 ns (a reference point), the flip-flop is still able to load $d = 1$ and change q to 1 at 5.922 ns after the rising edge of the clock. In this case, the flip-flop is said to have a negative hold time with respect to the input sources at the chip's boundary. The flip-flop also has $\tau_{cq} =$ 5.922 ns delay with respect to the chip's boundary. The flip-flop's simulation waveform illustrating a violation of negative hold time at −3.489 ns is shown in Fig. 4.17.

Likewise, it is possible for the flip-flop in Fig. 4.15 to have a negative setup time with respect to the *d* and *clock* signals at the chip's boundary. In this case, *d* may stabilize later with respect to the 0-1 *clock* transition at the chip's boundary, but still arrive in time, meeting the flip-flop's required setup time. The *d* may also be an input to a combinational circuit that has an output connected to the D-input of the flip-flop. In this case, the timing of when the *d* signal will affect the flip-flop depends on both signal routing delays and the propagation delay of the combinational circuit.

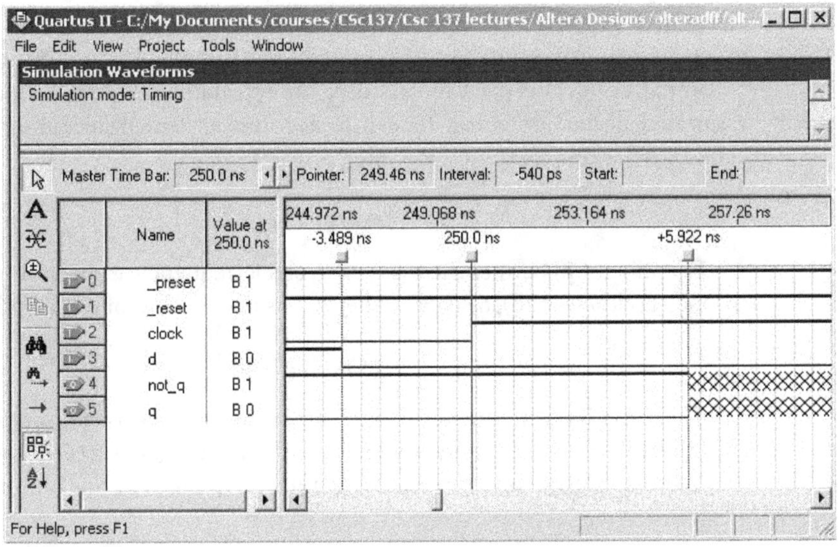

Figure 4.17 A D flip-flop timing diagram illustrating a negative hold-time violation at –3.489 ns.

Furthermore, an FPGA or an application-specific integrated chip (ASIC) may include fixed and prefabricated circuit modules. For example, consider an FPGA chip that contains the circuit in Fig. 4.1 designed with a CLA adder as a prefabricated module. In addition, the module may include some internal clock signal routing delay. Such modules may have negative hold/setup times with respect to the input signals at their boundary. It has been shown that a more accurate timing simulation can be achieved when negative hold/setup time requirements of prefabricated modules are modeled in HDL simulation software [1].

4.6 Clock Frequency Estimation without Clock Skew

Figure 4.18 illustrates the minimum clock period (τ) required to operate a D flip-flop. The period is the duration of one clock cycle and includes the amount of time that the clock is 1 and the amount of time that the clock is 0. The period is

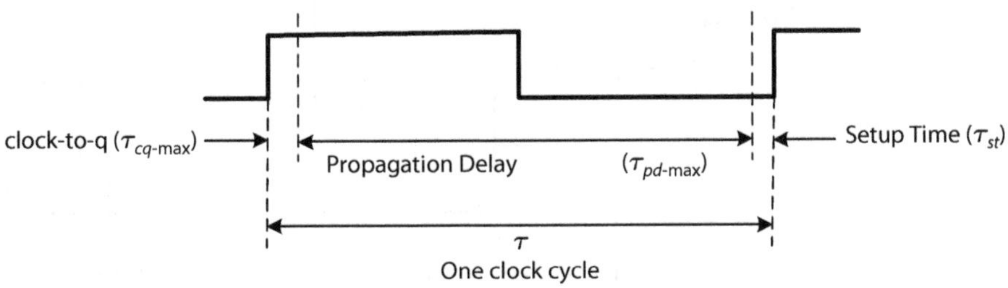

Figure 4.18 The relationship between the clock period and different delays.

calculated in terms of signal *d*'s maximum propagation delay (τ_{pd-max}), the maximum clock-to-q delay (τ_{cq-max}), and the flip-flop's setup time (τ_{st}).

Equation (4.1) is used to calculate the estimated minimum clock period. The estimated period does not take into account a sequential circuit phenomenon called clock skew discussed in Chap. 5.

$$\tau \geq \tau_{cq-max} + \tau_{pd-max} + \tau_{st} \tag{4.1}$$

Equation (4.2) defines the maximum clock frequency as the number of cycles per second (cycles/second), also called hertz. It is the number of clock cycles in 1 second.

$$f \leq \frac{1}{\tau} \text{ cycles/second or hertz} \tag{4.2}$$

Typically, large frequency values are converted into thousands, or kilohertz (KHz); millions, or megahertz (MHz); or billion, or gigahertz (GHz), by dividing the number of cycles/second by one thousand, one million, or one billion, respectively.

4.7 Flip-Flop with Enable

A complex digital circuit typically contains hundreds or thousands of flip-flops. The flip-flops are not all active (sampling) at the same time. Some of the flip-flops may be individually activated, and some may be activated as a group. For instance, a processor that has 16 32-bit registers contains 512 (16 * 32) flip-flops. A group of, say, 32 flip-flops that make up a 32-bit register would be all selected (enabled) at the same time to load a 32-bit result generated, for example, by an adder during the execution of an ADD instruction. Therefore, an additional control signal is necessary to select a flip-flop or a group of flip-flops during a particular clock cycle.

Figure 4.19 illustrates the design of a D flip-flop with an enable signal *e* that controls a 2-to-1 multiplexer (MUX). When $e = 0$, the MUX selects *q* and causes the flip-flop to reload *q* and thus retain its stored value. When $e = 1$, the MUX selects the *d_in* input and causes the flip-flop to load and change *q* to *d_in*. The flip-flop is said to be enabled, or selected, when $e = 1$, and disabled, or not selected, when $e = 0$.

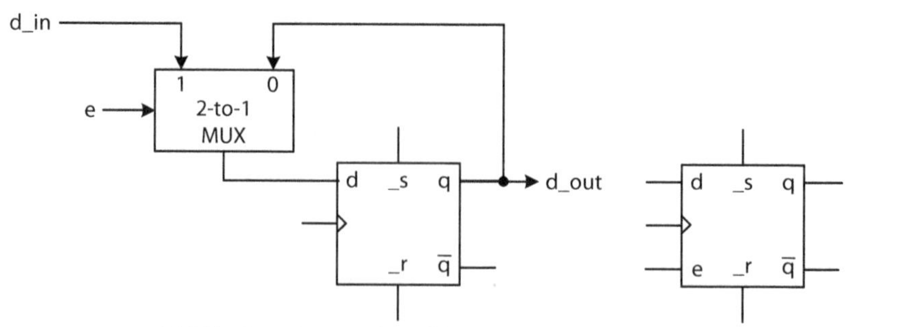

(a) D flip-flop circuit with enable (b) Symbol of a D flip-flop with enable

FIGURE 4.19 A D flip-flop circuit with enabling signal and its symbol.

4.8 Other Flip-Flops

A historically popular flip-flop is called the **JK flip-flop**. It uses two input signals j and k as shown in Fig. 4.20(*a*). In the figure, the *SET* and *CLR* signals indicate active-high asynchronous set and reset signals, respectively. The outputs of the flip-flop toggle; q becomes $\overline{q}$ and $\overline{q}$ becomes q during each clock cycle when j and k inputs are both 1. JK flip-flops have the disadvantage of requiring two control signals. However, in general, circuits that use JK flip-flops have the advantage of requiring simpler circuits for the j and k signals as compared to the d signals of a circuit that uses D flip-flops. An edge-triggered JK flip-flop can be designed using a D flip-flop, as illustrated in Fig. 4.21.

In Fig. 4.20(*b*), another flip-flop called a **T flip-flop** is shown. This flip-flop is designed using a JK flip-flop with both j and k inputs connected to a single signal t, which stands for toggle. A T flip-flop performs only one of two functions: retain its current state (q) or toggle. If $t = 0$, a T flip-flop retains the value of its q. If $t = 1$, its outputs toggle: q becomes $\overline{q}$ and $\overline{q}$ becomes q every clock cycle. A T flip-flop, similar to a D flip-flop, also has the advantage of requiring only a single input (t) to synchronously either set or reset its q.

In general, any flip-flop can be designed systematically from any other flip-flop. For example, the behavior of a JK flip-flop, like many other sequential circuits,

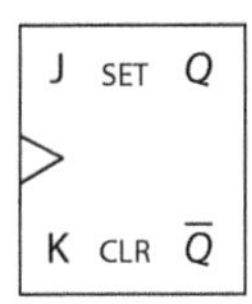

j	k	q(t + 1)
0	0	q(t)
0	1	0
1	0	1
1	1	!q(t)

(a) JK Flip-flop

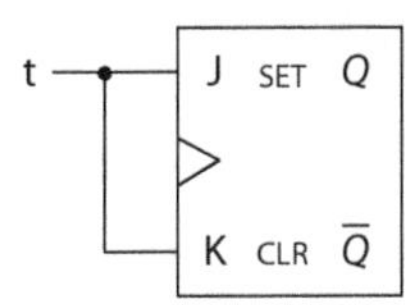

t	q(t + 1)
0	q(t)
1	!q(t)

(b) T flip-flop

Figure 4.20 Other types of flip-flops.

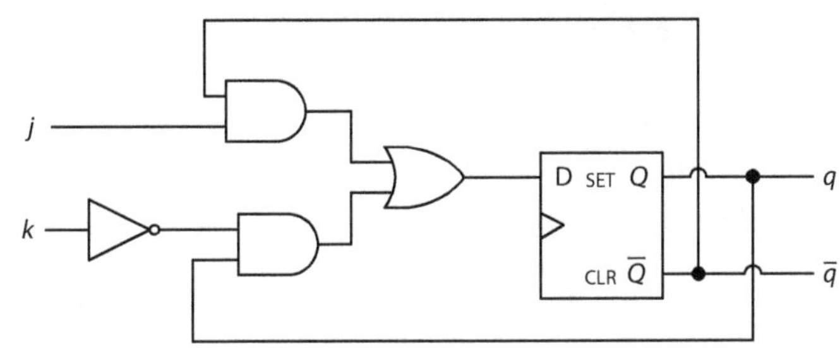

Figure 4.21 A circuit for an edge-triggered JK flip-flop.

can be expressed by a finite state diagram (FSD). FSDs and their implementations, known as finite state machines (FSMs), are discussed in Chap. 5.

4.9 Hardware Description Language Models

Example 4.1 presents a Verilog behavior model for a positive-level D latch with active-low asynchronous *_reset* and *_preset* signals. With latches and flip-flops, the nonblocking or concurrent assignment operator "<=" guarantees simultaneous clocking of multiple latches and flip-flops that operate with the same clock. The "always" block in the example includes the four signals—*clock*, *_reset*, *_preset*, and *d*—in its sensitivity list. In the description, the *_reset* is assigned the highest priority and the *clock* is the lowest. When *_reset* = 1 (not active), *_preset* = 1 (not active), and *clock* = 1, a change in the *d* signal will affect the *q* signal, as expected, as long as *clock* stays at 1 (sampling level). The Verilog code has a missing "else" statement for when *clock* = 0, and this creates an implicit latch that retains *q* when *clock* = 0.

Example 4.1. A behavior model of a positive-level, asynchronous active-low reset (*_reset*) and active-low preset (*_preset*) for a D-latch, a test-bench, and the simulation output are listed here. As shown in the output, when *clock* = 0, *d* = 1 at simulation time 3, *q* still 0 indicates the latch is retaining its 0 state as expected. At time 4, when *clock* = 1 and *d* = 1, *q* changes to 1 as expected, but while *clock* is still 1 at time 5 and *d* changes to 0, *q* changes to 0, as it should, illustrating a latch behavior.

HDL Model:

```
module d_latch

 (
   input clock, _reset, _preset, d,
   output reg q,
   output nq
 );

assign nq = ~q;
always@(clock or !_reset or !_preset or d)
begin
  if(!_reset)
       q <= 0;
  else if(!_preset)
       q <= 1;
  else if(clock)
       q <= d;
end
endmodule
```

Test-bench:

```
'include "d_latch.v"
module tester();
reg clock, _reset, _preset, d;
wire q, nq;
d_latch    dlatch1(clock, _reset, _preset, d, q, nq);
initial begin
$monitor("%4d clock = %b _reset = %b, _preset = %b d = %b q =
%b nq = %b\n", $time, clock, _reset, _preset, d, q, nq);
end
initial begin
clock = 0; _reset = 1; _preset = 1;
#1 _reset = 0;
#1 _reset = 1;
#1 d = 1;
#1 clock = 1;
#1 d = 0;
#10 $finish;
end
endmodule
```

Simulation Output:

```
Chronologic VCS simulator copyright 1991-2009
Contains Synopsys proprietary information.
Compiler version D-2009.12; Runtime version D-2009.12;

  0 clock = 0 _reset = 1, _preset = 1     d = x q = x nq = x
  1 clock = 0 _reset = 0, _preset = 1     d = x q = 0 nq = 1
  2 clock = 0 _reset = 1, _preset = 1     d = x q = 0 nq = 1
  3 clock = 0 _reset = 1, _preset = 1     d = 1 q = 0 nq = 1
  4 clock = 1 _reset = 1, _preset = 1     d = 1 q = 1 nq = 0
  5 clock = 1 _reset = 1, _preset = 1     d = 0 q = 0 nq = 1

$finish called from file "tester.v", line 21.
$finish at simulation time          15
```

VCS Simulation Report

```
Time: 15
CPU Time:   0.460 seconds;    Data structure size:  0.0Mb
```

Example 4.2 presents a Verilog D flip-flop behavior model. The posedge or negedge keyword stands for rising- or falling-edge triggered behavior, respectively. The Verilog code describes a positive-edge triggered D flip-flop with asynchronous active-low *_reset* and *_preset* signals, with priority given to *_reset*. These two signals operate the circuit asynchronously because they are listed as part of the "always" block sensitivity list; otherwise, if they are not included in the sensitivity list, they would operate the circuit synchronously and only when *clk* makes (in this case) a rising-edge transition.

Example 4.2. A behavior model for a positive-edge triggered D flip-flop with asynchronous active-low *_reset* and *_preset* and active-high enable (*e*) signals and a test-bench and simulation output are listed. Note that, in the model, the *d* and *e* signals are not included in the sensitivity list of the "always" block; thus, as expected, they will affect the D flip-flop synchronously. When *_reset* = 1 and *_preset* = 1, a 0-1 *clock* transition will cause the flip-flop to load and make $q = d$ if the flip-flop is enabled ($e = 1$). Otherwise, if $e = 0$ and the flip-flop is disabled, the value of q is retained. A flip-flop also retains q when *clk* is not making a 0-1 transition; that is, *clk* is either 0 or 1. A 1-0 level change in signal *_reset* asynchronously resets q ($q = 0$), and a 1-0 level change in signal *_preset* asynchronously would set q ($q = 1$). Again, the concurrent, nonblocking operator "<=" causes simultaneous clocking of multiple instantiated flip-flops (if any).

As shown in the simulation output, when *clock* = 0 at simulation time 3, *d* changing 1 will not change q as expected. When *clock* makes a 0-1 transition at time 4, $d = 1$ also changes q to 1 as expected. When *clock* is still 1 at time 5 and *d* changes to 0, as expected, q should not change, illustrating a flip-flop behavior.

HDL Model:

```
module dff

(
  input clock, _reset, _preset, d, e,
  output reg q,
  output nq
);

assign nq = ~q; //nq indicates not q
always@(posedge clock, negedge _reset, negedge _preset)
begin
  if(!_reset)
       q <= 0;
  else if(!_preset)
       q <= 1;
  else if(e)
       q <= d;
end
endmodule
```

Test-bench:

```
`include "dff.v"
module tester();
reg clock, _reset, _preset, d, e;
```

```
wire q, nq;
dff   dff1(clock, _reset, _preset, d, e, q, nq);
initial begin
$monitor("%4d clock = %b _reset = %b _preset = %b e = %b d = %b
q = %b nq = %b\n", $time, clock, _reset, _preset, e, d, q, nq);
clock = 0; _reset = 1; _preset = 1; e = 0;
#1 _reset = 0;
#1 _reset = 1;
#1 e = 1; d = 1;
#1 clock = 1;
#1 d = 0;
#10 $finish;
end
endmodule
```

Simulation Output:

```
Chronologic VCS simulator copyright 1991-2009
Contains Synopsys proprietary information.
Compiler version D-2009.12; Runtime version D-2009.12;
  0 clock = 0 _reset = 1 _preset = 1 e = 0   d = x q = x nq = x
  1 clock = 0 _reset = 0 _preset = 1 e = 0   d = x q = 0 nq = 1
  2 clock = 0 _reset = 1 _preset = 1 e = 0   d = x q = 0 nq = 1
  3 clock = 0 _reset = 1 _preset = 1 e = 1   d = 1 q = 0 nq = 1
  4 clock = 1 _reset = 1 _preset = 1 e = 1   d = 1 q = 1 nq = 0
  5 clock = 1 _reset = 1 _preset = 1 e = 1   d = 0 q = 1 nq = 0
  $finish called from file "tester.v", line 19.
  $finish at simulation time          15
```

VCS Simulation Report

```
Time: 15
CPU Time:  0.460 seconds;  Data structure size:  0.0Mb
```

References

1. Mi-Sook Jang and Hoi-Jin Lee, "Methods of HDL simulation considering hard macro core with Negative Setup/Hold time," US Patent 7,213,222 B2, May 1, 2007.

Exercises

4.1 Part 1: Given the circuit in Fig. 4.22 where q_a is the output of a positive-level D latch, complete the timing waveform for q_a given in Fig. 4.23. Assume $\tau_{st} = \tau_{cq} = 0$.

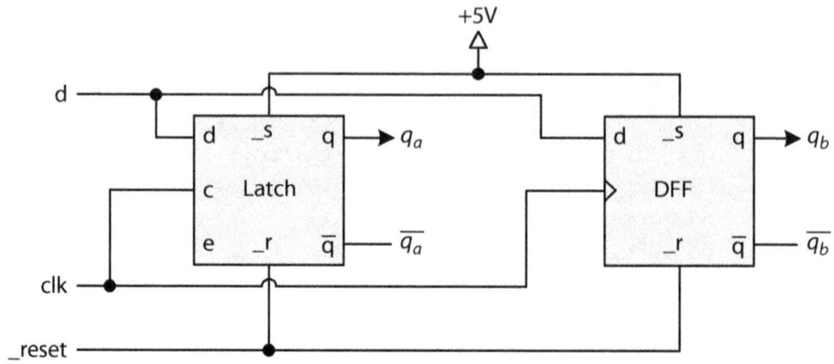

FIGURE 4.22 Circuit for Exercises 4.1 and 4.2.

4.2 Part 2: Given the circuit in Fig. 4.22 where q_b is the output of a positive-edge triggered D flip-flop, complete the timing waveform for q_b given in Fig. 4.23. Assume $\tau_{st} = \tau_{cq} = 0$.

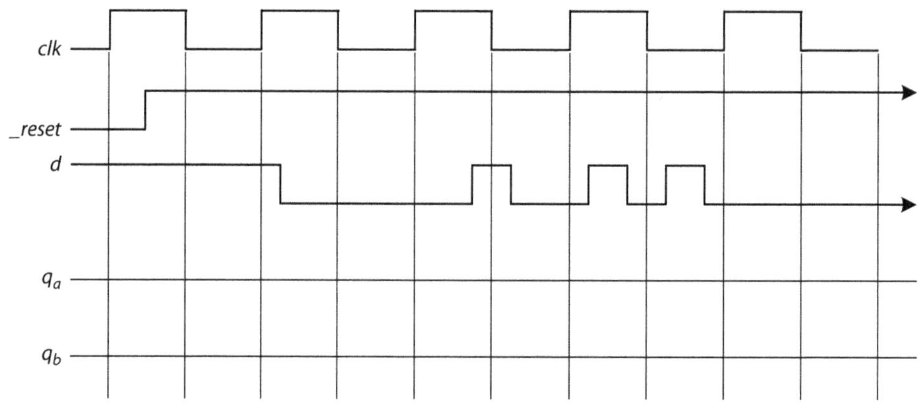

FIGURE 4.23 Waveform for Exercises 4.1 to 4.4.

4.3 Part 1: Given the circuit in Fig. 4.24 where q_a is the output of a negative-level D latch, complete the timing waveform for q_a given in Fig. 4.23. Assume $\tau_{st} = \tau_{cq} = 0$.

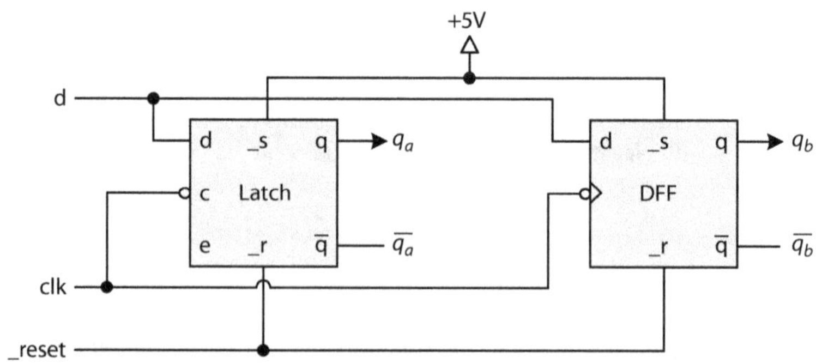

FIGURE 4.24 Circuit for Exercises 4.3 and 4.4.

4.4 Part 2: Given the circuit in Fig. 4.24 where q_b is the output of a negative-edge triggered D flip-flop, complete the timing waveform for q_b given in Fig. 4.23. Assume $\tau_{st} = \tau_{cq} = 0$.

4.5 Given the circuit in Fig. 4.22 where q_b is the output of a positive-edge triggered D flip-flop, complete the timing waveform for q_b given in Fig. 4.25. Assume $\tau_{st} = 0.15$ ns and $\tau_{cq} = 0.1$ ns.

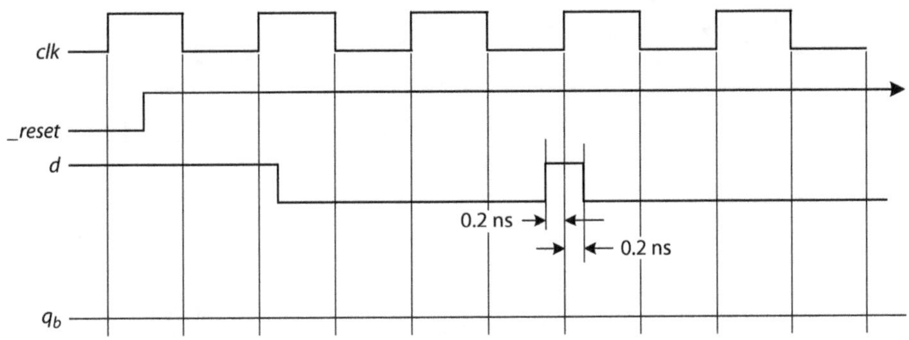

FIGURE 4.25 Waveform timing diagram for Exercise 4.5.

4.6 Given the circuit in Fig. 4.22 where q_b is the output of a positive-edge triggered D flip-flop, complete the timing waveform for q_b in Fig. 4.26. Assume τ_{st} = 0.15 ns and $\tau_{cq} = 0.2$ ns.

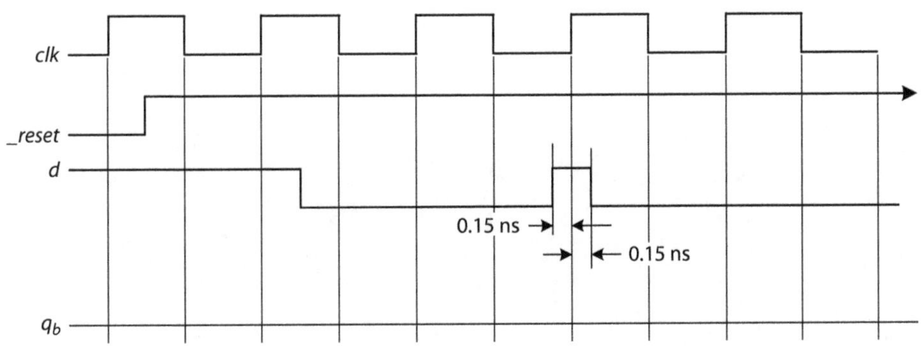

FIGURE 4.26 Waveform for Exercises 4.6.

4.7 Given that a D flip-flop has 0.1 ns setup time and 0.1 ns hold time and the maximum propagation delay for *d* is 0.3 ns, determine the maximum clock frequency for the proper operation of the D flip-flop.

4.8 Create and simulate a Verilog structural model of a D flip-flop using NAND gates with reset and preset signals.

CHAPTER 5

Sequential Circuits: *Small Designs*

5.1 Introduction

Like combinational circuits (CCs), sequential circuits can also be classified as small and large circuits. An up-counter, for example, that generates the sequence 0, 1, 2, 3, etc., as output would be considered a small sequential circuit. On the other hand, a processor core (a CPU) is a very large sequential circuit that may execute multiple instructions simultaneously. Each instruction would require a set of operations and involves one or more registers and possibly memory.

All small and large sequential circuits are made of flip-flops and a set of CCs, such as the one shown for a 2-bit up-counter in Fig. 5.1. The counter circuit includes two flip-flops and a set of CCs associated with the 2-bit adder shown in the figure. Contrary to CCs, a sequential circuit has states and transitions from a current state to a next state every clock cycle. A current state is determined from the flip-flops' q bits. In the figure, if the 2-bit $Q = q_1q_0 = (00)_2$, then the counter is said to be currently in state 0; if the $Q = q_1q_0 = (01)_2$, then the counter is said to be currently in state 1, etc. The counter also outputs the 2-bit state number $Q = q_1q_0$ as count every clock cycle.

The adder (i.e., the set of the counter's CCs) generates the counter's next state as a 2-bit number $D = d_1d_0 = q_1q_0 + 1$. If the counter's current state is $Q = q_1q_0 = (00)_2$, then its next state is $D = d_1d_0 = (01)_2$. The d_1 and d_0 are saved in the flip-flops during the next clock cycle.

The design of sequential circuits requires additional methodologies. A sequential circuit design problem is typically modeled as a finite state diagram (FSD). An FSD consists of circles as states and arcs (arrows) as transitions. It formally specifies the behavior of a target sequential circuit. The 2-bit counter has four states, numbered 0, 1, 2, and 3. It transitions from state 0 to state 1, then from state 1 to 2, then from state 2 to 3, and finally from state 3 back to 0. An FSD is systematically converted into a circuit called a finite state machine (FSM), such as the one in Fig. 5.1.

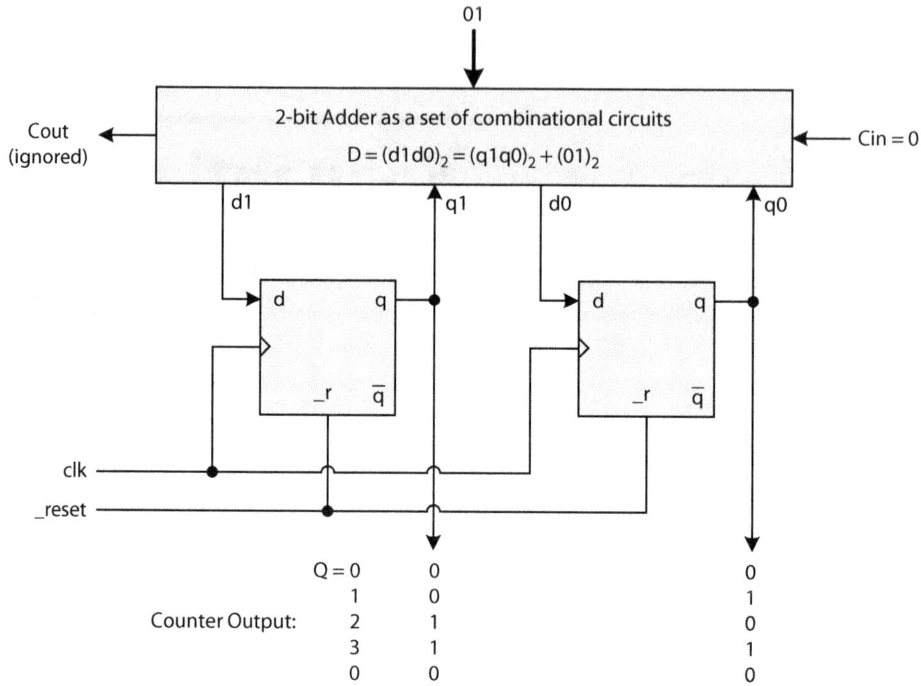

FIGURE 5.1 An FSM as a 2-bit counter with two flip-flops.

A large sequential circuit design problem is typically partitioned into the design of a data path and a control unit. The data path would contain both CC modules, such as arithmetic logic units (ALUs), multiplexers (MUXs), and decoders, and small sequential circuits, such as registers and counters. The design of large sequential circuits is discussed in Chap. 6, and CPU design, specifically, is discussed in Chap. 8.

Occasionally, it is possible to design a sequential circuit without first constructing an FSD. This is an important design concept and, in many cases, simplifies the design of some small and large sequential circuits, including CPUs. For example, the CC module (i.e., an adder) in Fig. 5.1 performs addition, which is a known function, and thus the counter can be designed without an FSD.

On the other hand, the design of, for example, a **sequence recognizer** that inputs a sequence of 1's and 0's, one at a time, and outputs a 1 each time it encounters a prespecified subsequence would require an FSD. In this case, as opposed to the 2-bit counter example, it would be nearly impossible for a designer to determine in advance what specific known function the CCs of a sequence recognizer would be performing.

Sequential circuits are also subject to environmental hazards, such as **transient faults** that occur at random and can change the state of a sequential circuit, causing a malfunction. One way to protect sequential circuits from environmental hazards is through fault-tolerant design.

We will start this chapter with a small FSM design problem and then formally present the design of FSMs in general, including the design of a fault-tolerant FSM. We will also present timing requirements of sequential circuits and discuss examples of FSM descriptions in Verilog.

5.2 Introduction to FSM: Register Design

A register, as a small sequential circuit, is used as a storage module to save the output of a CC. As a shift register, it would have the capability of shifting its content either to the right or to the left by a number of bits. We have selected a simple register as the first design problem for the following reasons:

- To illustrate design partitioning
- To first design a simple FSM
- To provide a formal design for a flip-flop that was discussed in Chap. 4

Figure 5.2 illustrates a 4-bit **parallel-load register** with input $X = x_3 ... x_0$ and output $Z = z_3 ... z_0$. Assuming that *_reset* = 1 (i.e., not active), the register would load X, making $Z = X$ on the next rising edge of the *clk* if the register is enabled (i.e., *enable* = 1). Otherwise, if *enable* = 0, the register is disabled and retains its content.

In the figure, the asynchronous preset signal *_s* of each flip-flop is assumed to be 1 (disabled), and thus is not shown. However, if *_reset* is supposed to initialize some of the register bits to 1 and others to 0, then *_reset* signal would be connected to the *_s* inputs of those flip-flops that initialize to 1 and the *_r* inputs of the remaining flip-flops. In addition, all the unused *_s* and *_r* inputs would need to be disabled, and being active-low signals in this case, they would be connected to 1.

5.2.1 Register Model

As illustrated in Fig. 5.2 for $n = 4$, the design of an n-bit register can be partitioned using n 1-bit register slices. The behavior of the register slice can be formally modeled as an FSD, shown in Fig. 5.3. The circles represent a set of unique but

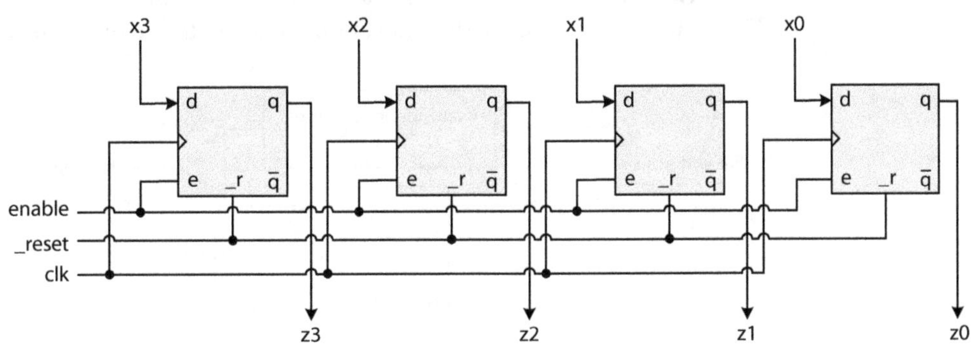

FIGURE 5.2 A 4-bit, parallel-load register.

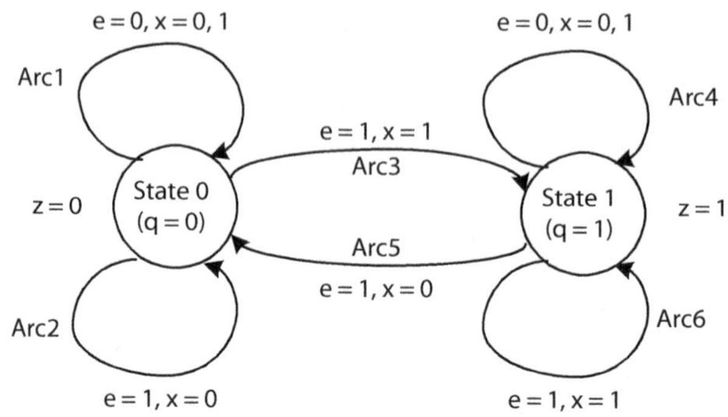

Figure 5.3 One-bit register slice modeled as an FSD.

finite number of states determined by analyzing the design problem. Because the 1-bit register slice can store either a 0 or a 1 as its content, its FSD has two possible states (i.e., two circles). The arcs represent transitions from one state to the next, and there are conditions for each transition.

As shown in the figure, if the register slice is not enabled (i.e., $e = 0$), as expected, the register slice will remain in its current state 0 or 1. This is indicated by an arc (Arc1) from the state 0 back to itself and again a second arc (Arc4) from the state 1 back to itself. The signal values associated with each transition are listed next to its corresponding arc. For example, the signals associated with Arc1 and Arc4 are $e = 0$ and x as don't-care (0 and 1). If the register slice is in state 0 and $e = 1$ and $x = 1$, then the register slice will transition from state 0 to state 1 on the next clock cycle. This is illustrated by Arc3. The other arcs are similarly drawn.

Figure 5.4 illustrates a detailed block diagram of the 1-bit register slice as an FSM. It includes a single flip-flop that holds the 1-bit register state as 0 or 1 and a CC that inputs the current state, indicated by the q signal, and the external inputs x and e to output the next state, indicated by the d signal. The output of the register slice is defined as $z = q$.

The logic expression for the d signal is determined from a truth table, also known as a **transition table**. The table is a tabular representation of information in an FSD. Table 5.1 presents the transition table of the 1-bit register slice.

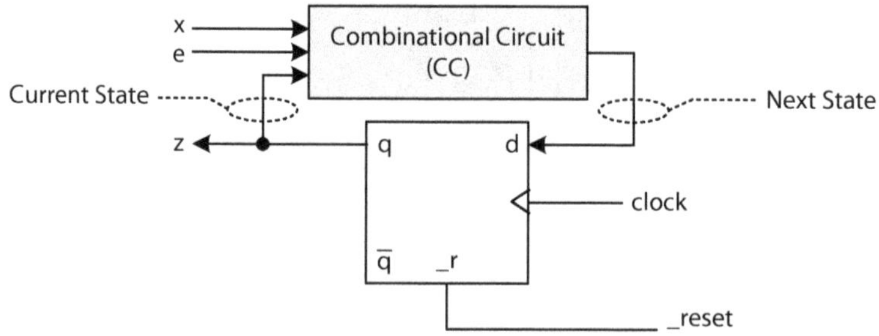

Figure 5.4 The detailed block diagram of a 1-bit register slice.

Current State	External Inputs		Next State
q	*e*	*x*	*d*
0	0	0	0
0	0	1	0
0	1	0	0
0	1	1	1
1	0	0	1
1	0	1	1
1	1	0	0
1	1	1	1

TABLE 5.1 Transition (Truth) Table for the 1-Bit Register Slice Determined from Its FSD

For example, the first two rows in the table define Arc1 (Fig. 5.3); row 3 defines Arc2; row 4 defines Arc3; etc. From the transition table, one can determine the minimal sum of product (SOP) expression $d = \bar{e}q + ex$.

Note that the expression defines a 2-to-1 MUX. As expected, if $e = 0$, then $d = q$, else if $e = 1$, then $d = x$. The circuit for the 1-bit register slice is shown in Fig. 5.5. Recall that the exact same circuit was initially introduced in Chap. 4 as a flip-flop with an enable signal (Fig. 4.19). However, the circuit was formally designed here using an FSD.

5.2.2 Multifunction Registers

Occasionally, a register may need to perform one of several functions. Figure 5.6 shows the block diagram of a four-function register with an n-bit input X, an n-bit output Z, and a 2-bit function code $F = f_1 f_0$. The register is enabled when its active-high *enable* signal is asserted. The register also requires an active-low *_reset* signal to initialize it asynchronously to 0. The four functions of the register are synchronous reset (clear), parallel load, arithmetic right shift that repeats the sign bit, and right shift that enters a 0 from left.

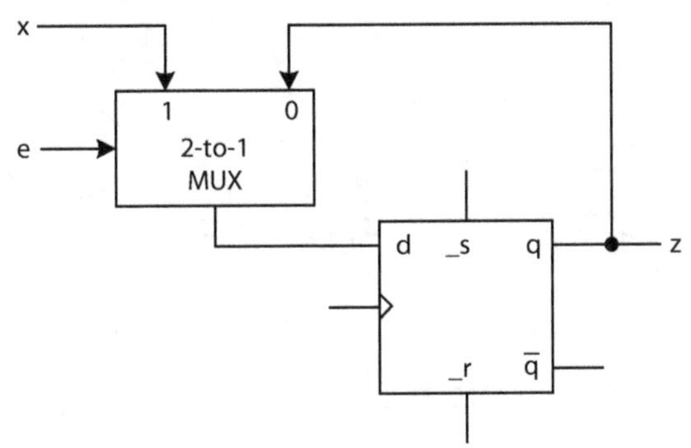

FIGURE 5.5 One-bit register-slice (also shown in Fig. 4.19).

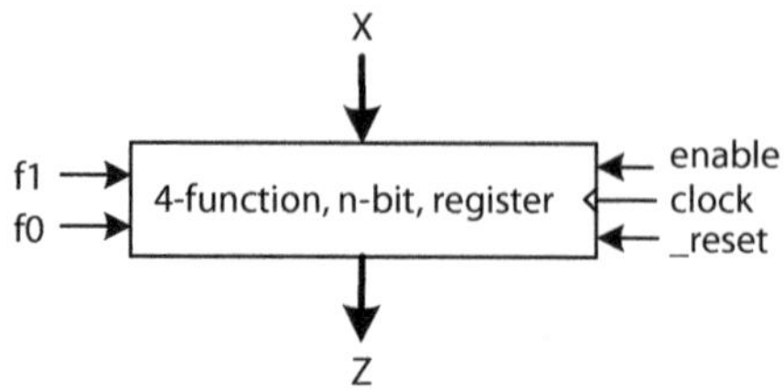

f1	f0	Action
0	0	Clear (synchronous reset)
0	1	Load X
1	0	Arithmetic shift (shifts right repeating the sign bit)
1	1	Right shift (shifts right entering 0 from left)

FIGURE 5.6 The block diagram of a four-function register.

Bit-Serial Design

The detailed block diagram and the FSD of a 1-bit four-function register slice are given in Fig. 5.7. Contrary to the previous example, the flip-flops in this case are assumed to include an enable signal *e*. Also, for clarity and convenience, don't-care signal values are not shown and omitted from the arcs in the FSD. For instance, *i* and *x* are don't-care and thus are not shown for the arcs for which $F = 0$ ($f_1 f_0 = 00$). When $F = 0$, the register slice must be synchronously initialized to 0 (cleared), independent of the values of its *i* and *x* inputs.

In Fig. 5.7, the combinational circuit has five inputs—q (the current state), f_1, f_0, i, and x—and one output d (the next state). Its truth table would consist of 32 rows and thus is not shown. However, using the Espresso minimization software yields the expression $d = \overline{f_1} f_0 x + f_1 i$. The final circuit and the symbol for the 1-bit four-function register slice are shown in Fig. 5.8. Using four copies of the register slice, a 4-bit four-function register can be designed as illustrated in Fig. 5.9. For each register slice in the figure, the input i, except the leftmost one, is connected to the z output from its preceding connected slice. The i input for the leftmost slice is defined as $i = f_1 \overline{f_0} z$. It repeats the sign-bit z_3 when $F = 2$ that specifies an arithmetic right shift.

Note that the combination circuit in Fig. 5.8 defines a 1-bit 4-to-1 MUX with the inputs 0, x, i, and i (the i is used twice), as shown in Fig. 5.10. The SOP expression of the MUX simplifies to $d = \overline{f_1} f_0 x + f_1 i$, which is the exact same

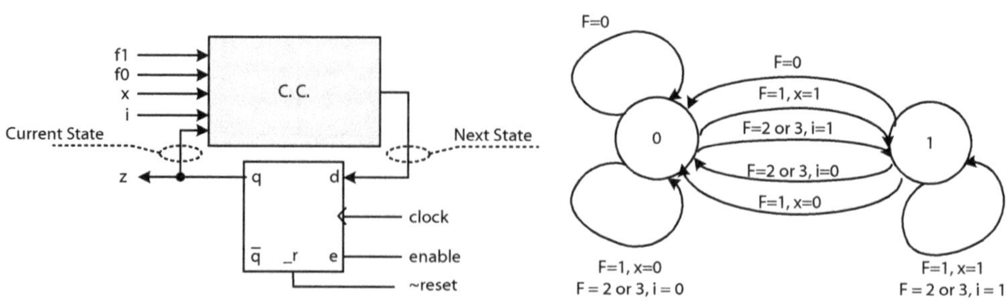

FIGURE 5.7 A detailed block diagram and the FSD of a 1-bit four-function register slice.

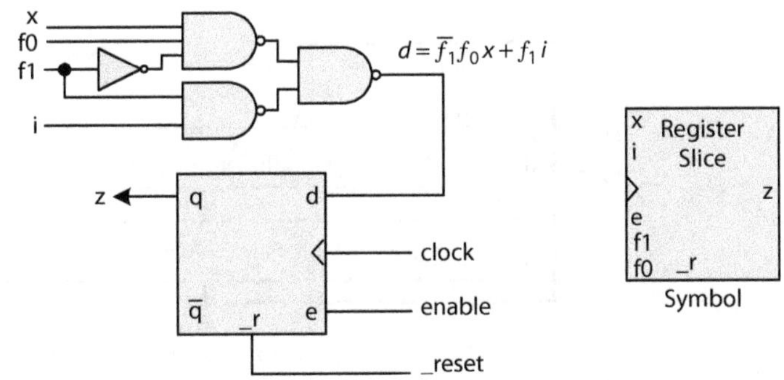

Figure 5.8 One-bit four-function register slice and its symbol.

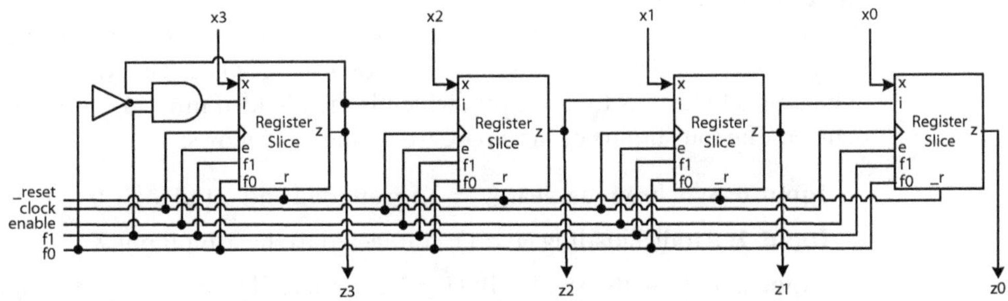

Figure 5.9 A 4-bit 4-function bit-serial register design.

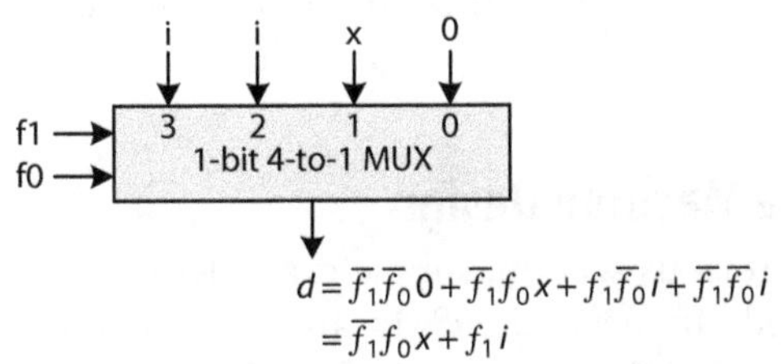

Figure 5.10 A 1-bit 4-to-1 MUX with inputs, 0, *x*, and *i* used twice.

expression shown in Fig. 5.8 obtained using the FSD. This is a valuable technique that can be used to design a bit-serial or a bit-parallel FSM without actually requiring an FSD model. The technique does, however, require analyzing the design problem to determine whether a known CC module, such as the MUX in this case, can be used in the design. This is illustrated next by using the bit-parallel technique to design the 4-bit four-function register. However, if one cannot identify a known module by analyzing the design problem, then one must use an FSD to formally model the design problem.

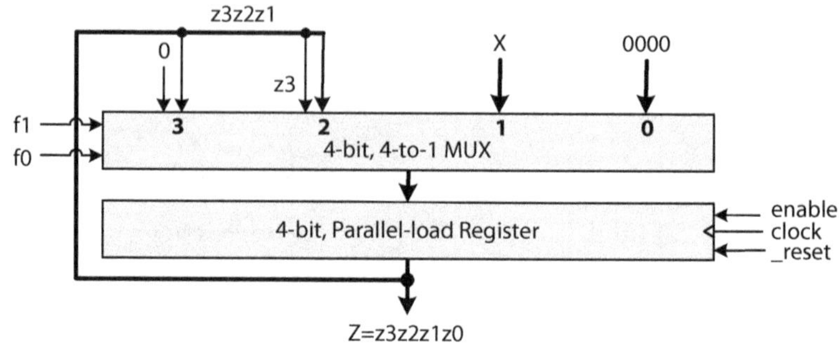

Figure 5.11 Bit-parallel design of a 4-bit four-function register.

Bit-Parallel Design

Figure 5.11 illustrates the design of the four-function register given in Fig. 5.6 using a 4-bit parallel-load register and a 4-bit 4-to-1 MUX. The MUX routes one of its four 4-bit inputs to its output, which is then loaded into a 4-bit parallel-load register. The four inputs of the MUX are defined as follows:

Input-0: Synchronous clear ($F = 0$), grounded and set to logic 0

Input-1: Parallel loading ($F = 1$), connected to the 4-bit input X

Input-2: Arithmetic right shift ($F = 2$), connected to $\{z_3, z_3z_2z_1\}$, the sign-bit z_3 is concatenated (indicated by {}) with the upper three bits of register output Z to create a 4-bit number

Input-3: Right shift ($F = 3$), connected to $\{0, z_3z_2z_1\}$, a 0 is concatenated with the upper three bits of register output Z to create a 4-bit number

5.3 Finite State Machine Design

In the previous section, simple examples of bit-serial and bit-parallel FSMs were presented. In general, FSMs are categorized into **Mealy**, **Moore**, or **hybrid** machines. In addition, the combinational circuits of an FSM can be grouped into two sets: those forming a **next-state generator** (NSG) and those forming an **output generator** (OG). An NSG determines the next state, and an OG generates output signals.

Figure 5.12 illustrates an FSM **detailed block diagram**. An FSM is called a Mealy machine if its outputs, known as **Mealy outputs**, are determined using its current state as well as using its current (external) inputs—not counting the clock, reset, and preset signals that directly connect to flip-flops and the enable signal used with flip-flops. A change in the values of one or more of the external inputs could change the value of a Mealy output independent of the clock signal. In the figure, this is shown by a wire (a bold and dashed line) connecting the external input signals to the inputs of the OG. Mealy outputs are said to asynchronously dependent on the external input values. Unlike Mealy outputs, **Moore outputs**

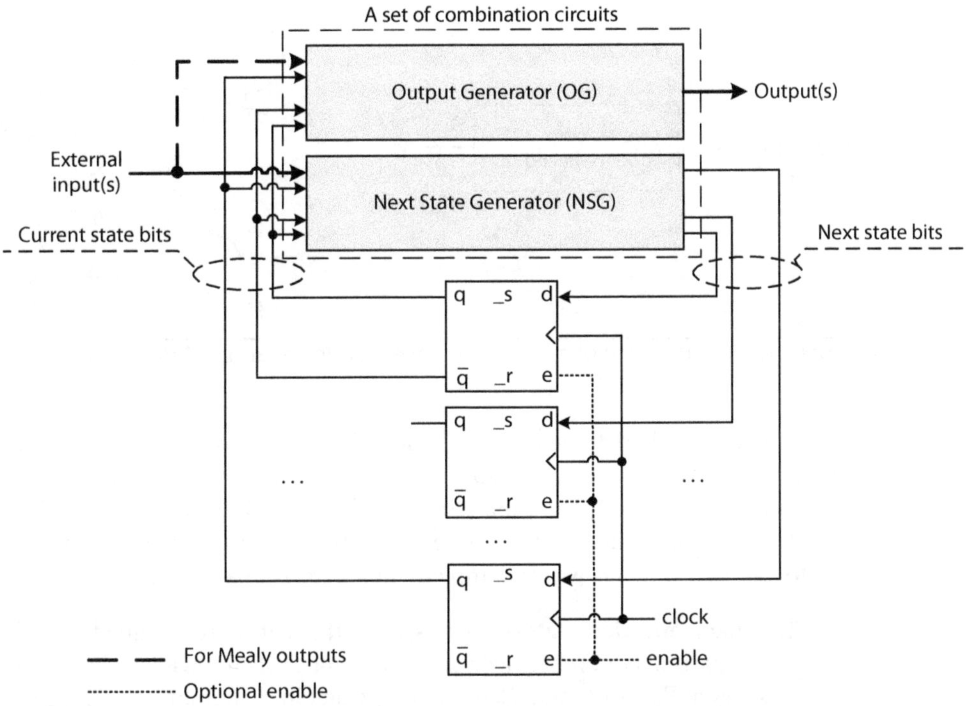

Figure 5.12 A detailed block diagram illustrating Mealy and Moore FSMs.

synchronously depend on the external input values. A hybrid machine has both Mealy and Moore outputs.

We use a sequence recognizer as an example to further discuss these topics, as well as the design decisions that affect circuit size. A sequence recognizer operates much like the controller of a combinational digital lock. It monitors inputs one bit at a time and outputs a 1 (assuming active high) each time that the recognizer encounters a target sequence, for example, a 3-bit sequence "101." A recognizer can be designed to recognize either an **overlapping** or **nonoverlapping** sequence. For example, the input sequence "10101" contains two overlapping sequences of "101," where the "1" in the center of the input sequence is shared. On the other hand, there are two nonoverlapping "101" sequences in the input sequence "101101."

Example 5.1 The design of a Moore FSM that detects the overlapping sequence "101":

Solution Figure 5.13 shows the top-level block diagram of the sequence recognizer with the external input x and the output z. Its Moore FSD is also shown with four states labeled A, B, C, and D. The active-high *reset* signal is used to asynchronous initialize the machine to a known state A, as illustrated by an arrow labeled *reset* in the FSD. An input sequence is processed one bit at a time. The recognizer makes a transition to a new state each time that it counters the next bit in the target sequence. For example, if the recognizer is in the state C, it indicates that it has received the first 2-bits of the target sequence. The output z is shown below each state and becomes 1 when the recognizer receives the last bit of the target sequence and enters state D. The z is 0 in all other states. The recognizer would reject all other 3-bit sequences that it inputs and start over each time. The details of alternative solutions are discussed next.

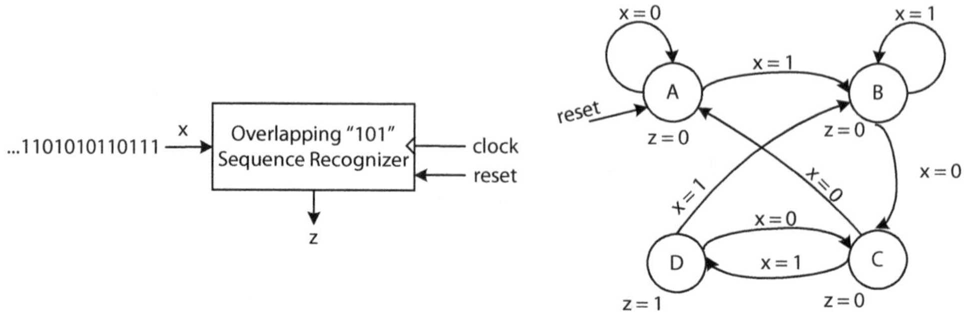

Figure 5.13 A block diagram of a "101" sequence recognizer and its Moore FSD.

An FSD is called **deterministic** if each of its states has a unique set of transitions (arcs). However, if there was also a second transition from state A in Fig. 5.13, say, to state C when $x = 1$, then the FSD would be **nondeterministic**. In this case, if $x = 1$, then it could not be determined if the transition should be from state A to B or to C. There are two ways to implement a deterministic FSD:

1. **Binary encoded states**—In this case, the states are assigned (labeled) with unique binary numbers with the least number of bits possible. For example, four states in Fig. 5.13 can be labeled with 2-bit binary numbers: 00, 01, 10, and 11.
2. **One-hot encoded states**—In this case, the states are labeled with unique binary numbers, each consisting of only a single 1 (one-hot); the remaining bits are 0. For example, the 4-bit one-hot numbers 0001, 0010, 0100, and 1000 can be used to label the four states in Fig. 5.13.

5.3.1 Binary Encoded States

The minimum number of bits required to encode the states of an FSD is determined from Eq. (5.1), where k is the total number of states. The symbols $\lceil \; \rceil$ indicate the ceiling function.

$$\text{Number of bits} = \lceil \log_2(k) \rceil \tag{5.1}$$

For instance, if the number of states is more than 4 and less than 8 (i.e., $4 < k < 8$), then 3-bit numbers are needed to label anywhere between five and eight states. Which one of the numbers should be assigned to each state is a logic optimization problem. Figure 5.14 illustrates a detailed block diagram for the sequence recognizer in Fig. 5.13 using binary encoded states.

The following two transition (truth) tables (Tables 5.2 and 5.3) are determined from the FSD, where binary numbers 00 is assigned to the state A, 01 to state B, 10 to state C, and 11 to state D. The truth tables are used to determine minimum SOP expression (Eq. (5.2)) for each of the state variables d_1, d_0, and the output variable z. A completed circuit using positive-edge triggered flip-flops and active-high reset is shown in Fig. 5.15.

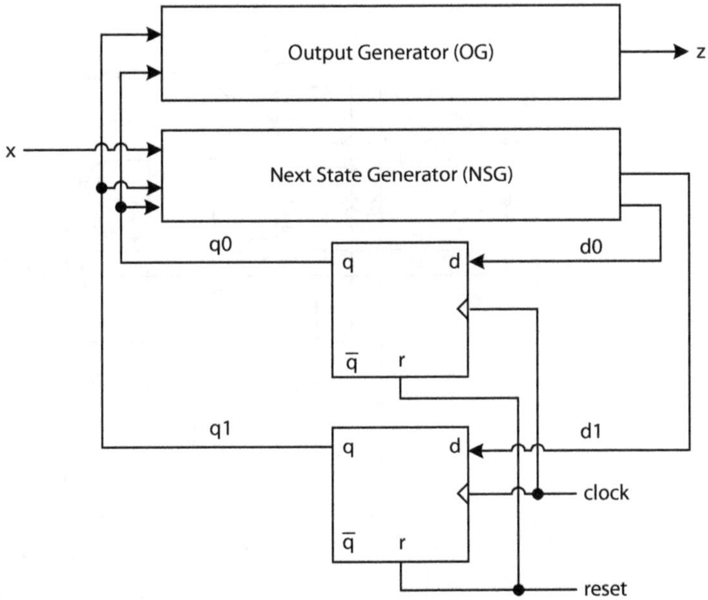

FIGURE 5.14 The detail block diagram of a "101" sequence recognizer.

	Current State		Input	Next State		
	q_1	q_0	x	d_1	d_0	
A	0	0	0	0	0	A
	0	0	1	0	1	B
B	0	1	0	1	0	C
	0	1	1	0	1	B
C	1	0	0	0	0	A
	1	0	1	1	1	D
D	1	1	0	1	0	C
	1	1	1	0	1	B

TABLE 5.2 The NSG Transition Table Compiled from the FSD in Fig. 5.13

Current State		Output
q_1	q_0	z
0	0	0
0	1	0
1	0	0
1	1	1

TABLE 5.3 The OG Truth Table Complied from the FSD in Fig. 5.13

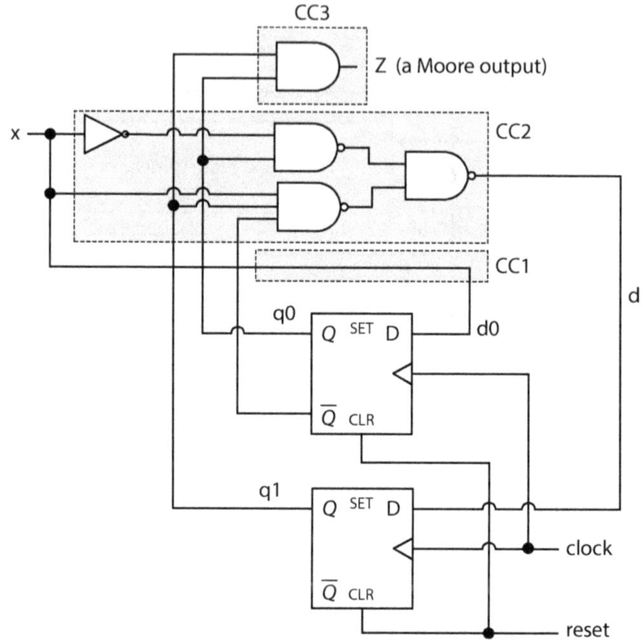

FIGURE 5.15 A Moore FSM to detect the overlapping sequence "101."

An alternative circuit schematic (layout) with distributed CCs is illustrated in Fig. 5.16. In this case, each of the CCs—CC1 and CC2—are schematically shown next to its corresponding flip-flop. This is how circuits are generally implemented.

$$
\begin{aligned}
d_1 &= q_0\bar{x} + q_1\bar{q}_0 x \\
d_0 &= x \\
z &= q_1 q_0
\end{aligned}
\tag{5.2}
$$

Upon reset, the sequence recognizer initializes to state 0 (i.e., $q_1q_0 = 00$) and, as expected, outputs $z = 0$. Assuming $x = 1$, the expressions yield $d_1 = 0$ and $d_0 = 1$, given $q_1 = 0$ and $q_0 = 0$. The recognizer transitions from state 0 to state 1 as soon as d signals are loaded into the flip-flops, making $q_1q_0 = 01$.

Suppose the next input is 0 (i.e., $x = 0$). This time, the expressions yield $d_1 = 1$ and $d_0 = 0$, and thus would cause the recognizer to transition to state 2 ($q_1q_0 = 10$). Finally, suppose the next input is 1 ($x = 1$). This yields $d_1 = 1$ and $d_0 = 1$, and the recognizer would transition to state 3 ($q_1q_0 = 11$). Once in state 3, z becomes 1 and recognizes the sequence "101." In general, testing every transition of an FSM, especially a large one, is difficult and can require a prohibitively large number of tests.

The amount of hardware required for implementing an NSG and an OG depends on the binary label assigned to each state. The circuit in Fig. 5.15 was designed using 00, 01, 10, and 11 to label the states A through D, respectively. Suppose we decide to

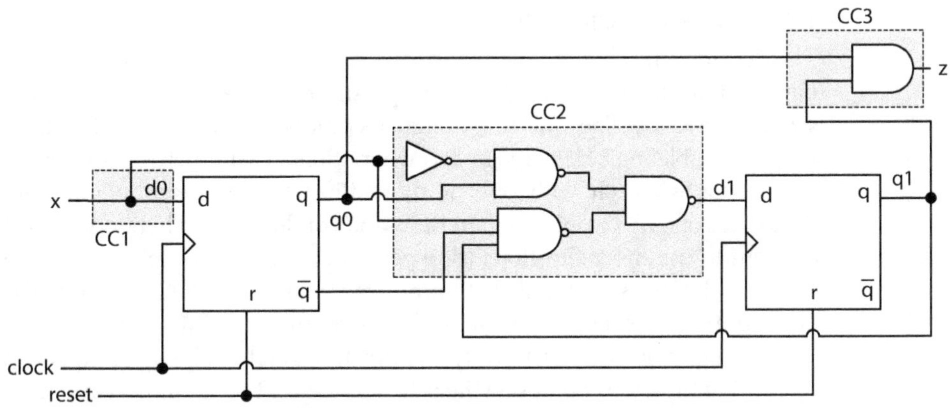

Figure 5.16 An alternative and typical layout for the circuit shown in Fig. 5.15.

change the assignments and instead use the labels 00, 11, 10, and 01 in order for states A to D. This would yield a circuit requiring less total hardware. Equation (5.3) lists a minimal SOP expression for each of the state variables d_1 and d_0 and the output z using the new binary state labels. Note that, compared to the circuits in Fig. 5.15, the circuit for d_1 is simpler, as shown in Fig. 5.17.

$$\begin{aligned} d_1 &= q_0 + \overline{q_1}x = \overline{(\overline{q_0})(\overline{\overline{q_1}x})} \\ d_0 &= x \\ z &= \overline{q_1}q_0 \end{aligned} \tag{5.3}$$

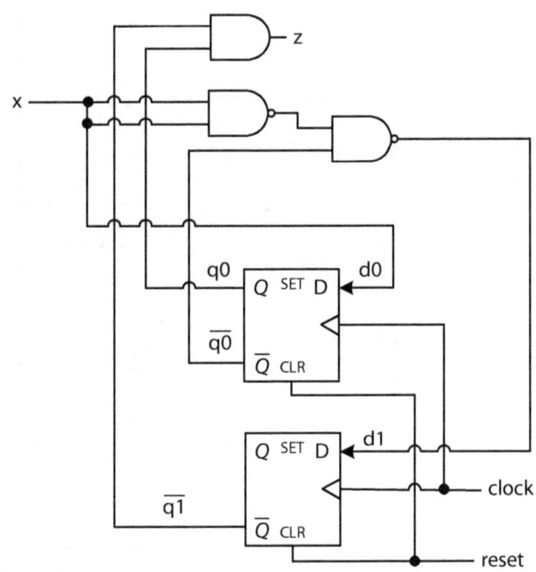

Figure 5.17 A Moore FSM to detect the overlapping sequence "101" with alternative encoded states.

5.3.2 One-Hot Encoded States

The design technique of binary encoded states minimizes the number of flip-flops. On the other hand, the design technique of one-hot encoded states minimizes the size of the CCs. The one-hot design technique is especially advantageous with programmable logic devices (PLDs) where there are numerous configuration logic blocks (CLBs), each with one or more flip-flops. For example, an FPGA with thousands of CLBs would contain thousands of flip-flops. Therefore, it may be more efficient to use more flip-flops (one per state) and less complex CCs. The one-hot design technique is more likely to produce circuits with the least propagation delay, not counting wire delays, which can be longer in some PLD designs.

During each clock cycle, only one flip-flop is set (i.e., only one q is 1), while the remaining flip-flops are reset (the other q's are 0). For example, suppose instead of using 2-bit numbers to label the four states of the FSD in Fig. 5.13, 4-bit one-hot labels 0001, 0010, 0100, and 1000 are used. Figure 5.18 illustrates a detailed block

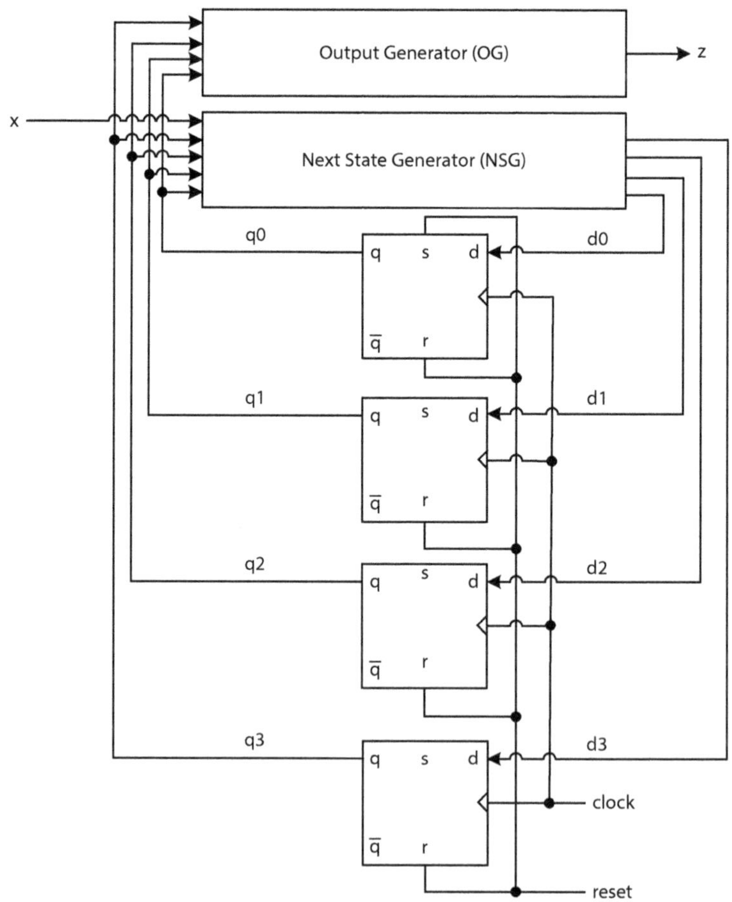

FIGURE 5.18 The detailed block diagram of a one-hot FSM for detecting the sequence "101."

	Current State				Input	Next State				
	q_3	q_2	q_1	q_0	x	d_3	d_2	d_1	d_0	
A	0	0	0	1	0	0	0	0	1	A
	0	0	0	1	1	0	0	1	0	B
B	0	0	1	0	0	0	1	0	0	C
	0	0	1	0	1	0	0	1	0	B
C	0	1	0	0	0	0	0	0	1	A
	0	1	0	0	1	1	0	0	0	D
D	1	0	0	0	0	0	1	0	0	C
	1	0	0	0	1	0	0	1	0	B

TABLE 5.4 One-Hot Design NSG Truth Table Compiled from the FSD in Fig. 5.12

diagram of the corresponding one-hot FSM. Note that, upon reset, the machine must start in state A ($q_3q_2q_1q_0 = 0001$) with one of the flip-flops set. Therefore, the *reset* signal must be connected to the preset (s) input of the flip-flop associated with q_0 and to the reset (r) input of the remaining flip-flops, as illustrated in the figure. Table 5.4 presents the truth table for the NSG, and Table 5.5 presents the truth table for the OG. The missing table entries are don't-cares and are not shown. However, the don't-care output values should be entered in Karnaugh maps (K-maps), Espresso files, and in HDL models to further minimize the circuits. In addition, don't-care values are especially important in HDL models to avoid creating implicit latches.

Equation (5.4) lists the minimal expressions for the next state variables d_3 through d_0 and the output variable z.

$$\begin{aligned} d_3 &= q_2 x \\ d_2 &= \overline{q_2}\,\overline{q_0}\,\overline{x} \\ d_1 &= \overline{q_2}\,x \\ d_0 &= \overline{q_3}\,\overline{q_1}\,\overline{x} \\ z &= q_3 \end{aligned} \tag{5.4}$$

	Current State				Output
	q_3	q_2	q_1	q_0	z
A	0	0	0	1	0
B	0	0	1	0	0
C	0	1	0	0	0
D	1	0	0	0	1

TABLE 5.5 One-Hot Design OG Truth Table Compiled from the FSD in Fig. 5.12

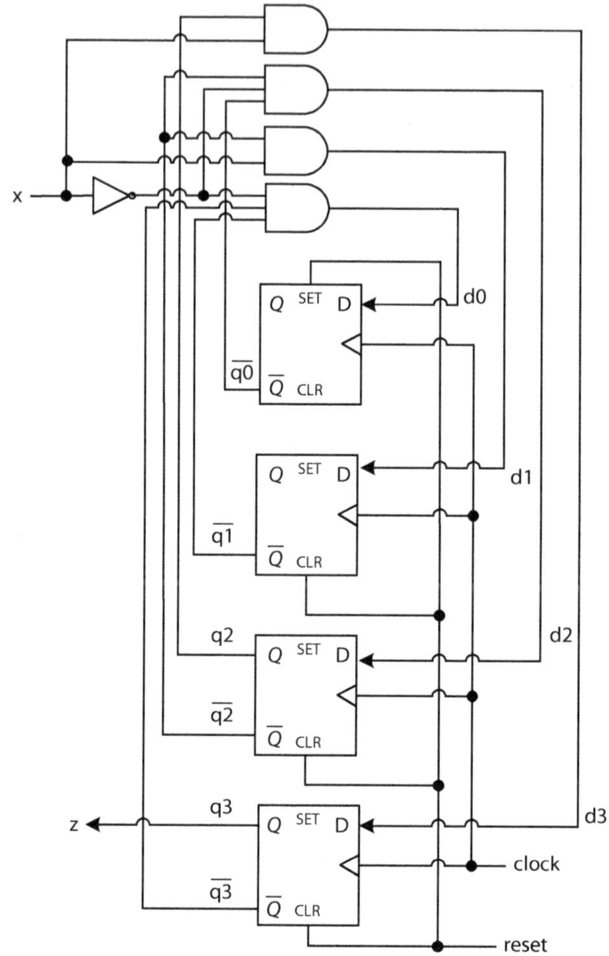

FIGURE 5.19 A one-hot design FSM to detect the sequence "101."

The final circuit is shown in Fig. 5.19. When compared to the binary encoded FSM in Fig. 5.17, the one-hot FSM requires more gates, but simpler circuits for each next state-bit, and output z requires no circuit.

Example 5.2 The design of a Mealy FSM that detects an overlapping sequence "101":

Solution Figure 5.20 shows the block diagram and a Mealy FSD for the sequence recognizer. Note that, in this case, the z being a Mealy output is assigned to the arcs and not to the states. The label of each arc has two parts separated by a slash (/), the inputs (only x in this case) are listed to the left of the slash, and the outputs (only z in this case) are listed to the right of the slash. Before a Mealy solution is presented, additional, Mealy and Moore design issues are discussed next using this example.

The detailed block diagram for the Mealy FSM is illustrated in Fig. 5.21. Note that if the FSM is in state C, then either $z = 0$ if $x = 0$ or $z = 1$ if $x = 1$. Therefore, as expected, the Mealy output z (z-Mealy) depends asynchronously on external input x.

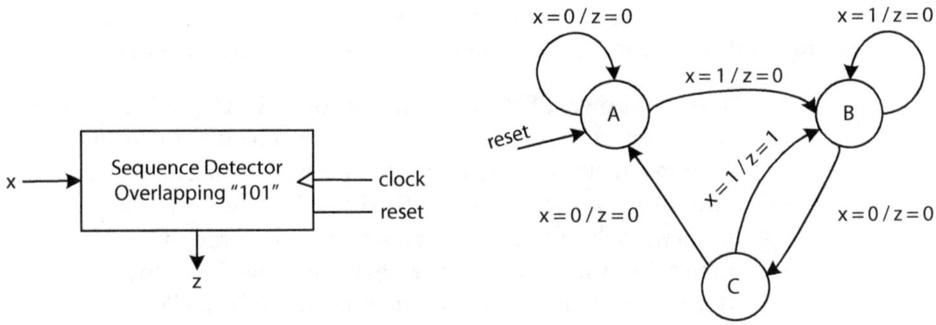

Figure 5.20 The block diagram of the "101" sequence recognizer and its Mealy FSD.

This is a typical behavior of a Mealy FSM. As soon as the x signal changes, the z signal could change, independent of the clock signal. The FSD has three states. It can be implemented using either two flip-flops with binary encoded state labels or three flip-flops with one-hot state labels.

Assuming that states A, B, and C are encoded with 2-bit binary codes, one of the four possible binary state labels would not be used. For example, suppose, the binary label 00 is assigned to state A, 01 to state B, and 10 to state C. The binary label 11 will not be used and would constitute an unknown/undefined machine state. An environmental hazard (e.g., a transient fault) that causes a state change in one or more of the flip-flops could alter the state of the FSM. For example, a 1-bit accidental change in the value of q_1 or q_0 would switch state 01 (C) either to the

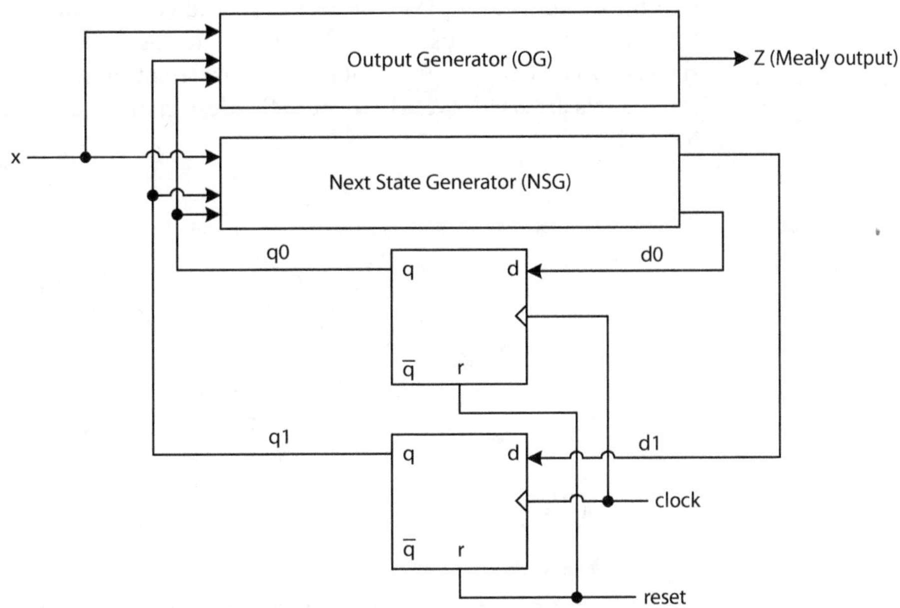

Figure 5.21 A detail block diagram of the "101" Mealy sequence recognizer.

known state 00 (A) or to the unknown state 11 (e.g., D). In general, there are several choices for how to handle the unknown states in a design:

- **The unknown FSM states are ignored in the design**—In this case, the unknown states are considered unimportant, for example, if the FSM operates a simple toy. If the FSM enters into one of its unknown states and thus malfunctions, the FSM needs to be reset. More specifically, the binary labels of the unknown states are entered in the NSG and OG truth tables, but the values for the next state and the output variables are set to don't-care. This helps to reduce the size of the NSG and OG circuits.
- **The unknown FSM states are transitioned to a known state**—In this case, each time that an FSM enters into an unknown state, the machine is transitioned into a known state during the next clock cycle. For example, Fig. 5.22 shows a binary encoded FSD with one unknown state D. If, due to an environmental hazard, the FSM accidentally enters the unknown state D, it would not only transition to the known state A on the next clock cycle, as shown in the figure, but also would not generate invalid outputs. In this case, the label for the unknown state D would be entered in both the NSG and OG truth tables, but the label for the state A would be entered as the next state for both when $x = 0$ and $x = 1$. Also, the output z would be set to 0.
- **The sequential circuit is designed as a fault-tolerant FSM**—In this case, the FSM would be able to recover from an unknown state or from an accidental transition to a known state and would continue operating normally. For example, a single bit fault, which would set or reset a single flip-flop, can be detected and corrected by including additional hardware that implements a single error detection and correction scheme. A fault-tolerant FSM would use extra flip-flops and logic to detect and correct the errors caused by faults. Note that error detection is simpler if one-hot state labels are used. We will discuss the design of a fault-tolerant FSM in Sec. 5.5.

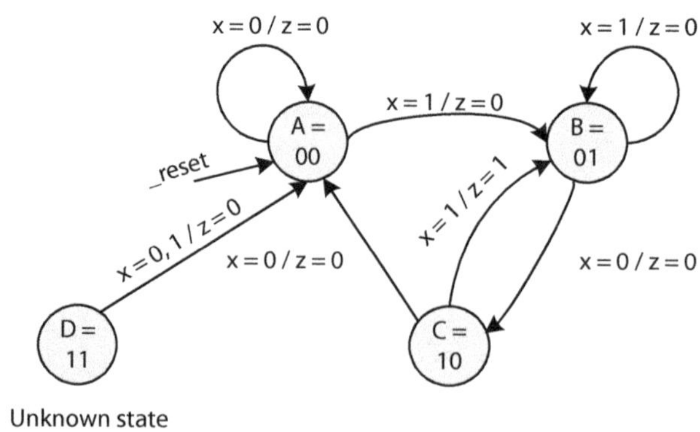

Figure 5.22 A binary encoded FSD with one unknown state D.

	Current State		Input	Next State		
	q_1	q_0	X	d_1	d_0	
A	0	0	0	0	0	A
	0	0	1	0	1	B
B	0	1	0	1	0	C
	0	1	1	0	1	B
C	1	0	0	0	0	A
	1	0	1	0	1	B
D	1	1	0	d	d	
	1	1	1	d	d	

TABLE 5.6 The NSG Truth Table Determined from the FSD Given in Fig. 5.20

Assuming that the first option (ignoring the unknown states) is used to design the Mealy FSD in Fig. 5.20, Table 5.6 presents the truth table for the NSG with the unknown state D ignored in the design; when the current state is D, the next state is defined as don't-care. Likewise, Table 5.7 presents the truth table for the OG with the output z set to don't-care when the current state is D. Equation (5.5) lists the minimal expressions for the next state variables d_1 and d_0 and the output variable z.

$$\begin{aligned} d_1 &= q_0\overline{x} \\ d_0 &= x \\ z &= q_1 x \end{aligned} \tag{5.5}$$

The corresponding FSM circuit is given in Fig. 5.23. Mealy machines, in general, require fewer flip-flips than their equivalent Moore machines. However, sometimes Moore outputs are preferred. In such cases, additional flip-flops,

	Current State		Input	Output
	q_1	q_0	x	z
A	0	0	0	0
	0	0	1	0
B	0	1	0	0
	0	1	1	0
C	1	0	0	0
	1	0	1	1
D	1	1	0	d
	1	1	1	d

TABLE 5.7 The OG Truth Table Determined from the FSD Given in Fig. 5.20

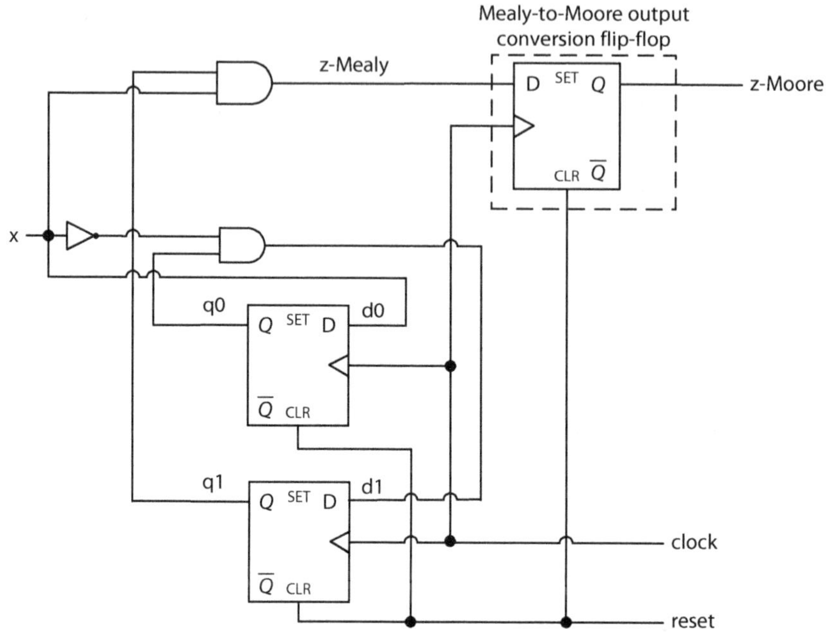

Figure 5.23 The circuit for the Mealy "101" sequence recognizer. Also illustrated is the conversion of a Mealy output to a Moore output using a synchronizing flip-flop.

one per Mealy output, are used to convert a Mealy output to its corresponding Moore output. In the figure, a flip-flop converts the *z*-Mealy to its equivalent *z*-Moore. While the *z*-Mealy asynchronously depends on the external input *x*, the *z*-Moore does not. However, the *z*-Moore would lag the *z*-Mealy by one clock cycle.

5.4 Counters

In the previous sections, various FSM design techniques and sample applications were discussed. A counter is a sequential circuit that outputs a finite set of prespecified values. For example, a 2-bit binary counter, also known as **mod-4 counter**, outputs the binary numbers 00, 01, 10, and 11 in order and then it repeats, $(3 + 1) \bmod 4 = 0$. In general, a mod-k counter outputs k values starting at 0 and ending at $k - 1$ and then it repeats.

There are many other counter examples. A **binary-coded-decimal** (BCD) counter would output the sequence 0 through 9 and then it would repeat. A **Gray** code counter would output a sequence of numbers, where each number is different in only 1-bit when compared with an immediately preceding number. For example, the numbers 000, 001, 011, 010, 110, 111, 101, 100, and then 000, repeating, would be the outputs of a 3-bit Gray code counter.

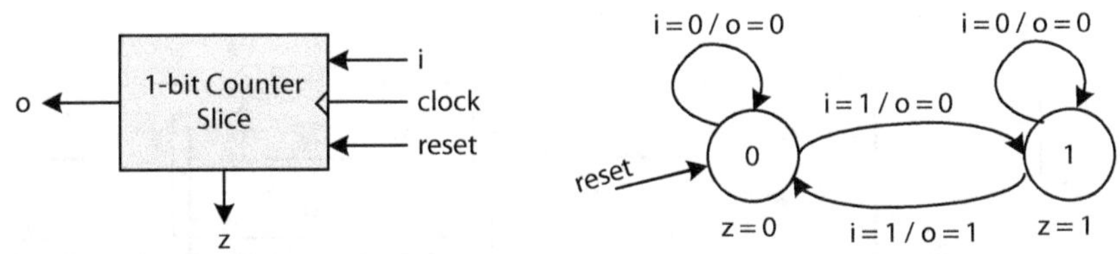

FIGURE 5.24 The block diagram and the FSD of a 1-bit counter slice.

The fact that two consecutive Gray codes differ in only 1-bit helps to limit the number of possible bit errors to only one when a Gray code is accessed as external data by a totally separate sequential circuit. For an example of why Gray code counters would be necessary for two independently functioning sequential circuits to access a first-in-first-out (FIFO) buffer, refer to Exercise 5.30 (also see Sec. 5.6.2).

Example 5.3 The design of a bit-serial mod-8 counter with asynchronous active-high reset that uses three copies of a 1-bit counter slice. For $k = 8$, the counter outputs 0 through 7 and then repeats. This would require the design of a hybrid FSM with both Mealy and Moore outputs.

Solution A k-bit binary counter slice must perform one of two operations: It should either retain its current count or increment it by 1. The slice associated with the least significant digits must increment every clock cycle. The other slices only increment when they receive a signal indicating "increment." Figure 5.24 illustrates the block diagram and the FSD of a 1-bit binary counter slice. The FSD defines a hybrid FSM with a Moore output z-Moore and a Mealy output o-Mealy. The i input is used as an enabling input signal in each slice. If $i = 1$, the slice increments its current value; otherwise, it retains its current value. Each o-Mealy output connects to the i input of its immediately succeeding slice.

Table 5.8 presents a combined NSG and OG transition (truth) table determined from the FSD of the 1-bit counter slice. The minimal expressions for the z-Moore and o-Mealy outputs and the next state variable d are listed in Eq. (5.6). The circuit for the 1-bit counter slice and a corresponding mod-8 counter are shown in Fig. 5.23. The 1-bit counter slice can be used to design any mod-k counter as long as $k \geq 2$ is power of 2 (i.e., $k = 2^m$, where $m \geq 1$).

$$\begin{aligned} z &= q \\ o &= q \cdot i \\ d &= q \oplus i \end{aligned} \tag{5.6}$$

Current State	Input	Next State	Outputs		
q	i	d	z-Moore	o-Mealy	Comment
0	0	0	0	0	Retain
0	1	1	0	0	Increment
1	0	1	1	0	Retain
1	1	0	1	1	Increment

TABLE 5.8 A Combined NSG and OG Truth Table Complied from the FSD in Fig. 5.24

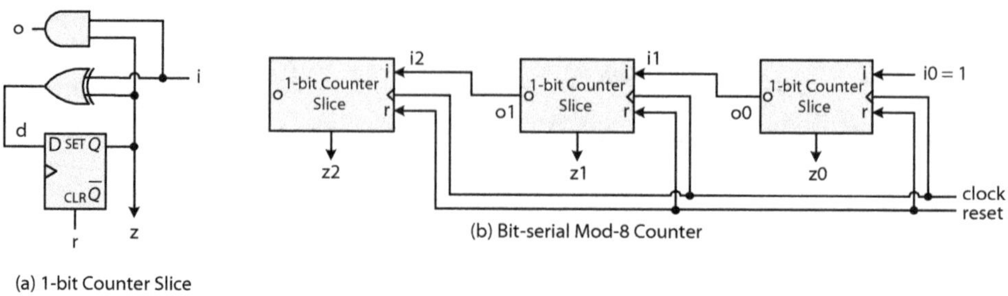

Figure 5.25 A bit-serial mod-8 counter: (*a*) 1-bit counter slice; (*b*) mod-8 counter with three slices.

Example 5.4 A bit-serial mod-6 counter with active-high reset; note that 6 is not a power of 2.

Solution A mod-6 counter outputs 0, 1, 2, 3, 4, 5, and then it repeats. Like a mod-8 counter, it also requires three flip-flops to store a value between 0 and 5 as its current states. However, a mod-6 counter, contrary to a mod-8 counter, must reinitialize when it reaches 5 and starts the count from 0 on the next clock cycle. For this, a simple but less preferred solution is to use the output of a simple CC that asynchronously resets the counter as soon as count reaches 6, but before the next clock cycle. A signal labeled *aclear* ("a" for asynchronous) is defined as $aclear = z_2 z_1 \bar{z}_0$ and then combined with the master asynchronous signal *reset* to asynchronously reset each slice, as illustrated in Fig. 5.26.

However, since $r = aclear + reset$ is used to asynchronously reset all the flip-flops, Δr (delay of signal r) would not be included in the calculation of the minimum operating clock period. Thus, it is very likely (due to signal routing delays) that all the flip-flops may not reset to the initial value 0. That is, it is possible that when $aclear = 1$, some of the flip-flops may reset quickly, causing *aclear* to become 0 (deasserted) before all the flip-flops are able to reset. Hence, the counter may produce an incorrect output on the next clock cycle.

A preferred solution is to reset the flip-flops synchronously each time the counter outputs 5 and reserve the asynchronous reset only for counter initialization during startup. To do this, the 1-bit counter slice must include an additional external input, for example, *sc* (synchronous clear). Figure 5.27 illustrates the block diagram of the modified 1-bit counter slice. Its FSD is also shown. When $sc = 1$, the next state is always 0 and the *o*-Mealy is a don't-care independent

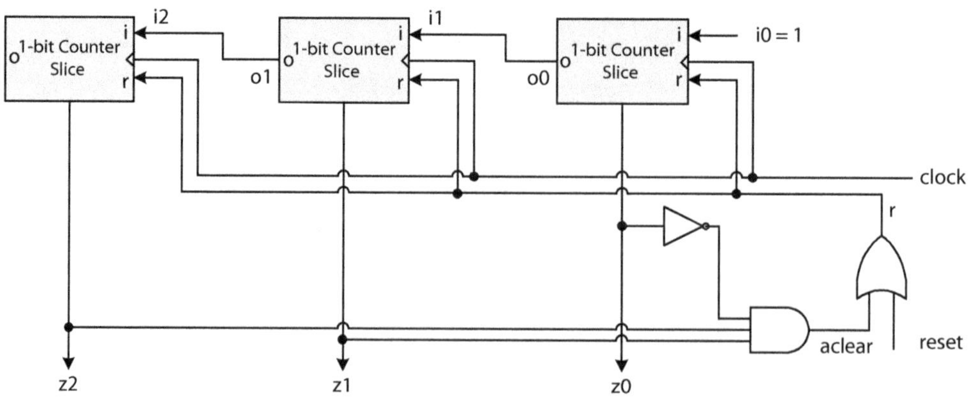

Figure 5.26 An asynchronously cleared bit-serial mod-6 counter (not a preferred solution).

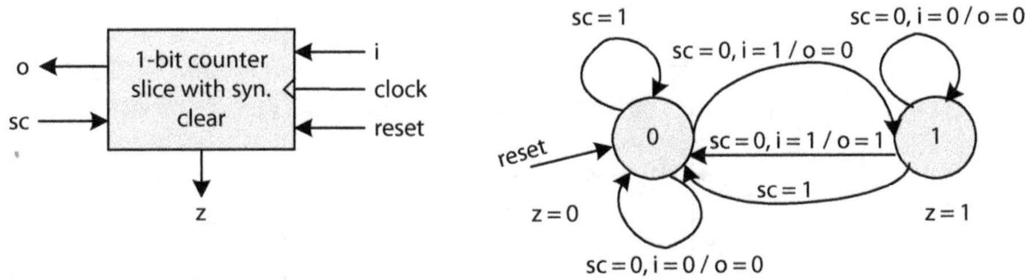

Figure 5.27 The block diagram and the FSD of a synchronously cleared (*sc*) 1-bit counter slice (a preferred solution).

of the value of the *i* input. For convenience and clarity, the *i* and the *o* signals are omitted and not shown on those arcs that $sc = 1$.

Table 5.9 presents the combined NSG and OG truth table. The minimized logic expressions for the next state variable *d* and the *z*-Moore and *o*-Mealy outputs are given in Eq. (5.7). The circuits for the bit slice and the mod-6 counter are illustrated in Fig. 5.28. The signal *sc* is defined as $sc = z_2\bar{z}_1z_0$ and is asserted each time that the counter outputs 5 and thus synchronously clears all the flip-flops during the next clock cycle.

$$\begin{aligned} d &= \overline{sc}(q \oplus i) \\ o &= qi \\ z &= q \end{aligned} \tag{5.7}$$

Also note that often an enabling signal is also required in the design of a counter so that the counter operates only when it is enabled. Two design options are available: (1) include a counter enable signal in its FSD model, or (2) use flip-flops with enable. The former counter, however, would operate with a faster clock.

Example 5.5 The design of a bit-serial mod-16 up/down counter using two copies of a 2-bit counter slice is presented. An up/down counter either counts up or down, based on the value of a control signal. The direction to count up or down can change at any time.

Current State	**Input**		**Next State**	**Outputs**	
q	***sc***	***i***	***d***	***z*-Moore**	***o*-Mealy**
0	0	0	0	0	0
0	0	1	1	0	0
0	1	0	0	0	d
0	1	1	0	0	d
1	0	0	1	1	0
1	0	1	0	1	1
1	1	0	0	1	d
1	1	1	0	1	d

Table 5.9 The Combined Truth Table of the NSG and OG for the 1-Bit Counter Slice in Fig. 5.27

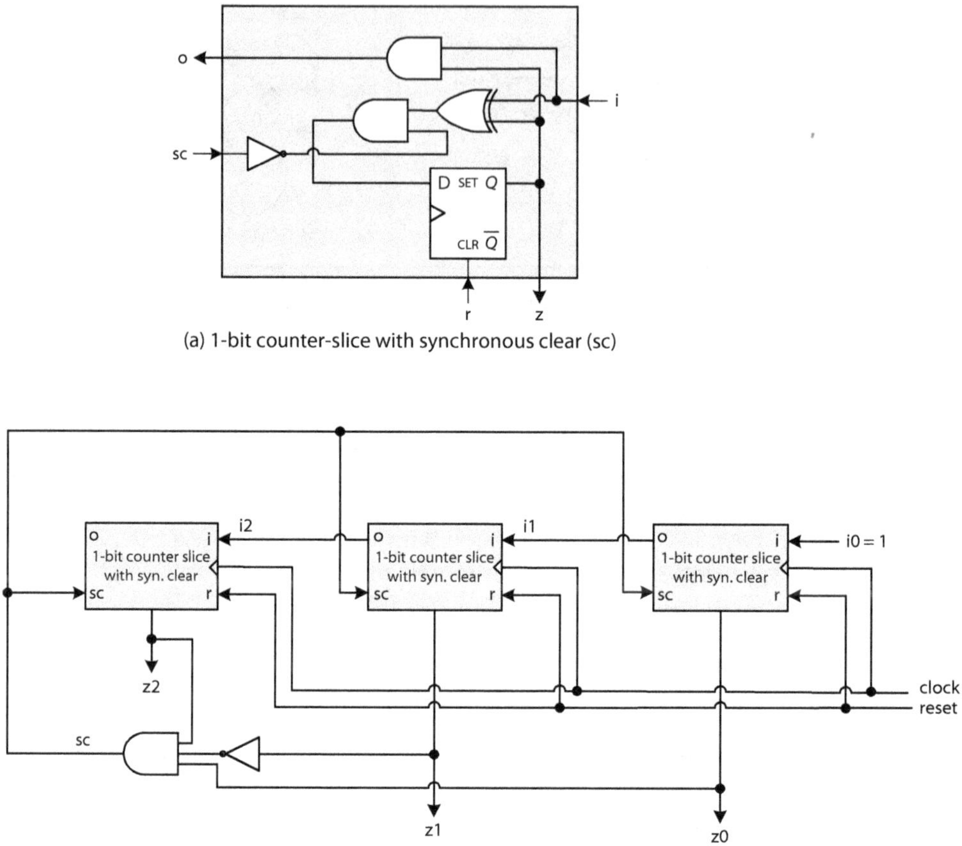

(a) 1-bit counter-slice with synchronous clear (sc)

(b) Bit-serial synchronously cleared Mod-6 Counter

Figure 5.28 A synchronously cleared bit-serial counter: (*a*) modified 1-bit counter slice; (*b*) circuit for the mod-6 counter (a preferred solution).

Solution Figure 5.29 illustrates the block diagram and the FSD of a mod-4 up/down counter slice. The slice generates the sequence 0, 1, 2, 3 and then repeats when counting up, and generates the sequence 3, 2, 1, 0 and then repeats when counting down. The signal u indicates the direction of the counter. If $u = 1$, the counter counts up; otherwise, if $u = 0$, the counter counts down. For a mod-16 counter, the first slice, which is responsible for the least two significant bits, counts every clock cycle, while the second slice counts only once every four clock cycles. The input i, if asserted, enables the corresponding slice during the next clock cycle. Disabling the first slice will automatically disable the second slice. However, the second slice is only enabled when the first slice reaches its maximum value = 3 when counting up or its minimum value = 0 when counting down. For example, when current $count = (0011)_2$, both the slices must be enabled to generate the next $count = (0100)_2$ while counting up. That is, the first slice must count up to produce $0 = (00)_2$, $(3 + 1) \bmod 4 = 0$, and the second slice must count up to produce $1 = (01)_2$, $(0 + 1) \bmod 4 = 1$.

The design requires two flip-flops and thus there are two current-state variables q_1 and q_0 and two next-state variables d_1 and d_0. Table 5.10 is the combined truth tables of the NSG and OG. The corresponding Espresso minimized SOP logic terms are listed next. Figure 5.30 is the final

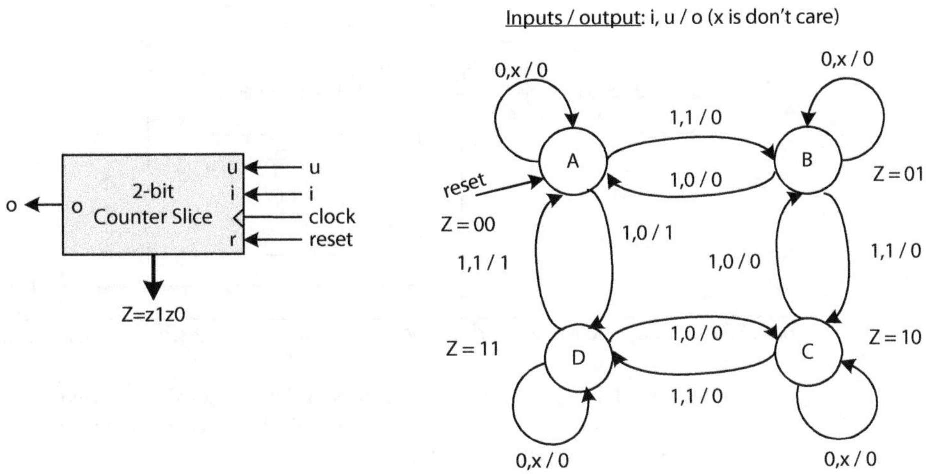

Figure 5.29 The block diagram and the FSD of a 2-bit up/down counter slice.

	Current State		Inputs		Next State		Outputs		
							Moore		Mealy
	q_1	q_0	i	u	d_1	d_0	z_1	z_0	o
A	0	0	0	0	0	0	0	0	0
	0	0	0	1	0	0	0	0	0
	0	0	1	0	1	1	0	0	1
	0	0	1	1	0	1	0	0	0
B	0	1	0	0	0	1	0	1	0
	0	1	0	1	0	1	0	1	0
	0	1	1	0	0	0	0	1	0
	0	1	1	1	1	0	0	1	0
C	1	0	0	0	1	0	1	0	0
	1	0	0	1	1	0	1	0	0
	1	0	1	0	0	1	1	0	0
	1	0	1	1	1	1	1	0	0
D	1	1	0	0	1	1	1	1	0
	1	1	0	1	1	1	1	1	0
	1	1	1	0	1	0	1	1	0
	1	1	1	1	0	0	1	1	1

Table 5.10 The Combined Truth Table for the NSG and OG of the 2-Bit Counter Slice in Fig. 5.29

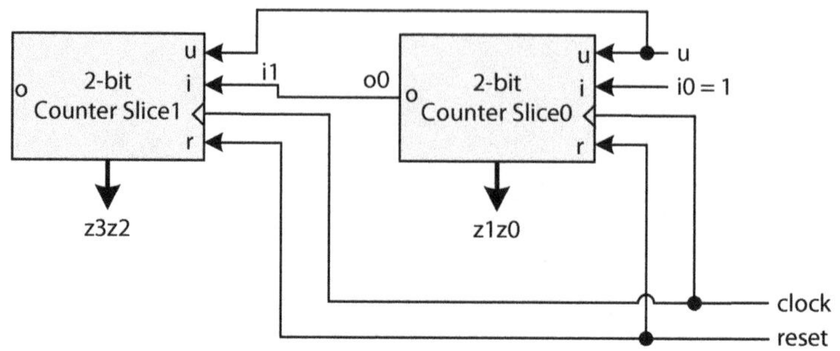

FIGURE 5.30 An up/down bit-serial mod-16 counter using 2-bit up/down counter slices.

circuit for the mod-16 counter using two copies of the counter slice, labeled Slice1 and Slice0. Slice0 is shown always enabled (i.e., $i_0 = 1$), while Slice1 is enabled only when Slice0 outputs 3 when counting up or 0 when counting down.

```
#mod-4 counter slice NSG and OG modules
#Input signal labels
#output bit label
.i  4
.o  5
.ilb  q1  q0  i  u
.ob  d1  d0  z1  z0  o
.p 10
0111 10000
0010 10001
1111 00101
10-1 10000
-10- 01000
-01- 01000
11-0 10100
10-- 00100
1-0- 10100
-1-  00010
.e
```

Alternatively, because all counters require some kind of an adder that performs a known function, they could be designed without requiring an FSD. This is illustrated in the following example.

Example 5.6 The bit-parallel design of a mod-8 up-counter using a set of known combinational circuit (CC) modules and a parallel-load register is presented.

Solution Figure 5.31 illustrates the data path of a bit-parallel mod-8 up-counter. The data path includes a 3-bit binary adder, a 2-to-1 3-bit MUX, and a 3-bit parallel-load register with an active-high asynchronous reset. The register is also designed using flip-flops with enable. The adder always outputs the register content plus 1. The signal *sc* (synchronous clear) controls the MUX and is used to synchronously initialize the counter to 0. On every clock cycle, the register, if enabled, loads the 3-bit output of the MUX, either the quantity $Z + 1$ if $sc = 0$ or zero if $sc = 1$.

The speed of the counter depends on the propagation delay of the adder. A carry propagate adder (CPA), for example, would introduce a longer signal delay, similar to that of a bit-serial circuit.

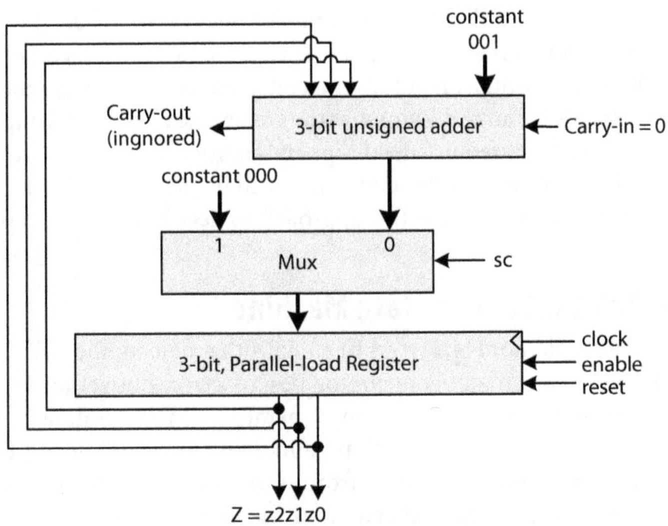

FIGURE 5.31 A synchronously cleared bit-parallel mod-8 up-counter.

On the other hand, one may use a faster adder, such as, a carry look-ahead (CLA) adder, to design a high-speed counter. Figure 5.32 illustrates the simplified circuit of a synchronously cleared mod-8 up-counter using a 3-bit CLA adder. The circuit in Fig. 5.32 is the result of further simplifications of the CLA's and the MUX's logic expressions using the constant operand $(001)_2$ for the CLA adder and the constant input $(000)_2$ for the MUX in Fig. 5.31.

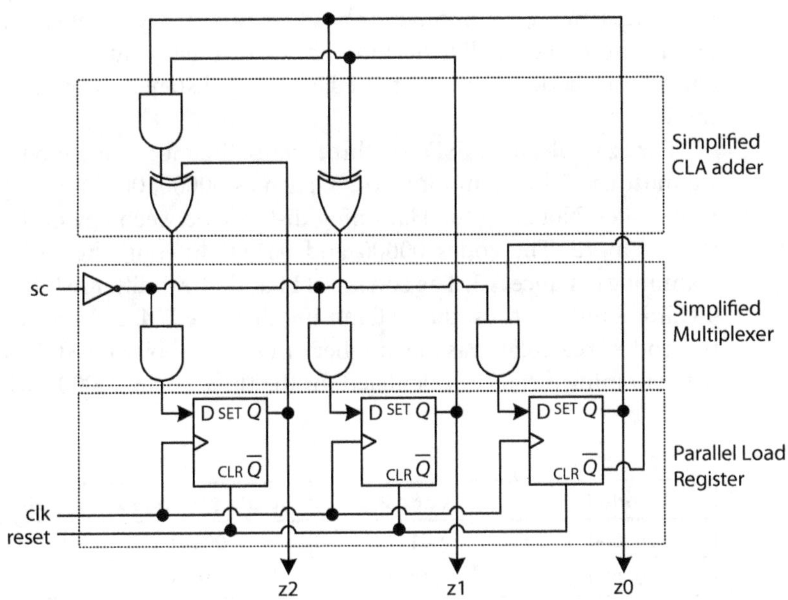

FIGURE 5.32 A synchronously cleared bit-parallel mod-8 up-counter using a simplified CLA adder and a simplified MUX.

In order to design a bit-parallel k-bit counter slice, the slice must include both the i and o interface signals that were discussed earlier in the design of the bit-serial counters. In addition, a bit-parallel (including a bit-parallel slice) solution, especially for large designs, has the advantage of not requiring an FSD. The technique, however, does require the designers' ability to determine, from the description of the design problem, the functions that will be performed by the combinational circuits. In Chap. 8, we will use this approach to design a CPU data path.

5.5 Fault-Tolerant Finite State Machine

A fault-tolerant FSM refers to an FSM that detects and corrects faults that occur, not because of a manufacturing or design error, but rather because of some random environmental hazard during operation. In general, such hazards affect storage elements like latches, flip-flops, and memory. The state of a flip-flop, indicated by its q bit, can suddenly change from 0 to 1 or vice versa. A fault can cause an FSM to transition to an invalid state and therefore cause circuit malfunction. A fault can cause a counter to suddenly output a wrong value, a sequence detector to recognize a wrong sequence, or skip and not recognize a right sequence, etc.

In general, faults can affect a single bit or multiple bits. However, single-bit faults are more common. A fault-tolerant FSM requires extra hardware to implement redundancy in the circuit and be able to detect and correct errors caused by faults. For example, a fault-tolerant FSM requires extra flip-flops and extra combinational circuits. The number of extra flip-flops depends on how many states there are originally in the FSD. For example, to design a single-bit fault-tolerant FSM, each binary state label must be different in at least 3-bits when compared to the other labels. In general, the number of bits that two binary numbers differ is called their **Hamming distance**. Furthermore, a set of binary numbers is called **Hamming code** if each code is at least a three Hamming distance away from any other code in the set.

For example, if an FSD has three states, then for a single-bit fault-tolerant FSM we must use 5-bit Hamming codes, such as 00000, 00111, and 11001, to label the three states. Note that the Hamming distance between any of the three 5-bit codes is 3 or more. The codes 00000 and 00111 differ in the first 3-bits; thus, their Hamming distance is 3. The codes 00111 and 11001 differ in bit numbers 1, 2, 3, and 4 (Little Endian); thus, their Hamming distance is 4. A Hamming distance of any two codes is calculated as the number of 1's in their bitwise XOR. Table 5.11 presents the Hamming distances calculated for the three codes 00000, 00111, and 11001.

Code 1	Code 2	Code 1 ⊕ Code 2	Hamming Distance
00000	00111	00111	3
00000	11001	11001	3
00111	11001	11110	4

Table 5.11 Hamming Distance between a Pair of Codes

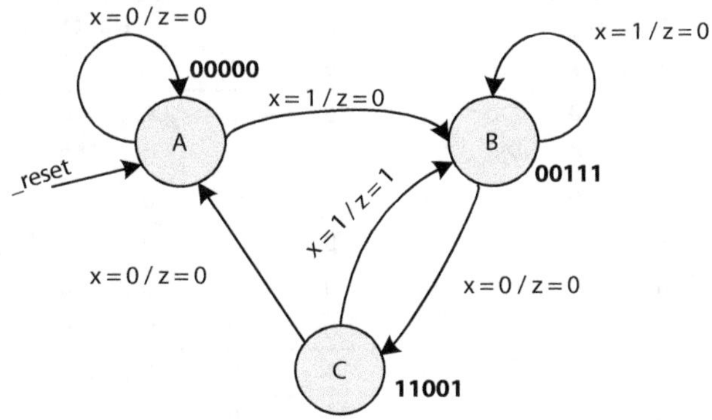

Figure 5.33 An FSD using 5-bit Hamming state codes as the state labels.

Suppose a fault causes bit 2 in the code 00000 to change from 0 to 1 and produce an invalid code 00100. The new code is 1 Hamming distance away from the valid 00000, but 2 or more distances away from each of the valid codes 00111 and 11001. Thus, the invalid code is closer to the valid code 00000 than the other two valid codes. Therefore, the error caused by the fault can be detected and corrected by replacing the invalid code 00100 with the valid code 00000.

Example 5.7 The design of a fault-tolerant FSM using the Mealy FSD given in Fig. 5.33 is presented. As shown in the figure, each state is labeled with a 5-bit Hamming code.

Solution Table 5.12 presents a combined truth table for the NSG and OG of the fault-tolerant FSM using five flip-flops. In the table, all the single-bit invalid current-state labels are interpreted the same as the corresponding valid-state label. For example, because each of the invalid labels 00001, 00010, 00100, 01000, and 10000 is 1 Hamming distance away from the valid code 00000, they all are interpreted as state A. In the first six rows in the table, if $x = 0$, then the next state label is 00000 (state A). Therefore, a single-bit error in any of the q_4, q_3, q_2, q_1, and q_0 signals will not alter the state of the FSM. The final circuit is not shown. However, one can use Espresso minimization software to minimize the truth table and obtain the required logic expressions for the circuit. Alternatively, the circuit can be modeled in Verilog using a "case" statement to enter the truth table (see Exercise 5.25).

A single fault can be simulated if one uses individually controlled input signals to reset or preset each flip-flop. For example, assuming that the current state is 00000, a fault can be introduced, say, in the q_2 bit by using the corresponding preset signal to change the state to 00100 (an invalid state). As indicated in Table 5.12, the FSM should correctly transition from this invalid state either to the valid state 00000 if $x = 0$ or to the valid state 00111 if $x = 1$. Thus, the circuit will detect and correct the single-bit error. Note that in this case, the detecting and correcting mechanism would be embedded in the NSG and OG modules. A technique that would convert a nonfault-tolerant sequential circuit to a single-bit fault-tolerant circuit without requiring the construction of a large truth table is discussed later.

In addition, in general, it is easy to come up with a small number of Hamming codes to design a small fault-tolerant FSM. However, it is more appropriate to use the **Hamming coding scheme** described next to generate as many Hamming codes as necessary for a given FSD of any size.

	Current State					Input	Next State					Output	
	q_4	q_3	q_2	q_1	q_0	x	d_4	d_3	d_2	d_1	d_0	z	
A	0	0	0	0	0	0	0	0	0	0	0	0	A
	0	0	0	0	1	0	0	0	0	0	0	0	
	0	0	0	1	0	0	0	0	0	0	0	0	
	0	0	1	0	0	0	0	0	0	0	0	0	
	0	1	0	0	0	0	0	0	0	0	0	0	
	1	0	0	0	0	0	0	0	0	0	0	0	
	0	0	0	0	0	1	0	0	1	1	1	0	B
	0	0	0	0	1	1	0	0	1	1	1	0	
	0	0	0	1	0	1	0	0	1	1	1	0	
	0	0	1	0	0	1	0	0	1	1	1	0	
	0	1	0	0	0	1	0	0	1	1	1	0	
	1	0	0	0	0	1	0	0	1	1	1	0	
B	0	0	1	1	1	0	1	1	0	0	1	0	C
	0	0	1	1	0	0	1	1	0	0	1	0	
	0	0	1	0	1	0	1	1	0	0	1	0	
	0	0	0	1	1	0	1	1	0	0	1	0	
	0	1	1	1	1	0	1	1	0	0	1	0	
	1	0	1	1	1	0	1	1	0	0	1	0	
	0	0	1	1	1	1	0	0	1	1	1	0	B
	0	0	1	1	0	1	0	0	1	1	1	0	
	0	0	1	0	1	1	0	0	1	1	1	0	
	0	0	0	1	1	1	0	0	1	1	1	0	
	0	1	1	1	1	1	0	0	1	1	1	0	
	1	0	1	1	1	1	0	0	1	1	1	0	
C	1	1	0	0	1	0	0	0	0	0	0	0	A
	1	1	0	0	0	0	0	0	0	0	0	0	
	1	1	0	1	1	0	0	0	0	0	0	0	
	1	1	1	0	1	0	0	0	0	0	0	0	
	1	0	0	0	1	0	0	0	0	0	0	0	
	0	1	0	0	1	0	0	0	0	0	0	0	
	1	1	0	0	1	1	0	0	1	1	1	1	B
	1	1	0	0	0	1	0	0	1	1	1	1	
	1	1	0	1	1	1	0	0	1	1	1	1	
	1	1	1	0	1	1	0	0	1	1	1	1	
	1	0	0	0	1	1	0	0	1	1	1	1	
	0	1	0	0	1	1	0	0	1	1	1	1	

TABLE 5.12 A Fault-Tolerant Combined NSG and OG Truth Table Determined from the FSD in Fig. 5.33

5.5.1 Hamming Coding Scheme

Hamming codes are useful for detecting and correcting a single-bit error or detecting a double-bit error. The errors can occur during data transmission in digital communication or in a data storage module such as a flip-flop or memory. Each Hamming code includes a certain number of **parity** bits and a certain number of data bits. A parity bit can be calculated as **even parity** or **odd parity**. For example, consider a 7-bit Hamming code. The bits would be numbered right to left from 1 to 7 with bits 1, 2, and 4 reserved for three even parity bits and bits 3, 5, 6, and 7 for four data bits. Equation (5.8) is used to compute the even parity bits p_1, p_2, and p_4 from the data bits d_3, d_5, d_6, and d_7. Now suppose $d_7 = 1$, $d_5 = 1$, and $d_3 = 0$. The parity bit p_1, as an even parity, must be 0 so that the number of 1's among d_7, d_5, d_3 and p_1 is an even number. If $d_7 = 1$, $d_5 = 1$, and $d_3 = 1$, then p_1 must be a 1.

$$\begin{aligned} p_1 &= d_3 \oplus d_5 \oplus d_7 \\ p_2 &= d_3 \oplus d_6 \oplus d_7 \\ p_4 &= d_5 \oplus d_6 \oplus d_7 \end{aligned} \tag{5.8}$$

Given the 4-bit data 1101, its 7-bit Hamming code is determined as follows using Eq. (5.8).

$$\text{Data bits: } d_3 = 1,\ d_5 = 0,\ d_6 = 1,\ d_7 = 1$$

Even parity bits:

$$\begin{aligned} p_1 &= 1 \oplus 0 \oplus 1 = 0 \\ p_2 &= 1 \oplus 1 \oplus 1 = 1 \\ p_4 &= 0 \oplus 1 \oplus 1 = 0 \end{aligned}$$

The 7-bit Hamming code is organized as $d_7d_6d_5p_4d_3p_2p_1 = 1100110$. The parity bit p_1 is determined from the data bits d_3, d_5, and d_7. These data bits are located in the bit positions $01\underline{1}$ (3), $10\underline{1}$ (5), and $11\underline{1}$ (7) in the Hamming code, respectively. Note that the first bit in each of the position numbers is 1, indicated by the underline. Likewise, the parity bit p_2 is determined from the data bits located in the bit positions $0\underline{1}1$ (3), $1\underline{1}0$ (6), and $1\underline{1}1$ (7). In general, a parity bit p_2k for $k = 0, 1, 2$, etc., is generated by XORing all the data bits that have a 1 in the kth bit of their respective position numbers. For instance, for $k = 2$, p_4 is determined by XORing the data bits at bit positions 5 ($\underline{1}01$), 6 ($\underline{1}10$), and 7 ($\underline{1}11$) because 5 is 4 + 1, 6 is 4 + 2, 7 is 4 + 3, where 4 is the common value among them.

The parity bits are used to determine the location of an error bit (if any). For example, consider the 7-bit Hamming code $d_7d_6d_5p_4d_3p_2p_1 = 1001100$. Suppose this Hamming code is transmitted wirelessly to a remote destination. Furthermore, suppose, the received Hamming code indicated as $d'_7d'_6d'_5p'_4d'_3p'_2p'_1 = (1\underline{1}01100)_2$ includes a single-bit error in bit d_6 (underlined). The received parity and data bits are then $p'_4p'_2p'_1 = 100$ and $d'_7d'_6d'_5d'_3 = 1\underline{1}01$. Using Eq. (5.8), the new parity bits,

p''_4, p''_2, and p''_1 are computed from the received data bits $d'_7d'_6d'_5d'_3 = 1\underline{1}01$ as follows:

$$p''_1 = d'_3 \oplus d'_5 \oplus d'_7 = 1 \oplus 0 \oplus 1 = 0$$

$$p''_2 = d'_3 \oplus d'_6 \oplus d'_7 = 1 \oplus 1 \oplus 1 = 1$$

$$p''_4 = d'_5 \oplus d'_6 \oplus d'_7 = 0 \oplus 1 \oplus 1 = 0$$

Equation (5.9) is used to determine the location of the bit in error as a 3-bit number $E = e_2e_1e_0$.

$$E = e_2e_1e_0 = p''_4\, p''_2\, p''_1 \oplus p'_4\, p'_2\, p'_1 \tag{5.9}$$

That is,

$$\begin{aligned} E = e_2e_1e_0 &= 010 \oplus 100 \\ &= 110 \\ &= 6 \text{ (the position of the bit in error)} \end{aligned}$$

The 6 indicates that the bit in location 6 (i.e., d_6) in the received Hamming code is in error and should be changed from 1 to 0 to yield the correct Hamming code 1001100 that was transmitted.

With an additional overall parity bit c, a Hamming code can also be used to detect, but not correct, a double-bit error. Equation (5.10) defines an overall even parity bit. The four parity bits c, p_4, p_2, and p_1 and the four data bits d_7, d_6, d_5, and d_3 create an 8-bit Hamming code. Table 5.13 presents some examples of Hamming code sizes.

$$c = d_7 \oplus d_6 \oplus d_5 \oplus p_4 \oplus d_3 \oplus p_2 \oplus p_1 \tag{5.10}$$

Table 5.14 presents the rules of the Hamming coding scheme (HCS) using the E, c' (the received overall parity bit), and c'' (the computed overall parity bit from the received data bits). When c' and c'' are equal and $E = 0$, then there are no faults in the received Hamming code. If the c' and c'' are the same but $E \neq 0$, then there is a double and uncorrectable error in the received code. If $c' \neq c''$ and $E = 0$, then the error is in c'. Finally, if $c' \neq c''$ and $E \neq 0$, then the E identifies the bit in error. If 3 or more bits are in error, the HCS will interpret the error (incorrectly) either as a single-bit or a double-bit error.

In Example 5.7, we selected the three Hamming codes 00000, 00111, and 11001 to encode the three states of the FSD in Fig. 5.33. Here, we will illustrate

Number of Parity Bits	Maximum Hamming Code Size
4 as p_1, p_2, p_4, and c	8 with 4 maximum data bits
5 as p_1, p_2, p_4, p_8, and c	16 with 11 maximum data bits
6 as p_1, p_2, p_4, p_8, p_{16}, and c	32 with 26 maximum data bits
7 as p_1, p_2, p_4, p_8, p_{16}, p_{32}, and c	64 bits with 57 maximum data bits
8 as p_1, p_2, p_4, p_8, p_{16}, p_{32}, p_{64}, and c	128 bits with 120 maximum data bits

TABLE 5.13 Sample Hamming Code Sizes

Received Overall Parity Bit vs. Computed Using the Received Data Bits $c' \oplus c''$	Error Indicator	Meaning
0	$E=0$	There are no errors
0	$E \neq 0$	A double-bit but uncorrectable error
1	$E=0$	The received overall parity bit c' is in error
1	$E \neq 0$	A single-bit correctable error at bit position E

TABLE 5.14 The Rules of the HCS (Hamming Code Scheme)

how the HCS can be used to systematically encode the states of an FSD with Hamming codes. Suppose the states A, B, and C in Fig. 5.33 are initially encoded with 2-bit binary numbers $s_1s_0 = 00$ for A, 01 for B, and 10 for C. By interpreting the 2-bit state labels as two data bits, Eq. (5.11) is used to compute the parity bits using the data bits $d_7 = 0$, $d_6 = 0$, $d_5 = s_1$ and $d_3 = s_0$.

$$
\begin{aligned}
p_1 &= s_0 \oplus s_1 \\
p_2 &= s_0 \\
p_4 &= s_1
\end{aligned}
\tag{5.11}
$$

For example, for $s_1s_0 = 01$, $p_1 = 1$, $p_2 = 1$, and $p_4 = 0$, the corresponding Hamming code is $s_1p_4s_0p_2p_1 = (00111)_2$. Table 5.15 presents a summary of the parity bit calculations. Note that the resultant Hamming codes are the same as those used in Example 5.7.

An alternative fault-tolerant FSM design to the one was discussed in Example 5.7 involves first designing the FSM as a nonfault-tolerant circuit and then incorporating in the circuit the additional circuits required to implement the Hamming error detection and correction mechanism. Consider, for example, the fault-tolerant FSM design problem in Example 5.7. First, the FSM is designed as a nonfault-tolerant FSM. This will require two flip-flops (two state bits), an NSG, and an OG. Suppose the two next state bits generated by the NSG are labeled s_1 and s_0. A fault-tolerant circuit would require six flip-flops that are associated with the two state bits s_1 and s_0, three parity bits p_1, p_2, and p_4, and an overall parity bit c.

Suppose the six flip-flop q bits are labeled q_0 to q_5. For a fault-tolerant design, the flip-flops would be interpreted as a transmission medium and a receiver. The d bits are "transmitted" through the flip-flops and then are "received"

	Initial Binary State Labels		Parity Bits			5-Bit Hamming Codes
State	s_1	s_0	p_1	p_2	p_4	$s_1p_4s_0p_2p_1$
A	0	0	0	0	0	00000
B	0	1	1	1	0	00111
C	1	0	1	0	1	11001

TABLE 5.15 The 5-Bit Hamming Codes Generated From the 2-Bit Numbers 00, 01, and 10

Flip-Flops' *d* Bits ("transmitted")	Flip-Flops' *q* Bits ("received")	Computed Parity Bits from the Received State Bits	Error Position
$d_0 = p_1$	$p'_1 = q_0$	$p''_1 = q_2 \oplus q_4$	$E = p''_4 p''_2 p''_1 \oplus p'_4 p'_2 p'_1$
$d_1 = p_2$	$p'_2 = q_1$	$p''_2 = q_2$	
$d_2 = s_0$	$s'_0 = q_2$	$p''_4 = q_4$	
$d_3 = p_4$	$p'_4 = q_3$	$c'' = q_4 \oplus q_3 \oplus q_2 \oplus q_1 \oplus q_0$	
$d_4 = s_1$	$s'_1 = q_4$		
$d_5 = c$	$c' = q_5$		

TABLE 5.16 Converting a FSM to a Fault-Tolerant FSM Using the Hamming Error Detection and Correction Mechanism

as the q bits. A single fault can switch one of the q bits and thus cause an error. On the transmission side, the six state bits (d's) would be connected to the s_1 and s_0 signals and the four parity bits p_1, p_2, p_4, and c. On the receiver side, the q's represent two current state bits labeled s'_1 and s'_0 and four received parity bits labeled p'_1, p'_2, p'_4, and c'. The relationships between the Hamming codes and the flip-flop inputs and outputs are summarized in Table 5.16.

An explicitly designed fault-tolerant FSM would require two additional modules: a parity generator (PG) module and an error detection and correction (EDC) module. The PG module would input s_1 and s_0 and would generate four even parity bits p_1, p_2, p_4, and c. The EDC would input the bits s'_1 and s'_0 (the "received" next state bits) and would generate the parity bits p''_4, p''_2, and p''_0. The bits p'_4, p'_2, and p'_0 and the p''_4, p''_2, and p''_0 would be used to compute the error bit position E (Eq. (5.9)) if $c' \neq c''$.

Using a 3-to-8 decoder, the E is decoded into one of seven output signals (1 through 7) for which only one could be active if $E \neq 0$. The active decoder signal (if any) would subsequently be used to correct the bit in error using an XOR gate if $c' \neq c''$. The resultant FSM, however, will have a longer propagation delay than the one designed using the truth table in Example 5.7 and SOP or POS expressions.

5.6 Sequential Circuit Timing

Flip-flops that share a common clock signal are expected to receive the sampling edge of the clock at approximately the same time so that all the flip-flops can sample their respective inputs simultaneously and before the arrival of the next sampling edge. Otherwise, as illustrated in Fig. 5.34, if there are delays in the transmission of the clock signal to the flip-flops, then it is possible that some of the flip-flops will receive the sampling clock edge much later than other flip-flops. For example, as shown in the figure, *clk*1 arrives after some delay (due to signal routing delay) as *clk*2 at the second flip-flop (FF2). This variation in the arrival times of a sampling clock edge at different flip-flops is known as **clock skew** (τ_{cs}).

A clock skew can cause many problems. The flip-flops that receive a sampling clock edge earlier will be able to sample their inputs, and thus change their outputs,

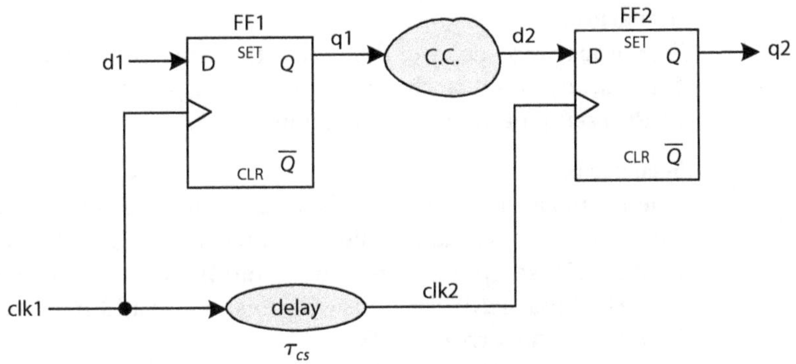

Figure 5.34 Illustrating a potential clock skew problem within one clock cycle; the d_2 signal could be changing if the sampling edge of *clk*2 arrives late at FF2.

before others will. As a result, this could cause the newly sampled inputs to modify all or some of the inputs for those flip-flops that have not yet completed sampling their inputs. This may, in turn, either cause a timing (setup-time or hold-time) violation or cause an invalid state transition resulting in a circuit malfunction.

In the figure, when FF1 receives the sampling edge of *clk*1, it samples d_1^{current} and changes q_1^{current} to q_1^{new}. When FF2 receives the sampling edge of *clk*2, it is supposed to sample d_2^{current} and change q_2^{current} to q_2^{new}. However, because of a clock skew, one of three scenarios may occur, as stated next and illustrated in Fig. 5.35.

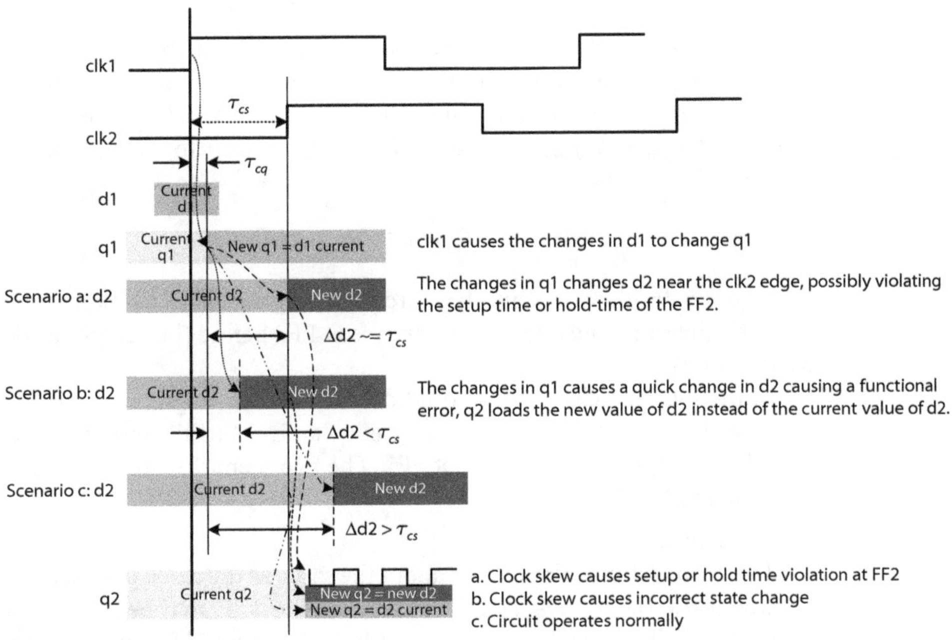

Figure 5.35 A timing diagram illustrating the effect of a clock skew within one clock cycle (Fig. 5.34 circuit).

Scenario a:
The propagation delay of d_2 is about the same as the clock skew ($\Delta d_2 \cong \tau_{cs}$). In this case, d_2 would be changing while FF2 is still sampling and, therefore, may cause a setup or hold-time violation at FF2.

Scenario b:
The propagation delay of d_2 is smaller than the clock skew ($\Delta d_2 < \tau_{cs}$). In this case, q_1^{new} would be able to change $d_2^{current}$ to d_2^{new} prior to the arrival of the *clk*2's sampling edge at FF2, and thus would cause FF2 to load d_2^{new} while FF1 loads $d_1^{current}$. This will cause an invalid state transition, resulting in a functional error.

Scenario c:
The propagation delay of d_2 is larger than the clock skew ($\Delta d_2 > \tau_{cs}$). In this case, the *clk*2's sampling edge will arrive at the FF2 before q_1^{new} can change the $d_2^{current}$ to d_2^{new}; thus, FF2 would load $d_2^{current}$ as it should for normal circuit operation.

The circuit in Fig. 5.34 includes only two flip-flops. In general, a circuit may contain many flip-flops, and the *d* input of each flip-flop may depend on one or more *q*'s. In this case, for the circuit to operate normally, the earliest time that a q^{new} can change a $d^{current}$ is the sum of the minimum clock-to-*q* time ($\tau_{cq\text{-min}}$) and the minimum circuit propagation delay ($\tau_{pd\text{-min}}$). This implies that the relationship in Eq. (5.12) must hold for the circuit to operate correctly:

$$\tau_{cs} < \tau_{cq\text{-min}} + \tau_{pd\text{-min}} \tag{5.12}$$

Otherwise, a $d^{current}$ would change too quickly, such as in the scenarios a and b in Fig. 5.35, and would either cause a setup or hold time violation (scenario a) or cause a functional error (scenario b). How close τ_{cs} can be to quantity $\tau_{cq\text{-min}} + \tau_{pd\text{-min}}$ can be defined in relation to the τ_{ht} (hold time) as shown in Eq. (5.13):

$$\tau_{cq\text{-min}} + \tau_{pd\text{-min}} - \tau_{cs} \geq \tau_{ht} \tag{5.13}$$

Figure 5.36 illustrates another circuit with possible clock skew problems. In this case, the sampling edge of *clk*1 arrives at FF2 before the sampling edge of *clk*2 arrives at FF1.

In the figure, a clock skew could create the following two possible scenarios between the time that *clk*2's sampling edge arrives at FF1 and the time that the next *clk*1's sampling edge arrives at FF2. The two scenarios are described next and illustrated in Fig. 5.37:

Scenario a:
The propagation delay of d_2 is about the same as the clock period (τ) minus the clock skew (i.e., ($\Delta d_2 \cong \tau - \tau_{cs}$). In this case, the d_2^{new} may be changing when the next sampling edge of *clk*1 arrives at FF2; therefore, d_2^{new} may cause a setup or hold-time violation at the FF2.

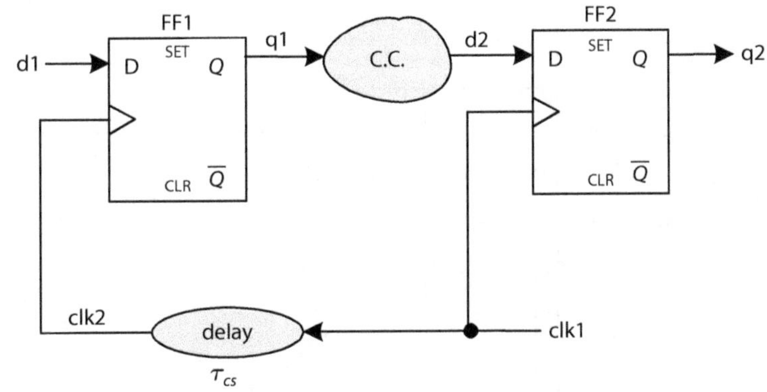

FIGURE 5.36 Illustrating a potential clock skew problem between a current sampling edge and the next sampling edge of *ckl*1; the d_2 signal could be changing when the next sampling edge of *clk*1 arrives at FF2.

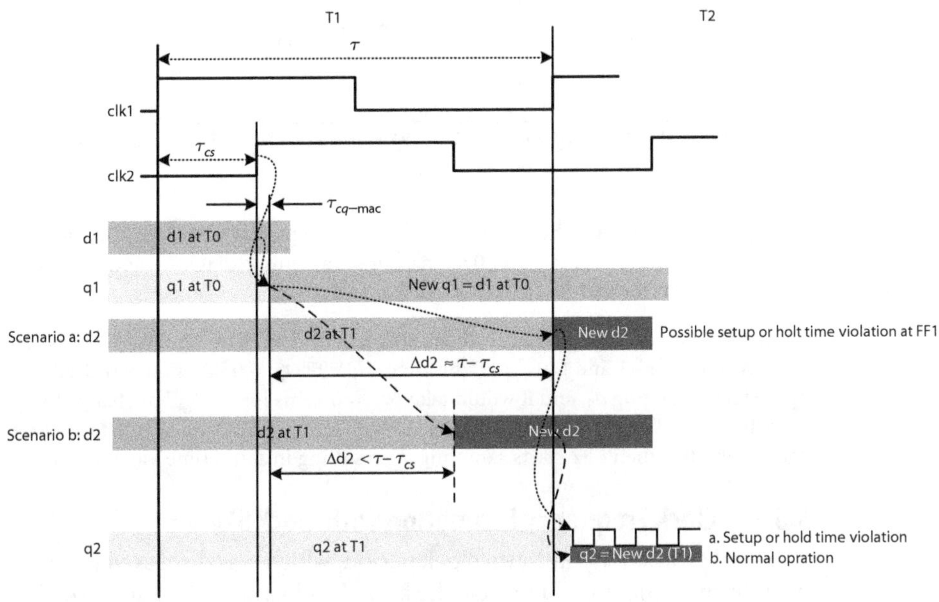

FIGURE 5.37 A timing diagram illustrating the effect of clock skew between the current and the next sampling edge of the *clk*1 (Fig. 5.36 circuit).

Scenario b:

The propagation delay of d_2 is smaller than the clock period (τ) minus the clock skew (i.e., $\Delta d_2 < \tau - \tau_{cs}$). In this case, d_2^{new} stabilizes prior to the arrival of the next *clk*1's sampling edge, and thus, FF2, as expected, would load d_2^{new}. Therefore, the circuit would operate normally.

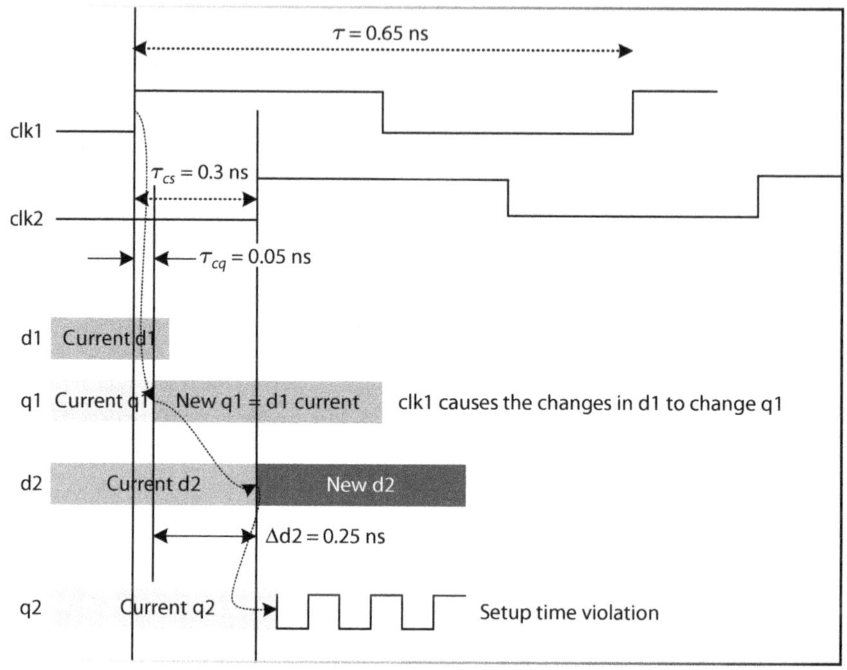

FIGURE 5.38 A timing diagram illustrating a setup or hold-time violation due to clock skew.

Example 5.8 Consider the circuit in Fig. 5.34. Suppose $\tau = 0.65$ ns, $\Delta_{CC} = 0.25$ ns, $\Delta_{delay} = 0.3$ ns, $\tau_{st} = 0.05$ ns, and $\tau_{cq} = 0.05$ ns. Draw a timing diagram and discuss if there is a problem due to clock skew.

Solution The timing diagram is illustrated in Fig. 5.38. The signal $d_1^{current}$ is sampled at the positive edge of *clk*1 and thus, q_1^{new} becomes $d_1^{current}$ $\tau_{cq} = 0.05$ ns after *clk*1 edge. At this time, q_1^{new} starts changing d_2, and it would take $\Delta_{CC} = 0.25$ ns for $d_2^{current}$ to change to d_2^{new} from the time that $q_1^{current}$ changes to q_1^{new}. This time is $\Delta_{CC} + \tau_{cq}$, or, in this case, 0.3 ns. Therefore, d_2 will change exactly when FF2 starts sampling d_2, resulting in setup time violation at FF2.

5.6.1 Clock Frequency Estimation with Clock Skew

The total estimated minimum clock period, which was discussed in Chap. 4, did not include the time lost due to clock skew. As illustrated in Fig. 5.37, the total time needed for a signal to stabilize before the arrival of the next clock edge can be reduced by an amount equal to the clock skew. This implies that the minimum clock period must include the delay caused by clock skew, as given in Eq. (5.14) [1].

$$\tau \geq \tau_{cq-\max} + \tau_{pd-\max} + \tau_{st} + \tau_{cs} \tag{5.14}$$

5.6.2 Asynchronous Interface

Sequential circuits that depend on external inputs, such as d_1 in Fig. 5.36, expect that an external signal always changes at the right time with respect to the

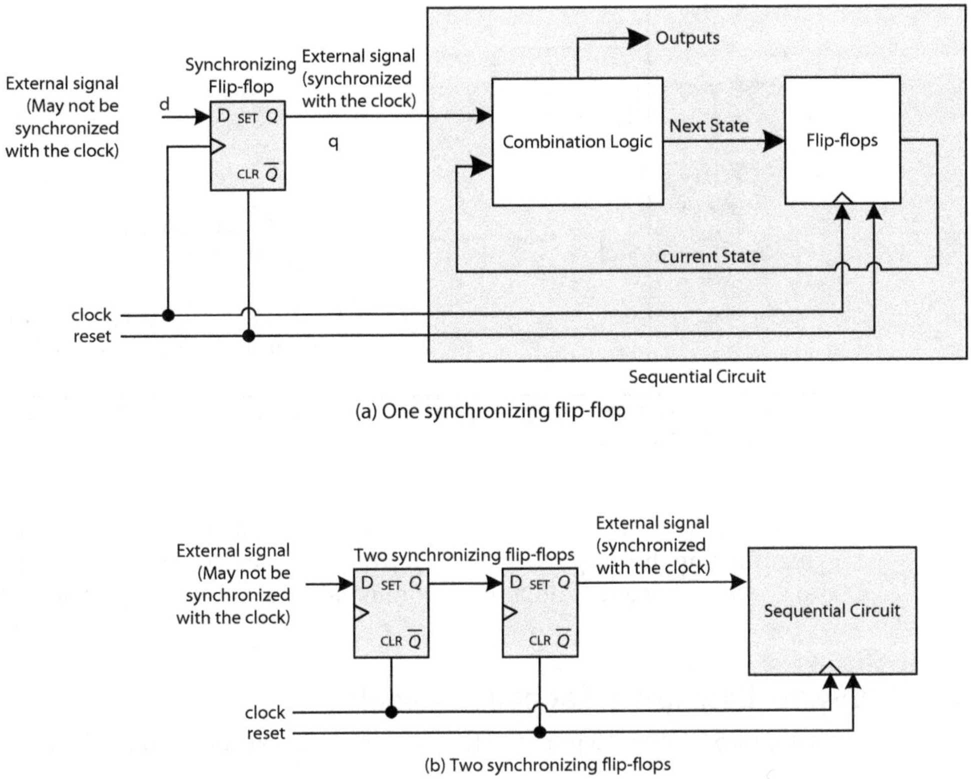

Figure 5.39 External input synchronization [1]: (*a*) using one synchronization flip-flop; (*b*) using two flip-flips to allow the maximum metastability resolution time.

sampling edge of the clock. However, an external signal may change at any time if it is generated by an input device, for example, a keyboard, or it is the output of another sequential circuit that uses a different clock source. As a result, this may violate the setup or hold time of the sequential circuit's flip-flops, causing metastability and possibly a malfunction.

A recommended solution [2–3] for resolving this problem is to sample the external inputs before they are fed into the target sequence circuit, as illustrated in Fig. 5.39(*a*). In the figure, the external input is fed into a **synchronizing flip-flop**. It is assumed that any possible metastability caused by the input will be resolved by the synchronizing flip-flop before the next clock edge. That is, if the input violates the setup or the hold time of the synchronizing flip-flop and causes the flip-flop's output to oscillate (metastability), it is expected that the oscillating output would stabilize to 1 or 0 before the arrival of the next clock edge. Therefore, the sequential circuit will input the synchronized signal and potentially will avoid a metastability of its own. This is illustrated by an example timing diagram shown in Fig. 5.40 using the circuit in Fig. 5.39(*a*). Furthermore, a synchronizing flip-flop may be designed so its outputs stabilize quickly when the external input value changes at the wrong time, violating the flip-flop's setup or hold time [4].

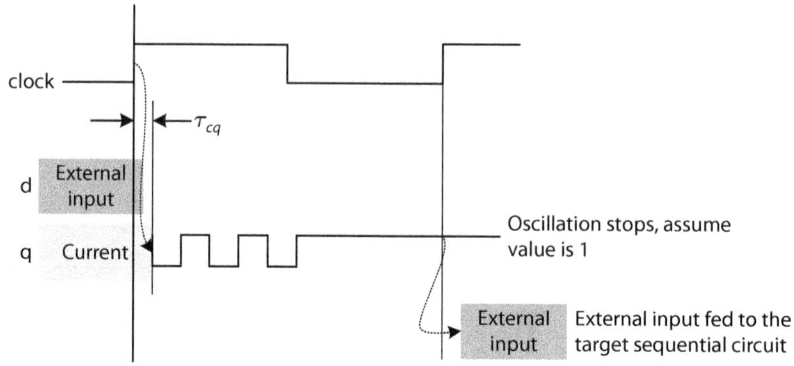

FIGURE 5.40 An example timing diagram illustrating the resolution of a metastability due to an external input.

However, in order to avoid a potential metastability at the output of the synchronizing flip-flop in Fig. 5.39(*a*) from ever entering the sequential circuit, two synchronizing flip-flops are recommended, as illustrated in Fig. 5.39(*b*).

5.7 Hardware Description Language Models

Examples 5.9 and 5.10 present HDL models for Mealy and Moore FSMs.

Example 5.9 A Verilog behavior model for the Moore sequence recognizer in Example 5.1 that detects the overlapping sequence "101" is presented, where the code is divided into three sections as follows:

Code section 1: A behavior description of the NSG. It describes the arcs of the recognizer's FSD. The FSD consists of four states labeled A to D.

Code section 2: A behavior description of the OG. It describes the state in which the recognizer outputs a 1, signaling the detection of a "101" sequence.

Code section 3: A behavior description of the flip-flops with asynchronous reset capability. Upon reset, the FSM is initialized to the initial state A.

HDL Model

```
//A Moore sequence recognizer that detects the overlapping
//sequence "101".
//Using binary encoded state labels
module moore_seq
(
        input clock, reset, x,
        output reg z
);
```

```
//assign binary encoded codes to the states A through D
parameter     A = 2'b00,
              B = 2'b01,
              C = 2'b10,
              D = 2'b11;

reg [1:0] current_state, next_state;

//Section 1: Next state generator (NSG)
always@(*)
begin
    casex  (current_state)  //ignore  unknown  and  high
                            //impedance (Z) inputs

    A:  if (x == 1)
                next_state = B;
        else
                next_state = A;
    B:  if (x == 1)
                next_state = B;
        else
                next_state = C;
    C:  if (x == 1)
                next_state = D;
        else
                next_state = A;
    D:  if (x == 1)
                next_state = B;
        else
                next_state = C;
    endcase
end

//Section 2: Output generator (OG)
always@(*)
begin
    if(current_state == D)
        z = 1;
    else
        z = 0;
end
```

```
//Sections 3: The flip-flops
always@(posedge clock, posedge reset)
begin
     if (reset == 1)
           current_state <= A;
     else
           current_state <= next_state;
end
endmodule
```

Simulation Test-Bench

```
`include "moore_seq.v"
module tester();
reg clock, reset, x;
wire z;
moore_seq      u1(clock, reset, x, z);
initial begin
$monitor("%4d: z = %b", $time, z);
clock = 0;
reset = 1;     //reset the flip-flops
x = 0;
#10 reset = 0; //end reset
end
always
begin
#5clock = ~clock;  //generates a clock signal with period 10
end
initial begin  //one input per clock cycle
#10 x = 1; $display("%4d: x = %b", $time, x);
#10 x = 1; $display("%4d: x = %b", $time, x);
#10 x = 1; $display("%4d: x = %b", $time, x);
#10 x = 0; $display("%4d: x = %b", $time, x);
#10 x = 1; $display("%4d: x = %b", $time, x);
#10 x = 0; $display("%4d: x = %b", $time, x);
#10 x = 1; $display("%4d: x = %b", $time, x);
#10 x = 1; $display("%4d: x = %b", $time, x);
#10 x = 0; $display("%4d: x = %b", $time, x);
#10 x = 0; $display("%4d: x = %b", $time, x);
#10 $finish;
end
endmodule
```

Simulation Output

The functional simulation output for the Moore FSM is shown next. The Moore signal *z* becomes 1 at the simulation time slots 55 and 75, indicating that there were two "101" sequences in the test vector.

```
Chronologic VCS simulator copyright 1991-2009
Contains Synopsys proprietary information.
Compiler version D-2009.12; Runtime version D-2009.12;

  0:             z = 0
 10: x = 1
 20: x = 1
 30: x = 1
 40: x = 0
 50: x = 1
 55:             z = 1
 60: x = 0
 65:             z = 0
 70: x = 1
 75:             z = 1
 80: x = 1
 85:             z = 0
 90: x = 0
100: x=0
$finish called from file "tester.v", line 32.
$finish at simulation time        110
```

Example 5.10 A Verilog behavior model for the Mealy sequence recognizer in Example 5.2 that recognizes the overlapping sequence "101" is presented, where the code consists of two sections as follows:

Code section 1: A behavior description for the combined NSG and OG. However, note that a combined code for some larger designs may create synthesizing problems, especially when PLDs with restricted hardware resources are used. The code describes the Mealy FSD. The FSD consists of only three states, labeled A, B, and C.

Code section 2: A behavior description of the flip-flops with asynchronous reset capability. Upon reset, the FSM is initialized to its initial state A.

HDL Model

```
//A Mealy sequence recognizer that detects the overlapping
//sequence "101"
//Using binary encoded state labels
module mealy_seq
(
        input clock, reset, x,
        output reg z
);
```

```
parameter    A = 2'b00,
             B = 2'b01,
             C = 2'b10;

reg [1:0] current_state, next_state;

//Section 1: A combined next state generator (NSG) and
output generator (OG)
//unknown states are ignored
always@(*)
begin
    casex(current_state)
    A:  if (x == 1) begin
                next_state = B;
                z = 0;
        end
        else begin
                next_state = A;
                z = 0;
        end
    B:  if (x == 1) begin
                next_state = B;
                z = 0;
        end
        else begin
                next_state = C;
                z = 0;
        end
    C:  if (x == 1) begin
                next_state = B;
                z = 1;
        end
        else begin
                next_state = A;
                z = 0;
        end
    default: begin
                next_state = 2'bxx;
                z = 1'bx;
        end
    endcase
end
```

```
//Section 2: flip-flops
always@(posedge clock, posedge reset)
begin
     if (reset == 1)
           current_state <= A;
     else
           current_state <= next_state;
end
endmodule
```

Simulation Test-Bench

The test-bench is the same as the one given in Example 5.9, with the exception that "mealy_seq.v" is instantiated instead.

Simulation Output

The functional simulation output for the Mealy FSM is given next. Note that, in this case, the Mealy output *z* depends on the current state as well as the input *x*. If *x* changes, so might *z*. On the other hand, in the case of the Moore machine, the input *x* affects the Moore output *z* on the next clock cycle. Here, Mealy-*z* is 1 at the same time as when the last bit of the target sequence "101" is entered at the simulation times 50 and 70.

```
Chronologic VCS simulator copyright 1991-2009
Contains Synopsys proprietary information.
Compiler version D-2009.12; Runtime version D-2009.12;

  0:           z = 0
 10: x = 1
 20: x = 1
 30: x = 1
 40: x = 0
 50:           x = 1
 50:           z = 1
 55: z = 0
 60: x = 0
 70: x = 1
 70:           z = 1
 75:           z = 0
 80: x = 1
 90: x = 0
100: x = 0
$finish called from file "testmealy.v", line 32.
$finish at simulation time                  110
```

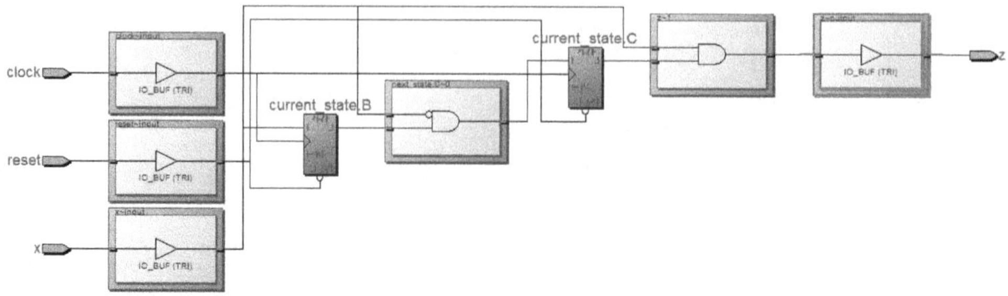

Figure 5.41 The synthesized circuit of the Mealy sequence recognizer of Example 5.10.

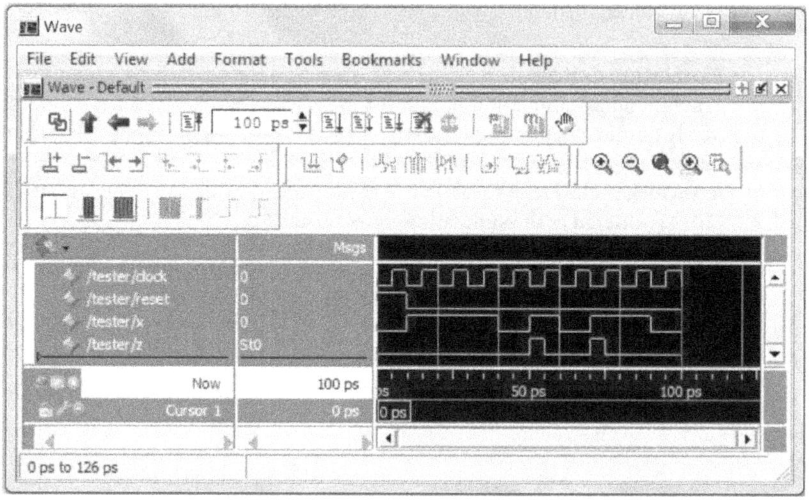

Figure 5.42 A simulation waveform for the synthesized Mealy sequence recognizer in Fig. 5.41.

5.7.1 Synthesis and Simulation

The Verilog model for the Mealy sequence recognizer in Example 5.10 was synthesized and simulated using the Altera Quartos II and ModelSim design and simulation tools. The synthesized circuit is shown in Fig. 5.41 and its simulation waveform is shown in Fig. 5.42. As illustrated in the timing diagram, signal z becomes 1 each time x inputs indicates a "101" sequence.

References

1. E. G. Friedman, ed., *Clock Distribution Networks in VLSI Circuits and Systems*, IEEE Press, 1995.
2. Wakerly, J. F., *Digital Design: Principles and Practices*, 4th ed., Prentice Hall, 2006.
3. Cypress Semiconductor, "Are Your PLDs Metastable?" May 1992, Revised March 6, 1997.

4. Ryan Donohue, Synchronization in digital logic circuits, Lecture notes, Stanford University.
5. Shomit Das, Comparison of synchronization techniques in pointer FIFOs, Technical report, University of Utah, 2009.

Exercises

5.1. Use Espresso to minimize the truth tables for the FSD in Fig. 5.7 and draw the minimized circuit. Is your solution the same as that shown in Fig. 5.8?

5.2. Consider a 1-bit 4-to-1 MUX like the one given in Fig. 5.10. Use the MUX and design a 1-bit four-function register slice that performs synchronous clear when $F = f_1 f_0 = 0$, parallel-load when $F = 1$, arithmetic right shift when $F = 2$, and right shift with a left input (li) when $F = 3$. Then, use four copies of the slice to draw the details of a 4-bit four-function register showing all the signal connections to each MUX.

5.3. Determine the minimum clock period for the bit-serial multifunction register given in Fig. 5.9 where τ_{st} and τ_{cq} are each assumed to be 0.1 ns and delay for NOT and NAND gates are each 0.1 ns.

5.4. Determine the minimum clock period for the bit-parallel multifunction register given in Fig. 5.11 in terms of τ_{st}, τ_{cq}, and delay of a NAND gate, and assuming only 2-to-1 MUXs are available.

5.5. Refer to Exercise 5.2. This time, design the 4-bit register directly from a 4-bit 4-to-1 MUX and a 4-bit parallel-load register.

5.6. Design an 8-bit multifunction register with asynchronous reset that performs parallel load, circular right shift, or circular left shift.

5.7. Verify that the circuit in Fig. 5.16 works correctly by constructing a truth table with current state bits q_1 and q_0, input x, next state bits d_1 and d_0, and output z. After reset, the flip-flops initialize to $q_1 = 0$ and $q_0 = 0$. Enter these values in the table as they indicate the current state of the FSM. Next, set $x = 1$ and determine the values of d_1, d_0, and z using the expressions in Eq. (5.2). Enter these values in the table. Now assume the clock signal makes a 0-1 transition and thus changes the values of q_1 and q_0 to those of d_1 and d_0. Enter the new values of q_1 and q_0 in the table, and repeat the process for the next values of x in order as 1, 0, 1, 0, 1, and 0. From the z values, determine if the circuit works correctly.

5.8. Design a Moore sequence recognizer that detects the nonoverlapping sequence "101." Use binary encoded state labels and design and draw the circuit schematic similar to the one shown in Fig. 5.16.

5.9. Design a Mealy sequence recognizer that detects the nonoverlapping sequence "101." Use binary encoded state labels and draw the circuit schematic similar to the one shown in Fig. 5.16.

5.10. Design a Moore sequence recognizer that detects the overlapping sequence "1001." Use binary encoded state labels.

5.11. Design a Mealy sequence recognizer that detects the overlapping sequence "1001." Use binary encoded state labels.

5.12. Consider the FSD in Fig. 5.13. Use the binary encoded state labels 11, 01, 10, and 00, in order, for the states A to D. Draw the circuit, making sure that upon reset, the FSM starts at state 11 (A). Compare the size of the combinational circuits with those in Fig. 5.15 and Fig. 5.17.

5.13. Design a Mealy sequence recognizer that detects the overlapping sequence "1001." Use one-hot state labels and use Espresso to minimize the combined truth table.

5.14. Formally design a JK flip-flop using a D flip-flop (also see Chap. 4).

5.15. Simulate the following circuits modeled in Verilog as specified:

a. Design the FSM in Example 5.1, but use the expressions given in Eq. (5.2).
b. Design the FSM in Example 5.1 by directly describing the FSD.

5.16. Consider the FSD in Fig. 5.22. Design the corresponding FSM using binary encoded labels and compare the circuit size with the one in Fig. 5.15.

5.17. Design a bit-serial mod-11 counter (also refer to Fig. 5.28) with asynchronous active-low reset.

5.18. Design a bit-parallel mod-11 counter (also refer to Fig. 5.31) with asynchronous active-low reset.

5.19. Simulate the following circuits modeled in Verilog as specified:

a. Model the 1-bit binary counter slice given in Fig. 5.28 and then use it to design the counter in Exercise 5.17.
b. Use behavioral models for an adder, an MUX, and a parallel-load register and then use them to design the counter in Exercise 5.18.
c. A complete behavioral model for the counter in Exercise 5.18.

5.20. Design a mod-4 up/down counter (not a counter slice) with asynchronous active-low reset.

5.21. Design a mod-4 up/down counter (not a counter slice) with both synchronous and asynchronous reset capabilities.

5.22. Design a 3-bit gray-code counter with both synchronous and asynchronous reset signals.

5.23. Non-return-to-zero inverted (NRZI) is a data coding scheme used to communicate with universal serial bus (USB) devices. The output signal (z) of an NRZI generator transitions when the input bit (x) is 0 and remains at the constant previous value (0 or 1) when the input bit is 1. That is, from right to left, when the input to the NRZI generator is 0 0 0 0 0 0, its output from right to left will transition as 1 0 1 0 1 0. Its output for consecutive 1's at the input, however, will remain at the previous output value. For example, the NRZI generator outputs from right to left z: 0 0 0 0 0 1 0 1 1 1 1 0 1 0 1 1 for input X: 1 1 1 1 0 0 0 1 1 1 0 0 0 0 1 1 read from right to left. Likewise, for X B 0xCF0C, Z: 0xEFAE. Design the NRZI generator. (Hint: design a Mealy FSM).

5.24. Suppose an FSD has five states. Use the Hamming coding scheme and generate five Hamming codes to label the states. Each pair of labels should have a Hamming distance of 3 or more.

5.25. Create and simulate a Verilog model for the fault-tolerant FSM in Example 5.7. Specifically, copy the entries of Table 5.12 to an Excel sheet. Then use the Excel "concatenate" function to concatenate bits q_4 to q_0 and x in each row into a 6-bit binary number. Do the same for bits d_4 to d_0 and z. (You may also use the Excel "bin2hex" function to convert the concatenated 6-bit numbers to a 6-bit hex). Then sort the table so the rows of the table for the current state bits and x in the Excel sheet are in ascending order. Then create two columns and use concatenation to convert the two 6-bit entries to a syntactically correct Verilog statement to be used in a "case" statement. For example, 000000 for the

q's and x is written as "6'b000000:" and 000000 for the d's and z is written as "{d, z} = 6'b000000;" where d would be declared as a 5-bit next state variable in Verilog. Copy the two columns from Excel into a Verilog text editor and model the FSM using a combined model for the NSG and OG modules. For simplicity, model the flip-flops with a common reset but separate preset signals. In the test-bench, use the preset signals to cause a 1-bit fault; that is, change a $q = 0$ to 1. The FSM should continue operating as if there were no faults.

```
...
always@(*) //NSG and OG
begin
case({q, x}) //q is declared as 5-bit current state variable
6'h000000: {d, z} = 6'h000000;   //1st row of the table
...
default: {d, z} = 6'hx;
endcase
end
```

5.26. Design a fault-tolerant mod-4 up-counter (not a counter slice).

5.27. Consider the circuit given in Fig. 5.15. Without altering its combinational circuits, use the Hamming error detection scheme and add modules to the circuit so it would operate as a single-bit fault-tolerant FSM.

5.28. A serial adder inputs 2-bits x and y and outputs their sum-bit s every clock cycle. It keeps the presence or the absence of a carryout bit internally. Assuming that the initial carry-in is zero, do the following:

a. Draw a Mealy FSD for the serial adder.
b. Design the Mealy serial adder FSM.
c. Draw a Moore FSD for the serial adder.
d. Design the Moore serial adder FSM.
e. Design a fault-tolerant Mealy serial adder.

5.29. Consider the circuit in Fig. 5.34. Suppose Δ_{CC} = 0.3 ns, τ_{sc} = 0.2 ns, τ_{st} = 0.05 ns, τ_{cq} = 0.05 ns. Assuming that τ = 0.6 ns, draw a timing diagram and discuss if there can be problems in operating the circuit due to clock skew.

5.30. Consider the circuit in Fig. 5.36. Suppose Δ_{CC} = 0.3 ns, τ_{sc} = 0.2 ns, τ_{st} = 0.05 ns, and τ_{cq} = 0.05 ns. Assuming that τ = 0.6 ns, draw a timing diagram and discuss if there can be problems in operating the circuit due to clock skew.

5.31. Consider the circuit in Fig. 5.36. Suppose Δ_{CC} = 0.3 ns, τ_{sc} = 0.2 ns, τ_{st} = 0.05 ns, and τ_{cq} = 0.05 ns. Assuming τ = 0.7 ns, draw a timing diagram and discuss if there can be problems in operating the circuit due to clock skew.

5.32. Consider a hardware FIFO buffer with input and output ports accessed by two sequential circuits A and B that perform concurrent computations [4–5]. Circuit A generates IN as the next buffer location to write a value to, and circuit B generates OUT as the next buffer location to read a value from. Two counters, X and Y, each controlled by one of the circuits, are used to generate the IN and OUT values, respectively. The buffer operates in a circular fashion. The two circuits, and thus their respective counters, also operate with two different clocks. Circuits A and B also need to input IN and OUT values in order to determine when the buffer is full or empty. Do the following:

a. Draw the block diagrams of circuits A and B, counters, and the buffer with labels and signal names. IN is an asynchronously input to circuit B, and OUT is an asynchronous input to circuit A. Use two synchronizing flip-flops for each bit.
b. Suppose the buffer size is 8 and X and Y are designed as mod-8 counters. Assuming that IN is 6, list the possible values circuit B may input as IN from the synchronizing flip-flops. Discuss how the buffer empty flag can be generated if input IN to circuit B is not 3.
c. Repeat part (b), but this time assume X and Y are designed as Gray code counters.

5.33. Computer security (hardware Trojans): See Exercise 11.12 to understand single-input trigger computational malicious circuits (also see Sec. 11.2).

5.34. Computer security (hardware Trojans): See Exercise 11.13 to understand timer attack malicious circuits (also see Sec. 11.2).

5.35. Computer security (confidentiality): See Exercise 11.14 to design a hardware encryption circuit (also see Sec. 11.5).

5.36. Computer security (computer security threats): See Exercise 11.15 to understand computer security threats (also see Sec. 11.1.3).

5.37. Computer security (hardware developmental threats): See Exercise 11.16 to understand homomorphic encryption as a hardware developmental security policy mechanism (also see Secs. 11.1.3 and 11.2).

CHAPTER 6

Sequential Circuits: *Large Designs*

6.1 Introduction

A large sequential circuit is made of a data path and a control unit, as illustrated in Fig. 6.1. The data path consists ofboth sequential and combinational circuit modules, such as registers, counters, multiplexers (MUXs), decoders, arithmetic logic unit (ALUs), and others, that are either standard or problem specific. The modules collectively implement a list of simple operations, such as adding the content of two registers and storing the result in a third register. The control unit is responsible for asserting the necessary control signals for the data path to carry out an operation. A data path may perform one or more operations during each clock cycle.

Each data path operation requires one or more inputs as data and generates one or more outputs, where each is then stored in a register or memory. An operation may also use one or more combinational circuit modules to compute an output. Two or more operations may share some or all of the combinational circuit modules in the data path. For example, two operations that compute a sum but use data values read from different sources may share an adder to reduce hardware.

A data path may also perform one or more operations conditionally, depending on the value of a signal either internal to the data path or generated by another module external to the data path. Arithmetic overflow flag, a specific register bit value, and a specific counter value are examples of signals internal to the data path that indicate a condition. External event-triggering signals, such as a signal starting the control unit, a signal generated by a keystroke, and a signal indicating a memory data is available to read, are examples of conditions indicated by external signals to the data path.

There are alternative architectures to design data paths and control units. Many factors, including operating clock frequency and how often results should be generated, affect the architecture of a data path. A high clock frequency implies that the data path has a short maximum propagation delay, which determines how

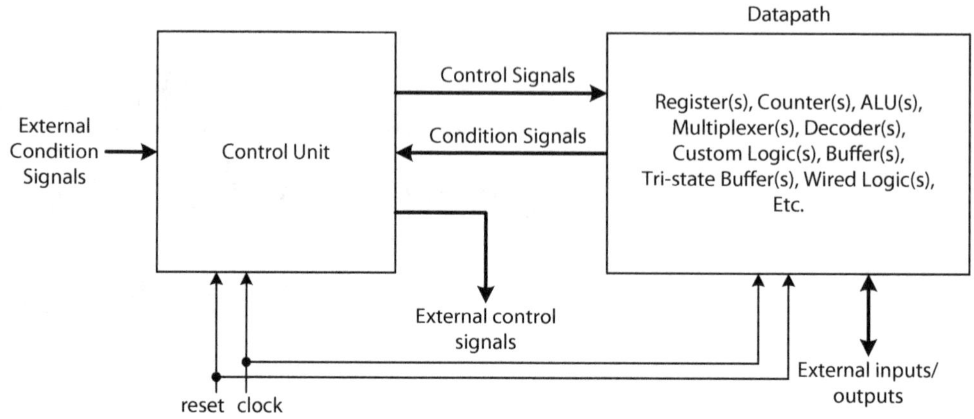

FIGURE 6.1 A diagram block of a large sequential circuit.

fast results can be generated. However, there is a relationship between a circuit's clock frequency and its number of transistors with the amount of power the circuit consumes, which determines how much heat the circuit can dissipate. As was mentioned in Chap. 1, there is a limit to how fast a sequential circuit may operate and still remain within the allowable temperature range using a fan cooling system. In practice, the amount of heat an integrated circuit (IC), such as a processor, can dissipate on average sets a limit on how high its clock frequency can be and how many transistors it can contain using current chip technologies.

In this chapter, we examine different data path and control unit architectures, illustrate their organizations, and estimate performance parameters. We also present power and energy usage models for a complex sequential circuit, discuss how such models may be used to reduce power consumption, and how to estimate the energy efficiency of a complex sequential circuit.

6.1.1 Register Transfer Notation

A register transfer notation (RTN) is used to formally describe an operation of a data path. Each operation generates a result that must be stored in a storage module such as a register or memory. The syntax for RTN is arbitrary. For example, $R3 \leftarrow R1 + R2$ that involves three registers, $R1$, $R2$, and $R3$, is an RTN. The left arrow ($\leftarrow$) indicates that the sum of the two register contents will be stored in register $R3$ during the next clock cycle. Table 6.1 presents some RTN syntax examples used in this book. Some of the syntax is borrowed from the Verilog hardware description language (HDL).

6.2 Data Path Design

The architecture of a data path can be classified as **single-cycle**, **multicycle**, or **pipelined**. A single-cycle data path requires more hardware but a simpler control unit. A multicycle data path requires less hardware but generates results in steps

RTN Example	Meaning
$R \leftarrow value$	Describes a register loading. The register *R* loads the *value*.
$CNTR \leftarrow CNTR + 1$ Or *CNTR* <= *CNTR* + 1; (Verilog)	Describes incrementing a counter named CNTR.
$R \leftarrow 0//R[7..1]$ Or *R* <= *R* >> 1; (Verilog) Or *R* <= {0, *R*[7:1]}; (Verilog)	Describes a register right shift with 0 fill. The symbol "//" is used here to indicate a concatenation.
$R \leftarrow R[7]//R[7..1]$ Or *R* <= *R* >>> 1; (Verilog) Or *R* <= {*R*[7], *R*[7:1]}; (Verilog)	Describes an arithmetic right shift.
$R3 \leftarrow R1 + R2$ Or *R3* <= *R1* + *R2*; (Verilog)	Describes adding the contents of two registers R1 and R2 and storing the sum in R3.
$M[X] \leftarrow R$;	Describes a memory transfer. The content of register R is stored in memory location X.

Table 6.1 RTN Syntax Examples Using Both an Arbitrary Syntax and Verilog HDL Syntax

using several clock cycles. A pipelined data path also requires more hardware but can operate on multiple inputs concurrently. Pipelining is only efficient when there are many inputs to process.

Example 6.1. The design and performance of single-cycle, multicycle, and pipelined data paths performing one or more RTN operations of the type $R \leftarrow A + B + C \pm D$ is presented. *A* to *D* represent values read simultaneously from some registers or memory. The pipeline data path, however, will perform $R_i \leftarrow A_i + B_i + C_i \pm D_i$ for $i = 0, 1, 2, 3$, etc.

The RTNs $R \leftarrow A + B + C + D$ and $R \leftarrow A + B + C - D$ do not represent operations a typical CPU instruction performs, as we will see in Chap. 8. Such RTNs would be considered fused operations, requiring computations performed on three or more data values. A typical arithmetic instruction operates on two data values. Here, the RTNs are used to illustrate and compare single-cycle, multicycle, and pipelined data path designs. However, some CPUs (e.g., [1]) have instructions that perform **fused operations** on three data values, such as multiply-add ($R \leftarrow A + B * C$), a common operation used in computations involving matrices. If *A*, *B*, and *C* are floating-point (FP) numbers, a fused operation has the advantage of producing a result in memory with only one rounding error (see Sec. 3.8.3). Done separately with two instructions, the result of $B * C$, if stored in memory, will result in one rounding error, and the result of adding *A* to the memory content, again if stored in memory, will result in another rounding error.

Other examples include custom instructions that perform fused operations as proposed in the design of configurable CPUs [2]. In this case, the dependent operations performed by a set of instruction sequences within a program loop can be combined into a single custom instruction with fused operations. The new instruction replaces the instruction sequence within the loop, and thus increases performance by reducing the number of instructions that must be fetched from memory.

The SIMD architecture, which was discussed in Chap. 1, is another variation of instructions that operate on multiple data values. In this case, however, each SIMD instruction specifies only one operation that is performed simultaneously and with no data dependency on multiple data values. The design of control units for each of the three data paths is discussed in Sec. 6.4.

6.2.1 Single-Cycle

Figure 6.2 illustrates a single-cycle data path that computes either the quantity $A + B + C + D$ or $A + B + C - D$ and stores the result in register R within one clock cycle. The data path contains two adder (+) modules and one adder/subtractor (+/–) module. The signal *mode* controls the functions of the adder/subtractor module. If $mode = 0$, the data path performs $R \leftarrow A + B + C + D$; otherwise, it performs $R \leftarrow A + B + C - D$.

Equation (6.1) estimates the minimum clock period required to run the data path. The period is proportional to the propagation delay of the longest signal path that starts from the inputs of the first adder and ends at the input of the register.

$$\tau \geq 2\Delta_{\text{ADD}} + \Delta_{\text{ADD/SUB}} + \tau_{st} + \tau_{cq} + \tau_{cs} \tag{6.1}$$

In general, if a single-cycle data path implements several simple and complex operations, its minimum clock period would be proportional to the time required to complete the most complex operation. Therefore, both simple and complex operations would each require the same amount of time to complete. This will increase the total time needed to complete a task that requires both simple and complex operations.

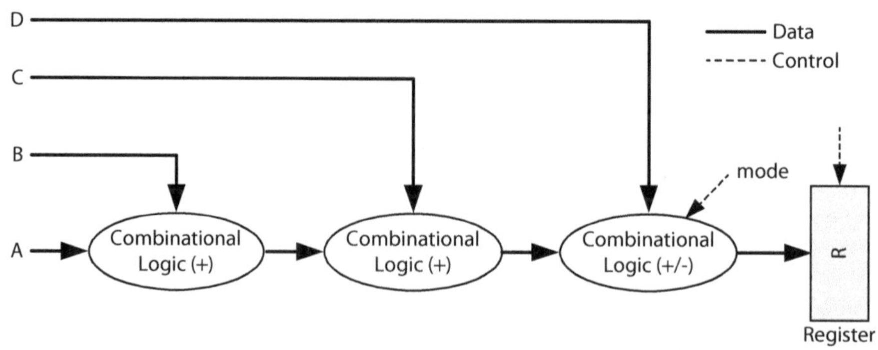

Figure 6.2 A single-cycle two-function data path that computes either $A+B+C+D$ or $A+B+C-D$ in one clock cycle.

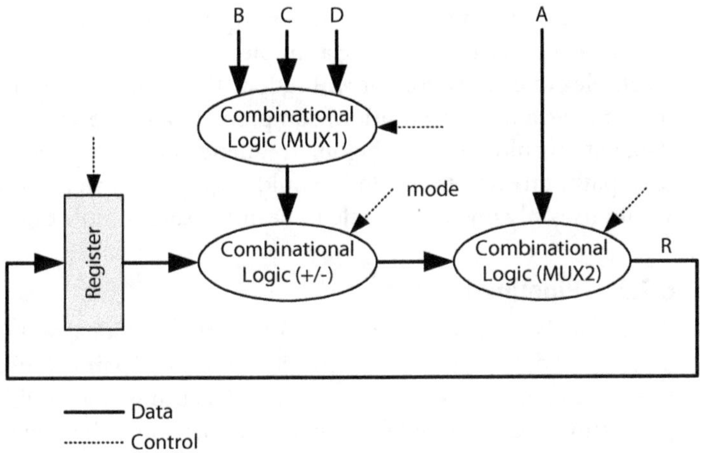

FIGURE 6.3 A multicycle data path requiring four clock cycles to compute $A+B+C+D$ or $A+B+C-D$.

6.2.2 Multicycle

A multicycle data path requires that a computation be divided and completed in steps, each requiring a simple data path operation. Figure 6.3 illustrates a multicycle data path with a single adder/subtractor and two multiplexer (MUX) modules. The data path can perform five possible simple operations as $R \leftarrow A$, $R \leftarrow R + B$, $R \leftarrow R + C$, $R \leftarrow R + D$, or $R \leftarrow R - D$. The following algorithm implements $R \leftarrow A + B + C \pm D$ using four clock cycles:

A multicycle algorithm to implement $R \leftarrow A + B + C \pm D$:
Cycle 1: $R \leftarrow A$
Cycle 2: $R \leftarrow R + B$
Cycle 3: $R \leftarrow R + C$
Cycle 4: If $mode == 0$ then $R \leftarrow R + D$; otherwise, $R \leftarrow R - D$

The clock period of a multicycle data path is also proportional to the delay of its longest signal path. In this case, the longest path starts from the inputs of the MUX1 through the adder/subtractor module and ends at the input of the register. Equation (6.2) estimates the minimum clock period of the data path. Note that the propagation delay of a MUX is less than that of an adder. Therefore, the estimated minimum clock period of the multicycle data path is shorter than that of the single-cycle data path. However, the multicycle algorithm requires four clock cycles to complete the task versus one clock cycle required by the single-cycle data path.

$$\tau \geq \Delta_{\text{MUX1}} + \Delta_{\text{ADD/SUB}} + \Delta_{\text{MUX2}} + \tau_{st} + \tau_{cq} + \tau_{cs} \quad (6.2)$$

A multicycle data path has the advantage of reducing the required total hardware. In the figure, a single adder/subtractor module is used several times to produce the final result. In general, a multicycle data path is also advantageous if it performs both simple and complex computations. In this case, fewer clock cycles would be required to complete a simpler computation and more cycles to complete a complex one. Therefore, it will reduce the total number of clock cycles required to complete a task. In addition, compared to a single-cycle data path, a multicycle data path would require a higher-frequency (shorter period) clock. A single-cycle data path, on the other hand, would require a slower (longer period) clock but would use only one clock cycle to perform each simple or complex computation.

6.2.3 Pipelined

A pipelined data path, or **pipeline** for short, is the ideal architecture for processing a stream of data. For example, consider adding N pairs of FP numbers one pair at a time, to generate N sums, or consider executing N assembly instructions. When processing in a pipelined fashion, a computation is divided into a set of dependent operations, much like the ones used for the multicycle data path, where each is performed in a separate subdata path called a **pipeline stage**.

The stages do not share any modules and are separated by parallel-load registers forming an assembly line, like a car-manufacturing assembly line discussed in Chap. 1. All the stages operate concurrently to process a stream of data. For example, consider the problem of computing N quantities, each $A_i + B_i + C_i \pm D_i$ for $i = 0$ to $N - 1$. Each A_i to D_i identifies four data items, such as the ith element from four different arrays. One way to divide each computation $A_i + B_i + C_i \pm D_i$ to its set of dependent operations is as follows:

$X_i \leftarrow A_i + B_i$ //performed by a pipeline stage 1

$Y_i \leftarrow X_i + C_i$ //performed by a pipeline stage 2

$R_i \leftarrow Y_i \pm D_i$ //performed by a pipeline stage 3

Figure 6.4 illustrates the architecture of a pipelined data path with three stages labeled 1 to 3. Three sets of registers are used to separate the result generated by each stage. In the figure, the pipeline receives a set of four values, A_i, B_i, C_i, and D_i, from an external module and sends the result R_i, also to an external module. During each clock cycle, stage 1 takes its inputs from an external data source, stage 2 takes its inputs from stage 1, and stage 3 takes its inputs from stage 2. Therefore, all three stages operate concurrently during each clock cycle.

Figure 6.5 shows two different **pipeline chart** styles for illustrating pipelining. Note the charts do not include the one-cycle delay caused by the interfacing registers shown in Fig. 6.4. The pipeline chart in Fig. 6.5(a) has a horizontal organization, with the clock cycles shown on the x-axis. On the other hand, the chart in Fig. 6.5(b) has a vertical organization, with the clock cycles shown on the negative y-axis.

As illustrated in Fig. 6.5, during cycle 3, while stage 3 is generating R_0, stage 2 is generating the intermediate result Y_1 required for R_1, and stage 1 is generating the intermediate result X_2 required for R_2. Therefore, the pipeline performs three simple operations at the same time, and thus concurrently operates on multiple data values. This helps to complete tasks quickly.

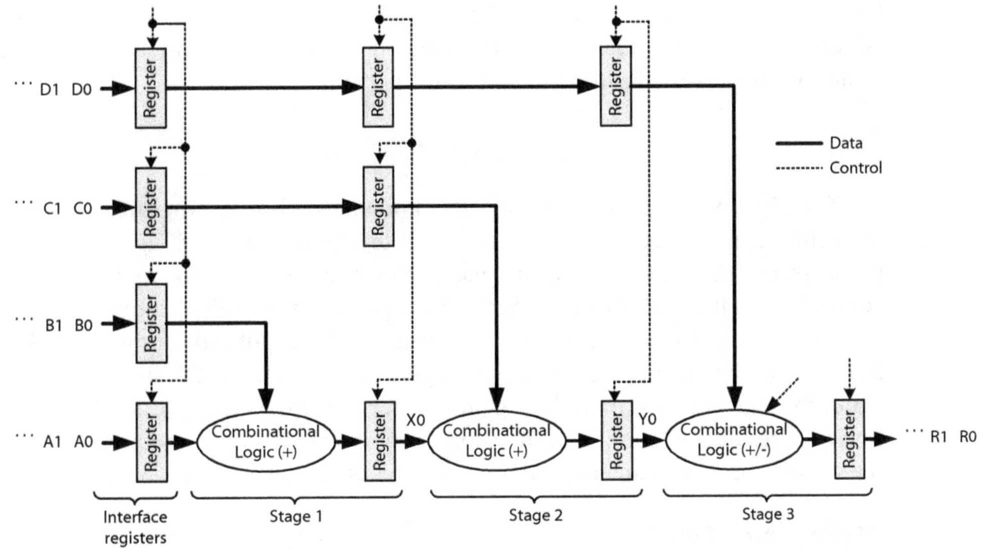

Figure 6.4 A two-function pipelined data path computing a stream of quantities $A_i + B_i + C_i \pm D_i$ for $i = 0, 1, 2$, etc.

A pipeline uses more hardware, similar to a single-cycle data path, but operates with a higher-frequency clock, similar to a multicycle data path. Furthermore, it can process a stream of data a lot faster than the other two data paths. The clock period of a pipelined data path is proportional to the propagation delay of its

Stage 3			R0 ← Y0 ± D0	R1 ← Y1 ± D1	...
Stage 2		Y0 ← X0 + C0	Y1 ← X1 + C1	Y2 ← X2 + C2	...
State 1	X0 ← A0 + B0	X1 ← A1 + B1	X2 ← A2 + B2	X3 ← A3 + B3	...
Cycle	1	2	3	4	...

(a) Horizontal Organization

Cycle	Stage 1	Stage 2	Stage 3
1	X0 ← A0 + B0		
2	X1 ← A1 + B1	Y0 ← X0 + C0	
3	X2 ← A2 + B2	Y1 ← X1 + C1	R0 ← Y0 ± D0
4	X3 ← A3 + B3	Y2 ← X2 + C2	R1 ← Y1 ± D1
5	X4 ← A4 + B4	Y3 ← X3 + C3	R2 ← Y2 ± D2
...	...	...	...

(b) Vertical Organization

Figure 6.5 Two alternative pipeline charts: (*a*) from left to right; (*b*) from top to bottom; showing the results R_0, R_1, etc.

longest stage. In Fig. 6.4, stage 3 has the longest propagation delay; it uses an adder/subtractor module, while the other two stages use an adder module. Equation (6.3) estimates the pipeline clock period.

$$\tau \geq \Delta_{\text{ADD/SUB}} + \tau_{st} + \tau_{cq} + \tau_{cs} \tag{6.3}$$

The pipelined data path in Fig. 6.4 is known as a **linear pipeline**, where each of its three stages is used only once to compute a final result (e.g., R_0). A **nonlinear pipeline**, on the other hand, would use one or more of its stages multiple times to compute a final result. Designs of nonlinear pipelines are referred to elsewhere.

During a clock cycle, each stage in a linear pipeline inputs one or more data items from its immediately preceding stage. In general, the data for the first stage is read either from an external module (e.g., memory) or from memory located internally. The final result is either stored in a storage module (register or memory) in one of the stages or is sent to an external module. This is further discussed in Chap. 8.

Pipeline Performance

A careful study of the pipeline chart (*a*) or (*b*) in Fig. 6.5 reveals that R_0, being the first result, requires three clock cycles to compute (not including the clock cycle required to load the interface registers), whereas the results R_1, R_2, etc., each requires only one clock cycle to compute. This reduces the total time required to compute N final results. In general, a k-stage (linear) pipeline requires k clock cycles to produce its first output. Equation (6.4) estimates the total time required to process a data stream of size N using a k-stage linear pipeline in terms of its clock period τ_{pipeline}.

$$T_{\text{pipeline}} = k * \tau_{\text{pipeline}} + (N - 1)\, \tau_{\text{pipeline}} \tag{6.4}$$

For example, when $N = 3$ and $k = 3$, the pipeline will require a total of 5 τ_{pipeline} to produce three outputs, such as the three outputs R_0, R_1, and R_3 shown in Fig. 6.5(*b*). Assuming that the clock period of a single-cycle data path $\tau_{\text{single-cycle}}$ is approximately equal to $k * \tau_{\text{pipeline}}$, where τ_{pipeline} is the clock period of a corresponding pipeline, Eq. (6.5) estimates the total time required to process a data stream of size N using a single-cycle data path.

$$T_{\text{single-cycle}} = N * k * \tau_{\text{pipeline}} \tag{6.5}$$

However, note that, in general, even if both a single-cycle data path and its corresponding k-stage pipeline use identical combinational circuit modules with identical propagation delays, $\tau_{\text{single-cycle}}$ would be slightly less than $k * \tau_{\text{pipeline}}$. The quantity $k * \tau_{\text{pipeline}}$ includes the sum of k register setup time and k clock-to-q delays, whereas $\tau_{\text{single-cycle}}$ would include the sum of only one register setup time and one clock-to-q delay. The $k * \tau_{\text{pipeline}}$ is also an upper bound estimate because τ_{pipeline} is proportional to the propagation delay of the longest stage in the pipeline. In Fig. 6.4, τ_{pipeline} is calculated based on the propagation delay of stage 3, which uses an adder/subtractor module, while the other two stages each use an adder. However, the difference between $\tau_{\text{single-cycle}}$ and its $k * \tau_{\text{pipeline}}$ approximation is ignored to simplify the performance analysis of a pipeline as compared to a corresponding single-cycle data path.

Speedup is a performance parameter that, in general, measures the performance of a faster system as compared to a slower system when performing the same task. It is defined as the ratio of the time required to complete a task by a slower system over that of a faster system. It indicates how much faster a faster system is as compared to an equivalent slower system. For example, Eq. (6.6) defines the speedup between a faster pipelined data path as compared to a corresponding slower single-cycle data path when processing a data stream of size N with $\tau = \tau_{\text{pipeline}}$. For $N = 3$ and $k = 3$, a pipeline is approximately 1.8 ($3 * 3 * \tau_{\text{pipeline}} / 5\ \tau_{\text{pipeline}} = 1.8$) times faster than the single-cycle data path. For $N = 1000$ and $k = 3$, the speedup is approximately 2.99. Note that the speedup increases and approaches k (the number of stages) as N approaches infinity (∞). The more stages there are in a pipeline, the larger the speedup would be, provided that it processes a large data stream. As N increases, the processing of a data stream becomes more efficient as the time required to fill the pipeline becomes negligible as compared to the total time required to process the entire stream.

$$\text{Speedup} = \frac{T_{\text{single-cycle}}}{T_{\text{pipeline}}} = \frac{Nk\tau}{k\tau + (N-1)\tau} = \frac{Nk}{k+N-1} \tag{6.6}$$

Efficiency is a performance parameter that measures how well a system's resources were utilized to complete a task. The efficiency of a pipeline data path is said to be 100% if all its stages were busy all the time; that is, there were no idle stages. For example, consider the pipeline chart in Fig. 6.5(*a*). The efficiency of the pipeline reaches 100% starting with clock cycle 3 when all the three stages become busy for the rest of the computations. However, we need an overall efficiency value and not just when all the resources are utilized. From Eq. (6.6), the speedup of the pipeline approaches K, the number of stages, when the number of computations N approaches infinity (∞). Therefore, an overall efficiency of a system can be defined as the ratio of its speedup to its maximum possible speedup. Equation (6.7) defines the efficiency of a pipeline data path with K stages.

$$\text{Efficiency} = \frac{\text{Speedup}}{k} \tag{6.7}$$

The substitution of Eq. (6.6) in Eq. (6.7) yields,

$$\text{Efficiency} = \frac{N}{k+N-1}$$

As N approaches infinity (∞), the efficiency of the pipeline in Fig. 6.4, as expected, approaches 1 or 100%.

Throughput is another performance parameter that measures a system's rate of processing. It indicates the number of items (N) performed per second. It is calculated as the ratio of the total number of items (tasks, calculations, operations, Google search, etc.) performed over the total required time (T). Equation (6.8) defines the throughput of a linear pipeline with k stages. For $N = 3$ and $k = 3$, the throughput is about $0.6\tau^{-1}$ ($3/5\tau$). For $N = 1000$ and $k = 3$, it is about $0.99\tau^{-1}$ ($1000/1002\tau$). In general, the throughput of a linear pipeline could approach to τ^{-1} (the operating clock frequency) as N approaches infinity (∞). For example, if the clock frequency ($f = 1/\tau$) of the pipeline in Fig. 6.4 is 1 GHz (one billion cycles per second), its peak throughput (τ^{-1}) would be one billion computations (each $A + B + C \pm D$), or three billion arithmetic operations (each + or −) per second, not including the delays required for reading the input data, for example, from memory and writing the outputs back to memory.

$$\text{Throughput} = \frac{N}{T_{\text{pipeline}}} = \frac{N}{k\tau + (N-1)\tau} \tag{6.8}$$

As discussed in Chap. 1, the CPU data path is pipelined as it executes many instructions, including floating-point (FP) instructions that operate on FP numbers. For a data path to perform FP arithmetic, it must perform several operations, such as initialization, lining up decimal points, integer arithmetic, normalization, and rounding, as discussed in Chap. 3. These FP operations are typically divided into several pipeline stages to increase throughput. For example, consider the following for-loop where an FP ADD instruction (e.g., "FADD") would be executed 1000 times to add 1000 elements of array A with 1000 elements of array B to produce 1000 elements of array C. With a pipelined floating-point unit (FPU), the 1000 ADD instructions would execute in less time as compared to, say, a single-cycle FPU.

```
float A[1000], B[1000], C[1000];
int i;

for(i = 0; i < 1000; i++)
       C[i] = A[i] + B[i];
```

The **FLOPS** (floating-point operations per second) or the less popular MIPS (millions of instructions per second) are two examples of throughput measurement units that are typically reported by processor designers. However, these throughput units reported by the designers often assume ideal conditions and may represent peak performance values. In addition, MIPS may be based on executing a set of some random instruction mix. In general, a more realistic performance measurement requires the execution of some benchmark (existing standard) programs, such as the compute-intensive workload called standard performance evaluation corporation 2006 (SPEC CPU2006) benchmark for measuring the performance of a computer system, or the graphic-intensive workload SPECviewperf benchmark for measuring the performance of a computer-graphic system [3].

6.3 Control Unit Design Techniques

A control unit is a finite state machine (FSM). As a **hardwired control unit**, the control signals are generated using a set of combinational circuits. For maximum speed, each control signal can be the output of an AND-OR (SOP expression) or OR-AND (POS expression) circuit with the maximum $3\Delta_{\text{NAND}}$ or $3\Delta_{\text{NOR}}$.

A control unit may be modeled as a finite state diagram (FSD) or designed using the bit-parallel methodology discussed in the previous chapter. However, once a hardwired control unit is built, it cannot be repaired if there are design errors, especially if it implements a set of very complex algorithms. A high-performance pipelined data path is typically controlled by a hardwired control unit.

On the other hand, a memory-based control unit, called a **microprogrammed control**, keeps the values of the control signals in memory inside the IC. The content of the memory can be updated in the future in case some design errors are discovered after manufacturing. A memory-based control unit, however, can be slow, depending on the size of the memory. It takes a lot longer than $3\Delta_{NAND}$ to perform a memory read/write operation, which will be discussed in Chap. 7.

The application of a microprogrammed control has diminished over the years, especially due to the advantage of reduced instruction set computer (RISC) versus complex instruction set computer (CISC) architecture. With an RISC, as opposed to a CISC, the CPU has simpler and fewer instructions. Therefore, it is easier to design a hardwired control unit to control an RISC data path. RISC and CISC will be discussed in Chap. 8.

The application of microprogrammed control has also diminished due to the availability of modern HDL synthesis tools today. The tools have simplified the design and verification of hardwired control units. However, microprogrammed controls would still be used when designing control units that have a large number of states [4], or when it is necessary to translate legacy CISC instructions to a list of simple operations that would be performed on a more efficient RISC data path.

6.3.1 Hardwired Control: FSD

Figure 6.6 shows the multicycle data path given in Fig. 6.3 with an FSD model of its control unit. The FSD has four states and defines the data path operations in RTN. The control unit generates control signals for the data path to compute either the quantity $A + B + C + D$ if the external input signal *mode* = 0 or the quantity $A + B + C - D$ if *mode* = 1 in four clock cycles. The result will be stored in the register within the data path. An external input signal *start,* when asserted, triggers the start of the computation. In contrast, Fig. 6.7 illustrates the FSD with the actual control signals. Note that it is much easier to verify an FSD if data path operations are specified in RTN than if they are specified with actual control signals.

Figure 6.8 illustrates a detailed block diagram of the corresponding FSM with two flips: a next-state generator (NSG), and an output generator (OG). The OG generates the data path control signals. During each clock cycle, only the control signals associated with a specific data path operation are asserted. For example, initially, when *start* = 0, the register *R* is disabled and the data path is said to be "doing nothing." When *start* becomes 1, the control unit asserts *e* (i.e., *e* = 1), enabling the register *R*, and makes $s_0 = 0$ so the MUX selects the input *A*. In each clock cycle, only control signals that are required to perform a specific data path operation are asserted, while other control signals would be deasserted or set to

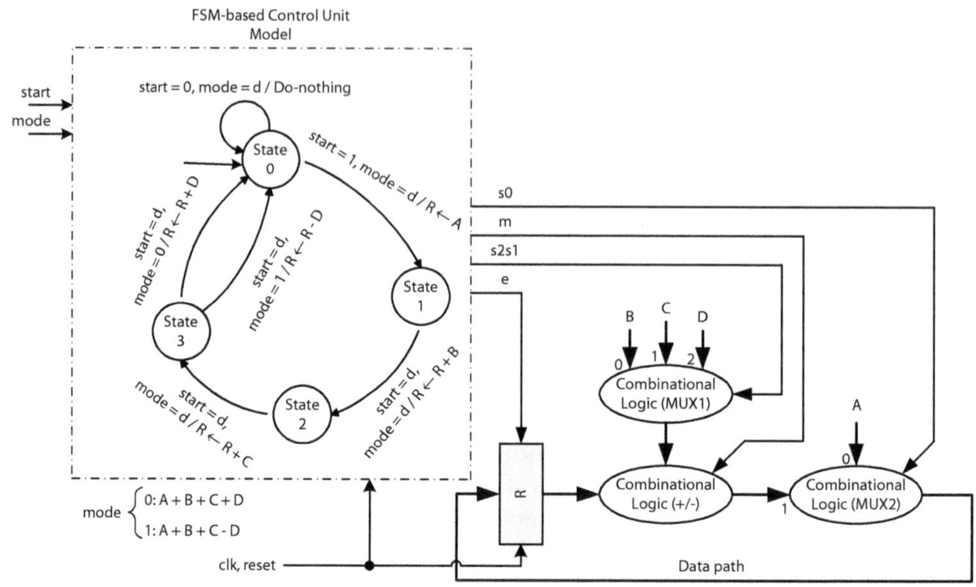

FIGURE 6.6 Illustrating an FSM-based control unit for the data path shown in Fig. 6.3 with RTNs, where "d" stands for don't-care.

don't-care (d) as necessary. The NSG module implements the specific order in which the data path must perform operations as dictated by the algorithm. The design is completed by following the FSM design steps discussed in Chap. 5.

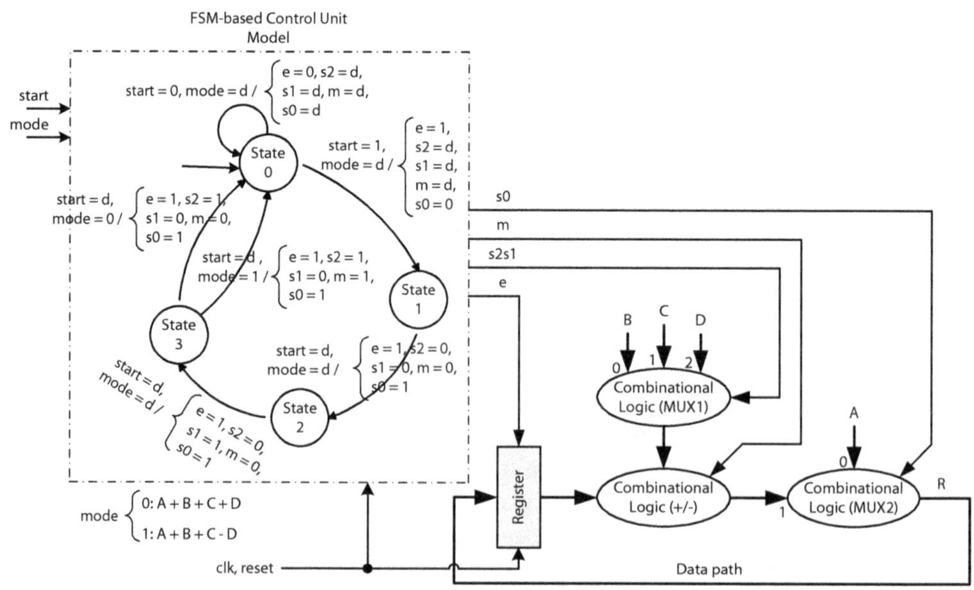

FIGURE 6.7 The FSD in Fig. 6.6 with the actual control signals; d stands for don't-care.

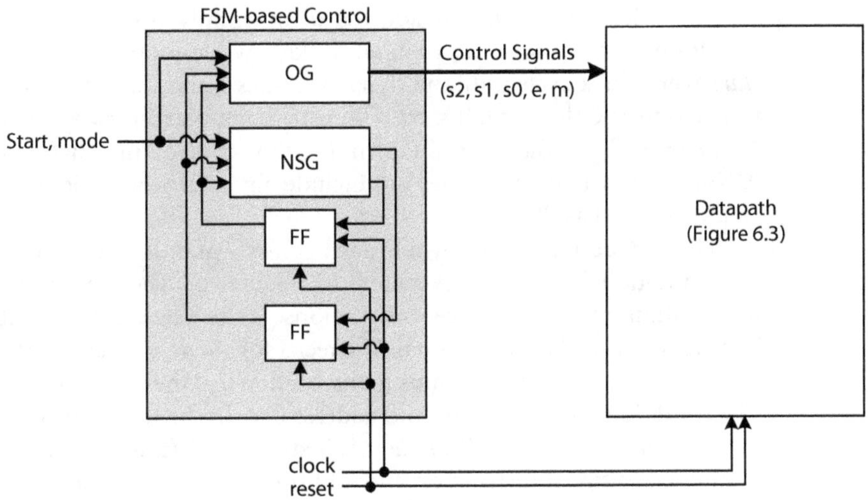

FIGURE 6.8 The detailed block diagram of the FSM-based control unit.

6.3.2 Microprogrammed Control

A microprogrammed control unit uses a memory called **control memory** (CM) for storing a description of an algorithm called a **microprogram**. The program is made of a set of **microinstructions**, where each specifies one or more data path operations called **micro-operations**. Each microinstruction also includes **microprogram flow control** information in the form of "jump" or "no-jump" that decides which microinstruction will execute next.

Figure 6.9 shows the microprogram for computing $R \leftarrow A + B + C \pm D$ using the multicycle data path in Fig. 6.3. Note, the microprogram reads like a program. It consists of five microinstructions that would be translated into binary, called a **microcode**, and would be stored in CM at locations (i.e., addresses) 0 to 5. The microprogram consists of three types of microinstructions.

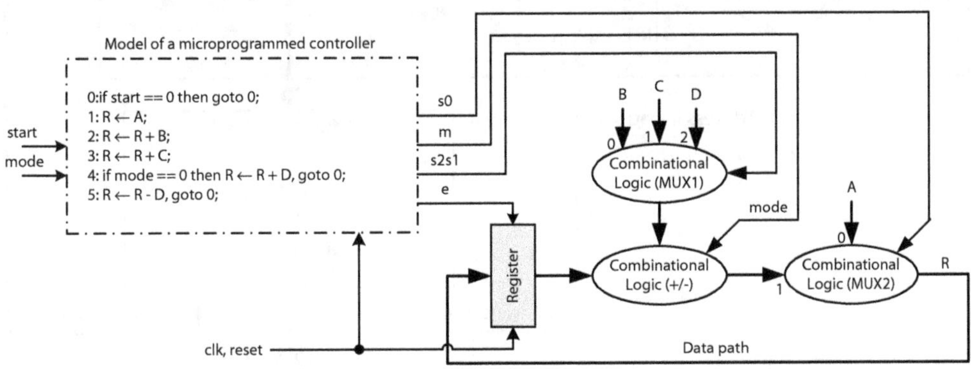

FIGURE 6.9 The multicycle data path in Fig. 6.3 is controlled using a microprogram.

The microinstruction in address 0, or simply instruction 0, includes the condition "if *start* == 0" and operates like a while-loop that checks the value of *start* every clock cycle until the signal becomes 1 and the control is transferred to instruction 1. Otherwise, if *start* = 0, instruction 0 executes again. The condition "if *start* == 0" and the address 0 form the flow control information of instruction 0. Note that instruction 0 does not include any micro-operations that, otherwise, would be listed as RTNs.

The instructions 1 to 3, where each consists of only one micro-operation, are not conditional and are executed in sequence. The program flow control information for these three instructions is to execute the next instruction. Instruction 4 is also conditional. It checks for *mode* = 0, and if 0, it performs a micro-operation and then jumps to instruction 0 at the start of the microprogram. The condition "if *mode* == 0" and address 0 form the flow control information of instruction 4. Otherwise, if *mode* = 1, instruction 5 (the next instruction), which is an unconditional instruction and additionally performs a micro-operation (an RTN), executes. Instruction 5 also performs "go to 0"—an unconditional jump to address 0—where 0 is the flow control information of the instruction.

A detailed block diagram of a microprogrammed control unit is illustrated in Fig. 6.10. It consists of a CM and a **next-address-generator** (NAG) that generates the address of the next microinstruction in CM. It also consists of a multifunction counter called a **microprogram counter** (MPC) and a 1-bit k-to-1 MUX, where k is the number of condition signals (e.g., *start* and *mode*) plus 2; $k = 4$ in this case. The MPC holds the address of the currently executing microinstruction. In each clock cycle, the MPC either increments the address of the current instruction it

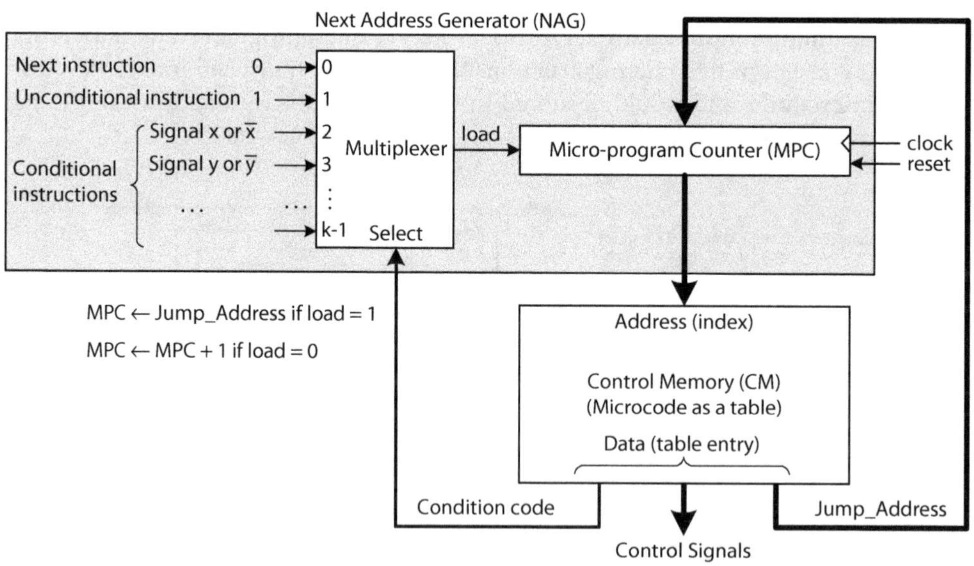

Figure 6.10 A detailed block diagram of a microprogrammed control unit.

Condition Code (c_1c_0)	NSG (Next Address Generator)	
	MPC	**MUX Output**
00	$MPC \leftarrow MPC + 1$	$load = 0$
01	$MPC \leftarrow Jump\_Address$	$load = 1$
10	If $start == 0$ then $MPC \leftarrow Jump\_Address$ else $MPC \leftarrow MPC + 1$	$load = \overline{start}$
11	If $mode == 0$ then $MPC \leftarrow Jump\_Address$ else $MPC \leftarrow MPC + 1$	$load = \overline{mode}$

Table 6.2 Condition Codes for the Microprogram in Fig. 6.9

holds or loads a new (jump) address from CM. The MUX decides which function, increment or load, the MPC will perform next. In the figure, the MPC is designed to load a jump address if *load* = 1 or to increment its content if *load* = 0.

A microcode is organized as a table and stored in CM. Each table entry consists of a condition code, a set of control signals, and may be a jump address. Table 6.2 lists four arbitrarily assigned condition codes to implement the microprogram control shown in Fig. 6.9 as illustrated in Fig. 6.11. Each condition code selects one of the four inputs of the 4-to-1 MUX as *load* signal value.

Condition code 0 ($c_1c_0 = 00$) is assigned to the microinstructions that are not conditional, such as instructions 1 to 3 in Fig. 6.9. The code represents a "no-jump" statement (making *load* = 0), which causes the MPC that points to the current microinstruction to increment the next clock cycle. Code 1 ($c_1c_0 = 01$) represents a

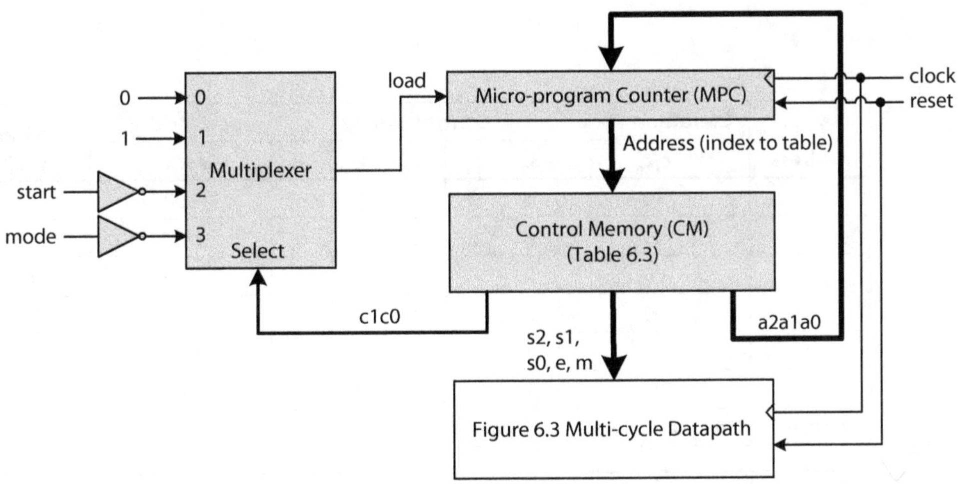

Figure 6.11 Microprogrammed control unit for the multicycle data path in Fig. 6.9.

"jump" statement (making *load* = 1). It is assigned to the microinstructions that are unconditional, such as instruction 5. Code 2 (c_1c_0 = 10) is assigned to the microinstruction 0. It represents the "if *start*==0" condition and, via the MUX, makes $load = \overline{start}$. If *start* = 0, then *load* = 1 (jump); otherwise, *load* = 0 (no-jump). Last, code 3 (c_1c_0 = 11) is assigned to the microinstruction 4. It represents the "if *mode* == 0" condition and, via the MUX, makes $load = \overline{mode}$.

Table 6.3 lists the microcode for the microprogram in Fig. 6.9. It has six rows, each a 10-bit binary representation of a microinstruction as $\{c_1c_0, s_2, s_1, s_0, e, m, a_2a_1a_0\}$. The binary representations are also shown in hex in the last column.

6.3.3 Hardwire Control: Pipeline

The control signals for all the stages of a pipeline are generated at once, but applied to each stage at the right time. A stage that performs only a single operation requires no control signals. Figure 6.12 illustrates the pipelined data path in Fig. 6.4 with a pipelined control unit. The data path performs N computations each $A_i + B_i + C_i \pm D_i$ for $i = 0, 1, 2, \ldots N{-}1$. The signal $mode_i$ decides if the last arithmetic operation that the data path will perform is an addition or a subtraction. All the registers in the figure are enabled by the incoming *start* signal. It is assumed that *start* will remain at logic 1 (active) for $N+4$ clock cycles, the required number of clock cycles to perform N computations (also counting the one clock signal required by the interface registers). The $mode_i$ signal along with four data values A_i to D_i enter the pipeline on every clock cycle for N cycles, but the $mode_i$ is passed from one stage to the next until it is used in stage 3 to generate the final quantity $Y_i + D_i$ if $mode_i = 0$ or $Y_i - D_i$ if $mode_i = 1$. The values of $mode_i$ for the last four final cycles are set to don't-care. Note that, in this case, the pipeline control unit uses only a set of registers and no combinational circuits, and stages 1 and 2 require no control signals other than the *start*.

CM Address	Condition Code	Control Signals					Jump Address	In Hex (10-Bits)
	c_1c_0	s_2	s_1	s_0	e	m	$a_2a_1a_0$	
0	10	d	d	d	0	d	000	200
1	00	d	d	0	1	d	ddd	010
2	00	0	0	1	1	0	ddd	030
3	00	0	1	1	1	0	ddd	070
4	11	1	0	1	1	0	000	3B0
5	01	1	1	1	1	1	000	1F8

d: Stands for don't care and is set to 0 when stored in the CM.

TABLE 6.3 Microcode for the Microprogram in Fig. 6.9

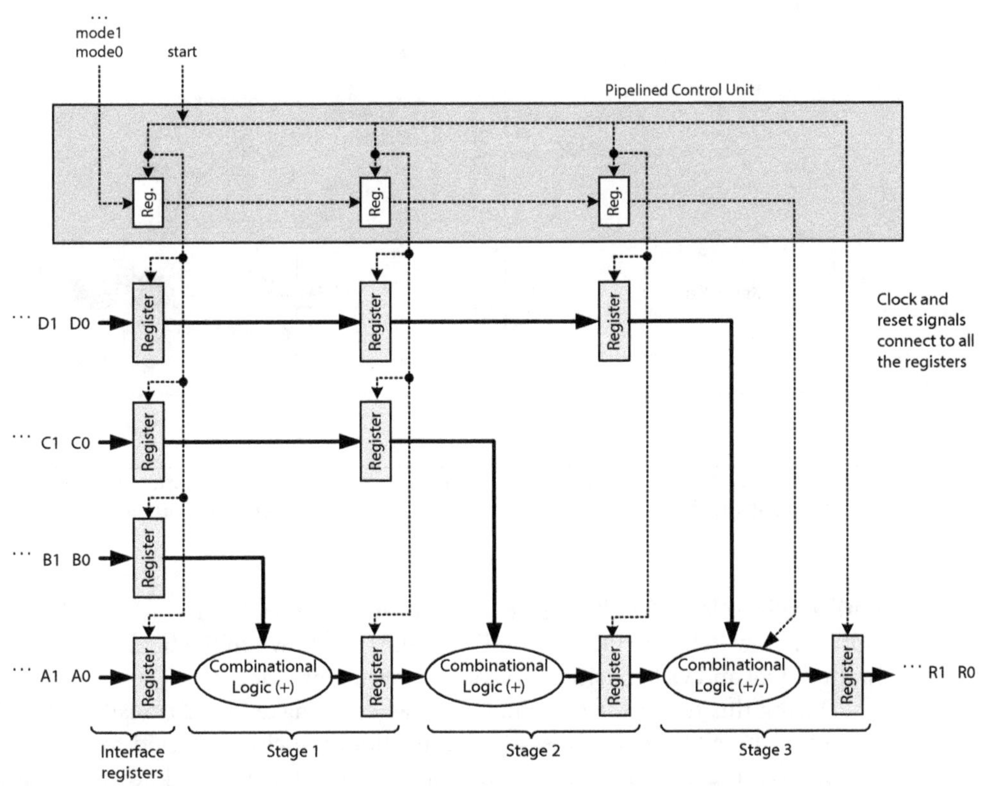

FIGURE 6.12 A pipeline control unit and the pipelined data path in Fig. 6.4.

6.4 Energy and Power Consumption

As stated in Chap. 1, as the number of transistors and the operating clock frequency of ICs increase, they use more power and also dissipate more heat. Consider the CMOS NOT gate circuit shown next that was discussed in Chap. 1. Recall that in CMOS circuits, pMOS and nMOS transistors are complementary; one transistor remains in ON position and the other in OFF position once the gate output stabilizes to either logic 1 or logic 0 voltage level.

In Fig. 6.13, the circuit is also shown with a capacitive load C, where its size determines the amount of **dynamic energy** necessary to charge the capacitance and generate logic 1 as the gate output. It is called dynamic energy because inputs and outputs of a gate do not instantly change from logic 1 to logic 0 or from logic 0 to logic 1, as was discussed in Sec. 2.6 (Chap. 2). For example, when input $x = 1$, the pMOS and nMOS transistors remain OFF and ON (not shown), respectively. As x starts to change from logic 1 voltage level to logic 0 voltage level, both the transistors start switching. Each transistor becomes partially ON or partially OFF

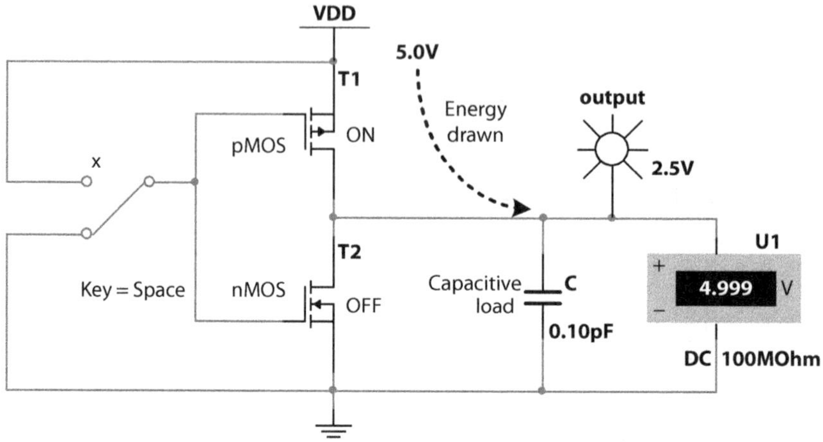

FIGURE 6.13 CMOS NOT gate circuit from Chap. 1 illustrating capacitance charging when x transitions from 1 to 0 that results in a 0-1 transition at the output.

until x falls to logic 0 voltage level. At that time, the pMOS and nMOS that have completely switched to ON and OFF positions, as shown in the figure, will remain in those ON and OFF positions as long as x remains at logic 0.

During this transistor-switching time where x makes a 1-0 transition and the output of the NOT gate makes a 0-1 transition, the circuit is said to be a **short circuit**. During this time, a certain amount of current, also known as **shoot-through current**, flows from V_{DD} to ground. The total amount of energy drawn from the power source for 0-1 transition at the NOT gate output is CV_{DD}^2. From this total energy, one half dissipates as heat and the other half is stored in the capacitance, as specified in its simplified form by Eq. (6.9).

Energy stored in the capacitance:
$$E_{\text{dynamic}}^{0-1} = \int_{v=0}^{v=V_{DD}} Cv\,dv = \frac{1}{2}CV_{DD}^2 \text{ Joules} \tag{6.9}$$

where "Joules" is the unit for energy. The 1-0 transition at the output of the NOT gate, however, does not draw energy from the power source. Instead, the charge ($\frac{1}{2}CV_{DD}^2$) that was stored in the capacitance discharges to ground, as illustrated in Fig. 6.14. This is called the 1-0 transition dynamic energy, or $E_{\text{dynamic}}^{1-0} = \frac{1}{2}CV_{DD}^2$.

The amount of total dynamic energy that a NOT gate dissipates as heat due to a single transition, 0-1 or 1-0, at the gate output is shown in Eq. (6.10):

Energy dissipated as heat by a NOT gate per 0-1 or 1-0 transition at its output:
$$E_{\text{dynamic}} = \frac{1}{2}CV_{DD}^2 \text{ Joules} \tag{6.10}$$

Recall that in sequential circuits, signals make 0-1 or 1-0 transitions during each clock cycle; some signals transition from 0 to 1 and others from 1 to 0. Each

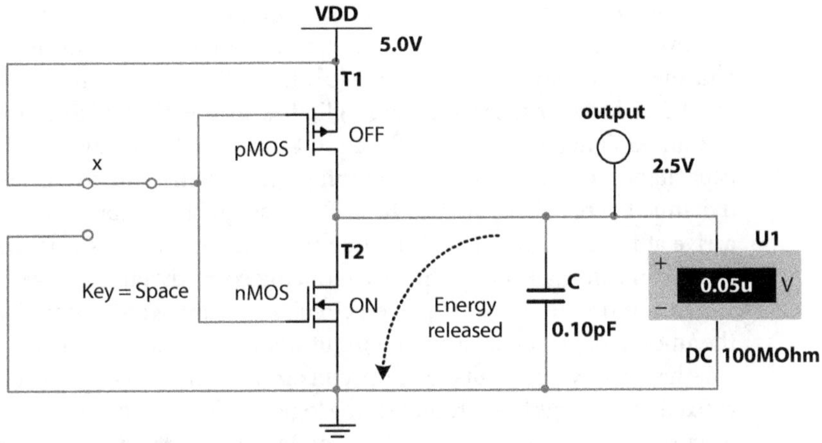

FIGURE 6.14 CMOS NOT gate circuit from Chap. 1 illustrating capacitance discharging when *x* transitions from 0 to 1 that results in a 1-0 transition at the output.

signal then remains at its final logic value 0 or 1 until the next clock cycle. Equation (6.11) defines the total **dynamic power** (in watts) a sequential circuit consumes during one clock cycle. It is determined from the amount of total dynamic energy (in joules) the circuit consumes during one second. In the equation, τ and f represent the period and frequency of the clock signal, respectively and C_{total} represents the total equivalent capacitive load in the circuit; C_{total} is determined from the capacitive load of all the gates (NOT, NAND, etc.) and all the wire connections in the circuit.

$$P_{\text{dynamic}} = \frac{E_{\text{dynamic}}}{\tau} \quad \text{watts, defined as joules per second} \tag{6.11}$$

Or,

$$P_{\text{dynamic}} = E_{\text{dynamic}} * f \quad \text{watts, where } f = \frac{1}{\tau}$$

$$P_{\text{dynamic}} = \frac{1}{2} C_{\text{total}} V_{DD}^2 f \quad \text{watts}$$

Equation (6.11) indicates that as the clock frequency increases, so will the number of 0-1 and 1-0 transitions in the circuit. This, in turn, will increase the total dynamic power consumed and the heat generated by the circuit. There are three ways that one may be able to reduce the dynamic power consumption of a complex circuit:

- Reduce total capacitance, C_{total}
- Reduce supply voltage, V_{DD}
- Reduce clock frequency, f

The size of the capacitive load or more precisely the effective capacitive load, however, depends on many parameters including the circuit topology (the gates and they way they are connected) [5]. In addition, dynamic power consumption can be reduced if there are fewer glitches in the circuit. Recall that glitches are unwanted signal transitions in the circuit and, therefore, they would cause unwanted capacitance charging and discharging that contribute to the amount of total dynamic energy dissipated as heat. Because glitches happen when signals do not arrive at the gates' inputs at the same time, designing gates with equal rise and fall times can eliminate some glitches, reducing some dynamic power usage.

In addition to dynamic power, circuits consume **static** (standby) power. This is the amount of power used by the circuit when no transistor switching is taking place. This happens when inputs to the circuit are static (not changing) and the outputs are at fixed DC voltage levels representing logic 1 or logic 0. In this case, a certain amount of current, called DC current (I_{DD}) or **leakage current**, would flow through the transistors that are off. Equation (6.12) defines a circuit's static power consumption.

$$P_{\text{static}} = V_{DD} I_{DD} \tag{6.12}$$

While energy and power consumption in a circuit are related, energy is a preferred metric to compare the efficiency of two complex circuits (e.g., processors) [6]. From Eq. (6.11), if we increase the clock frequency, the amount of dynamic power consumed to perform a task also increases. However, the amount of dynamic energy consumed remains unchanged (constant). Consider, for example, two processors A and B. Now suppose, during the execution of a program, processor A's dynamic power consumption P_A is greater than P_B consumed by processor B (i.e., $P_A > P_B$). However, processor A is able to execute the program faster than processor B. That is, $t_A < t_B$, where t_A and t_B are, respectively, the program execution times by processors A and B. In this case, it is possible that processor A could be more energy efficient than processor B.

Suppose, for a given program, P_A is 20% more than P_B and t_A is 40% less than t_B. That is, processor A consumes 20% more power than processor B, but executes the program faster, requiring only 60% of the time it takes processor B to execute the program. In another word, $P_A = (1 + 0.2)P_B$ and $t_A = (1 - 0.4)t_B$. Therefore, according to Eq. (6.13), processor A consumes 72% of the energy consumed by processor B.

$$P_A = \frac{E_A}{t_A} \text{ from Eq. (6.11) for the duration of execution time, or} \tag{6.13}$$

$$\begin{aligned} E_A &= P_A * t_A \\ &= (1+0.20)P_B * (1-0.40)t_B \\ &= (1.2)P_B * (0.60)t_B \\ &= (0.72)P_B * t_B \\ &= 0.72 E_B \end{aligned}$$

Even though during the execution of the program, processor A consumes more dynamic power than processor B, processor A is better than processor B because it consumes 28% less total energy. Processor B consumes less dynamic power, but since it takes a longer time to execute the program, overall, processor B consumes more dynamic energy than processor A. As also discussed in Chap. 1, the power and cooling requirements (i.e., thermal design power) of a complex IC can be incorporated into its operations so that its clock frequency may be increased on occasion to improve performance, subject to meeting its cooling requirements.

6.5 Design Examples

In Chap. 3, an unsigned multiplier was designed as a combinational circuit with several adder modules. As a sequential circuit, a multicycle multiplier can reduce hardware, and a pipelined multiplier can increase throughput. Table 6.4 lists a set of design examples. Section 6.5.1 presents the design of a multicycle unsigned multiplier using a hardwired control unit designed from an FSD. Section 6.5.2 presents the design of a multicycle signed multiplier using a microprogrammed control unit. The signed multiplier uses a single adder/subtractor module and iteratively multiplies two 2's complement numbers. Section 6.5.3 presents the design of a rudimentary graphic pipeline that implements a two-dimensional (2-D) CORDIC (COordinate Rotation Digital Computer) rotation algorithm. In general, CORDIC algorithms can be used to implement elementary complex functions, including trigonometric, hyperbolic, logarithmic, exponential, and square root. The last two large sequential circuit designs in the table will be covered in Chap. 8.

6.5.1 Unsigned Sequential Multiplier

The advantage of a sequential multiplier is that it computes the product of two numbers in steps using a multicycle data path with only one adder. In each step,

Design Example	Data Path Type	Control Type	
Unsigned sequential multiplier	Multi-cycle	Hardwired, deigned using a FSD	Section 6.5.1
Signed sequential multiplier	Multi-cycle	Microprogrammed	Section 6.5.2
Computer Graphics: 2D CORDIC rotation	Pipeline	Pipelined	Section 6.5.3
A simple CPU (processing-core)	Single-cycle	Hardwired, designed using bit-parallel design methodology (Chap. 3, Chap. 5)	Chapter 8
A simple pipelined CPU	Pipeline	Pipelined	Chapter 8

Table 6.4 A Set of Large Sequential Circuit Design Examples

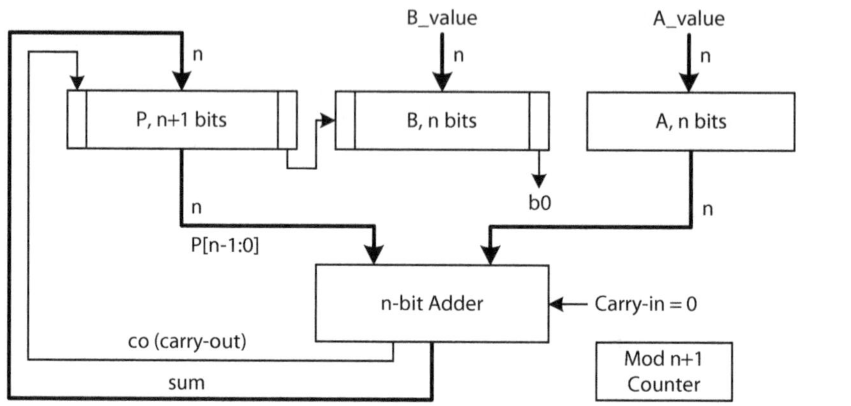

FIGURE 6.15 A multicycle unsigned multiplier data path.

the next addend is added to the accumulated sum of previously generated addends. As discussed earlier, a multicycle data path has the disadvantage of being slow but uses less hardware. This section presents the design of the unsigned multiplier data path and its FSD-based controller. The section also discusses alternative hardware description language (HDL) design models, and the code for an all-behavioral Verilog model for the multiplier is provided and simulation results are discussed.

Data Path

Figure 6.15 illustrates the data path of an unsigned multiplier using a single adder, three registers, and a mod $n+1$ counter. The A and B registers are n bits each and are used to load an n bits multiplicand *A_value* and n bits multiplier *B_value*. The P register is used to hold an $n+1$ bits partial sum (including the carry-out bit) each time that two multiplication addends are added. Recall that an addend is the result of a bitwise AND of all the A register bits with a single B register bit. Here, only the addends that are not zero are added to reduce the total computation time. Therefore, the wired-AND circuits that were used in the design of the multiplier as a combinational circuit in Chap. 3 are not necessary. If $b_i = 1$, $addend_i = A\_value$; otherwise, $addend_i = 0$ and the step to add the addend to the partial sum is skipped. Furthermore, the register B will be shifted right after each multiplication step so that its least significant bit (LSB) b_0 is used to determine the value of the next addend. The counter is used to keep track of the n iterations required to produce the final product result.

In the first clock cycle, all three registers and the counter are initialized. Each time that a partial sum is generated, the sum is loaded to P register. The P and B registers are then both shifted right. This simplifies the algorithm and reduces hardware. Specifically, the shifts allow (1) to replace the current b_0 with the next higher bit in B; (2) to line up the partial sum bits for the next multiplication step; and (3) to store the P's LSB in B as P and B registers are simultaneously shifted right. The final product will be stored in both P and B registers. The aforementioned steps are summarized as the multiplier algorithm as follows:

Sequential unsigned multiplication algorithm:

```
A ← A_value,
B ← B_value,               //Initialization, all done in
CNTR ← 0,                  //one clock cycle
P ← 0;
Repeat
    If (b0 == 1) P ←       //add the next non-zero addend
P[n-1:0] + A;              //to the current partial sum to
                           //produce a new partial sum. It is
                           //assumed that adding and
                           //loading the sum into the P takes
                           //one clock cycle

{P, B} ← {P, B} >> 1;      //right shift P and B registers
CNTR ← CNTR + 1;           //simultaneously. This will
                           //lineup the P bits for the next
                           //cycle; and also replace b0 with
                           //the next higher bit in B. The
                           //counter is incremented to keep
                           //track of number of iterations.
                           //One clock cycle is needed to do
                           //both the shifting and the
                           //incrementing
Until CNTR < n;            //repeat n times
```

Table 6.5 presents a step-by-step illustration of the unsigned multiplier algorithm using $A\_value = 7 = (111)_2$ and $B\_value = 5 = (101)_2$. After three steps, the final 6-bit result in {P[2:0], B} is $(100{,}011)_2$, or 35 in decimal.

Control Unit Design: FSD

There are several ways to design a large sequential circuit using HDL and/or an schematic design tool. The following outlines three general design practices for all types of data paths and controllers:

I. All structural—The design would use a hierarchy of interconnected modules, where all the modules at the leaf of the hierarchy would be modeled with Boolean expressions (i.e., using "assign" statements) or circuits (i.e., using primitive gates). The modules would then be interconnected operating with explicitly declared control signals. This option is not recommended for very large designs. In addition, one may use a schematic design tool.

II. Hybrid—The design would use both structural and behavioral models. In this case, a hierarchical model uses behavioral models (e.g., "always" blocks) for the leaf modules that would then be interconnected and operated with explicitly declared control signals. One may also use an schematic design tool that has an HDL interface.

III. All behavior—The design would describe the behavior of the circuit modeled as an FSD with data path operations indicated in RTNs. The design would require no explicitly RTN-related control signals.

Mod-3 CNTR	P[3], P[2:0]	B	A	Comment
0	0,000	101	111	Initialization
	0,111	101	111	b0 = 1, thus P ← P[2:0] + A
1	0,011	110	111	Right shift {P, B} with 0 fill
	0,011	110	111	b0 = 0, thus P remains unchanged
2	0,001	111	111	Right shift {P, B} with 0 fill
	1,000	111	111	b0 = 1, thus P ← P[2:0] + A
3	0,100	011	111	Right shift {P, B} with 0 fill

TABLE 6.5 Step-by-Step Illustration of Multiplying Unsigned Numbers *A_value* = $(111)_2$ and *B_value* = $(101)_2$

We first present the design requirements for the unsigned multiplier circuit, starting with the explicitly declared control signals that would be needed to operate each of the registers and the counter in Fig. 6.15. Table 6.6 presents the list of functions each of the registers A, B, and P and the counter (CNTR) must perform. The A is a single-function parallel-load register; B is a dual-function parallel-load and right-shift register; and P is a three-function parallel-load, right-shift, and synchronous-clear register. The CNTR is a dual-function counting-up and synchronous-clear counter.

For all structural (Option I) or hybrid (Option II) designs, all the registers and the counter must be assigned a set of control signals, as shown in Fig. 6.16. A simple combinational circuit (CC) is used to convert a multibit output from the CNTR to a single signal, *flag*, used by the control unit. If count = *n*, *flag* = 1; otherwise, *flag* = 0. Table 6.7 lists the specific control signal values and the corresponding data path operation for each. The registers and the CNTR are assumed to be implemented with flip-flops with no enable signals.

Figure 6.17 shows the detailed block diagram of the unsigned multiplier circuit. The external triggering signal *start*, which is the output of a synchronizing flip-flop (FF1), starts the multiplier control unit. The *done* (Mealy) signal is asserted at the end of the computation and is saved as *done_moore* in another flip-flop (FF2). FF1 is reset if either *_reset* = 0 or *done* = 1, and FF2 is reset if either *_reset* = 0 or *start_asyn* = 1.

Function	Meaning
A ← A_value;	Parallel load
B ← B_value;	Parallel load
B ← {P[0], B[n-1:1]};	Right shift with a left-input (Li)
P ← 0;	Synchronously initialized to 0
P ← {co, sum};	Parallel load
P ← {0, P [n:1]};	Right shift with zero fill
CNTR ← 0;	Synchronously initialized to 0
CNTR ← CNTR + 1;	Increment

TABLE 6.6 Register and Counter Functions in RTN

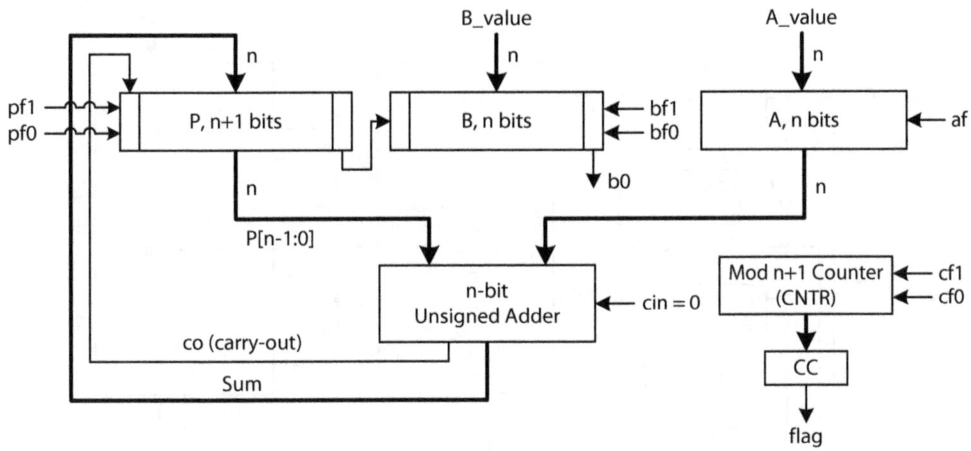

Figure 6.16 Data path of unsigned multiplier with control signals.

Figure 6.18 presents the multiplier control unit FSD with data path operations indicated in RTNs. The FSD consists of three states labeled "Idle," "Check," and "Add". Upon reset, the control unit initializes to the Idle state as shown in the FSD. Once in Idle state, the control unit monitors the *start* signal until the signal

Control Signals		Description
af		The single control signal of register A. Enables or disables the register.
1		Enables register A for parallel load
0		Disables register A
bf1	**bf0**	Control signals of register B
0	0	Disables register B
0	1	Enables register B for parallel load
1	0	Enables register B for a right shift
1	1	Not used
pf1	**pf0**	Control signals of register P
0	0	Disables register P
0	1	Enables register P for parallel load
1	0	Enables register P for a right shift
1	1	Enables and synchronously clears the P register
cf1	**cf0**	
0	0	Disables the CNTR
0	1	Enables the CNTR to increment
1	0	Not used
1	1	Enables and synchronously clears the CNTR

Table 6.7 The Control Signals of the Data Path in Fig. 6.16 for Structural HDL Models

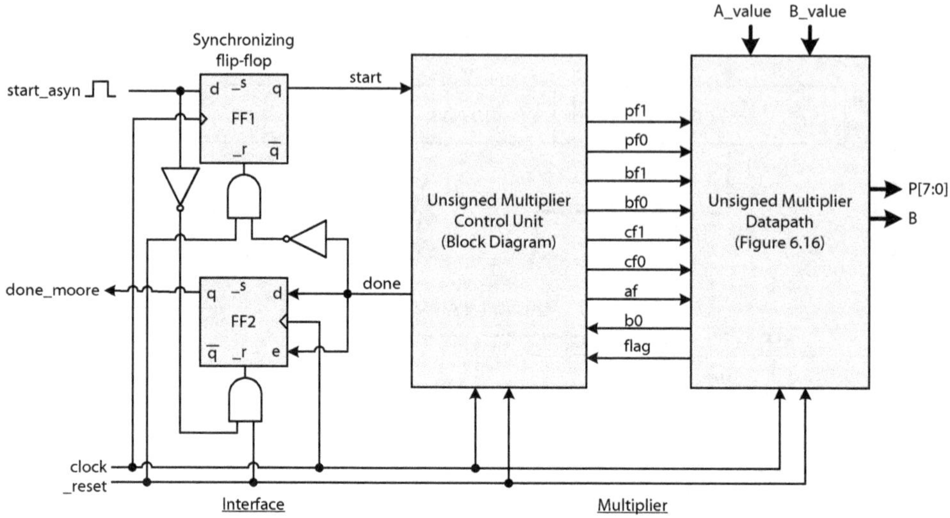

FIGURE 6.17 Unsigned multiplier block diagram, control signals, and interfacing signals.

becomes 1 and triggers the control unit that multiplies *A_value* and *B_value* according to the unsigned multiplication algorithm discussed earlier.

HDL Model

The multiplier all-structural and hybrid designs are deferred to the Exercises section. However, for an example of a design that uses explicitly declared control signals refer to Sec. 6.5.2. Next is an all-behavioral (Option III) HDL code for the unsigned multiplier and its interface module shown in Fig. 6.17. The HDL code models the FSD in Fig. 6.18 with no explicitly declared data path control signals. Specifically, the code describes an FSM with an NSG, an OG, and a set of flip-flips. The OG is responsible for generating the data path control signals as specified

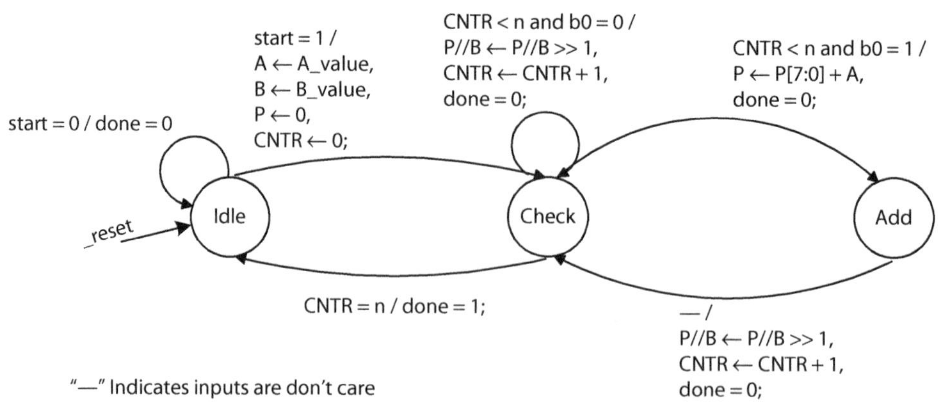

FIGURE 6.18 Unsigned multiplier controller FSD; adding only the non-zero addends.

implicitly by the RTNs. Therefore, an all-behavioral OG using the RTNs would model the multiplier data path with implicit control signals.

Example 6.2. A Verilog behavior model of the unsigned multiplier and its interface module in Fig. 6.17 is described. The description uses no explicitly declared data path control signals.
Solution: The multiplier is described exactly as specified in its FSD in Fig. 6.18 with the data path operations given in RTN.

```
module interface_unit(
  input clock, _reset, start_asyn, done,
  output reg start, done_moore
  );
  //------------- synchronization flip-flop -------------
  always@(posedge clock or negedge _reset or posedge done)
  begin
    if(_reset == 0 || done == 1)
         start <= 1'b0;
    else
         start <= start_asyn;
  end
  //------- Convert a Mealy output to a Moore output -------
  always@(posedge clock or negedge _reset or posedge start_asyn)
  begin
    if(_reset == 0  || start_asyn == 1)
         done_moore = 1'b0;
    else
         done_moore <= done;
  end
  endmodule
  //---------------- Unsigned Multiplier ----------------
  //A Mealy controller FSM
  module umult(
    input clock, _reset, start,
    input [7:0] a_value, b_value,
    output[15:0] result,
    output reg done
  );
  reg [8:0] p;
  reg [7:0] a, b;
  reg [3:0] cntr; //mod-16 counter
  reg co;
  reg [7:0] sum;
  reg [1:0] current_state, next_state;
  assign result = {p[7:0], b};
  //-------------------- The states --------------------
  parameter   Idle =        2'b00,
            Check =        2'b01,
              Add =        2'b10;
  /*---------------- Output Generator (OG) ----------------
```

```
The OG module defines the data path and also implicitly defines
the data path control signals using a behavior description*/
always@(posedge clock or negedge _reset)
begin
  if(!_reset)
  begin

       p <= 0;
       cntr <= 0;
  end
  else
  case(current_state)
       Idle: if(start == 1)
       begin // initialize
            a <= a_value;
            b <= b_value;
            p <= 0;
            cntr <= 0;
       end
  Check: begin
         if (cntr < 8)
             if(b[0] == 1)
                p <= {co, sum}; //or p <= p[7:0] + a; //without delay
         else
         begin
                {p, b} <= {p, b} >> 1;
                cntr <= cntr + 1;
         end
       end
       Add:begin
            {p, b} <= {p, b} >> 1;
            cntr <= cntr + 1;
         end
  endcase
end
//-------------- Next State Generator (NSG) -------------
always@(current_state or cntr or start or b[0])
begin
  case(current_state)
       Idle:begin
            done = 0;
            if(start == 1)
                 next_state = Check;
            else
                 next_state = Idle;
            end
  Check: if(cntr < 8)
            begin
              done = 0;
```

```
                if(b[0] == 0)
                   next_state = Check;
           else
              next_state = Add;
         end
         else
         begin
              done = 1;
              next_state = Idle;
         end
 Add:    begin
                   done = 0;
                   next_state = Check;
         end
 default:begin
              done = 0;
              next_state = Idle;
         end
   endcase
 end
 //------------------- The flip-flops -------------------
 always@(posedge clock or negedge _reset) //state transitions
 begin
   if (!_reset)
   current_state <= Idle;
 else
   current_state <= next_state;
 end

 //adder with delay
 always@(p or a)
 begin
 //assume 10ns delay for the adder
 #10 {co, sum} = p[7:0] + a;
 end
 endmodule
```

Simulation

The multiplier and its interface module were synthesized and simulated using Altera Quartus II and Altera ModelSim 10.1b. The tool also provides a "state machine viewer" verification feature that reconstructs the FSD from a given Verilog description. Figure 6.19 shows the reconstructed FSD of the multiplier control unit from the Verilog description in Example 6.2.

Example 6.3 is a test-bench with two test vectors. The corresponding simulation timing diagram is shown in Fig. 6.20. Note that because the algorithm skips addends that are zero, the multiplier takes a longer time to multiply *A_value* = 8'h03 with *B_value* = 8'h7F that has seven 1's, as compared to *A_value* = 8'h7F with *B_value* = 8'h03 that has only two 1's. Alternatively, a multiplier may be designed with a comparator that switches the operands *A_value* and *B_value* if there are fewer 1's in the *A_value*.

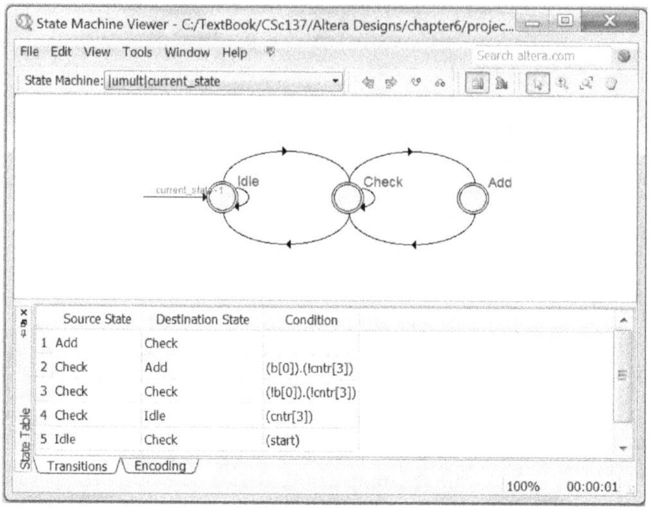

FIGURE 6.19 FSD reconstructed from the multiplier Verilog description in Example 6.2.

Example 6.3. The HDL model of a test-bench with two test cases 8'h03 × 8'b7F and 7'h7F × 8'h03 is described.

```
module tester();
reg clock, _reset, start_asyn;
reg [7:0] a_value, b_value;
wire [15:0] result;
wire done, done_moore, start;

interface u1(clock, _reset, start_asyn, done, start, done_moore);
umult  u2(clock, _reset, start, a_value, b_value, result, done);

initial begin
clock = 1;
#10 forever #10 clock = ~clock; //20ns clock period
end

initial begin
_reset = 0;
start_asyn = 0;
#15 _reset = 1;
#10 start_asyn = 1; a_value = 8'h03; b_value = 8'h7F;
#40 start_asyn = 0;

#400 start_asyn = 1; a_value = 8'h7F; b_value = 8'h03;
#40 start_asyn = 0;

#1000 $finish;
end

initial
    $monitor($stime,, _reset,, start_asyn,, clock,,, result,, done_
moore);
endmodule
```

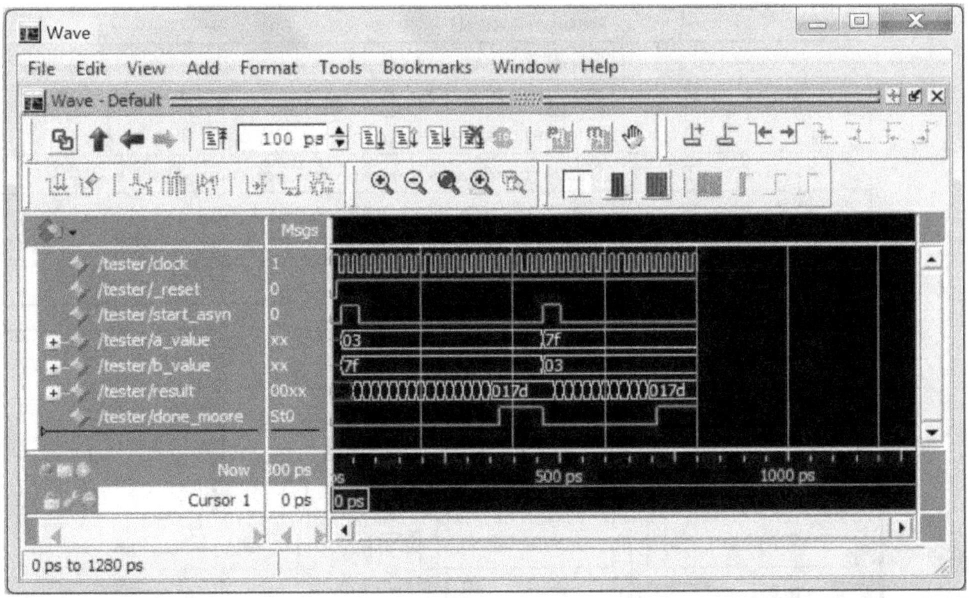

Figure 6.20 A simulation output for the multiplier in Fig. 6.17, using the test-bench in Example 6.3.

6.5.2 Signed Sequential Multiplier

A 2's complement multiplication algorithm, commonly known as the Booth's multiplier, uses both addition and subtraction to multiply two 2's complement positive or negative numbers. Its data path is similar to that of the unsigned multiplication discussed earlier with three registers A, B, and P. A 2's complement multiplicand *A_value* and a 2's complement multiplier *B_value* are loaded in the A and B registers, respectively, and the final product result is read from the P and B registers.

In the Booth's algorithm, a sequence of, for example, three 1's or $(111)_2$ is interpreted as $(100\overline{1})_2$; where $\overline{1}$ is used here to represent −1. Both $(111)_2$ and $(100\overline{1})_2$ represent 7 in decimal; $(111)_2 = 4 + 2 + 1$ is 7 and so is $(100\overline{1})_2 = 8 - 1$. This interpretation replaces the computations of three partial sum values that typically would be needed to multiply 7 by *A_value* by only two partial sum computations: a subtraction at the start of the sequence and an addition at the end of the sequence. The intermediate 1's would be interpreted as 0's and would be skipped.

This is done by examining the quantity {*B_value*, 0} (i.e., the *B_value* concatenated with a 0) two bits at a time but overlapping, starting from its LSB and using the rules specified in Table 6.8 to multiply *A_value* by *B_value*.

Data Path

A data path for the Booth's multiplier is illustrated in Fig. 6.21. The x and the b_1 signals are used by the control unit. If $x = 0$, both registers P and B (shown as {P, B}) are arithmetic right shifted at the same time, saving the LSB of P in B. If $x = 1$, then if

b_0	b_{-1}	Interpretation
0	0	Skip: Sequences of 0's; {P, B} ← {P, B} >>> 1.
0	1	Add: Indicates the end of a sequence of 1's; P ← P + A.
1	0	Subtract: Indicates the start of a sequence of 1's; P ← P − A.
1	1	Skip: Sequences of 1's are interpreted as 0's; {P, B} ← {P, B} >>> 1.
>>> indicates an arithmetic right shift with sign extension.		

Table 6.8 Bit Interpretations in the Booth's Multiplier

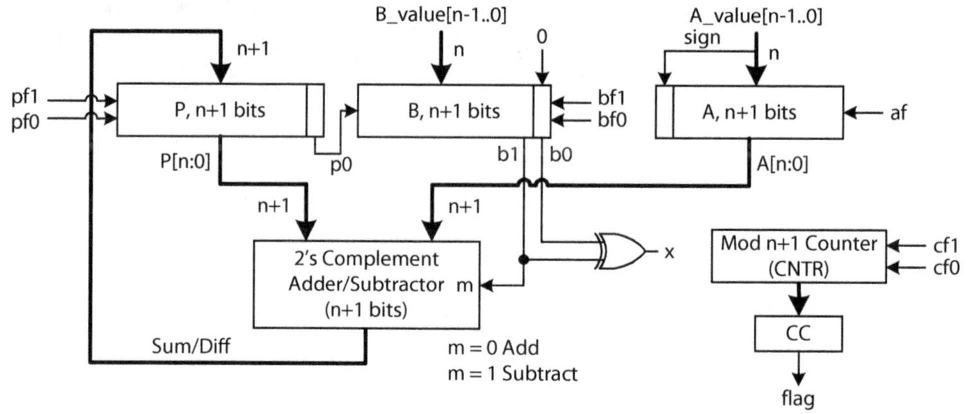

Figure 6.21 Sequential Booth's multiplier data path.

$b_1 = 0$, *A_value* is added to the content of P; otherwise, *A_value* is subtracted from the content of P. The combinational circuit (CC), as in the case of the unsigned multiplier, converts the counter output to a signal, *flag*. If count = n, *flag* = 1; otherwise, *flag* = 0.

In addition to the B register, which is an $n + 1$ bits register, both A and P registers are also $n + 1$ bits so the multiplier circuit can handle the largest magnitude n-bit 2's complement negative number. For example, consider *A_value* = −8 or $(1000)_{2s}$ as the smallest 4-bit 2's complement negative number and the content of P is 0 (i.e., *P_value* = 0). Now, the quantity *P_value* − *A_value*, which should equal to +8, would be represented incorrectly in 4-bits as $(1000)_{2s}$, which is −8 using 4-bit 2's complement representations. Therefore, making both the A and the P 5-bit registers resolves this problem. With 5-bits, −8 is represented as $(11000)_{2s}$ in the A register, and the quantity $(00000)_{2s} - (11000)_{2s}$ would be represented as $(01000)_{2s} = +8$ in the P register, correctly.

The B register stores the $n + 1$ bit quantity {*B_value*, 0}. Suppose the 4-bit *B_value* = $(1111)_{2s} = -1$. Note that *A_value* × *B_value* or *A_value* × −1 should result in −*A_value*, where *A_value* is an arbitrary 4-bit 2's complement number. Initially, as illustrated in the data path, the *B_value* would be stored in the

B register in 5-bits as $(11110)_{2s}$ with its two LSB bits $b_1b_0 = (10)_2$. Using the rules in Table 6.8, if $b_1b_0 = (10)_2$, then the content of the P register, initially 0, will become $0 - A\_value = -A\_value$. Next, {P, B} would be arithmetic shifted right, repeating the sign of P, which is now a 1. Because the remaining bits in the B register are all 1's, {P, B} will be arithmetic right shifted four times, each time repeating the sign of P. This will produce the final correct product result $-A\_value$ in $\{P_{n\text{-}1..0}, B_{n..1}\}$ as a $2n$ bits 2's complement negative number. Table 6.9 illustrates -8×-5 using the data path in Fig. 6.21 with $n = 4$. The result is $40 = (0010, 1000)_{2s}$ = 8'h28.

Multiplier Algorithm: Microprogram

Assuming that the interface module in Fig. 6.17 is also used to generate the two interfacing signals *start* and *done*, Table 6.10 lists the microprogram for controlling the signed multiplier in Fig. 6.21.

Mod-5 CNTR	P	B	A	Comment
0	00000	1011,0	1,1000	Initialization
1				b_1b_0= 10, subtract, load, and shift:
	01000	1011,0	1,1000	P ← P − A;
	00100	0101,1	1,1000	{P, B} >>> 1 and CNTR++;
2				b_1b_0 = 11, shift
	00010	0010,1	1,1000	{P, B} >>> 1 and CNTR++;
3				b_1b_0 = 01, add, load, and shift
	11010	0010,1	1,1000	P ← P + A;
	11101	0001,0	1,1000	{P, B} >>> 1 and CNTR++;
4				b_1b_0 = 10, subtract, load, and shift:
	00101	0001,0	1,1000	P ← P − A;
	00010	1000,0	1,1000	{P, B} >>> 1 and CNTR++;

Table 6.9 A 4-Bit Booth's Multiplication Example: $A\_value = -8 = (1000)_{2s}$ and $B\_value = -5 = (1011)_{2s}$

Address	Micro-Instruction	Comment
0	If *start* == 0 go to 0;	Wait to be started
1	P ← 0, A ← *A_value*, B ← *B_value*, CNTR ← 0;	Initialize
2	If CNTR == n go to 6;	Done if n iterations completed
3	If x == 0 go to 5	Skip if b_1b_0 = 00 or 11
4	P ← P ± A;	Compute and load the partial sum using addition if $b_1 = 0$ or subtraction if $b_1 = 1$
5	{P, B} >>> 1, CNTR ← CNTR + 1, go to 2;	Update P, B, and the CNTR and then repeat
6	*done* = 1, go to 0;	Done, branch to 0

Table 6.10 A Microprogram to Control the Signed Multiplier Data Path in Fig. 6.21

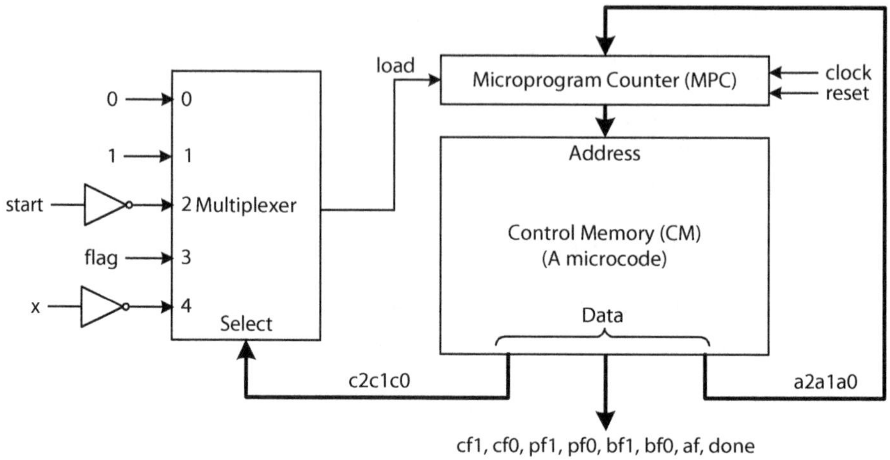

Figure 6.22 A microprogrammed control unit for the Booth's multiplier data path in Fig. 6.21.

Control Unit Design: Microprogrammed

A microprogrammed control unit of the signed multiplier is shown in Fig. 6.22. The conditions "if *start*==0," "if *CNTR*==n," and "if *x*==0" are arbitrarily assigned the condition codes 2, 3, and 4, respectively. The microprogram counter (MPC) loads a jump address ($a_2a_1a_0$) if *load* = 1 or increments its content if *load* = 0. The microcode for the microprogram is listed in Table 6.11.

HDL Model

Example 6.4 describes the microprogrammed control unit in Fig. 6.22 and the data path in Fig. 6.21 with explicitly specified control signals. The description of the data path includes the initialization of the CM.

CM Address	Condition Code	Control Signals					Jump Address	Hex
	$c_2c_1c_0$	cf_1cf_0	pf_1pf_0	bf_1bf^0	af	done	a2a1a0	(14 bits)
0	010	00	00	00	0	0	000	1000
1	000	11	11	01	1	0	ddd	07B0
2	011	00	00	00	0	0	110	1806
3	100	00	00	00	0	0	101	2005
4	000	00	01	0	0	0	ddd	0080
5	001	01	10	10	0	0	010	0B42
6	001	00	00	00	0	1	000	0808
The d stands for don't care and is replaced with a 0 to produce the hex values.								

Table 6.11 The Microcode of the Microprogram in Table 6.10

Example 6.4. A Verilog model of the 2's complement Booth multiplier using both HDL structural and behavioral descriptions is presented.

Solution: Assume the interface in Fig. 6.17 is used to interface with the control unit.

```
module smult(
    input clock, _reset, start,
    input [7:0] a_value, b_value,
    output [15:0] result,
    output done
);
wire flag, x;
wire [13:0] control;

controller  u2(clock, _reset, start, flag, x, control, done);
data path   u3(clock, _reset, a_value, b_value, control, flag, x,
result);

endmodule

module controller(
    input clock, _reset, start, flag, x,
    output reg[6:0] control,
    output reg done
);

reg [2:0] mpc; //control memory
(* ramstyle = "M512" *) reg [13:0] cm[0:7]; //control memory, using
                                            //a built-in memory
//reg [0:7][13:0] cm;
reg [2:0] ccode; //branch type
reg [2:0] jump_address; //branch address
reg load;

//------Initialize the CM,----------
initial begin
cm[0] = 14'h1000; //wait for start = 1
cm[1] = 14'h07B0; //initialize
cm[2] = 14'h1806; //if flag == 1 then goto 6
cm[3] = 14'h2005; //if x == 0 then goto 5
cm[4] = 14'h0080; //p <- sum_diff
cm[5] = 14'h0B42; //p//b <= p//b >>> 1, cntr++, goto 2
cm[6] = 14'h0808; //done = 1, goto 0
end

//-------- MUX ----------
always@(*)
begin
    case(ccode)
    0: load = 0; //next instruction
    1: load = 1; //unconditional jump
    2: if (start == 0) //conditional jump if start = 0
          load = 1;
    else
```

```
            load = 0;
        3: if (flag == 1) //conditional jump if done
            load = 1;
        else
            load = 0;
        4: if (x == 0) //conditional jump if b[1]b[0] =  00 or 11
            load = 1;
        else
            load = 0;
        default: load = 1;
        endcase
end

//--------------- MPC ---------------------
always@(posedge clock or negedge _reset)
begin
    if (_reset == 0)
        mpc <= 3'b000;
    else
        if(load == 0)
            mpc <= mpc + 1;
        else
            mpc <= jump_address;
end

//--------------  CM --------------------
always@(*)
begin
    ccode = cm[mpc][13:11];
    control = cm[mpc][10:4];
    done = cm[mpc][3];
    jump_address = cm[mpc][2:0];
end
endmodule

module data path(
    input clock, _reset,
    input [7:0] a_value, b_value,
    input [6:0] control,
    output reg flag,
    output x,
    output [15:0] result
);

reg [8:0] a, b;
reg [8:0] p;
reg [3:0] cntr; //mod-16 counter
reg [8:0] sum_diff;

wire af = control[0];
wire [1:0]     CF = control[6:5],
```

```
               PF = control[4:3],
               BF = control[2:1];
assign result = {p[7:0], b[8:1]};
assign x = b[1] ^ b[0];

always@(posedge clock or negedge _reset) //registers and cntr
begin
   if (_reset == 0)
   begin
        a <= 0;
        b <= 0;
        p <= 0;
        cntr <= 0;
   end
   else
   begin
   if (af == 1)
        a <= {a_value[7], a_value}; //a[8] is set to the sign bit
(a[7])
   case(BF)
        2'b01: b <= {b_value, 1'b0};
        2'b10: b <= {p[0], b[8:1]}; //shift right with left input
        2'b11: b <= 9'h000; //clear
        default: b <= b;
   endcase
   case(PF)
        2'b01: p <= sum_diff; //load
        2'b10: p <= {p[8], p[8:1]}; //arithmetic right shift
        2'b11: p <= 9'h000; //clear
        default: p <= p;
   endcase
   case(CF)
        2'b01: cntr <= cntr + 1; //increment
        2'b10: cntr <= cntr; //not used, retain
        2'b11: cntr <= 3'b000; //clear
        default: cntr <= cntr;
   endcase
   end
end

always@(*)
begin
   if (b[1] == 1'b0)
        sum_diff = p + a;
   else
        sum_diff = p - a;
end

always@(*) //condition flag
begin
```

```
    if (cntr < 8)
         flag = 1'b0;
    else
         flag = 1'b1;

end
endmodule

module interface_unit(
input clock, _reset, start_asyn, done,
output reg start, done_moore
);
//synchronization flip-flop
always@(posedge clock or negedge _reset or posedge done)
begin
    if(_reset == 0 || done == 1)
         start <= 1'b0;
    else
         start <= start_asyn;
end

//Moore output
always@(posedge clock or negedge _reset or posedge start_asyn)
begin
    if(_reset == 0  || start_asyn == 1)
         done_moore = 1'b0;
    else
    if(done == 1)
         done_moore <= done;
end
endmodule
```

Simulation

Example 6.5 describes a test-bench for the signed multiplier. Figure. 6.23 illustrates the simulation timing diagram illustrating –8 multiplied by –5. The result is 40 in decimal.

Example 6.5. A test-bench to compute -8×-5, indicated as 0xF8 × 0xFB in hex or `8'hF8` × `8'hFB` in Verilog is described. Assume that the interface module in Fig. 6.17 is used to interface with the control unit.

```
module tester();
reg clock, _reset, start_asyn;
reg [7:0] a_value, b_value;
wire [15:0] result;
wire done, done_moore, start, flag, x;
wire [6:0] control;

interface_unit  u1(clock,  _reset,  start_asyn,  done,  start,
    done_moore);
smult  u2(clock,  _reset,  start,  a_value,  b_value,result,
    done);
```

```
initial begin
clock = 1;
_reset = 0;
end

always #10 clock = ~clock; //20ns clock period

initial begin
start_asyn = 0;
#15 _reset  = 1;
#10 start_asyn = 1; a_value = 8'hF8; b_value = 8'hFB;
#40 start_asyn = 0;

#1000 $finish;
end

initial begin
   $monitor($stime, _reset, start_asyn, clock, result, done_moore);
end
endmodule
```

6.5.3 Computer Graphics: Rotation

In computer graphics, a virtual object is defined by several points in the Cartesian coordinate system. For example, as illustrated in Fig. 6.24, a 2-D virtual object "house" is defined by its five points, labeled *a* through *e* in the *x-y* coordinate

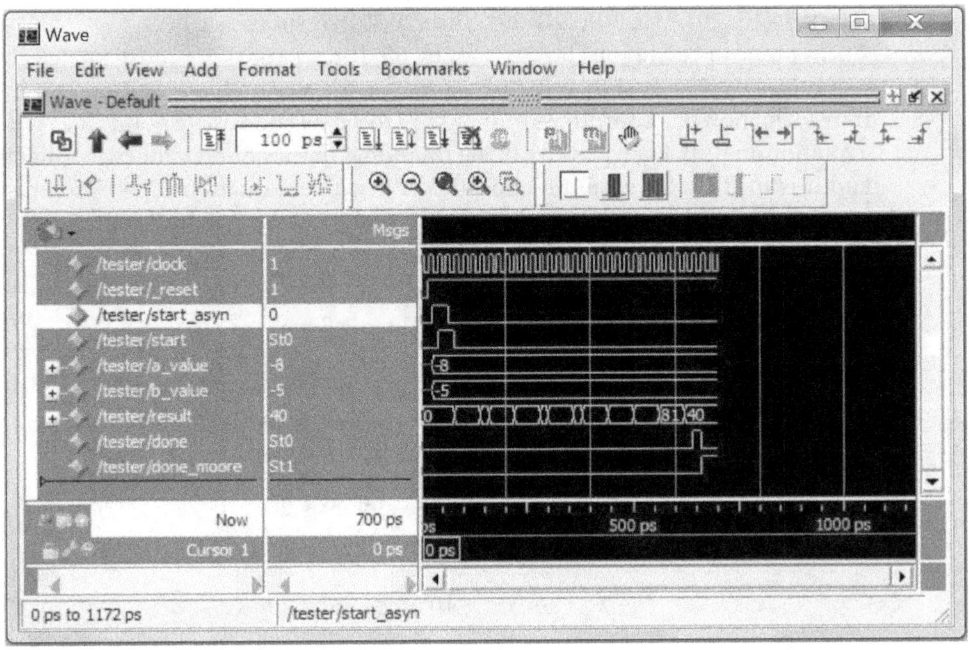

FIGURE 6.23 Simulation output of multiplying −8 by −5.

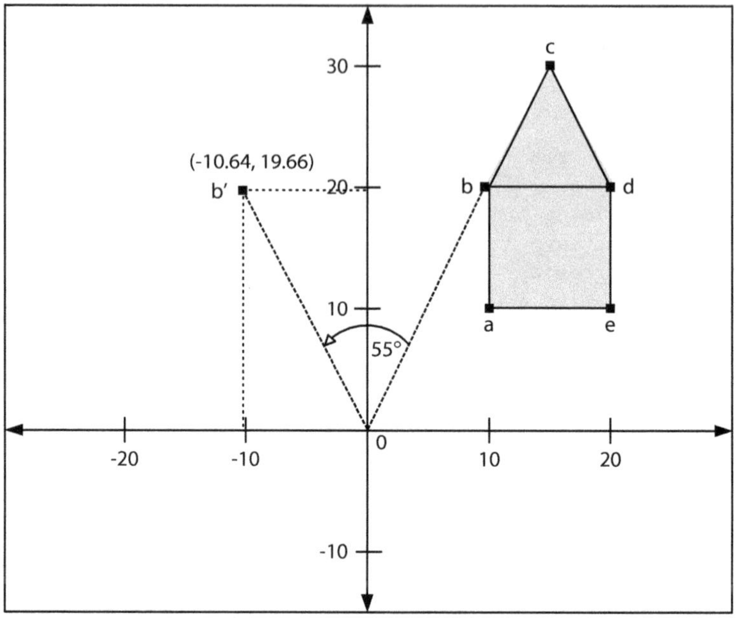

Figure 6.24 A 2-D virtual object with five *x-y* coordinate points. Also shown is the new vector *b′* as vector *b* rotated by 55°.

system. Each point is viewed as a **vector** connecting the origin coordinate point (0, 0) to a coordinate point (X, Y). The x-y coordinates for vectors a through e are (10, 10), (10, 20), (15, 30), (20, 20), and (20, 10), respectively.

In order to rotate the "house," say, by 55 degrees (55°), we must rotate each of its vectors by 55° as shown in the figure for the vector b. A rotation is a linear transformation of a coordinate point (X, Y) to a new coordinate point (X', Y') using a rotation angle β. For a positive β, the rotation is in the counterclockwise direction, and it is in the clockwise direction if β is negative. Equation (6.14) presents the expressions used for a 2-D rotation.

$$X' = \cos\beta * X - \sin\beta * Y \tag{6.14}$$

$$Y' = \sin\beta * X + \cos\beta * Y$$

In the figure, vector b, with coordinate points (0, 0) and $(X, Y) = (10, 10)$, is transformed to vector b' with the coordinate points (0, 0) and $(X', Y') = (-10.64, 19.66)$. The calculations for this transformation are given in Eq. (6.15).

$$X' = \cos 55° * 10 - \sin 55° * 20 \tag{6.15}$$

$$= -10.64$$

$$Y' = \sin 55° * 10 + \cos 55° * 20$$

$$= 19.66$$

CORDIC Algorithm

The CORDIC algorithms can be used to perform trigonometric, hyperbolic, logarithmic, exponential, square root, etc. functions. CORDIC algorithms may be used in the design of pocket calculators and 2-D/3-D graphic processors. Furthermore, it has been shown that one can develop simple CORDIC algorithms that are iterative and require only simple functions, such as integer addition, subtraction, and arithmetic right shift [7]. The right shift is used to perform an integer division by 2, 4, 8, etc. An iterative CORDIC algorithm, discussed next, can be used to perform a linear transformation (Eq. (6.14)). Also presented is the design of a data path for a simple graphic processor.

The expressions in Eq. (6.14) require complex cosine and sine functions. However, they can be simplified by factoring out the "cos β" term from both the expressions, as shown in Eq. (6.16). As you will see, it is easier to compute tan β and keep the factored-out "cos β" separate from the computations required for a linear transformation. The result of "cos β," however, will be used as a scaling factor to adjust each of the new computed coordinate points.

$$X' = \cos\beta * (X - \tan\beta * Y) \tag{6.16}$$
$$Y' = \cos\beta * (\tan\beta * X + Y)$$

The iterative and simple algorithm only uses a set of fixed rotation angles with predetermined tangent values. Table 6.12 lists seven angles, 45°, 27°, 14°, 7°, etc., with tangent values equal to, respectively, 1 and fractions 1/2, 1/4, 1/8, etc. For integer arithmetic, each angle in the table is rounded up to its nearest integer value.

A vector, such as the *b* in Fig. 6.24, can be rotated by an angle β = 55° in four steps by first rotating the vector by 45°, then by 7°, then by 2°, and finally by 1°, as illustrated in Eq. (6.17). The result in step 4 includes the scaling factor 0.701 = cos 1° * cos 2° * cos 7° * cos 45°. Without the scaling factor, the values $X' = -15.1$ and $Y' = 28.1$ are larger by a constant 1.427 (1/0.701). The final values of X' and Y' without this (1.427) gain are given in step 5. The small difference between these

i	$2^{-i} = \tan\beta_i$	$\beta_i = \tan^{-1} 2^{-i}$
0	1	45°
1	1/2	27°
2	1/4	14°
3	1/8	7°
4	1/16	4°
5	1/32	2°
6	1/64	1°

Table 6.12 Seven "Tan β" Values and Their Corresponding Approximate β Values

values and those obtained in Eq. (6.15) are due to rounding errors caused by manual calculations.

$$\begin{aligned}
\text{Step 1: } X' &= \cos 45^\circ * (10 - 1 * 20) \quad //(\tan 45^\circ = 1) \\
&= \cos 45^\circ * (-10) \\
Y' &= \cos 45^\circ * (1 * 10 + 20) \\
&= \cos 45^\circ * (30) \\
\text{Step 2: } X' &= \cos 7^\circ * \cos 45^\circ *(-10 - 1/8 * 30) \quad //(\tan 7^\circ = 1/8) \\
&= \cos 7^\circ * \cos 45^\circ * (-13.75) \\
Y' &= \cos 7^\circ * \cos 45^\circ *(1/8 * -10 + 30) \\
&= \cos 7^\circ * \cos 45^\circ * (28.75) \\
\text{Step 3: } X' &= \cos 2^\circ * \cos 7^\circ * \cos 45^\circ *(-13.75 - 1/32 * 28.75) \quad //(\tan 2^\circ = 1/32) \\
&= \cos 2^\circ * \cos 7^\circ * \cos 45^\circ * (-14.65) \\
Y' &= \cos 2^\circ * \cos 7^\circ * \cos 45^\circ *(1/32 * -13.75 + 28.75) \\
&= \cos 2^\circ * \cos 7^\circ * \cos 45^\circ * (28.32) \\
\text{Step 4: } X' &= \cos 1^\circ * \cos 2^\circ * \cos 7^\circ * \cos 45^\circ *(-14.65 - 1/64 * 28.32) \quad //(\tan 1^\circ = 1/64) \\
&= \cos 1^\circ * \cos 2^\circ * \cos 7^\circ * \cos 45^\circ * (\mathbf{-15.1}) \\
Y' &= \cos 1^\circ * \cos 2^\circ * \cos 7^\circ * \cos 45^\circ *(1/64 * -14.65 + 28.32) \\
&= \cos 1^\circ * \cos 2^\circ * \cos 7^\circ * \cos 45^\circ * (\mathbf{28.1}) \\
\text{Step 5: } X' &= 0.701 * (-15.1) \\
&= \mathbf{-10.59} \approx -10.54 \text{ (Eq. (6.15))} \\
Y' &= 0.701 * (28.1) \\
&= \mathbf{19.7} \approx 19.66 \text{ (Eq. (6.15))}
\end{aligned} \tag{6.17}$$

Although this example illustrates that the iterative process can eliminate the need for computing the tangent of an arbitrary angle, such as 55°, for a simple graphic data path, there exist some implementation complexities, as follows:

- How to select the next β_i in Table 6.12
- When to end the computation
- What scaling factor should be used for a given rotation angle

A solution that resolves all those implementation complexities is to use a fixed number of steps, independent of the target rotation angle [7]. This requires that some rotations in the opposite direction may be necessary if the previous step resulted in an over-rotation. For instance, a vector can be rotated, say, by 55°

in seven steps using 45°, 27°, −14°, −7°, 4°, 2°, and −1°. Furthermore, only one scaling factor = 0.6048 (Eq. (6.18)) would be needed for all target rotation angles. Note that since $cos\ \beta_i = cos\ -\beta_i$, the single scaling factor is not affected by the direction of the rotations. The more steps there are, the closer the scaling factor would become to its maximum 0.607 as the number of steps approaches infinity.

$$0.6048 = \cos 1° * \cos 2° * \cos 4° * \cos 7° * \cos 14° * \cos 27° * \cos 45° \quad (6.18)$$

For a simple graphic data path, each step of the algorithm requires integer arithmetic. Each of the products, 1/8 X, 1/32 X, etc., in Eq. (6.17) is implemented by an arithmetic right shift. For instance, the quantity 1/8 * 10, if converted to its nearest integer, is equal to $10 = (01010)_2$ being right shifted 3 times, as illustrated next. An arithmetic right shift is used to handle both positive and negative coordinate values.

$$1/8 * 10 = (01010)_{2s} / 8$$
$$= (01010)_{2s} >>> 3 \text{ (>>> indicates an arithmetic right shift operator)}$$
$$= 1$$

Or,

$$1/8 * -10 = (10110)_{2s} / 8$$
$$= (10110)_{2s} >>> 3$$
$$= (11110)_{2s}$$
$$= -2$$

The following describes the iterative rotation algorithm for β values that are between −90° and +90° (i.e., β ≤ |90°|). The final new coordinate values are determined by the two expressions in Eq. (6.19).

Iterative Rotation Algorithm for β ≤ |90°|:

$$X_0 = X$$
$$Y_0 = Y$$
$$\beta_0 = \beta$$
$$A_0 = 1;$$
$$d_i = +1\ if\ \beta_i \geq 0,\ or\ -1\ if\ \beta_i < 0$$
$$X_{i+1} = X_i - d_i\ (Y_i >>> i)$$
$$Y_{i+1} = d_i\ (X_i >>> i) + Y_i$$
$$\beta_{i+1} = \beta_i - d_i * tan^{-1}\ 2^{-i} \quad //\text{(Table 6.12)}$$
$$A_{i+1} = A_i * cos\ (tan^{-1}\ 2^{-i})$$

for $i = 0, 1, 2, \ldots k - 1$

i	B_{i+1}	X_{i+1}	Y_{i+1}	d_i	A_{i+1}	$X_i - d_i (Y_i >>> i)$	$d_i (X_i >>> i) + Y_i$
0	10° = 55° –45°	–10 =10 – 20	30 = 10 + 20	1	0.7071	10 – (1) (20 >>> 0)	(1) (10 >>> 0) + 20
1	–17° = 10° – 27°	–25 = –10 – 15	25 = 5 + 30	1	0.6300	–10 – (1) (30 >>> 1)	(1) (–10 >>> 1) + 30
2	–3° = –17° + 14°	–19 = –25 + 6	32 = 7 + 25	–1	0.6113	–25 – (–1) (25 >>> 2)	(–1) (–25 >>> 2) + 25
3	4° = –3° + 7°	–15 = –19 + 4	35 = 3 + 32	–1	0.6067	–19 – (–1) (32 >>> 3)	(–1) (–19 >>> 3) + 32
4	0° = 4° – 4°	–17 = –15 – 2	34 = –1 + 35	1	0.6053	–15 – (1) (35 >>> 4)	(1) (–15 >>> 4) + 35
5	–2° = 0° – 2°	–18 = –17 – 1	33 = –1 + 34	1	0.6049	–17 – (1) (34 >>> 5)	(1) (–17 >>> 5) + 34
6	–1° = –2° + 1°	–18 = –18 + 0	34 = 1 + 33	–1	0.6048	–18 – (–1) (33 >>> 6)	(–1) (–18 >>> 6) + 33

TABLE 6.13 Illustrating the Intermediate Results Obtained for Rotating the Vector b with the Coordinates (10, 20) in Figure 6.24 by 55°

$$X' = A_k * X_k \qquad (6.19)$$
$$Y' = A_k * Y_k$$

The quantity A_k is the scaling factor and is an FP number. For instance, for $k = 7$, $A_7 = 0.6048$. Therefore, for each vector, once its last computed coordinate point (X_k, Y_k) is determined, the X_k and Y_k are then each multiplied by the constant A_k to produce the final new coordinate point (X', Y'). This requires an FP multiplier, and thus it would be performed by CPU. Equation (6.20) shows the calculations of (X_{i+1}, Y_{i+1}) for $i = 0$, 1, and 2 where (X_0, Y_0) = (10, 20) and β_0 = 55°. Table 6.13 lists all the X_{i+1}, Y_{i+1}, and A_{i+1} values for i = 0, 1, and 6.

$$X_1 = cos\ 45° * (10 - (1)\ (20 >>> 0)),\ d = 1 //\ 55° - 45° = 10° \qquad (6.20)$$
$$= 0.7071 * (10 - 20)$$
$$= 0.7071 * (\mathbf{-10})$$
$$Y_1 = cos\ 45° * ((1)\ (10 >>> 0) + 20)$$
$$= 0.7071 * (10 + 20)$$
$$= 0.7071 * (\mathbf{30})$$
$$X_2 = cos\ 27° * cos\ 45° * (-10 - (1)\ (30 >>> 1)),\ d = 1 //10° - 27° = -17°$$
$$= 0.8910 * 0.7071 * (-10 - 15)$$
$$= 0.6300 * (\mathbf{-25})$$
$$Y_2 = cos\ 27° * cos\ 45° * ((1)\ (-10 >>> 1) + 30)$$
$$= 0.6300 * (-5 + 30)$$
$$= 0.6300 * cos\ 45° * (\mathbf{25})$$

$$X_3 = \cos(-14°) * \cos 27° * \cos 45° *(-25 - (-1)\,(25 >>> 2)),\ d = -1$$
$$//-17° - (-14°) = -3°$$
$$= 0.6113 * (-25 + 6)$$
$$= 0.6113 * (-19)$$
$$Y_3 = \cos(-14°) * \cos 27° * \cos 45° * ((-1)\,(-25 >>> 2) + 25)$$
$$= 0.6113 * (7 + 25)$$
$$= 0.6113 * (32)$$

The transformed coordinate point at the end of step 7 is $(X_7, Y_7) = (-18, 34)$ and includes a gain of $1/A_7 = 1.427$. The final new coordinate point (X', Y') is determined by multiplying X_7 and Y_7 by $A_7 = 0.6048$, as follows:

$$\begin{aligned} X' &= 0.6048 * (-18) \\ &= -10.88 \\ Y' &= 0.6048 * (34) \\ &= 20.56 \end{aligned} \tag{6.21}$$

The values $X' = -10.88$ and $Y' = 20.56$ given in Eq. (6.21) are slightly different from the -10.64 and 19.6 that were obtained in Eq. (6.15) due to integer arithmetic.

For rotation angles $\beta > |90°|$, an initial rotation by ±90° or ±180° is required. For example, for $\beta = 125°$, an initial rotation by 90° reduces the target rotation angle to 35°, which is < 90°. Since cosine of ±90° = 0 and sine of ±90° = ±1, an initial ±90° rotation changes the initial values X_0, Y_0, and β_0 as follows:

$$X_0 = -d * Y$$
$$Y_0 = d * X$$
$$\beta_0 = \beta - d * 90 \qquad \text{where } d = 1 \text{ if } \beta \geq 0 \text{ or } d = -1 \text{ if } \beta < 0$$

Alternatively, a 180° initial rotation would reduce a $\beta = 125°$ to $-55° > -90°$ and change the initial values X_0, Y_0, and β_0 as follows:

$$X_0 = -X$$
$$Y_0 = -Y$$
$$\beta_0 = \beta - d * 180 \qquad \text{where } d = 1 \text{ if } \beta \geq 0 \text{ or } d = -1 \text{ if } \beta < 0$$

For a $\beta > |180°|$ and $\leq |360°|$, an initial rotation by ±360° is required to reduce β to $< |180°|$. This, however, will keep the initial values the same as follows:

$$X_0 = X$$
$$Y_0 = Y$$
$$\beta_0 = \beta - d * 360 \qquad \text{where } d = 1 \text{ if } \beta \geq 0 \text{ or } d = -1 \text{ if } \beta < 0$$

Finally, for $\beta > |360°|$, the β is replaced with β mod 360. That is,

$$X_0 = X$$
$$Y_0 = Y$$
$$\beta_0 = \beta \; mod \; 360 \qquad \text{if } \beta > |360°|$$

The pseudo-code shown next specifies the steps necessary to rotate a virtual object by a given angle β.

Iterative Rotation Pseudo-Code:

```
object_transform(x[], y[], β, n, x' [], y' [])
    if β > |360| then
         β = β mod 360;
    endif
    if β > |180| then
         if β > 0 then
              β = β - 360;
         else
              β = β + 360;
    endif
    d = 1;
    if β > |90| then
         d = -1;
         if β > 0 then
              β = β - 180;
         else
              β = β + 180;
    endif
    //Rotate n vectors
    for i = 0 to n-1
         x0 = d * x[i];
         y0 = d * y[i];
         (x7, x7) = vector_transform(x0, y0, β);  //-90° ≤ β ≤ 90°
         x' [i] = 0.6048 * x7;  //apply the scaling factor A7
                                //= 0.6048
         y' [i] = 0.6048 * y7;
    endfor
end

vector_transform(x0, x0, β0)   //use the Iterative Rotation
Algorithm
         x = x0;
         y = y0;
         β = β0
         If (β >= 0)
              d = 1;
         else
              d = -1;
         for (i = 0; i < 7, i++)
```

```
            x = x - d * (y >>> i);
            y = d * (x >>> i) + y
            β = β - d * LUT[i];  //Look-up table <tn>Table 6.12,
                                 //column 3
        endfor
        return(x, y);
end
```

Pipelined Data Path and Control

Typically, a virtual object includes thousands or millions of coordinate points that all must be transformed to new coordinate points when the object is rotated by a given angle. A pipelined data path that implements the aforementioned "vector transform" function given in the pseudo-code can process many vectors in a short time or even in real time. A nonpipelined data path, such as a multicycle data path, would have a lower throughput as compared to a pipelined data path, but would require less hardware.

In a nonpipelined data path, these incremental angles must be stored in a look-up-table (LUT) and would be read one at a time to compute the rotation angle for the next step. In addition, a nonpipelined data path may need to use a combinational shifter (discussed in Chap. 3) to shift the X_i and Y_i values during iteration *i*.

Figure 6.25 illustrates a seven-stage pipelined data path implementing the "vector_transform" function. Note that $-90° \leq \beta \leq 90°$. Each pipeline stage performs one of the seven vector transformation steps, illustrated by an example in Table 6.13. Each stage includes three 2's complement adder/subtractor modules. The sign of an incoming 2's complement rotation angle (*Bin*) is used to determine the value of the direction signal *d* that is used to compute the rotation angle *Bout* and the new coordinate values *Xout* and *Yout* used for the next stage.

Each stage also inputs the arithmetic right shifted values *SXin* and *SYin* obtained by shifting *Xin* and *Yin* values. Note that in the pipeline data path, no circuits are used to generate the *SXin* and *SYin* values; they are hard-wired shifts. The first stage is responsible for transforming an initial coordinate point (*Xin*, *Yin*) by 45°; the second stage is responsible for transforming its input coordinate point by 27°; etc.

HDL Model

The HDL code in Example 6.6 describes the pipelined data path in Figure 6.25. The seven rotation angles are specified in Table 6.12, column 3. All the stages perform the same functions, and, therefore, only one control signal is needed to enable all the pipeline registers.

Example 6.6. An HDL behavior description of the seven-stage pipelined data path in Figure 6.25 is presented.

Solution: The "cordic" module is described structurally, while the "stage" and "register" modules are described behaviorally (i.e., using the Option II design model described in Sec. 6.5.1). Furthermore, in order to simply the description of the "cordic" module, the "stage"

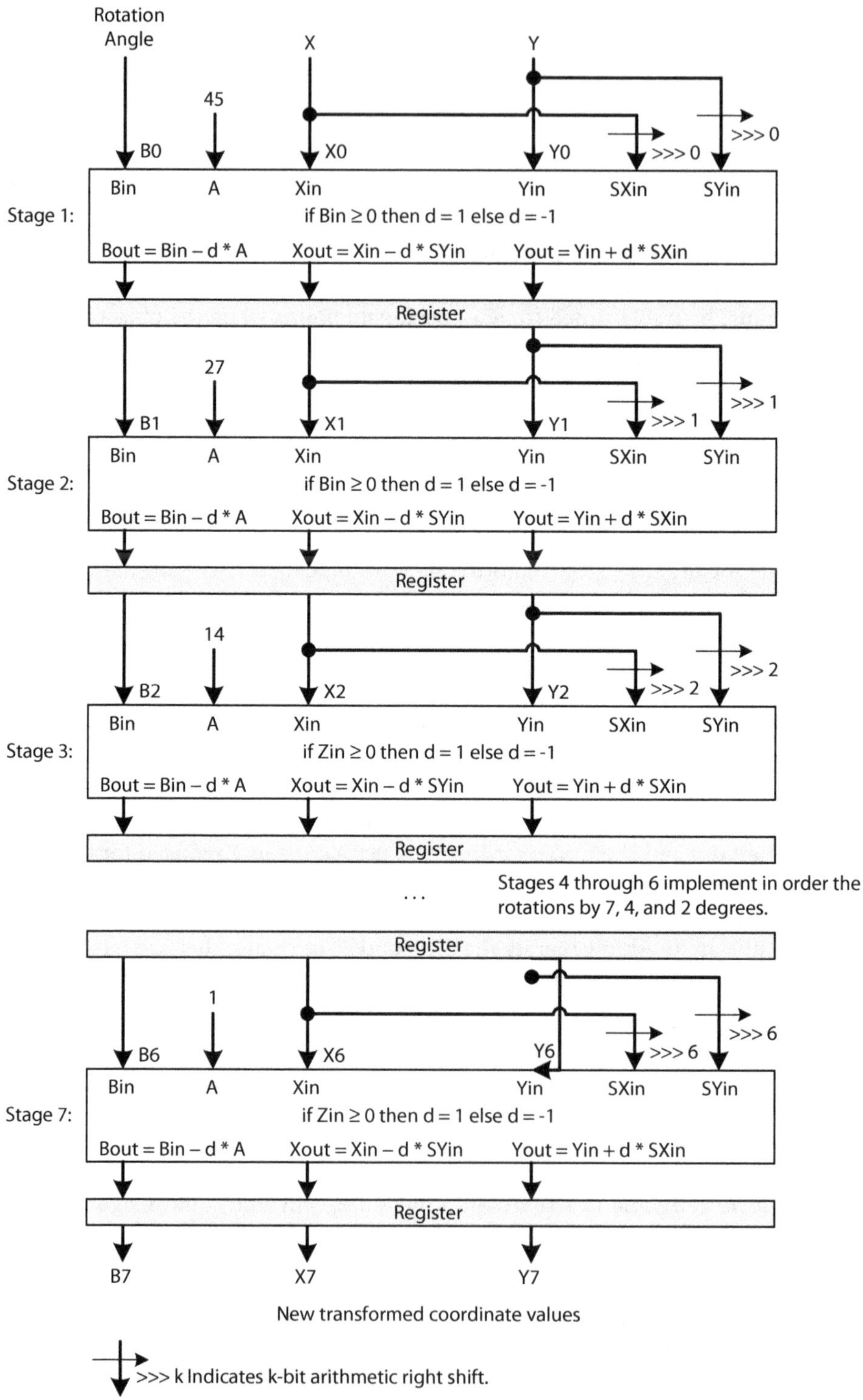

FIGURE 6.25 A seven-stage pipeline data path implementing a 2-D linear transformation.

and "register" modules are instantiated using their port names instead of by their port positions.

```
module cordic(
     input clock, reset, enable,
     input [7:0] xin, yin,
     input [7:0] degrees,
     output [7:0] cordxp, cordyp
  );

  wire [7:0] b0, x0, y0, b0o, x0o, y0o, sx0o, sy0o;
  wire [7:0] b1, x1, y1, b1o, x1o, y1o, sx1o, sy1o;
  wire [7:0] b2, x2, y2, b2o, x2o, y2o, sx2o, sy2o;
  wire [7:0] b3, x3, y3, b3o, x3o, y3o, sx3o, sy3o;
  wire [7:0] b4, x4, y4, b4o, x4o, y4o, sx4o, sy4o;
  wire [7:0] b5, x5, y5, b5o, x5o, y5o, sx5o, sy5o;
  wire [7:0] b6, x6, y6, b6o, x6o, y6o, sx6o, sy6o;
  wire [7:0] b7, x7, y7, b7o, x7o, y7o, sx7o, sy7o;

  assign b0 = degrees;     //target rotation angle in degrees
  assign x0 = xin;         //initial coordinate value X
  assign y0 = yin;         //initial coordinate value Y
  assign cordxp = x7o;     //final computed coordinate value X7
  assign cordyp = y7o;     //final computed coordinate value Y7

  //Stage 1: atan(2^0) = 45 degrees ------------------------
  stage   s1(
     .b(b0), .x(x0), .y(y0), .atan(8'd45), .sx(x0), .sy(y0),
     .bp(b1), .xp(x1), .yp(y1));
  pregregister1(
     .enable(enable),
     .clock(clock),
     .reset(reset),
     .bin(b1), .xin(x1), .yin(y1),
     .sxin({x1[7], x1[7:1]}), .syin({y1[7], y1[7:1]}),
     .bout(b1o), .xout(x1o), .yout(y1o), .sxout(sx1o), .syout(sy1o));

  //Stage 2: atan(2^-1) = 27 degrees -----------------------
  stage   s2(
     .b(b1o), .x(x1o), .y(y1o), .atan(8'd27), .sx(sx1o), .sy(sy1o),
     .bp(b2), .xp(x2), .yp(y2));

  pregregister2(
     .enable(enable),
     .clock(clock),
     .reset(reset),
     .bin(b2), .xin(x2), .yin(y2),
     .sxin({{2{x2[7]}}, x2[7:2]}), .syin({{2{y2[7]}}, y2[7:2]}),
     .bout(b2o), .xout(x2o), .yout(y2o), .sxout(sx2o), .syout(sy2o));
```

```
//Stage 3: atan(2^-2) = 14 degrees ------------------------
stage   s3(
    .b(b2o), .x(x2o), .y(y2o), .atan(8'd14), .sx(sx2o), .sy(sy2o),
    .bp(b3), .xp(x3), .yp(y3));
pregregister3(
    .enable(enable),
    .clock(clock),
    .reset(reset),
    .bin(b3), .xin(x3), .yin(y3),
    .sxin({{3{x3[7]}}, x3[7:3]}), .syin({{3{y3[7]}}, y3[7:3]}),
    .bout(b3o), .xout(x3o), .yout(y3o), .sxout(sx3o), .syout(sy3o));

//Stage 4: atan(2^-3) = 7 degrees -------------------------
stage   s4(
    .b(b3o), .x(x3o), .y(y3o), .atan(8'd7), .sx(sx3o), .sy(sy3o),
    .bp(b4), .xp(x4), .yp(y4));
pregregister4(
    .enable(enable),
    .clock(clock),
    .reset(reset),
    .bin(b4), .xin(x4), .yin(y4),
    .sxin({{4{x4[7]}}, x4[7:4]}), .syin({{4{y4[7]}}, y4[7:4]}),
    .bout(b4o), .xout(x4o), .yout(y4o), .sxout(sx4o), .syout(sy4o));

//Stage 5: atan(2^-4) = 4 degrees -------------------------
stage   s5(
    .b(b4o), .x(x4o), .y(y4o), .atan(8'd4), .sx(sx4o), .sy(sy4o),
    .bp(b5), .xp(x5), .yp(y5));
pregregister5(
    .enable(enable),
    .clock(clock),
    .reset(reset),
    .bin(b5), .xin(x5), .yin(y5),
    .sxin({{5{x5[7]}}, x5[7:5]}),
    .syin({{5{y5[7]}}, y5[7:5]}),
    .bout(b5o), .xout(x5o), .yout(y5o), .sxout(sx5o), .syout(sy5o));

//Stage 6: atan(2^-5) = 2 degrees -------------------------
stage   s6(
    .b(b5o), .x(x5o), .y(y5o), .atan(8'd2), .sx(sx5o), .sy(sy5o),
    .bp(b6), .xp(x6), .yp(y6));
pregregister6(
    .enable(enable),
    .clock(clock),
    .reset(reset),
    .bin(b6), .xin(x6), .yin(y6),
    .sxin({{6{x6[7]}}, x6[7:6]}),
    .syin({{6{y6[7]}}, y6[7:6]}),
    .bout(b6o), .xout(x6o), .yout(y6o), .sxout(sx6o), .syout(sy6o));
```

```
//Stage 7: atan(2^-6) = 1 degrees -------------------------
stage   s7(
    .b(b6o), .x(x6o), .y(y6o), .atan(8'd1), .sx(sx6o), .sy(sy6o),
   .bp(b7), .xp(x7), .yp(y7));
pregregister7(
    .enable(enable),
    .clock(clock),
    .reset(reset),
    .bin(b7), .xin(x7), .yin(y7),
    .sxin({7{x7[7]}}),
    .syin({7{y7[7]}}),
    .bout(b7o), .xout(x7o), .yout(y7o), .sxout(sx7o), .syout(sy7o));

endmodule

module stage(
    input [7:0] b,
    input [7:0] x,
    input [7:0] y,
    input [7:0] atan,
    input [7:0] sx, sy,
    output reg [7:0] bp, xp, yp
);

//b is the 8-bit rotation angle between -90 and +90 degrees
//8-bit x-y coordinate point in a 2D space

wire mode = b[7]; //sign of rotation angle b

always@(*)
begin
    if (mode == 0)
         bp = b - atan; // b' = b – atan if b >= 0
    else
         bp = b + atan; //b' = b + atan if b < 0
end

always@(*)
begin
    if (mode == 0)
         xp = x – sy; //x' = x – y*(2^–i) if b >= 0 for the
                      //ith iteration
    else
         xp = x + sy; //x' = x + y*(2^–i) if b < 0
end

always@(*)
begin
    if (mode == 0)
```

```
            yp = y + sx; //y' = y + x * (2^-i) if b >= 0
        else
            yp = y - sx; //y' = y - x * (2^-i) if b < 0
    end
    endmodule

module preg(
        input enable,
        input clock, reset,
        input [7:0] bin, xin, yin,
        input [7:0] sxin, syin,
        output reg[7:0] bout, xout, yout, sxout, syout
    );

    always@(posedge clock or posedge reset)
    begin
        if(reset == 1)
        begin
            bout <= 0;
            xout <= 0;
            yout <= 0;
            sxout <= 0;
            syout <= 0;
        end
        else if(enable == 1)
        begin
            bout <= bin;
            xout <= xin;
            yout <= yin;
            sxout <= sxin;
            syout <= syin;
        end
    end
    endmodule
```

Simulation

The Verilog model of the pipeline in Example 6.6 was synthesized and simulated using the Altera Quartos II and ModelSim design and simulation tools. Example 6.7 describes a test-bench for transforming the five vectors of the virtual object "house" shown in Fig. 6.24 by 55°. The simulation waveform is shown in Fig. 6.26. For convenience, the simulation data is shown in decimal.

Example 6.7. A test-bench to simulate the pipelined model in Example 6.6 is presented.
Solution: There are only five vectors in the object "house," and thus they are listed as five test vectors in the code. One may read the test vectors from a file if there are many test vectors.

```
module tester();
  reg clock, reset, enable;
  reg [7:0] degrees, x, y;
  wire [7:0] xp, yp;
```

```
cordic   u1(clock, reset, enable, x, y, degrees, xp, yp);

initial begin
clock = 1;
reset = 1;
enable = 0;
#5 reset = 0;
end

always begin
      #5 clock = ~clock;
end

initial begin
enable = 1;
degrees = 55; //rotate by 55 degrees
    x = 10; y = 10; //point a
#10 x = 10; y = 20; //point b
#10 x = 15; y = 30; //point c
#10 x = 20; y = 20; //point d
#10 x = 20; y = 10; //point d
#80
enable = 0;
#20 $finish;
end
endmodule
```

Table 6.14 presents the original and the computed coordinate points captured from the simulation waveform in Fig. 6.26. The new coordinate values include a gain that is equal to 1.653 = 1/0.6048 and, therefore, the rotated virtual object

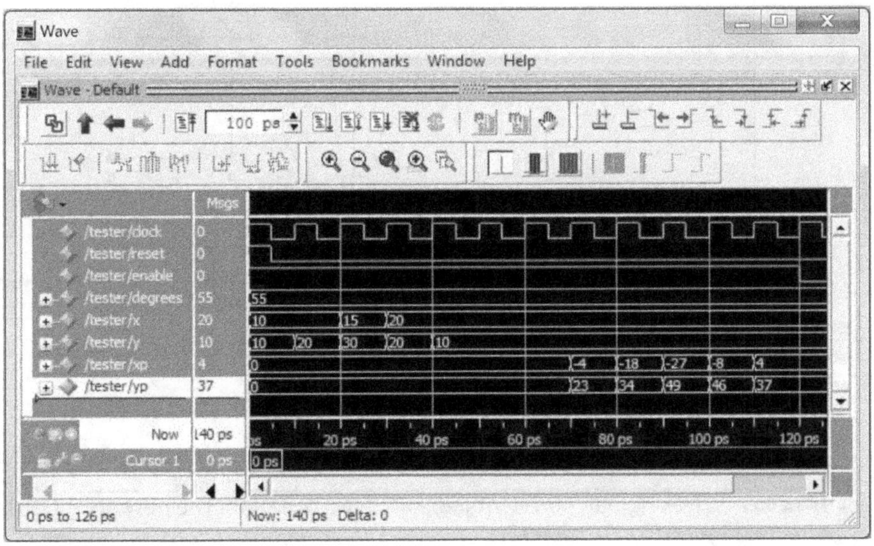

Figure 6.26 A waveform for simulating the pipeline description in Example 6.6; values are in decimal.

"house" Original Coordinate Points (in Decimal)	"house" Rotated Coordinate Points* (in Decimal)
a: (10, 10)	a7: (−4, 23)
b: (10, 20)	b7: (−18, 34)
c: (15, 30)	c7: (−27, 49)
d: (20, 20)	d7: (−8, 46)
e: (20, 10)	e7: (4, 37)
*The coordinate values include a gain of 1.653 = 1/0.6048.	

TABLE 6.14 The Simulation Data Summary Obtained from the Simulation Waveform in Fig. 6.26

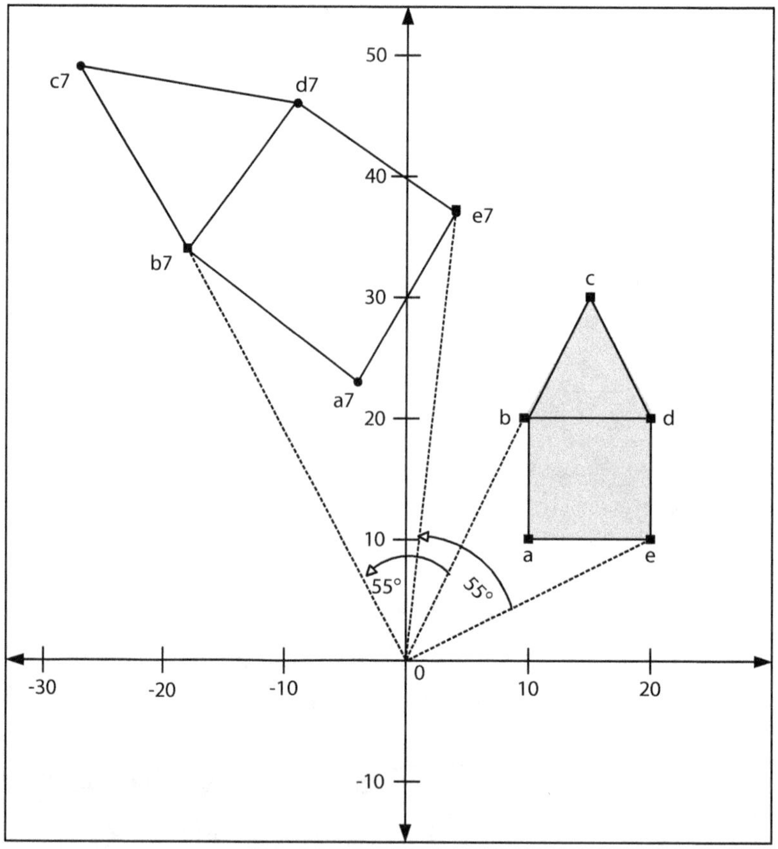

FIGURE 6.27 The original "house" object and its 55° rotation. The new object is shown enlarged by 1.653 = 1/0.6048.

Vector	Computed New Coordinates without the Gain = 1.653	Calculated New Coordinates
a	(−2.42, 13.91)	(−2.46, 13.93)
b	(−10.88, 20.56)	(−10.64, 19.66)
c	(−16.33, 29.64)	(−15.97, 29.49)
d	(−4.84, 27.82)	(−4.91, 27.85)
e	(2.42, 22.38)	(3.28, 22.11)

TABLE 6.15 The new Coordinate Values Computed Using the Iterative Rotation Algorithm versus Calculated Using the Transformation Expressions in Eq. (6.14)

looks bigger, as shown in Fig. 6.27. In order to remove the gain, each of the coordinate values must be multiplied by the constant 0.6048. This requires an FPU, and thus this rescaling must be performed by CPU.

The new computed coordinate values without the gain and those calculated using the transformation expressions in Eq. (6.14) are given in Table 6.15. The computed values are close, but not the same as the calculated ones. The reason for this difference is that the algorithm presented here uses integer division, which results in more rounding errors than if FP division is used.

The CORDIC rotation pipeline may be implemented as a simple 2-D graphic processor with two internal memory units: (1) to store a virtual object's initial coordinate points as input and (2) to store the computed new coordinate points as output. The CORDIC processor will be a co-processor much like a graphic processor unit (GPU). However, the co-processor, in this case, would perform the fixed CORDIC rotation task as outline earlier using its internal memories; the co-processor would have no instructions to execute. The CPU would start the co-processor by initiating the transfer of both the initial and final coordinate points of a virtual object between the main (system) memory and each of the input and output internal memory units. Once the co-processor is done computing the new coordinate points and the new points are transferred to the main memory, the co-processor will inform the CPU, which would then access the new coordinate points from the main memory and after multiply each new coordinate point by the constant scaling factor 0.6048 would display the rotated virtual object on the screen. For how to calculate the throughput of the CORDIC processor refer to Exercise 6.15. Memory design is presented in the next chapter and CPU initiated transfers of large memory data are discussed in Chap. 9.

References

1. Intel Architecture Instruction Set Extensions Programming Reference, www.intel.com.
2. Steven Leibson and James Kim, Configurable processors: a new era in chip design, *IEEE Computer*, 2005, pp. 51-59.

3. SPEC CPU2006 and SPECviewperf from the Standard Performance Evaluation Corporation, http://www.spec.org/.
4. B. W. Bomar, Implementation of microprogrammed control in FPGAs, *IEEE Transactions on Industrial Electronics*, Vol. 49, No. 2, Apr 2002, 415-422.
5. Anantha Chandrakasan and Robert Broderson, Minimizing power consumption in digital CMOS circuits, Proceedings of the IEEE, Vol. 83, No. 4, 1995, 498-523.
6. J. Hennessy and D. Patterson, *Computer Architecture: A Quantitative Approach*, 5th ed., Morgan Kaufmann, 2012.
7. Ray Andraka, A survey of CORDIC algorithms for FPGA based computers, In: *Proc. ACM/SIGDA 6th International Symposium on Field Programmable Gate Arrays*, 191-200, 1998.

Exercises

For Exercises 6.1 to 6.3: Suppose the propagation delay of an 8-bit adder is 0.8 ns, adder/subtractor is 1.1 ns, 2-to-1 MUX is 0.3 ns, and 4-to-1 MUX is 0.6 ns. Also, assume register setup time (τ_{st}), clock-to-q (τ_{cq}), and clock skew (τ_{cs}) are all 0.05 ns.

6.1 Calculate the required maximum clock frequency for each of the following data paths:
   a. Single-cycle data path in Fig. 6.2
   b. Multi-cycle data path in Fig. 6.3
   c. Pipelined data path in Fig. 6.4

6.2 Calculate the total time required to compute the quantity $A + B + C \pm D$ by the data paths in Exercises 6.1*(a)* and 6.1*(b)*.

6.3 Estimate the speedup between the following data paths when generating $N = 1000$ quantities $A_i + B_i + C_i \pm D_i$ for $i = 0, 1, 2, \ldots, 999$. Ignore the data reading and writing delays.
   a. Exercise 6.1(*a*) vs. 6.1(*c*)
   b. Exercise 6.1(*b*) vs. 6.1(*c*)

6.4 Suppose a new processor has 25% less capacitive load than the old processor and operates with 20% higher clock frequency. Determine the ratio of the dynamic powers consumed by the two processors. Comment on the result obtained.

6.5 Suppose the voltage source for a new processor is 50% of that used to operate the older processor, its total capacitive load is 15% less, and it operates with 40% higher clock frequency. Determine the ratio of the dynamic powers consumed by the two processors. Comment on the result obtained.

6.6 Show the register contents for multiplying 3-bit *A_value* = 6 with 3-bit *B_value* = 5 using the multicycle unsigned multiplier given in Fig. 6.15.

6.7 Design an 8-bit unsigned multiplier circuit modeled as follows:
   a. Use a schematic design tool (e.g., LogicWorks) or all structural HDL model to design to model the multiplier circuit. You may design a multifunction register to implement registers *A*, *B*, and *P* in the data path.
   b. Use a hybrid HDL model. Use behavioral models for registers *A*, *B*, and *P* and the mod-8 counter. Then combine the *A*, *B*, *P*, counter and an adder to complete the design.

6.8 Show the register contents for multiplying 4-bit 2's complement *A_value* = 5 and *B_value* = −2 using the 2's complement multiplier given in Fig. 6.21.

6.9 Show the register contents for multiplying 4-bit 2's complement *A_value* = −5 and *B_value* = −2 using the 2's complement multiplier given in Fig. 6.21.

For Exercises 6.10 and 6.11: Use the standard non-return-to-zero inverted (NRZI) generator FSM (see Chap. 5 Exercise section) and design an NRZI conversion system. Assume that the input stream is processed 16-bits at a time. Also, because no common clock is used between a source and a destination module and the bits are transmitted on a pair of twisted wires called D+ and D–, with D– being the opposite of D+, we must prevent the data synchronization problem between the source and destination modules by making sure the NRZI output does not remain at 1 or 0 for several clock cycles. This is done by making sure that for every six consecutive 1's at the input stream there is a transition at the output. This will ensure that an output stream can have maximum seven consecutive 1's or seven consecutive 0's. For example, for input $X = 1\,1\,0\,0\,0\,1\,1\,1\,1\,1\,1\,1\,0\,0\,1\,1$ (0xC7F3) processed, from right to left, the modified NRZI generator must output $Y = 1\,1\,1\,0\,1\,0\,\underline{0}\,1\,1\,1\,1\,1\,1\,1\,0\,1\,1$ (or 0x1D3FB from right to left); for $X =$ 0xFFFF, $Y =$ 0x3E03F; for $X =$ 0xCFF6, $Y =$ 0x123F8; and for $X =$ 0x0000, $Y =$ 0xAAAA. The NRZI system consists of a data path and a control unit. Do the following:

6.10 Design a data path that includes a 16-bit multifunction (parallel load and right shift) input register, a standard NRZI FSM, and a mod-17 counter that keeps track of the input bits. Design an FSM-based control unit for a data path that implements the NRZI and keeps track of six consecutive 1's at the input. (An 18-bit parallel-load and right-shift register may be used, if necessary, to capture the output bits.)

6.11 Design a data path that consists of a 16-bit multifunction input register, a standard NRZI FSM, a mod-17 counter (CNTR1) to keep track of the processed input bits, and a mod-7 counter (CNTR2) that keeps track of consecutive 1's at the input. Design an FSM-based control unit for a data path that implements the NRZI and keeps track of six consecutive 1's at the input. (An 18-bit parallel-load and right-shift register may be used, if necessary, to capture the output bits.)

6.12 Design a microprogrammed controller for the data path in Exercise 6.11.

6.13 Calculate the new coordinate points of the virtual "house" object in Fig. 6.24 rotated by 35°. Compare your results with those calculated using the expressions in Eq. (6.14).

6.14 Calculate the new coordinate points of the virtual "house" object in Fig. 6.24 rotated by −35°. Compare your results with those calculated using the expressions in Eq. (6.14).

6.15 Suppose a 2-D graphic processor implements the seven-step pipelined CORDIC rotation algorithm discussed in the text. Also, assume the delay of an adder or a subtractor is 0.8 ns and register setup time (τ_{st}), clock-to-q (τ_{cq}), and clock skew (τ_{cs}) are all 0.05 ns. What is the approximate maximum number of coordinate points the pipeline can process within 0.001 seconds? Ignore delays associated with reading and writing coordinate points.

6.16 Write a program in the language of your choice to implement the CORDIC rotation pseudocode described in the book.

6.17 The following defines an exponential function as a k-term Taylor series:

$$e^{x} = 1 + \frac{x}{1!} + \frac{x^{2}}{2!} + \ldots + \frac{x^{k-1}}{(k-1)!}$$

Each term in the series can be computed from the previous term, as illustrated here for the first four terms:

$$\text{1st term} = 1$$

$$\text{2st term} = \text{1st term} * \frac{x}{1}$$

$$\text{3rd term} = \text{2nd term} * \frac{x}{2}$$

$$\text{4th term} = \text{3rd term} * \frac{x}{3}$$

Using a combinational adder, a multiplier, a divider, and other modules as necessary, do the following:

a. Draw a multicycle data path to compute e^x for five terms for a given x. Also, determine the minimum clock period in terms of the delays of modules used and , τ_{st}, τ_{cq} and τ_{cs}.
b. Draw a minimum delay pipeline data path to compute e^x for five terms for a given x. Also determine the minimum clock period in terms of the delays of the modules used and τ_{st}, τ_{cq}, and τ_{cs}.

6.18 Computer security (confidentiality): Use Exercise 11.17 to design a stream cipher with control unit (also see Sec. 11.5.1).

6.19 Computer security (confidentiality): Use Exercise 11.18 and/or Exercise 11.19 to understand the RSA encryption algorithm (Sec. 11.5.3).

6.20 Computer security (confidentiality): Use Exercise 11.20 to understand asymmetric versus symmetric ciphers (also see Sec. 11.5).

6.21 Computer security (integrity, understanding cryptography hash): Use Exercises 11.21 to 11.23 (Sec. 11.6 and Sec. 11.7).

CHAPTER 7
Memory

7.1 Introduction

A register is designed to store single value that is readily available when needed. Memory, on the other hand, is designed to store the code and data of programs during execution. The storage technologies used to implement different types of memory require much less hardware than a latch or flip-flop. However, memory requires more time to store (write) or retrieve (read) data.

The storage size of memory is defined in terms of bytes, 8 bits per byte (B). Table 7.1 presents a list of commonly used memory storage sizes, and Fig. 7.1 illustrates two logical views of 1024 B (1 KB) memory. In Fig. 7.1(*a*), the memory is viewed as having 1024 locations, each with 1 B content. That is, a 1024 × 8 (i.e., 1024 by 8) memory requires a 10-bit address ($1024 = 2^{10}$) to identify each of the 1 B (8 bits) content. In Fig. 7.1(*b*), 1024 B memory is viewed as 512 × 16, with 512 locations, each with 2 B content. It requires a 9-bit address ($512 = 2^9$) to identify each of the 2B (16 bits) content.

The performance of a Von Neumann machine (Fig. 1.2 in Chap. 1) directly depends on how fast data can be read or written to memory. Over the years, as the speed of CPUs has increased at a higher rate than that of memory, a quest for faster memory technologies and better memory architectures and organizations have helped bridge this speed gap. Pipelining has been used to increase concurrency by overlapping memory operations and reduce average read/write time. Parallelism

Unit	Reads	Actual Size	Approximate Size
1 KB	One kilobyte	$2^{10} = 1024$ bytes	10^3 B
1 MB	One megabyte	$2^{20} = 1,048,576$ bytes	10^6 B
1 GB	One gigabyte	$2^{30} = 1,073,741,824$ bytes	10^9 B
1 TB	One terabyte	$2^{40} = 1,099,511,627,776$ bytes	10^{12} B

TABLE 7.1 Examples of Memory Sizes

Address		Data
Decimal	Binary (10 bits)	8-bit Content
0	0000000000	00010001
1	0000000001	10000111
2	0000000010	00111100
3	0000000011	11000000
	...	
	...	
	...	
	...	
	...	
	...	
	...	
	...	
	...	
1023	1111111111	10000001

(a) 1K × 8 Memory

Address		Data
Decimal	Binary (9 bits)	16-bit Content
0	000000000	0001000100010001
1	000000001	1000011111100001
2	000000010	0011110000001100
3	000000011	1100000011010100
	...	
	...	
511	111111111	1000000111001011

(b) 512 × 16 Memory

FIGURE 7.1 Two logical views of a 1-KB memory with arbitrary contents: (*a*) 1K × 1B; (*b*) 512 × 2B.

has been used to deliver more data in less time to improve the performance of multiprocessor systems and real-time applications.

This chapter introduces commonly used memory technologies and their applications, and covers memory cell structures, a cell schematic logic model, and the arrangements of memory cells within a memory chip to support various applications. The chapter also covers memory organization, timing, and communication protocols, including those of commonly used memory technologies today. Memory architectures of modern computer systems and an introduction to programming practices for reducing memory traffic and increasing performance are also discussed and examples provided. An example of the hardware description language (HDL) model for memory is also provided.

7.2 Memory Technologies

In general, memory technologies are categorized as read-only memory (ROM) or random access memory (RAM). The hardware used to store 1 bit of data is called a memory cell. In a ROM, the cells are **nonvolatile** and thus can retain their contents even when they are not powered. Other nonvolatile memory technologies used today are magnetic disks, flash memory [1], and optical discs (e.g., CD-ROM) where data is organized and accessed in blocks. Magnetic disks are discussed in Chap. 9.

On the other hand, the cells in a RAM are **volatile** and would lose their content when not powered. Both ROM and RAM are random access in the sense that the amount of time required to access the content of a location is the same. For this reason, ROM is sometimes called nonvolatile RAM (NVRAM).

7.2.1 Read-Only Memories

The content of a ROM cell is fixed at logic 0 or 1. A programmable ROM (PROM) uses fuse-based technologies that make them one-time programmable; each cell can be programmed to logic 0 (e.g., keeping the fuse) or 1 (e.g., burning the fuse). An ultraviolet erasable PROM (EPROM), not commonly used today, would erase its content when placed under an ultraviolet light source (e.g., for 30 minutes). An electrically erasable PROM (EEPROM), which is common today, uses electrically rewritable ROM memory cells to trap logic 0 or 1 values for a long time. EEPROM cells, however, can only be programmed a certain number of times (e.g., typically 100,000 minimum).

7.2.2 Random Access Memories

On the other hand, RAMs are designed to function as the main storage for programs and data during execution. A RAM cell is called **static** if logic 1 is stored as a static charge, and it can be retained as long as the memory is powered. The static-cell RAMs are called SRAMs. A RAM cell is called **dynamic** if logic 1 is stored as a dynamic charge, and it can only be retained for a very short time unless it is refreshed, typically once every few milliseconds (ms). The dynamic cell RAMs are called DRAMs.

SRAM versus DRAM Cells

Figure 7.2 illustrates circuit examples for SRAM and DRAM cells. The SRAM cell, which requires two transistors and two cross-coupled NOT gates, is much larger than the DRAM cell that requires one transistor and a small capacitor. The schematic of an nMOS **pass transistor** is shown in Fig. 7.2(*c*) with three pins labeled *a*, *b*, and *e*. The *e* input acts like an enable signal. If $e = 1$, the nMOS pass transistor conducts current in either direction from *a* to *b* or from *b* to *a* as if the pins *a* and *b* are connected by a wire. Otherwise, if $e = 0$, the transistor keeps the pins *a* and *b* electrically isolated (high impedance, Z), with a very small amount of current flowing in either direction as if there is no connection between the two pins.

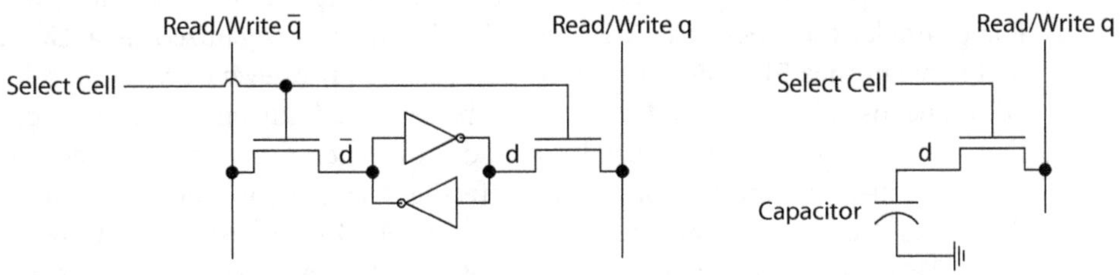

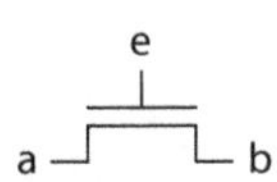

e	Action
0	a and b are disconnected (ellectrically isolated)
1	a and b are connected

(c) NMOS Transistor

Figure 7.2 Memory cells: (*a*) an SRAM cell; (*b*) a DRAM cell; (*c*) an nMOS pass transistor.

In Fig. 7.2(*a*), when the cell is selected, the two pass transistors connect d to q and $\overline{d}$ to $\overline{q}$. The cell content can now be either read or written as the q and $\overline{q}$ signals. Otherwise, when the cell is not selected, the two pass transistors keep the two cross-coupled NOT gates isolated from the q and $\overline{q}$ signals. During this time, the cell retains its stored value d as long as the cell is powered.

DRAM cells operate differently. In Fig. 7.2(*b*), when the cell is selected and the memory is performing a write operation, the capacitor is charged to a voltage level representing logic 1. Otherwise, the pass transistor keeps the capacitor isolated and not connected to the q signal. However, the isolated capacitor, if charged to logic-1 voltage level, retains its charge for only a short time, typically a few milliseconds (ms). Therefore, the charge in the capacitor must be refreshed periodically during a **refresh cycle**. Otherwise, the content would be lost over time due to a leakage current, much like batteries left in a flashlight that is turned off discharge over time.

The capacitor can be charged to a voltage required to represent logic 1 in about $\tau = R * C$ seconds when the cell is selected. The R is the size of the equivalent resistor of the transistor in ohms when the cell is selected. The C is the size of the capacitor in farads. The capacitor can discharge in about $t = R_z * C$ seconds, where R_z is the size of the equivalent resistor of the transistor when the cell is not selected (i.e., the cell is isolated). However, R_z is much greater than R.

[The equation $Vc = Vs\left(1 - e^{-1t/RC}\right)$ defines how a capacitor is charged. If the charging voltage source $Vs = 5.0\,\text{V}$ (V for volts), the voltage level in the capacitor would reach 3.16 V in about $t = 1RC$ seconds, $5.0(1 - e^{-1})\text{V} = 3.16\,\text{V}$. The 3.16 V is the voltage level for logic 1 when using the 0 to 5.0 voltage range. A capacitor would, however, be fully charged in about $t = 5RC$ seconds. The equation $\text{V} = Vc \times e^{-1t/R_zC}$ defines how a capacitor discharges [2].]

A DRAM refresh cycle must be performed before the cells lose their contents. For example, the Micron 64MB DRAM [3], which is organized as a 128M × 4 memory and has 512 million (128M * 4) cells, requires every cell to be refreshed within 64 ms. Therefore, many cells must be refreshed during each refresh cycle so that the DRAM can still be used for read/write operations between each refresh cycle. For instance, the Micron DRAM refreshes 64K (65,536) cells at the same time during a refresh cycle. This requires 8192 (512M/64K) refresh cycles to refresh all the 512M cells once every 64 ms. This implies that each refresh cycle can start once every 7.8 μs (64 ms/8192 cycles) to refresh 64K cells at the same time. The DRAM requires 7.5 ns to read or write its content. Therefore, the DRAM can support more than 1000 (7.8 μs/7.5 ns) read/write operations between each refresh cycle.

During a read operation, the charge in the capacitor is measured to determine the cell content as 1 or 0. If the capacitor's charge is at the logic-1 voltage level, some of its charge would be lost during this read operation, and thus the cell must be refreshed again. This is done by performing a write operation after each read to restore the contents of the cells that were just read.

7.2.3 Applications

There are many applications for ROMs. For example, EEPROM is used to store a startup (**bootloader**) program that begins to execute when the system is powered, or to store a configuration file for a ROM-based programmable logic device (PLD). EEPROM technologies are also used in the design of flash memory, for example, flash drives. However, because data in a flash drive is organized and accessed in blocks, like in magnetic disks and optical discs, flash drives are also relatively slow. Many portable devices today use flash drives in place of disk drives.

SRAMs require more hardware and thus are more expensive per byte, but they are faster than equivalent DRAMs because they do not require refresh and write-after-read cycles. Modern computers use DRAMs as main memory and SRAMs as cache memory to decrease the average time required to access data. Cache memory organizations are discussed in Chap. 10.

7.3 Memory Cell Array

All memories internally use two-dimensional (2-D) cell organizations in order (1) to reduce the total number of signals required to select a set of target cells, and (2) to refresh multiple cells at the same time if the cells are dynamic. A 2-D organization would require two selection signals per cell instead of only one that would be needed if the cells were organized in one dimension. However, a 2-D organization requires far fewer selection signals. For example, consider a 128B memory that contains 1024 ($128 * 8$) cells. If the cells are organized in one dimension, as illustrated in Fig. 7.3(*a*), 1024 selection signals would be needed to select 1024 cells, one at a time. The cells are said to be organized as a $1K \times 1 \times 1$ (i.e., 1K by 1 by 1) cell array consisting of 1024 rows, with one column and one cell at each row-column intersection.

On the other hand, as illustrated in Fig. 7.3(*b*), a $32 \times 32 \times 1$ cell array consists of 32 rows, 32 columns, and one cell at each row-column intersection. It requires only 64 ($32 + 32$) selection signals to select, one at a time, 1024 ($32 * 32$) cells. Likewise, a $1M \times 1$ memory would require 1M selection signals (a very large number) in one-dimensional organization versus only 2048 ($2 * 2^{10}$) selection signals in a two-dimensional organization. Clearly, a two-dimensional organization is advantageous.

A $32 \times 32 \times 1$ cell array would also require only 32 refresh cycles, each one refreshing all the 32 cells in one row, versus 1024 refresh cycles, each one refreshing only one cell in a $1K \times 1 \times 1$ cell array. When one of the row-selection signals is asserted, all the cells in that row are selected at the same time. This is called a **row activation**. Furthermore, because a row-selection signal is the output of an address decoding circuit (not shown), the signal does not have the required fan-out to directly activate a large number of cells. Instead, the signal enables a transistor that allows a power source to activate all the cells on that row.

The content of each cell on the activated row is determined ("sensed") as logic 0 or logic 1 using a special electronic circuit known as a **sense amplifier**. The amplifier compares a cell's voltage level with a reference voltage source, for example, 50% of

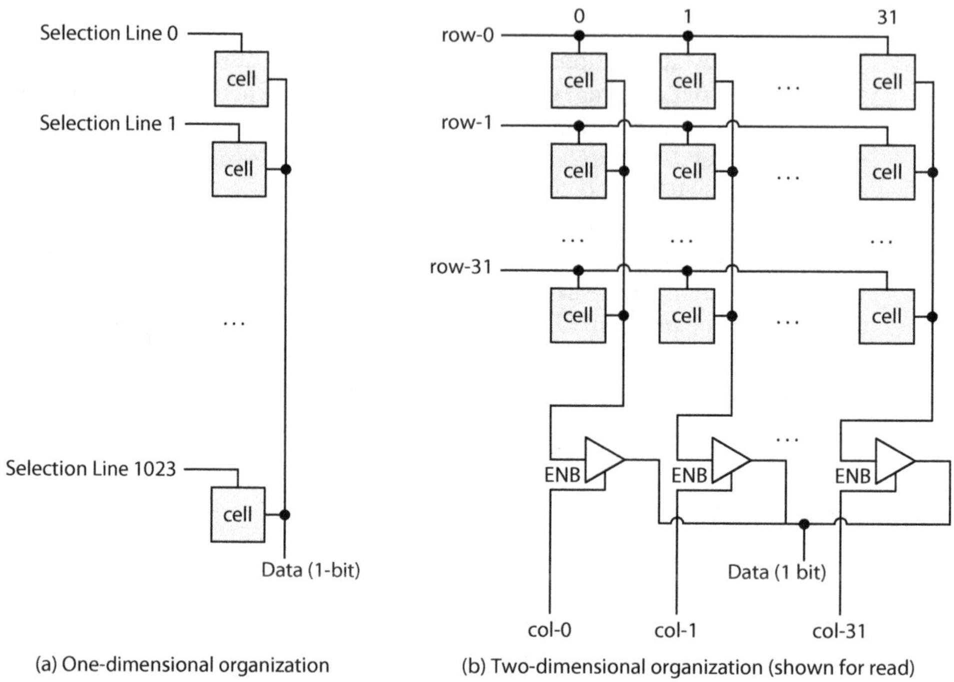

Figure 7.3 The internal organization of a 1024-cell memory: (*a*) one-dimensional organization requiring 1024 selection signals; (*b*) two-dimensional organization requiring only 64 selection signals.

the voltage used to represent logic 1. During a read operation, if the cell contains logic 0, it will pull down the reference voltage slightly, causing the sense amplifier to detect logic 0; otherwise, the voltage level for logic 1 in the cell will pull the reference voltage slightly up, causing the sense amplifier to detect logic 1.

Only one sense amplifier per column is needed. The logic 0 or 1 output from the sense amplifier is latched and is made available to the tri-state buffers controlled by the column-selection signals. In the case of DRAMs, the output of the latch is also used to refresh the target cell after each read. A more detailed discussion of sense amplifiers is outside the scope of this book and thus is referred to elsewhere.

7.3.1 Word Access

Figure 7.4 illustrates the organization of a 512 × 2 (a 2-bit word) memory. In the figure, the 1024 cells are organized as two separate 32 × 16 × 1 cell arrays, effectively creating a 32 × 16 × 2 cell array. The cells are now accessed two at a time, one from each of the 32 × 16 × 1 cell arrays. For example, the assertion of the row-0 and col-0 selection signals will select the cell at the intersection of row 0 and column 0, as well as the cell at the intersection of row 0 and column 16. The 32 × 16 × 2 cell array, which still consists of 32 rows and 32 columns, requires, as in Fig. 7.3, 32 refresh cycles to refresh a row of 32 cells at the same time.

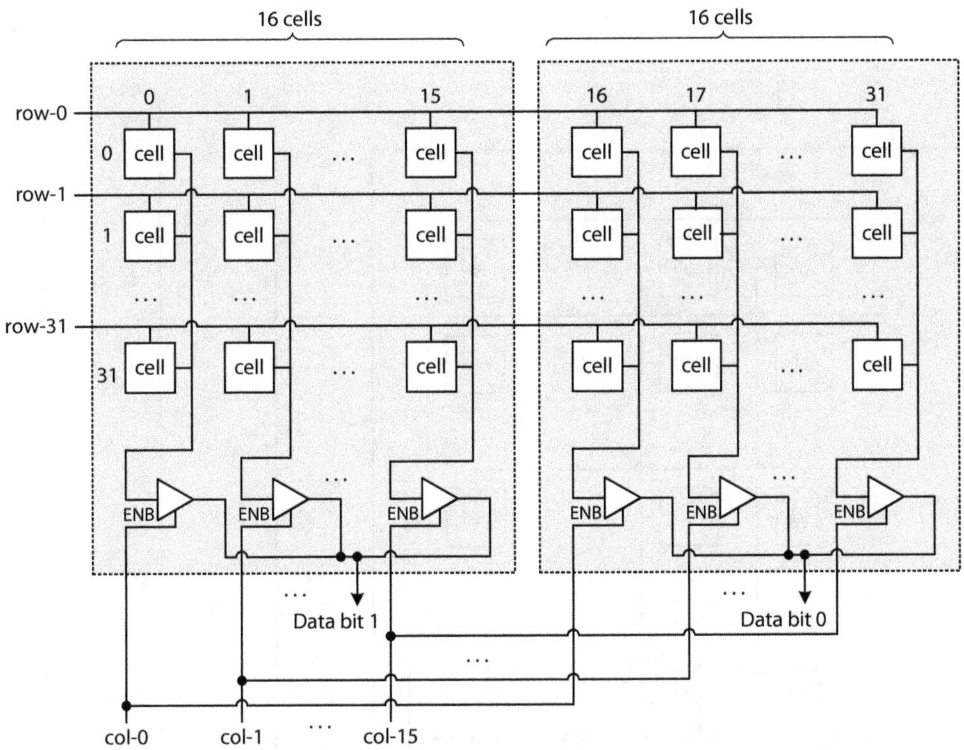

FIGURE 7.4 A 1K cell organized as a $32 \times 16 \times 2$ cell array to create a 512×2 memory (shown for read operation).

7.3.2 Burst Access

A burst access refers to the memory's ability to transfer a burst of data (one word each). A burst size may be small—a few bytes—or large—the size of a large block (e.g., 4 KB) called a **page**. Burst accessing is implemented by activating a row and then asserting the column-selection signals one at a time and in some specified order (e.g., sequentially) to either read or write a set of target cells. Each burst access must be first preceded by a row activation operation. If a page access, or **page mode access**, expands multiple rows, the memory may be enabled to automatically activate each succeeding row when the access from the current row is completed. A page mode access increases **memory efficiency** by reducing the memory idle time, and at the same time, it makes a page transfer seamless.

In order to make memory even more efficient, the cells can be organized into **banks**, each a cell array, as illustrated in Fig. 7.5. In the figure, a $4K \times 1$ memory is designed using four banks, each a $32 \times 32 \times 1$ cell array. In this case, a row activation operation in another bank can be started while the memory operation on the current bank is still in progress. Therefore, this makes bank-access turn around

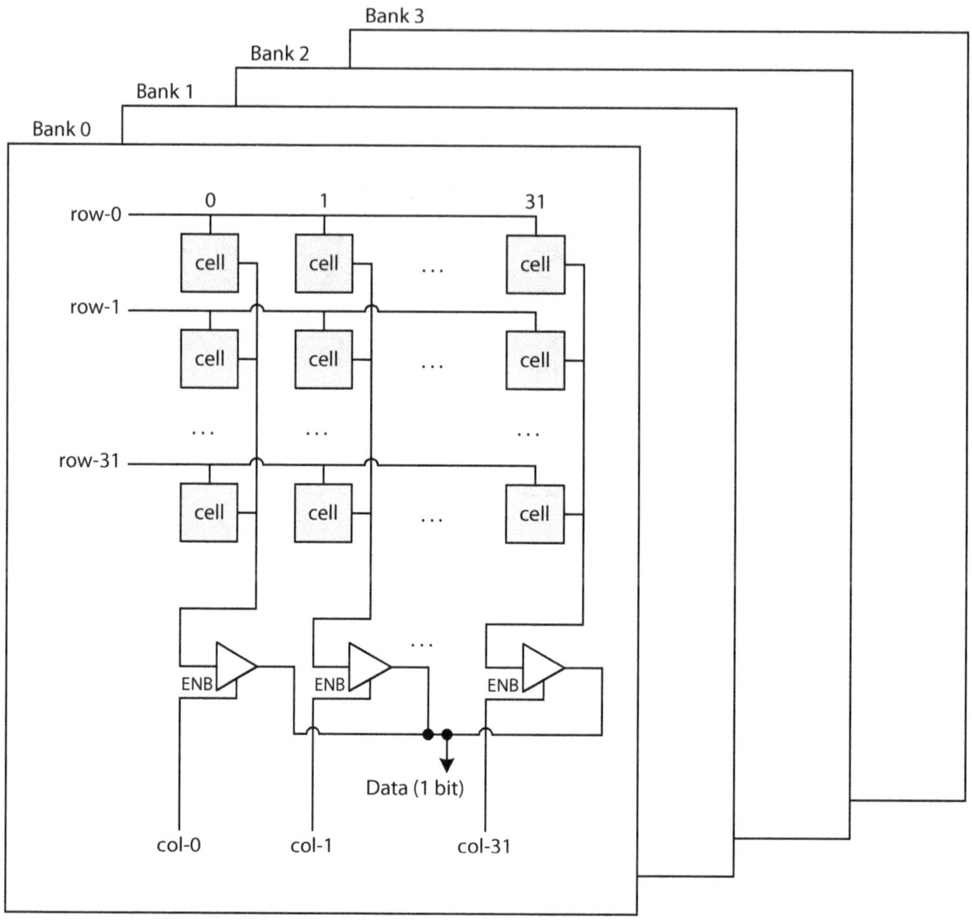

FIGURE 7.5 The organization of a 4K × 1 memory into four banks, each a 32 × 32 × 1 cell array.

time shorter when memory is required to perform operations that involve different banks.

A multibank memory may be designed to support intermittent short burst accesses from one bank while a block transfer is taking place from another bank. In this case, an intermittent access would momentarily interrupt an ongoing block transfer from one bank to allow a short burst access from another bank.

Most modern memory chips support word access and are also multibanked. They also support burst and page mode accesses. For example, consider the Micron 512 Mb (million bits) DRAM memory [3]. It is available as a 128M × 4 RAM with four banks, each a 8192 × 4096 × 4 cell array; a 64M × 8 RAM with four banks, each a 8192 × 2048 × 8 cell array; or a 8M × 16 RAM with four banks, each a 8912 × 1024 × 16 cell array. For instance, the Micron 128M × 4 RAM supports 1, 2, 4, or 8 burst word and 4096-bit page transfers, where a word is 4 bits.

7.4 Memory Organization

Memory organization refers to the internal components and their organization within a memory chip, using several memory chips to create a larger memory called a **memory unit**, and memory communication protocols. Three sets of signals—**address bus, data bus**, and **control bus**—are used to control the operations of a memory chip or unit. The address bus signals specify the address of a single memory location, which could be the starting address of a burst or page access. They are used to select a set of target cells for read or write operation. For a basic memory organization, the control bus is used to specify read or write operations, or neither in case the data bus is also used to communicate with other components in the system.

Figure 7.6 illustrates the logical view of a 1K × 1 memory, its block diagram as an SRAM, and its block diagram as a DRAM. The SRAM requires a 10-bit address bus with signals labeled a_0 *to* a_9; a 1-bit data bus labeled *d*; and three active-low control bus signals labeled *_ce* (chip enable), *_we* (write enable), and *_oe* (output enable). The *_ce*, when asserted, selects the memory chip and, therefore, a set of target cells to perform either a read operation if *_we* = 1 (not asserted) or a write operation if *_we* = 0 (asserted). The *_oe* signal, when asserted, causes the data from the SRAM to appear on the data bus during a read operation.

A DRAM, which typically has many more cells and also requires the cells to be refreshed, has additional control signals. The row address strobe (*ras*) and column address strobe (*cas*) are used to specify a single memory address in two parts, a row address and a column address, using the address bus. The *_ras* and *_cas* signals are also used to place the memory in a refresh cycle mode. The DRAM block diagram in the figure has a 5-bit address bus labeled a_0 *to* a_4, a 1-bit data bus also labeled *d*, and a 5-bit control bus, all active-low signals.

7.4.1 Modern DRAMs

A modern DRAM chip is designed to operate synchronously, and the chip is called a synchronous DRAM (SDRAM). The chip may contain one or more pipelined

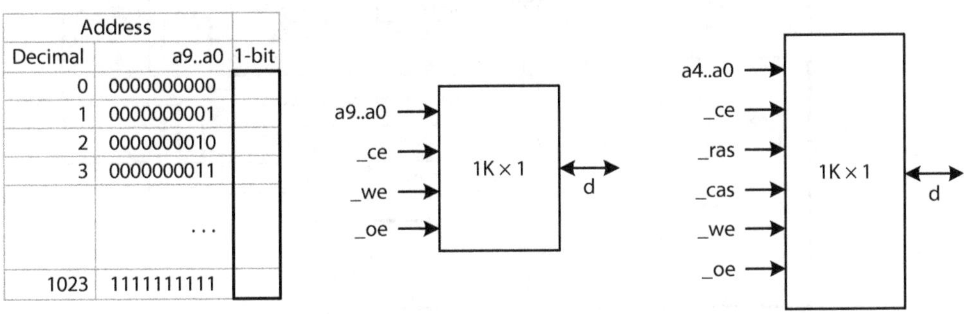

(a) Logical View (b) An SRAM block diagram (c) A DRAM block diagram

FIGURE 7.6 Logical view and block diagram of 1K × 1 memory with active-low control signals: (*a*) logical view; (*b*) SRAM block diagram; (*c*) DRAM block diagram.

data paths to increase memory bandwidth by processing multiple read/write requests concurrently.

Typically, a modern DRAM chip uses the interface signals *_ce*, *_ras*, *_cas*, and *_we* as a 4-bit instruction, also called a memory **command,** to select and send row and column addresses, and an **access mode** (e.g., a single or burst access) to the chip. Commands are also used to perform other memory tasks, such as to start a refresh cycle. Figure 7.7 shows the data path of a modern SDRAM consisting of two memory banks. It contains registers to load an access mode and row and column addresses. A bank number (0 or 1) is the same as one of the row address (e.g., the highest or the lowest) bit. The column address is stored in a counter and is incremented every clock cycle when the access mode indicates a burst or page transfer.

Table 7.2 presents a list of commands used in a Micron SDRAM. For example, if *command* = $(0000)_2$, which implies *_ce* = 0, *_ras* = 0, *_cas* = 0, and *_we* = 0, the SDRAM inputs an access mode using the address bus, which is then loaded into the mode register, as shown in the figure. If *command* = $(0011)_2$, a row address (including a bank number) is loaded into one of the row-address registers

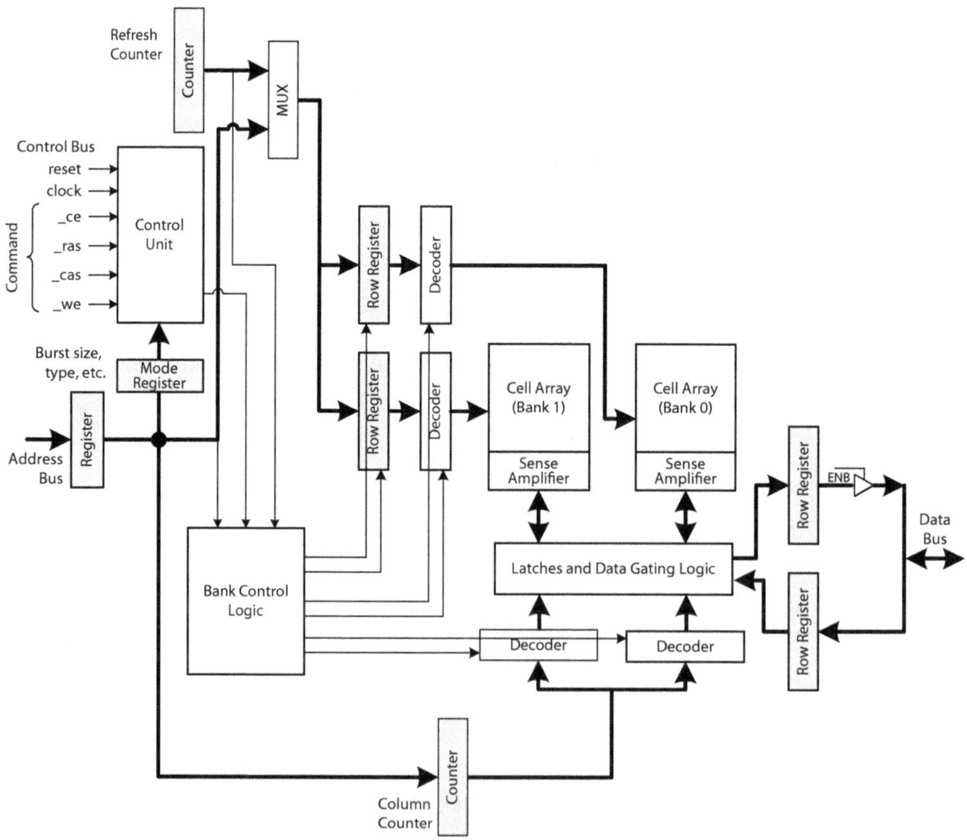

FIGURE 7.7 An internal organization of an SDRAM with two banks. A signal from the row address selects one of the banks. Not all signal connections are shown [3].

Action	Command			
	_ce	*_ras*	*_cas*	*_we*
Chip not selected	1	d	d	d
No operation (NOP)	0	1	1	1
Inputs next row address (row is activated)	0	0	1	1
Inputs next column address and start a read in burst mode	0	1	0	1
Inputs next column address and start a write in burst mode	0	1	0	0
Terminates burst mode (e.g., for a single read or write)	0	1	1	0
Deactivates current row (gets ready for a new row)	0	0	1	0
Starts a refresh cycle	0	0	0	1
Specifies access mode using the address lines (e.g., burst access or block access)	0	0	0	0

d: don't-care

Table 7.2 Examples of Commands Used by a Micron SDRAM

identified by the bank number. If *command* = $(0101)_2$, a column address is loaded into the column address counter. The row address is used to activate a row in the target bank. The counter is used to generate column addresses if the access mode indicates a burst access.

Over the years, the demands for greater memory bandwidth and various system requirements have produced various memory technologies. Advancements in memory chip internal organization and communication protocols have resulted in today's high-performance SDRAMs. For example, while an SDRAM operates at the speed of one data item per clock cycle, a double data rate (DDR, DDR2, DDR3, etc.) SDRAM operates at the speed of two data items per clock cycle; one data item is transferred on the positive edge of the clock signal and another on the negative edge of the clock signal.

Another example is the Rambus proprietary technologies, such as RDRAM and XDR DRAM high-speed, point-to-point communication using **packets** [4]. In this case, a packet is a short burst of, for example, 1-bit wide data transmitted from a source (e.g., memory) to a destination module (e.g., a processor). Packet communications work similar to how people communicate with letters. Each letter (a packet) contains a source address, a destination address, and a payload (content) that goes through one or more (point-to-point) post offices before arriving at the destination.

7.4.2 SRAM Cell Model

While a real memory cell cannot be modeled with logic gates, the behavior of an SRAM cell can be modeled with logic gates to illustrate memory design and its operation. Figure 7.8(*a*) illustrates a schematic logic model of an SRAM cell. It

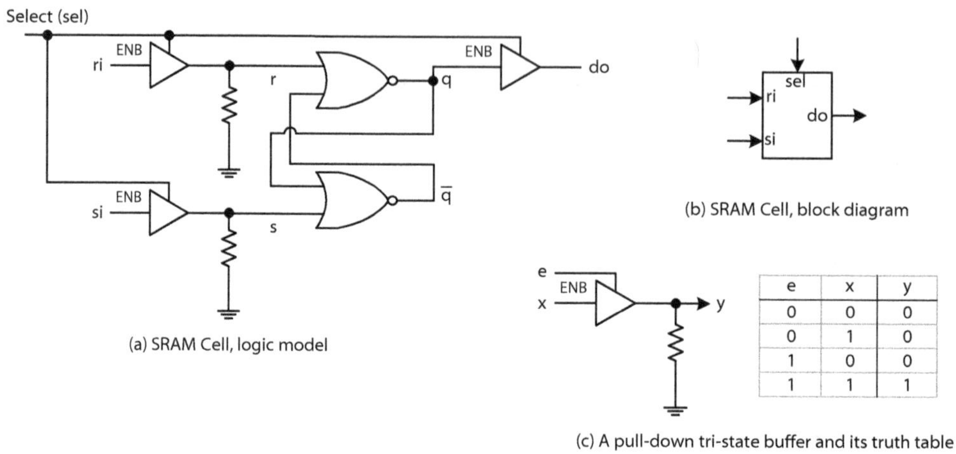

e	x	y
0	0	0
0	1	0
1	0	0
1	1	1

Figure 7.8 A logic model of an SRAM cell: (*a*) gate-level model; (*b*) the cell's block diagram; (*c*) a pull-down tri-state buffer and its truth table.

consists of an SR latch (without the clock) and three tri-state buffers. Two resistors that connect the outputs of the input tri-state buffers to ground are called pull-down resistors. They cause the outputs to become 0 instead of high impedance (Z) when the tri-state buffers are not enabled. This makes the *s* and *r* inputs of the SR latch both 0, causing the latch to retain its stored 1 or 0 value when the cell is not selected. The model uses tri-state buffers to mimic the functions of the pass transistors used in Fig. 7.2(*a*). The SR latch becomes electrically isolated when it is not selected, much like the cross-coupled NOT gates in a real SRAM cell. Figure 7.8(*b*) shows the cell's block diagram, and a pull-down tri-state buffer and its truth table are shown in Fig. 7.8(*c*).

7.4.3 Internal Organization: SRAM Chip

While a cell array is the core storage hardware, additional circuits are needed to translate a given memory address into one of many row and column selection signals and route data in and out of cell array, typically using a bidirectional data bus. A bidirectional bus reduces the number of wires required to transmit data in and out of memory.

Example 7.1. The design of 16 × 1 SRAM is presented. It requires a 4-bit address bus, a 1-bit bidirectional data bus, and a 3-bit control bus.
Solution: The logical and block diagrams of the SRAM are shown in Fig. 7.9 with four address bus signals labeled a_3 to a_0, a 1-bit bidirectional data bus signal labeled *d*, and three active-low control bus signals labeled *_ce*, *_we*, and *_oe*. The design detail and read/write operations are described next.

Figure 7.10 illustrates the internal organization of the SRAM using a 4 × 4 × 1 cell array and two 2-to-4 decoders. The row decoder translates the upper two address signals a_3 and a_2 into four row-selection signals. The column decoder translates the lower two address signals a_1 and a_0 to four column-selection signals.

Address		1-bit
Decimal	a3a2a1a0	
0	0000	
1	0001	
2	0010	
3	0011	
	...	
15	1111	

(a) Logical View

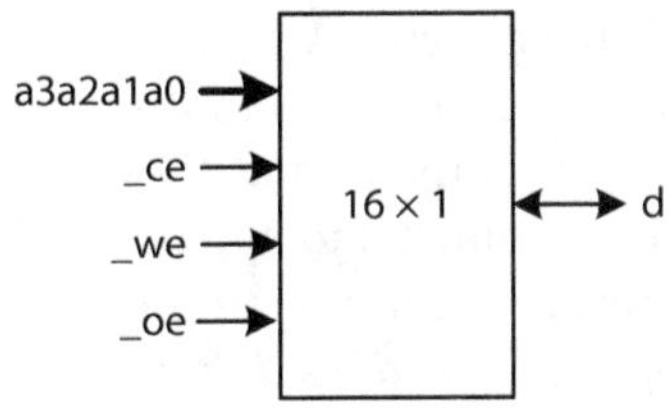

(b) Block diagram

Figure 7.9 Logical view and block diagrams of a 16 × 1 SRAM.

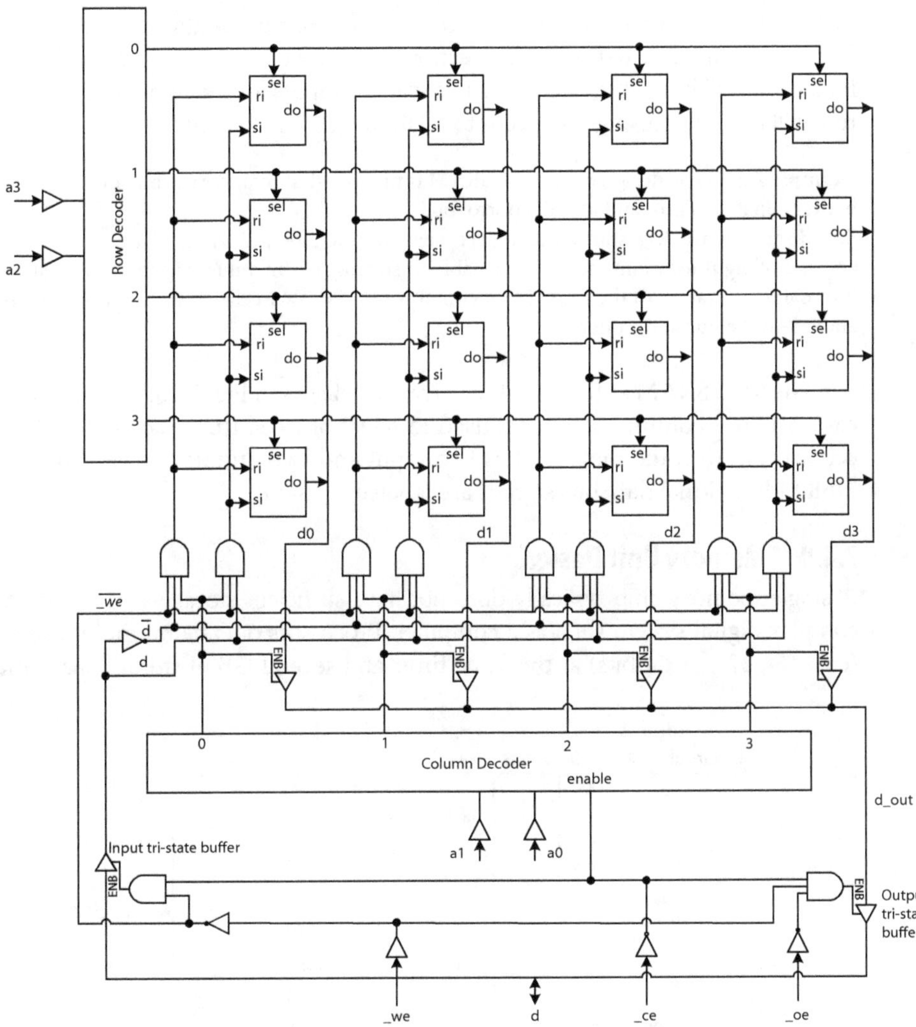

Figure 7.10 The internal organization of a 16 × 1 SRAM using a 4 × 4 × 1 cell array.

The row decoder may always be enabled to allow early row activation, but the column decoder must be enabled when *_ce* = 0, as shown in the figure. All the input signals are buffered to avoid possible fan-out violations at the source of the signals.

During a memory operation when *_ce* = 0, an active row-selection signal will enable all the four cells in the row associated with a given memory address. This will enable all the output tri-state buffers in each cell on that row, and will cause their 1-bit contents to appear on the four data lines d_0 to d_3 within the cell array. An active column-selection signal enables one of the column tri-state buffers to pass d_0, d_1, d_2, or d_3 as *d_out*. During a read operation when *_we* = 1, the *d_out* is placed on the data bus if *_oe* is asserted. The single output tri-state buffer is used to design a 1-bit bidirectional data bus labeled *d*.

A memory write operation works similar to the read operation, but this time, the single input tri-state buffer is enabled to route an incoming data *d* into the cell array. The tri-state buffer reduces the SRAM's power consumption during a read operation. The *_we*, *d*, and asserted column-selection signal are used to generate $si = d$ and $ri = \bar{d}$ for a target cell in the activated row. The *si* and *ri* inputs for the remaining cells on that row will be zero, causing these cells to retain their 1-bit contents.

Example 7.2. The design of an 8 × 2 SRAM is presented. It requires a 3-bit address bus, a 2-bit bidirectional data bus, and a 3-bit control bus.

Solution: Figure 7.11 shows the logical view and the block diagram of an 8 × 2 SRAM that consists of eight locations, each one capable of storing a 2-bit value. Figure 7.12 illustrates the internal organization of the SRAM that contains two 4 × 2 × 1 cell arrays, a 2-to-4 row decoder, and a 1-to-2 column decoder.

The 8 × 2 SRAM still uses a 4 × 4 array of cells, as in Fig. 7.10, except that in this case, a 1-to-2 column decoder is used to select one cell from each of the 4 × 2 × 1 cell arrays. Also, the memory has two input and two output tri-state buffers. The 2-bit bidirectional data bus signals are labeled d_1 and d_0.

7.4.4 Memory Unit Design

A single memory chip typically does not have sufficient storage space to support a complex digital system such as a computer. Processors operate on multiple data bits (e.g., 16-, 32-, or 64-bits) at the same time, and several GB of storage space may be

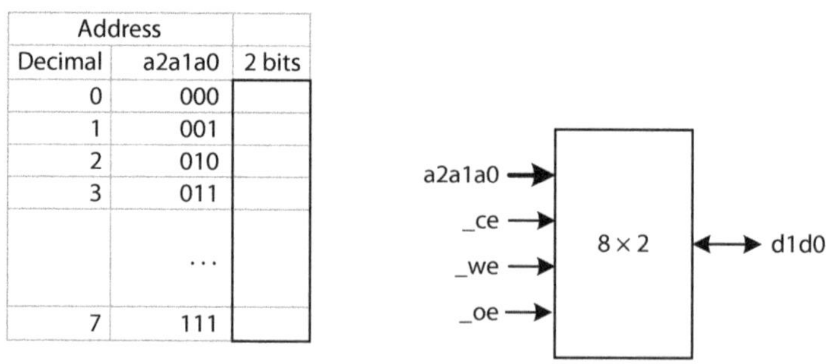

(a) Logical view (b) Block diagram

Figure 7.11 Logical view and block diagram of an 8 × 2 SRAM.

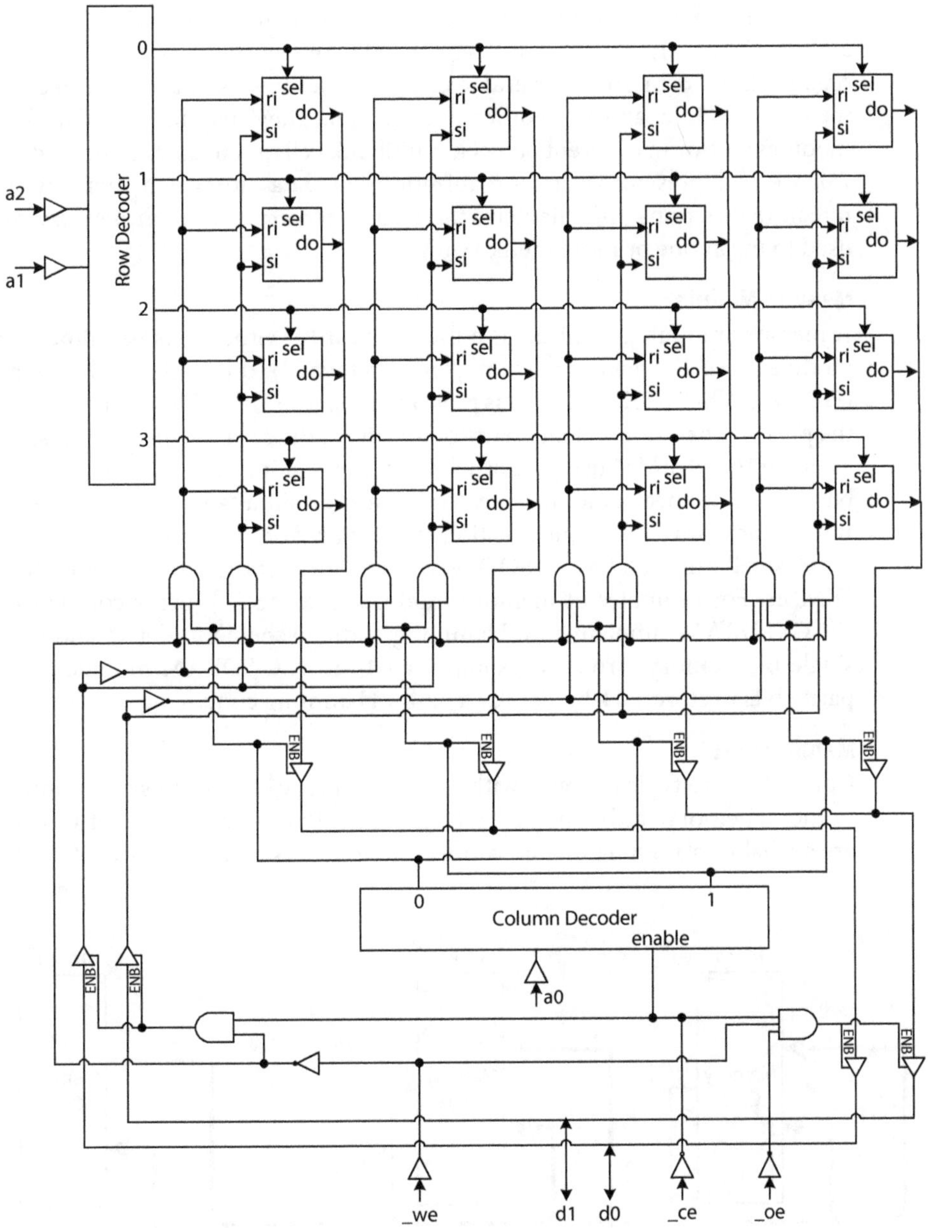

Figure 7.12 The internal organization of an 8 × 2 SRAM using two 4 × 2 × 1 cell arrays.

necessary to store program instructions and data, each as a memory data item, during execution. A memory unit refers to the physical organization of the storage space, known as the main memory that may be accessed by one or more processors. When there is more than one processor, each processor must take turns accessing the memory unit, which typically is designed using one or more memory modules (e.g., memory cards). Each module stores multiple bytes, generally 4B or 8B, per memory address.

The internal organization of a memory unit depends on how data is distributed among the many individually referenced storage spaces (i.e., cells). In one organization, data from several consecutive addresses may be stored in one memory module, and in another organization, they may be stored in different modules and/or in different banks if multibank cell organizations are used. In Sec. 7.6, we will present several organizations of data storage in memory. In the remaining sections and chapters, the terms *memory* and *main memory* may be used to mean one or more memory units.

Memory Module

A memory module organized as a memory card is either a single inline memory module (SIMM), which is not very common today, or dual inline memory module (DIMM). While a SIMM card has pins on one side, a DIMM is smaller and has half the pins on one side and half on the other side of the card. Figure 7.13 illustrates an example 16 × 8 SIMM using eight 16 × 1 memory chips. Address, data, and control bus signals connect to all the memory chips for simultaneous read/write access. Buffer gates prevent fan-out violation at the source of the bus signals.

A small outline DIMM (SODIMM) is about half the size of regular SDRAM DIMMs. For examples of memory modules, refer to [5]. Error correction code (ECC) SDRAMs implement a Hamming coding scheme to detect and correct a single-bit memory error. For example, a 64-bit ECC SDRAM module uses eight parity bits to store a 64-bit data as a 72-bit Hamming code.

Memory Unit

Computers are typically built with several memory expansion slots to install one or more memory cards, one per slot. Assuming that only 2-GB and 4-GB DIMMs are available, a 1G × 64 memory unit is designed using either four 2-GB DIMMs

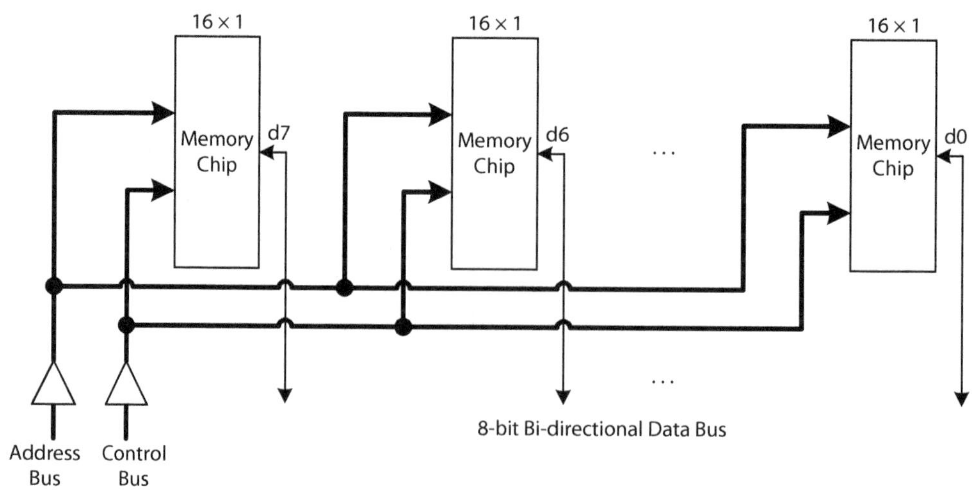

FIGURE 7.13 A 16 × 8 SIMM.

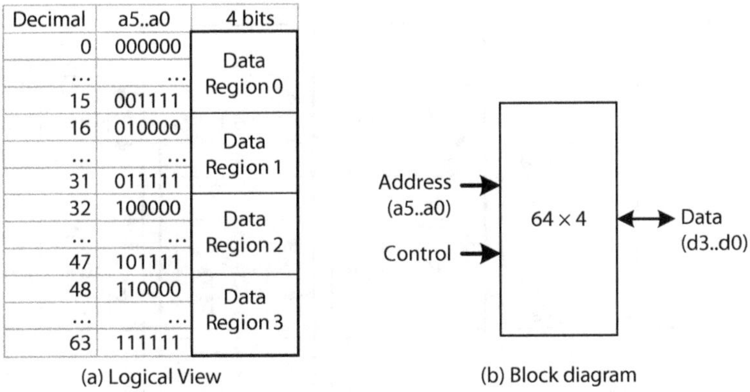

FIGURE 7.14 A 64 × 4 memory unit: (*a*) logical view divided into four data regions; (*b*) block diagram.

or two 4-GB DIMMs, depending on how many memory expansion slots are available.

Example 7.3. The design of a 64 × 4 (32 B) memory unit using 16 × 4 memory modules is presented, where a module uses 16 × 1 memory chips.
Solution: Figure 7.14 shows the logical view and block diagram of a 64 × 4 memory unit with six address lines labeled a_5 to a_0 and four data bus lines labeled d_3 to d_0. The memory unit requires four (64/16) modules each to store data from one-fourth of the address space. Two of the six address signals are used to identify a target memory module that will perform a read or write operation. The remaining four address signals are used to identify a 4-bit target memory content from the selected memory module. For example, the address bits a_5 and a_4 can be used to divide the memory address space into four data regions of size 16 each, as shown in Fig. 7.14(*a*). The data from each region is stored in one memory module.

The memory unit requires a 2-to-4 decoder to translate the address bits a_5 and a_4 to four chip enable signals $\_ce_0$ to $\_ce_3$, one for each of the memory modules, as illustrated in Fig. 7.15. The decoder itself is enabled when a master chip enable (i.e., *_ce*) is asserted. The address lines a_3 to a_0 and the remaining control signals, *_we* and *_oe* if SRAM or *_ras, _cas, _we, and _oe* if DRAM or SDRAM, are used to perform a read or write operation. Only the enabled memory module can transfer data using the 4-bit data bus.

The internal organization of the 64 × 4 memory unit in Fig. 7.15 stores each region's data, shown in Fig. 7.14(*a*), in one memory module. Other data storage organizations are presented in Sec. 7.6.

7.5 Memory Timing

A memory-timing diagram precisely illustrates memory communication protocol. It specifies the timing of SRAM or DRAM control signals or the timing of SDRAM commands. The total time required to select target cells and perform a read or write operation is called a **memory access time**. A **memory cycle** includes both an access time and a **data transfer time**.

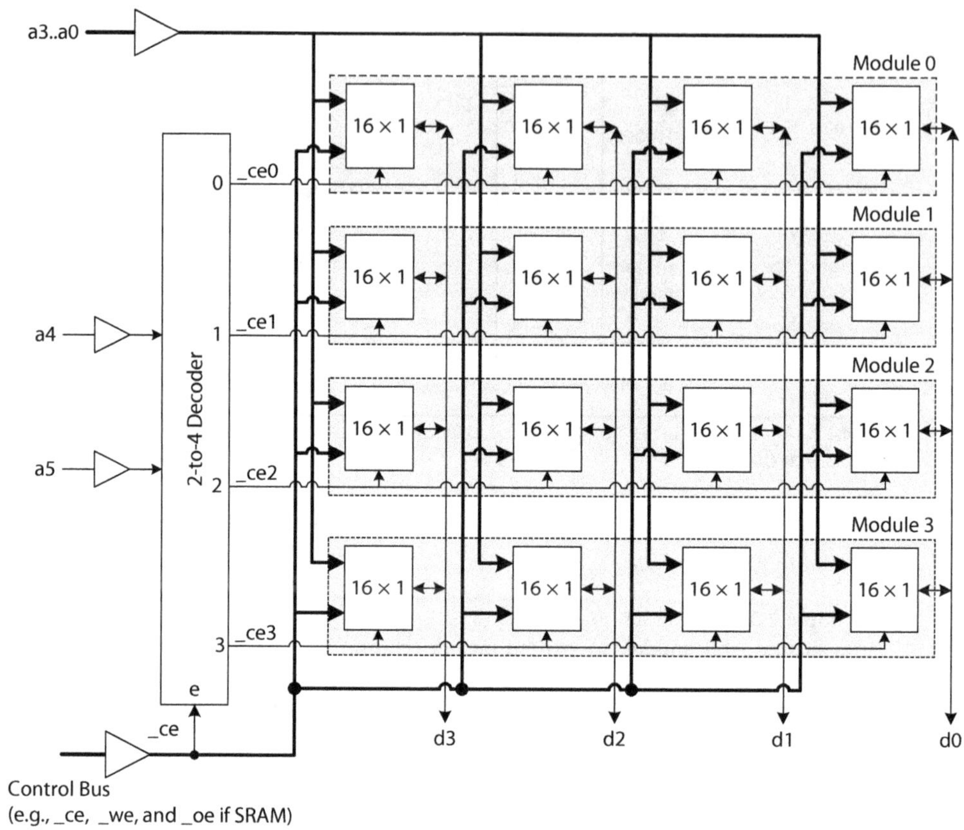

FIGURE 7.15 A 32-B memory organized as a 64 × 4 memory unit.

The access time is directly proportional to the size of the memory cell array, which determines the size of the row and column decoders inside the memory chip. The decoders in turn determine the time required to activate a target row and access target cells. Moreover, the decoders are multilevel and thus have long propagation delays. In addition, while the communication protocols of a memory unit, module, or chip are the same, the protocols differ with each memory technology.

7.5.1 SRAM

Figure 7.16 illustrates an SRAM read cycle from the memory point of view. A memory-timing diagram can also be drawn from the CPU point of view that will be discussed in Chap. 9.

At the start of a memory read cycle, a memory address is placed on the address bus prior to or at the same time that the *_ce* signal is asserted.

The *_oe* signal is used with the *_ce* to control one or more output tri-state buffers (e.g., Fig. 7.12). In order to minimize the duration of a read cycle, the *_oe* can be asserted at any time within a maximum time after the *_ce* is asserted, as illustrated in the timing diagram. The *_oe* allows the data bus to be used only when

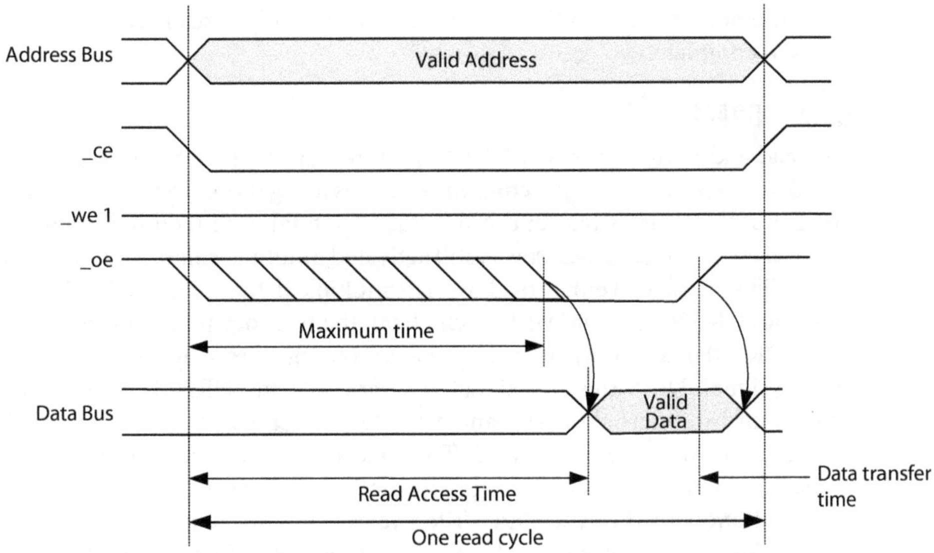

Figure 7.16 An SRAM read cycle from the memory point of view.

data from the cell array is available and not before. The *_ce* is deasserted last because it would disable the output tri-state buffer(s) as well as the column decoder.

A memory write cycle is similar to a read cycle, except that data must be placed on the data bus at the same time that *_ce* is asserted or within a maximum delay after *_we* is asserted to minimize the time the data bus is used. Figure 7.17 illustrates an SRAM memory write cycle.

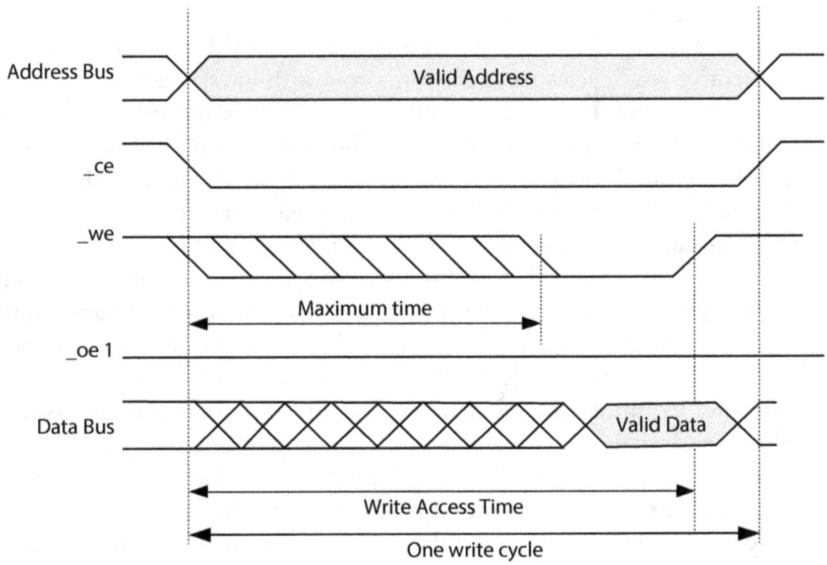

Figure 7.17 An SRAM write cycle from a memory point of view.

A memory cycle is initiated by CPU and typically takes multiple CPU clock cycles to complete.

7.5.2 DRAM

The read and write cycles of a DRAM require that the target memory address be issued in two parts as row and column addresses using the address bus. This reduces the size of the address bus, and can reduce the total time required to complete a burst access. That is, while a row is still activated, multiple column addresses can be applied in sequence to either read or write multiple data values quickly. Figure 7.18 illustrates DRAM read and write cycles from the memory point of view.

A DRAM read or write cycle starts by issuing a row address and asserting the *_ce* signal. Next, the *_ras* signal is asserted so the DRAM can load the row address into an internal register and activate a target row. Next, a column address is issued and the *_cas* is asserted. This selects one or more target cells on the activated row. The *_oe* and *_we* signals function as described earlier for SRAMs.

In addition to the read and write cycles, DRAMs require refresh cycles to restore the content of each cell. A **cas-before-ras refresh** cycle, for example, requires the assertion of the *_cas* signal before *_ras* to switch the DRAM into a refresh mode. DRAMs are used to build SDRAMs with standard communication protocols that have simplified the way computers are designed. This, therefore, has helped reduce the cost of computers.

7.5.3 SDRAM

In addition to standardized communication protocols, SDRAMs implement different refresh cycles, including the normal and partially powered-down modes [3]. A memory cycle starts with a row-activation command followed by one or more read/write commands for the same row.

Figure 7.19 illustrates a hypothetical SDRAM timing diagram for two consecutive read cycles from the same row with burst size = 4. It is assumed that one SDRAM bus clock cycle is required to enter a burst size, three clock cycles to activate a row, and two clock cycles to complete a read or write access. Each data value takes one clock cycle to be transmitted out of SDRAM. Furthermore, because an SDRAM's data path is pipelined, another read cycle for the same row may start while the data for the previous cycle is being transferred.

Specifically, in Fig. 7.19, a burst size of four is issued on the first clock cycle (burst type is not shown), followed by a row activation command on the second clock cycle, which also requires a row address. The target row will be activated by the end of the fourth clock cycle. On the fifth clock cycle, a read command is issued for column address x, and thus the four data values associated with column addresses x to $x + 3$ appear on the data bus starting at the seventh clock cycle for three more clock cycles. The column addresses $x + 1$ to $x + 3$ are generated internally. The second read command (from the same row) is issued at the ninth clock cycle. Its four data values appear on the data bus starting at the eleventh clock cycle as y to $y + 3$; however, only the data for address y is shown in the figure.

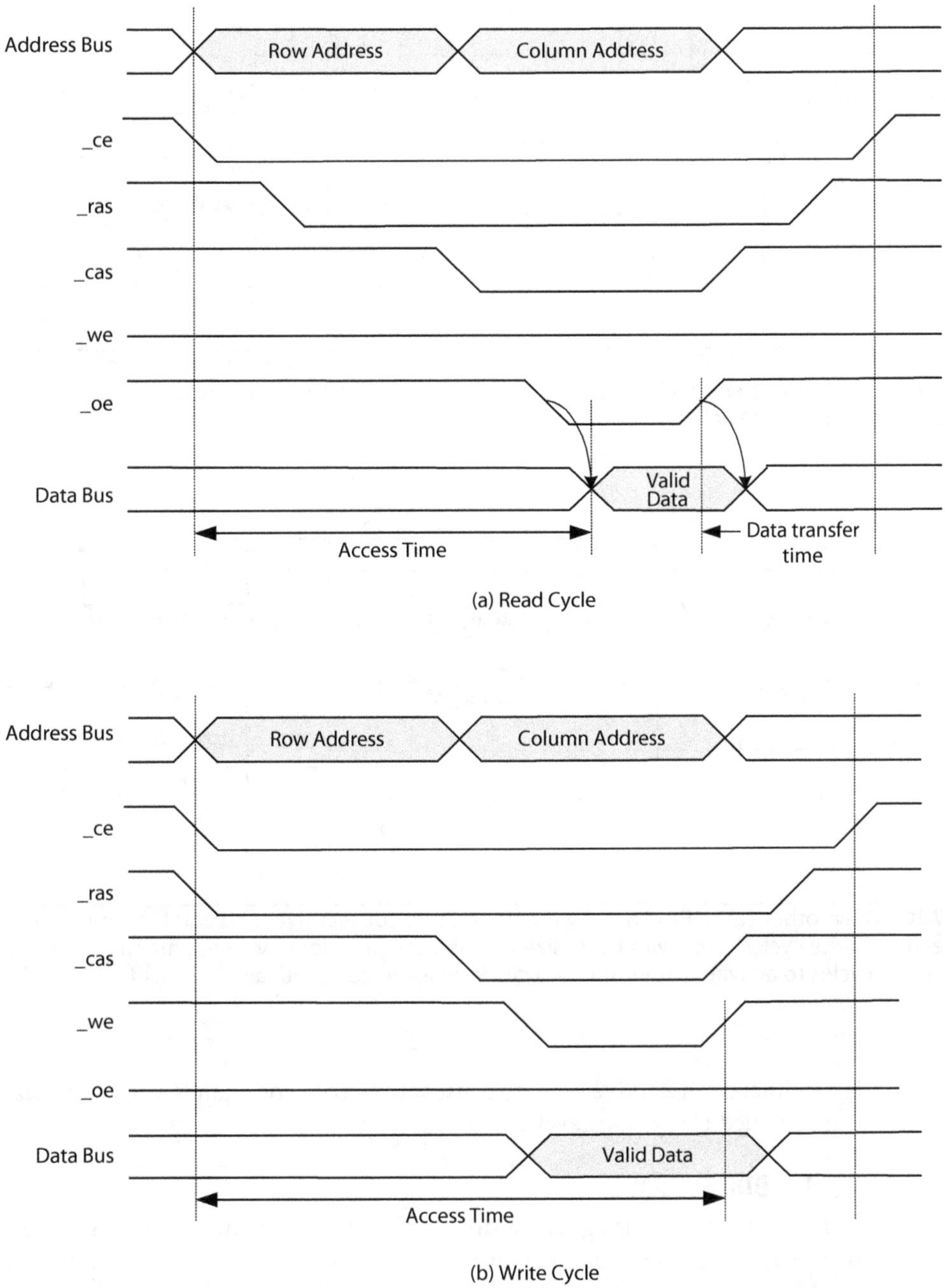

FIGURE 7.18 DRAM read and write cycles from the memory point of view: (*a*) read cycle; (*b*) write cycle.

Figure 7.20 illustrates a hypothetical SDRAM timing diagram with two consecutive read cycles with burst size = 1 followed by two consecutive write cycles also with burst size = 1 for the same row. Note that because the SDRAM data path

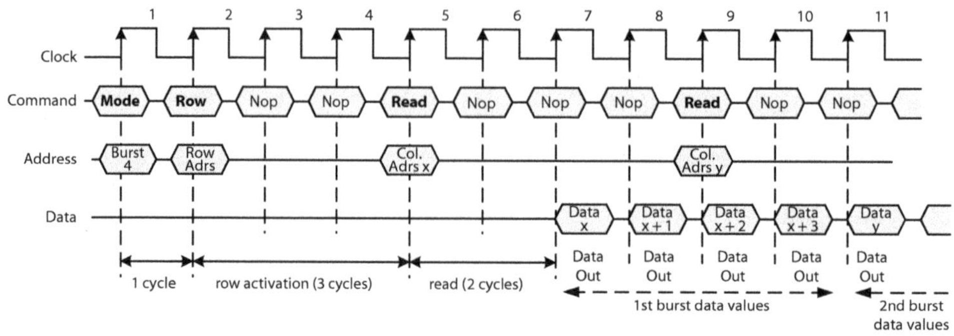

FIGURE 7.19 A hypothetical SDRAM access illustrating two consecutive read cycles with burst size = 4. Assume 1 clock cycle to enter a burst size, 3 clock cycles to activate a row, and 2 clock cycles to complete a read or write access.

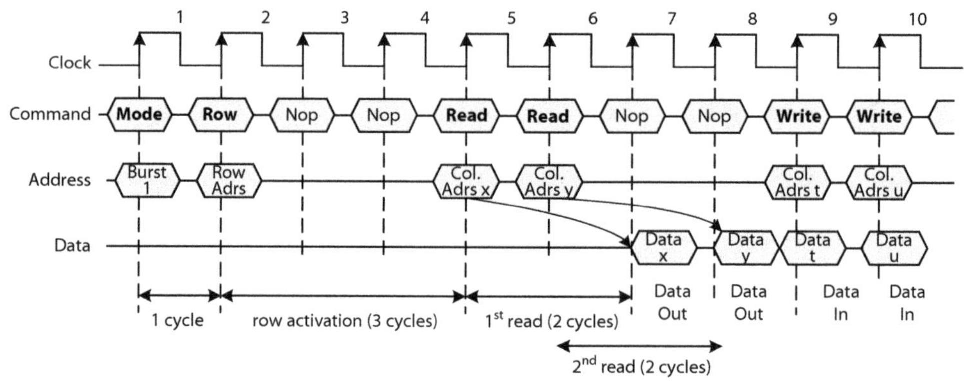

FIGURE 7.20 A hypothetical SDRAM access illustrating two consecutive read and then two consecutive write cycles, each with burst size = 1. Assume one clock cycle to enter a burst size, three clock cycles to activate a row, and two clock cycles to complete a read or write access.

is pipelined, consecutive read or consecutive write commands with burst size = 1 can be issued every clock cycle.

7.5.4 DDR SDRAM

DDR SDRAMs are designed to double the effective bandwidth of SDRAMs. In this case, data is transmitted on every positive edge and every negative edge of the clock. Figure 7.21 illustrates a hypothetical DDR SDRAM timing diagram with three read cycles for the same row with burst size = 4. As illustrated in the figure, the four data values of each read cycle are transferred in two clock cycles for a total of 6 clock cycles versus 12 in a SDRAM. Therefore, DDR effectively doubles **peak memory bandwidth**, which is defined as the maximum memory bandwidth when data bus is utilized 100% of the time (i.e., zero idle time).

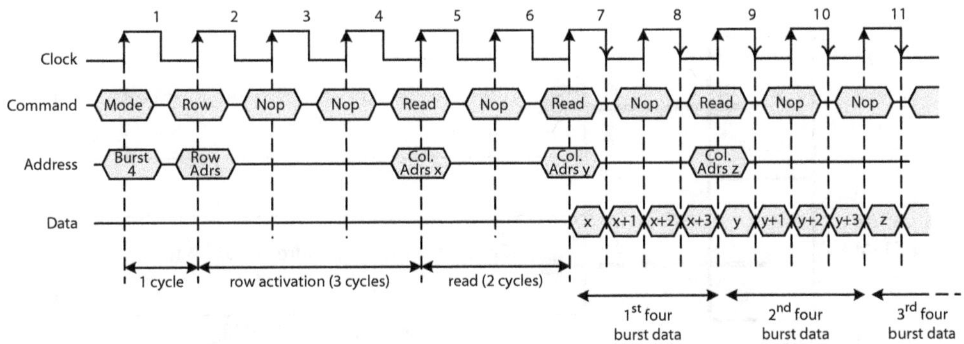

Figure 7.21 A hypothetical DDR SDRAM timing diagram illustrating three burst read cycles with burst size = 4. Assume one clock cycle to enter a burst size, three clock cycles to activate a row, and two clock cycles to complete a read or write access.

7.6 Memory Architecture

Memory architecture refers to various ways that data can be stored in (i.e., distributed to) two or more memory units, two or more memory modules, or two or more memory banks (cell arrays) in order to increase efficiency and thus achieve higher bandwidth. For example, in a two-processor system where each processor communicates with its own dedicated memory unit, it is more efficient if the data accessed by each processor is stored in a memory unit assigned to the processor. Moreover, the data in each of the memory units can be organized in ways to optimize and increase its bandwidth to satisfy the data rate required by the processing cores in each of the processors.

7.6.1 High-Order Interleaving

A high-order interleaving technique divides the total logical memory space into two or more continuous data regions. The data of each region is stored in separate memory modules, or even separate memory units. For example, consider the two-way high-order interleaving of 1 GB data to two memory units, as illustrated in Fig. 7.22. In this case, the two data regions are identified by byte addresses 0 to 512M – 1 (the top half, starting from address 0) and 512M to 1G – 1 (the bottom half).

In a k-way high-order interleaving, the k most significant address bits are used to partition the logical memory space into 2^k regions. For instance, the memory organization in Fig. 7.15 is an example of a four-way high-order interleaving of 32B memory space, organized as a 64×4 memory unit, into four regions, referenced by byte addresses 0 to 15 (region 0), 16 to 31 (region 1), 32 to 47 (region 2), and 48 to 63 (region 3). Data of each region, as shown, is stored in a separate memory module.

7.6.2 Low-Order Interleaving

Consider a multibank memory organization. A low-order (or fine) interleaving, also commonly known as simply interleaving, stores the first data item in one

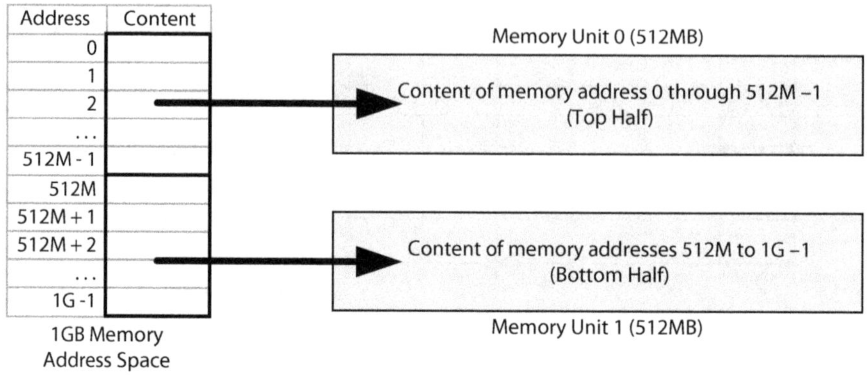

Figure 7.22 A two-way high-order memory interleaving of 1 GB data.

memory bank, the next sequentially accessed data item in another memory bank, and so on, and then it repeats. For example, assuming that each data item is stored as a 4B word, a four-way (low-order) interleaving stores word 0 in bank 0, word 1 in bank 1, word 2 in bank 2, word 3 in bank 3, and then starts over and stores word 4 in bank 0, word 5 in bank 1, etc. This is illustrated in Fig. 7.23 for a 512-MB memory space, organized as a 128M × 4B memory module, using four-bank memory chips. In this case, each four consecutively accessed 4B words are stored in four different banks.

Figure 7.24 illustrates the timing diagram of a hypothetical two-bank DDR SDRAM with interleaved data storage. In this case, the amount of data accessed is doubled from four in Fig. 7.21 to eight in Fig. 7.24, effectively doubling the peak bandwidth of the memory without changing the internal clock frequency of the SDRAM. Twice as many data items must now be transmitted to a destination

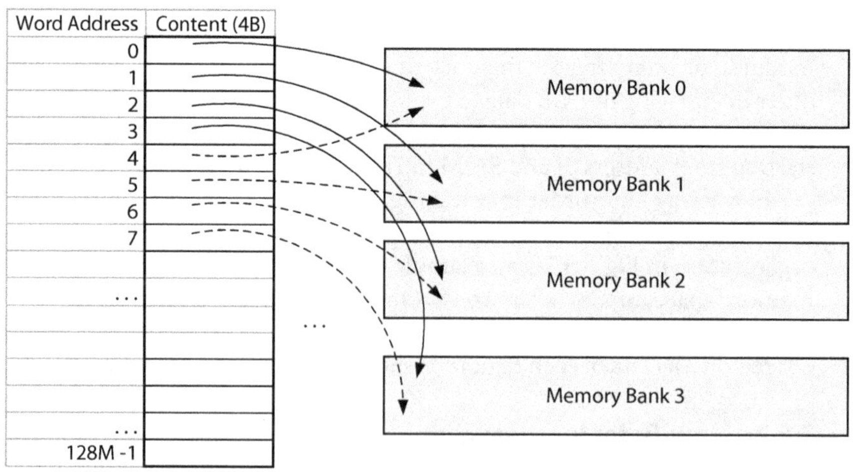

Figure 7.23 A four-way low-order interleaving of 512 MB data in a four-bank memory chips.

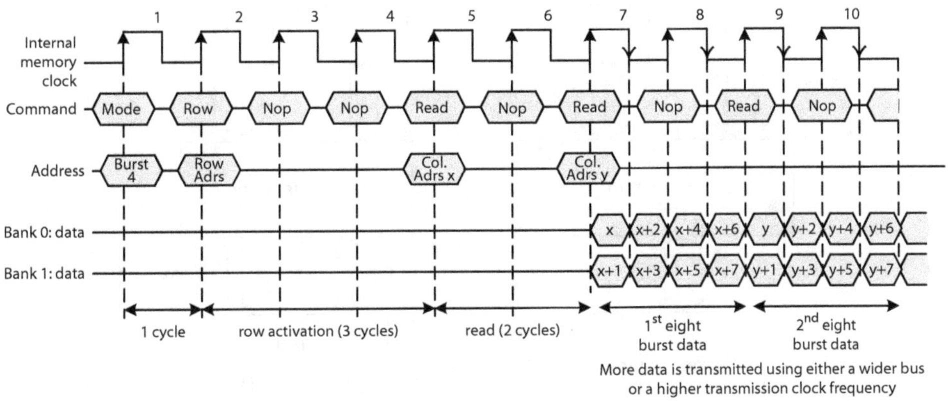

FIGURE 7.24 Illustrating a hypothetical two-bank interleaved (DDR2) SDRAM timing diagram. Assume one clock cycle to enter a burst size, three clock cycles to activate a row, and two clock cycles to complete a read or write access.

module using either a twice-as-wide data bus or a twice-as-fast bus clock. This is how DDR2, DDR3, etc., SDRAMs are designed, each doubling the size of data accessed.

Fine interleaving may also be used with memory modules to further increase bandwidth. For example, consider a memory unit with two fine interleaved memory modules. In addition, suppose the memory modules are made of two-bank DDR SDRAM (i.e., DDR2) chips. In this case, the memory unit would be capable of delivering twice as much data as compared to that shown in Fig. 7.24. That is, with each memory module capable of delivering four data items per SDRAM clock cycle—two on the rising edge (e.g., x and x + 1) and two on the falling edge (e.g., x + 2 and x + 3)—the memory unit would deliver eight data items (four from each module) per clock cycle, therefore doubling the bandwidth of one module.

7.6.3 Multichannel

A multichannel memory refers to memory architecture with two or more communication channels with the rest of the system [6, 7]. For example, Fig. 7.25 illustrates the architecture of two systems with a CPU and a digital signal processor (DSP) communicating with a single- or dual-channel memory. An interconnection network, as a set of buses or point-to-point connections (discussed in Chap. 9), is used to interconnect the CPU and the DSP with the memory. In this case, each channel potentially can serve a different processing unit.

In the single-channel architecture, the CPU and the DSP must take turns accessing the memory. In the dual-channel architecture, however, the two processors can simultaneously access the memory, each from a different channel, provided that their respective data is not stored in the same memory unit.

A multichannel memory unit becomes more efficient (i.e., less idle time) when larger burst sizes are used to deliver the same amount of data. In another words, each channel may continuously deliver data using larger and more efficient burst

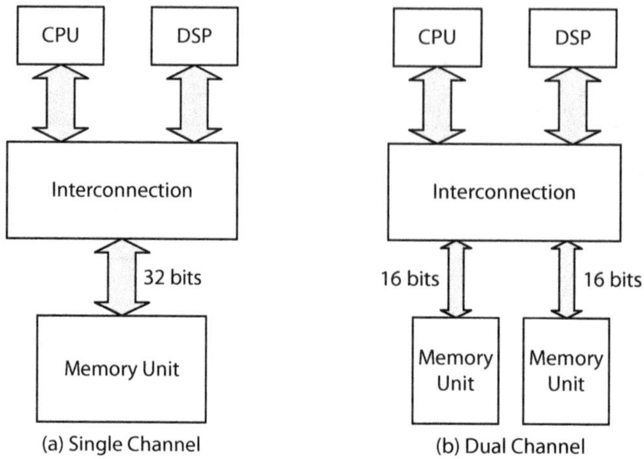

Figure 7.25 Illustrating single- and dual-channel memory architecture.

sizes without actually delivering too much data for a destination module to handle. This makes multichannel memory architecture more suitable for real-time systems where continuous delivery of data is important.

Example 7.4. Consider a 64-bit single-channel and 32-bit dual-channel memory organization as illustrated in Fig. 7.25. Assume the memory units are designed using DDR SDRAM modules. We would like to determine the efficiency of a channel when the CPU accesses 128 B data using four separate burst read cycles from the same row as illustrated in Fig. 7.21. Assume each read delivers 32 B data.

Solution: Equation (7.1) defines memory efficiency.

$$\text{Memory Efficiency (ME)} = \frac{\text{Data bus cycles with data}}{\text{Total number of data bus cycles}} \tag{7.1}$$

In order to access 32 B from the single-channel memory, the burst size must be 4 (32 B/64 b). As illustrated in Fig. 7.21, it takes six clock cycles before the first data item appears on the bus and two cycles per 32 B for a total of eight (2 * 4) cycles to access 128 B (4 * 32 B). Therefore, the memory efficiency is 57%, as calculated next:

$$ME_{\text{single-channel}} = \frac{8 \text{ cycles}}{(6+8) \text{ cycles}} = 0.57 \text{ or } 57\%$$

On the other hand, in order to access 32 B from one of the channels in the dual-channel memory, the burst size must be 8 (32 B/32 b). Again, it takes 6 clock cycles before the first data item appears on the bus, as in the single-channel memory, and 4 cycles per 32 B for a total of 16 (4 * 4) cycles to access 128 B. In this case, the channel efficiency increases to 73%, as calculated next:

$$ME_{\text{each-channel}} = \frac{16 \text{ data cycles}}{(6+16) \text{ total cycles}} = 0.727 \text{ or } 73\%$$

7.7 Design Example: Multiprocessor Memory Architecture

As discussed earlier a low-order memory interleaving is used to increase memory bandwidth by accessing multiple banks or multiple modules in parallel. On the other hand, a high-order memory interleaving is used to partition a memory space into multiple regions and store the data of each region in a separate memory module or memory unit.

7.7.1 UMA versus NUMA

A multiprocessor system with nonuniform memory access (NUMA) architecture is shown in Fig. 7.26(*a*). In this case, a memory unit is assigned to each processor, creating a processing node. An interconnection network interconnects the processing nodes, creating a shared memory system.

Each processor has a shorter connection to its local memory unit (or simply **local memory**) and longer connection to a **remote memory**, another processor's local memory. Therefore, the duration of a memory read/write cycle, called **memory latency**, is shorter when memory accesses are from local memory and longer when they are from remote memory. Therefore, read/write memory cycles in a NUMA system are nonuniform; some are short, requiring fewer clock cycles, and some are long, requiring many clock cycles.

On the other hand, in a uniform memory access (UMA) architecture multiprocessor system (Fig. 7.26(*b*)), all the processors communicate with a single memory unit that may, in addition, be organized as multi-channel. However, among memory accesses, there is less variation in memory latencies; they are all about the same (uniform). Moreover, in both the NUMA and UMA systems, memory units may be low-order interleaved to better serve the processing cores in each of the processors.

The Silicon Graphics' SGI Altix 4700 system, for example, is a NUMA system that can support 512 to 1024 processors and up to 128 TB shared memory [8]. However, the architecture in many smaller systems today is also NUMA, such as the AMD Quad FX platform with two dual-core processors [9].

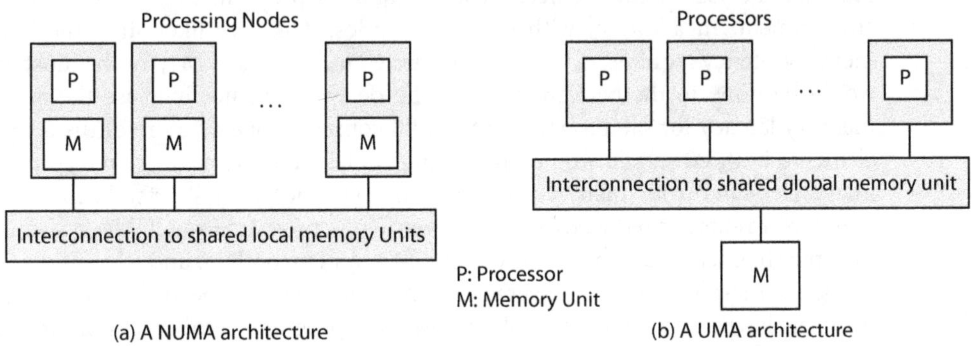

FIGURE 7.26 NUMA and UMA system architectures: (*a*) a NUMA system architecture with shared but locally distributed memory units; (*b*) an UMA system architecture with a shared global memory unit.

7.7.2 A NUMA Application

A low-order interleaving would also work with two or more memory units, called **node interleaving**, if two or more processors in a NUMA system cooperate to complete a task. For example, if two processors are performing an image-processing task, where one processor produces intermediate results to be processed by the second processor, it would be advantageous for the first processor (P0) to read the image data elements from its local memory unit (M0) but write the results to the second processor's (P1's) memory unit (M1). This can be done by creating, for example, an array of structures, each with two data items. For example, consider the following program code that defines an example data structure named "foo_t" with two elements *a* and *b*. Suppose P0 operates on the *a* elements and generates the *b* elements and P1 modifies the *b* elements. With node interleaving, all the *a* elements of the array would be stored in M0, and all the *b* elements would be stored in M1.

```
typedef struct{
        int a;
        int b;
}foo_t;
Main()
{
        foo_t array[100][100];
        ...
}
```

In this case, P0 would read the *a* image elements from its local memory (M0) with shorter latency and write the computed *b* image elements to M1, P1's local memory, with longer latency. P1 would then read the *b* image elements from its local memory (M1) with shorter latency. [As we will see in Chap. 10, P1 may access some recently computed *b* elements from P0's cache memories with longer latency. However, in order to simplify, P0 is assumed to write all the *b* elements to M1 with longer latency and P1 to read all the computed *b* elements from M1 with shorter latency. A more accurate analysis of the latencies would require a program execution simulation environment. In addition, with cache memories, it is very likely that the longer memory latency required to write the computed *b* elements in M1, for the most part, would be done in the background and the delay would not increase the average memory latency for the program.] Once, P0 computes one or more of the *b* image elements, both P0 and P1 would operate in parallel (at the same time) and access the image elements from their respective local memories with shorter latency. This, therefore, results in a reduced execution time for the program in a NUMA system. On the other hand, because in an equivalent UMA system, both P0 and P1 would access the same memory unit, the average memory latency for the program would be relatively longer. This in turn would increase the program's total execution time in a UMA system.

7.8 HDL Models

Example 7.5 describes a combined behavioral and structural model of a memory unit. Memory modules are modeled using SRAM chips. The memory unit implements high-order interleaving, similar to the one shown in Fig. 7.15. The SRAM chip HDL model implements bidirectional data lines. It also implements a simplified memory communication protocol; that is, the memory read/write control signals, when asserted or deasserted, are done at the same time. A more accurate model of the protocol would require generating memory control signals with precise timings, such as those shown in Figs. 7.16 and 7.17. This would require the memory chip (e.g., Fig. 7.10) either to be modeled using a schematic design tool or the HDL model (structural or behavioral) that includes signal timing delays.

Example 7.5. The design and simulation of a 64 × 8 memory unit using four 16 × 8 memory modules is illustrated. A memory module is structurally modeled using two 16 × 4 SRAM chips with bidirectional data lines. Note the use of the keyword "inout" to declare the variable *data* as bidirectional data lines. Also, note that in the SRAM chip model, the assign statement makes *data* high impedance (Z) when memory is not in the read mode. When writing to the memory unit (i.e., *_ce* = 0, *_we* = 0, and *_oe* = 1), *data* will be an input. When reading from the memory unit (i.e., *_ce* = 0, *_we* = 1, and *_oe* = 0), *data* will be set to the memory content identified by the 4-bit *adrs*.

HDL Model:

```
'include "ram16x8.v"
module ram64x8(
input [5:0] adrs,
inout [7:0] data,
input _ce, _we, _oe
);
reg [3:0] _cee;
ram16x8 u1(adrs[3:0], data, _cee[0], _we, _oe);
ram16x8 u2(adrs[3:0], data, _cee[1], _we, _oe);
ram16x8 u3(adrs[3:0], data, _cee[2], _we, _oe);
ram16x8 u4(adrs[3:0], data, _cee[3], _we, _oe);
//2-to-4 decode
always@(*)
begin
        if(_ce == 0)
        case(adrs[5:4])
        0: _cee = 4'b1110;
        1: _cee = 4'b1101;
        2: _cee = 4'b1011;
```

```
        3: _cee = 4'b0111;
        default: _cee = 4'hf;
        endcase
        else
            _cee = 4'hf;
end
endmodule
'include "ram16x4.v"
module ram16x8(
        input [3:0] adrs,
        inout [7:0] data,
        input _ce, _we, _oe
);
ram16x4    u1(adrs, data[7:4], _ce, _we, _oe);
ram16x4    u2(adrs, data[3:0], _ce, _we, _oe);
endmodule
module ram16x4(
        input [3:0] adrs,
        inout [3:0] data,
        input _ce, _we, _oe
);
reg [0:15][3:0] mem; // 16 X 4 RAM
assign data = ~_ce & _we & ~_oe ? mem[adrs]: 4'hz;
always@(*)
begin
if(_ce == 0)
        if(_we == 0 && _oe == 1)
            mem[adrs] = data;
end
endmodule
```

Simulation Test-Bench:

A simulation test model is shown here. In this case, an “assign” statement is used to place an 8-bit value as *content* on the bidirectional data lines when the memory unit is writing (i.e., *_ce* = 0, *_we* = 0, and *_oe* = 1). Four test cases are simulated: two memory write cycles followed by two memory read cycles. The test vectors simulate simplified memory read/write cycles. A write cycle begins with a target address *adrs*, a data value as *content*, and the activation of the control signals for a write. The control signals are deasserted at the end of the write cycle. A read cycle begins with an address *adrs* and the activation of the read control signals, which are then deasserted at the end of the read cycle.

This is similar to how a CPU accesses memory as it executes load and store instructions. A write cycle is initiated by a executing a store instruction, which provides a memory address and a value (i.e., a register content) to be stored in memory. A read cycle is initiated by a load instruction, which only provides a memory address and saves the retrieved memory content in a register. Because the CPU also executes arithmetic and other types of instructions and therefore does not access memory all the time, in the test-bench, this is simulated with some random delays between each memory read/write cycle.

```
'include "ram64x8.v"
module ram64x8test();
reg [5:0] adrs;
reg [7:0] content;
reg _ce, _we, _oe;
wire [7:0] data;
assign data = ~_ce & ~_we & _oe ? content : 8'hz;
ram64x8 u1(adrs, data, _ce, _we, _oe);
initial begin
$monitor ("%4d: adrs = %h  _ce = %b _we = %b _oe = %b data =
%h", $time, adrs, _ce, _we, _oe, data);
//two memory writes
adrs = 6'b001111; //memory module 0
content = 8'haa; //data as content
_ce = 0; _we = 0; _oe = 1; //begin a write cycle
#10
_ce = 1; _we = 1; _oe = 1; //end a write cycle
#50 //some delay
adrs = 6'b011111; //memory module 1
content = 8'hbb;
_ce = 0; _we = 0; _oe = 1;
#10
_ce = 1; _we = 1; _oe = 1;
#40
//two memory reads
adrs = 6'b001111;
_ce = 0; _we = 1; _oe = 0; //begin a read cycle
#10
_ce = 1; _we = 1; _oe = 1; //end a read cycle
```

```
#60
adrs = 6'b011111;
_ce = 0; _we = 1; _oe = 0;
#10
_ce = 1; _we = 1; _oe = 1;
#30
$finish;
end
endmodule
```

Simulation Result:

As expected, the memory data lines indicated as *data* become high impedance (Z) when memory is not accessed.

```
Chronologic VCS simulator copyright 1991-2009
Contains Synopsys proprietary information.
Compiler version D-2009.12; Runtime version D-2009.12; May
20 22:54 2014
   0: adrs = 0f  _ce = 0 _we = 0 _oe = 1 data = aa
  10: adrs = 0f  _ce = 1 _we = 1 _oe = 1 data = zz
  60: adrs = 1f  _ce = 0 _we = 0 _oe = 1 data = bb
  70: adrs = 1f  _ce = 1 _we = 1 _oe = 1 data = zz
 110: adrs = 0f  _ce = 0 _we = 1 _oe = 0 data = aa
 120: adrs = 0f  _ce = 1 _we = 1 _oe = 1 data = zz
 180: adrs = 1f  _ce = 0 _we = 1 _oe = 0 data = bb
 190: adrs = 1f  _ce = 1 _we = 1 _oe = 1 data = zz
$finish called from file "ram64x8test.v", line 41.
$finish at simulation time                  220
```

References

1. Flash Memory: An Overview, http://www.spansion.com/Support/TechnicalDocuments/Pages/TechnicalDocuments.aspx.
2. C. R. Nave, HyperPhysics, Georgia State University, http://hyperphysics.phy-astr.gsu.edu/.
3. Memory chips from Micron, http://www.micron.com/products/dram/.
4. Memory from Rambus, http://www.rambus.com/.
5. Memory modules, http://www.newegg.com/.
6. Gomony MD, Akesson B, Goossens K. Architecture and optimal configuration of a real-time multi-channel memory controller, Design, Automation & Test in Europe Conference & Exhibition (DATE), 2013, 1307–1312.
7. Siqueira HM, Silva IS, Kreutz ME, Correa EF. DDR SDRAM memory controller for digital TV decoders, *Symposium on Computing System Engineering (SBESC)*, 2011, 78–82.
8. NUMA SGI Altix 4700 system, http://www.sgi.com/products/.
9. AMD Quad FX Platform, http://support.amd.com/.

Exercises

7.1. Draw the logical view and the block diagram of a 128 KB memory. Also indicate the number of memory locations and number of address lines required for the following:

a. One byte wide

b. Two bytes wide

7.2. Show the cell array organizations for the following memory sizes:

a. 128×1

b. 64×2

c. 32×4

7.3. Using the SRAM cell model in Fig. 7.8, design a 32×2 SRAM (no need to draw every cell).

7.4. Using the cell model in Fig. 7.8, design a 16×4 SRAM (no need to draw every cell).

7.5. Design a 256-B memory organized as a 128×16 memory unit using memory modules with 32×4 SRAM chips.

7.6. Design a 256-B memory organized as a 128×16 memory unit using memory modules with 16×8 SRAM chips.

7.7. Design a 256-B memory organized as a 128×16 memory unit partitioned into 128-B ROM space for the lower addresses and 128-B RAM space for the upper addresses. Design the memory unit using 64×8 ROM chips and memory modules with 32×8 SRAM chips.

7.8. Research and write a short paper on each of the following memory technologies:

a. Rambus RDRAM

b. Rambus XDR

c. Rambus XDR2

d. EDO DRAM

7.9. Draw an SDRAM timing diagram for a read cycle followed by a write cycle, both with burst size = 4. Assume that it takes one clock cycle to enter a burst size, four clock cycles to activate a row, and three cycles to complete a read or write access.

7.10. Consider a 32-bit data bus SDRAM. Given that the clock frequency of the bus is 200 MHz, what is the peak memory bandwidth in megabytes per second (MBs)?

7.11. Consider a 64-bit data bus SDRAM. Given that the clock frequency of the bus is 200 MHz, what is the peak memory bandwidth in megabytes per second (MBs)?

7.12. Consider a 32-bit data bus DDR SDRAM. Given that the clock frequency of the bus is 200 MHz, what is the peak memory bandwidth in megabytes per second (MBs)?

7.13. Consider the SDRAM timing diagram in Fig. 7.19. Suppose there are four memory read cycles, as follows, where the data bus is 32 bits:

Row address x, burst size = 4, and column addresses $x1$ and $x2$ (32 B total)

Row address y, burst size = 4, and column addresses $y1$ and $y2$ (32 B total)

a. Draw the timing diagram.

b. Determine memory efficiency. Ignore row deactivation time.

7.14. Consider the DDR SDRAM timing diagram in Fig. 7.21. Suppose there are four memory read cycles, as follows, where data bus is 32 bits:

Row address x, burst size = 4, and column addresses $x1$ and $x2$ (32 B total)

Row address y, burst size = 4, and column addresses $y1$ and $y2$ (32 B total)

a. Draw the timing diagram.

b. Determine memory efficiency. Ignore row deactivation time.

7.15. Consider a 64-B memory unit high-order interleaved into four 16×8 memory modules, but data is low-order interleaved within each module. Suppose each module is made of eight SDRAM chips, each with four banks and organized as a $2 \times 2 \times 1$ cell array. Show/describe how a 2-bit burst data would be stored in the memory unit.

7.16. Consider a four-channel memory designed using 400 MHz DDR SDRAM modules. What is the peak bandwidth of the memory if each channel is 64 bits (8 B) wide?

7.17. Consider a NUMA system with local memory latency = 1τ and remote memory latency = 4τ. If a program execution results in 80% local accesses and 20% remote accesses, what is the average memory latency? Compare your result with an equivalent UMA system with 4τ average memory latency.

7.18. Show/describe how physically the elements of an array[8][8] of type "foo_t" would be stored in a two-node interleaved 128×32 memory units M0 and M1.

7.19. Computer security (memory authentication): Select Exercise 11.24 and/or 11.25 (also see Sec. 11.9). Note, the details of cache memory are covered in Chap. 10. However, here, first determine the number of blocks in memory, and then assign each block a block address starting with 0.

CHAPTER 8
Instruction Set Architecture

8.1 Introduction

The preceding chapters covered digital design concepts and techniques as well as memory organization and architecture. In this chapter, we discuss data paths for CPU. Unless explicitly stated, the terms *CPU* and *processor* are both used to refer to a processing core. Multicore processors will be discussed in Chap. 10.

An **instruction set architecture** (ISA) refers to a single-cycle, multicycle, or pipelined data path that executes a program. In this case, a data path is capable of executing many different instructions, each requiring a set of data path operations. The data path fetches an instruction from memory; decodes the instruction by generating the necessary data path control signals; executes the instruction by performing data path operations, which additionally may require retrieving data from memory, according to those control signals; and stores (writes back) the computed result (if any) or data read from memory in a register. Register content may be stored in memory.

In general, each different CPU has its own set of unique instructions. However, some Intel and AMD processors execute the same set of instructions. For example, X86 refers to a 32-bit instruction set in both Intel and AMD processors. Other commonly known instruction set examples are Intel's IA-64 (Itanium Architecture), AMD's X64 (64-bit instruction set), MIPS, Sparc, and ARM. While each instruction set is different, the instructions in each set are complete for developing any type of software, including systems software, stand-alone, and online application software.

With the increases in transistor count, modern CPUs implement pipelining and instruction-level parallelism (ILP) to increase performance as was briefly discussed in Chap.1. A pipelined data path in this case is also called an **instruction pipeline.** The efficiency of the data path increases as the stages of the pipeline are kept busy operating on multiple instructions concurrently. This increases instruction throughput, the number of instructions executed per second, and reduces a program's total execution time.

However, as data-dependent instructions go through the pipeline stages, additional hardware is needed to stall the pipeline, if necessary, and make sure data-dependency relationships are not violated. This, in turn, can reduce pipeline efficiency unless certain hardware is used and compiler optimizations are performed to eliminate or reduce such data dependencies.

In addition, branch instructions change execution flow, introducing bubbles that also decrease pipeline efficiency. However, modern CPUs implement branch prediction mechanisms to minimize this.

With ILP, an instruction pipeline is made of multiple parallel pipelines that execute several instructions at the same time. Which set of instructions can execute at the same time is program dependent and is determined either **statically** by compiler (i.e., in software) or **dynamically** in hardware. ILP also reduces a program's total execution time. However, as was discussed in Chap. 1, only a limited number of independent instructions in each program can be executed during each pipeline cycle. In addition, because instructions and data must come from memory, many modern CPUs (e.g., Intel Core i7) implement multithreading so they can switch to executing another program (thread) if the CPU must wait to receive instructions or data. While this does not reduce a program's total execution time, it helps to increase the overall efficiency of the processor, which would perform more tasks and idle less.

In this chapter, we start by providing some background information and examine different instruction set architectures, and then, in order to provide a better understanding of instruction set and data path design, we begin the discussion with a simple high-level language code example. Using this example, an instruction set and a single-cycle data path are presented. A hardware description language (HDL) description for the single-cycle data path is presented. Execution simulation results for both a single-cycle and pipelined data path for the example program are presented, and performance parameters are discussed.

The chapter then presents reduced instruction set computer (RISC) architecture and its advantages. It provides an introduction to RISC compiler optimization, as well as techniques used to improve instruction throughput. Specifically, we use examples to illustrate ways to increase pipeline clock frequency, branch prediction techniques, ILP, and multithreading. An introduction to multithreaded programs using an example is also provided.

8.1.1 Type of Instructions

A processor is typically designed for general-purpose programming. However, for better performance and some real-time applications, it is often necessary for computer systems to use special-purpose processors, such as a graphic processing unit (GPU) and digital signal processor (DSP). Each special-purpose processor has a set of instructions designed to efficiently perform computer graphics (e.g., object rotation), signal processing (e.g., audio compression), etc. DSPs are typically used in embedded systems, like cell phones and digital cameras, for signal and image processing tasks. Modern processors may also include certain special-purpose instructions, such as the single instruction multiple data (SIMD) (Chap. 1) instruction sets and computer security related instructions of the Intel and AMD processors.

Instructions that perform arithmetic and logic computations are generally referred to as **data-manipulation** instructions for performing calculations on data. Others are referred to as **program-flow control** instructions, such as conditional and unconditional branch (or jump) instructions, and **data-movement** instructions, such as those used for reading and writing memory. The program-flow control instructions alter a program's execution path, and are necessary in the execution of high-level language statements, such as "if-else," "for-loop," "while-loop," and subroutine procedure calls. The latter requires saving a return address and state (i.e., register contents) of the processor, either in a special set of registers inside the processor (e.g., Sparc's register windows) or in memory (e.g., memory stack). (For more information on register windows, refer to Exercise 9.14 in Chap. 9.)

8.1.2 Program Translation

As shown in Fig. 8.1, a software program is typically written in a high-level language, such as C/C++ or Java, and is translated by a compiler into assembly

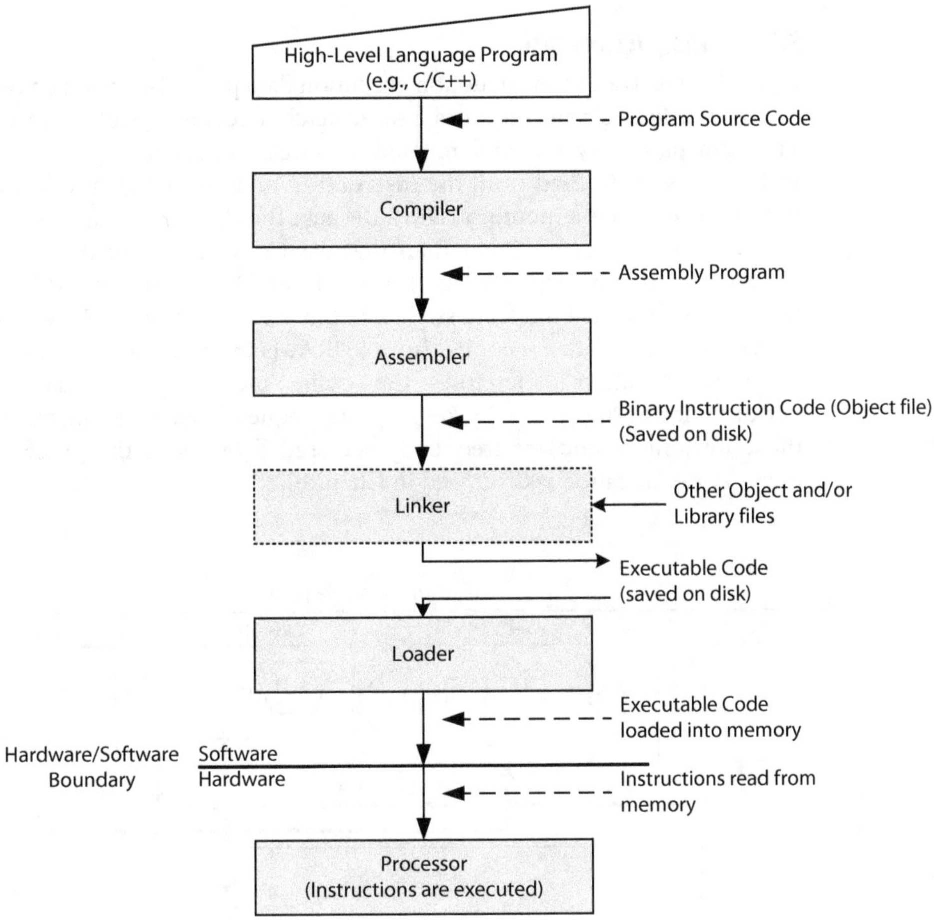

Figure 8.1 Basic program translation and execution process.

instructions (C/C++) or bytecode (Java). A bytecode is converted to instructions at run time. An assembly instruction is defined by its **mnemonic**, an easy-to-remember operation-code (**op-code**) name, such as ADD for addition and SUB for subtraction. For more information on mnemonics and assembly language conventions, refer to IEEE Standard for Microprocessor Assembly Language [1].

Unique binary numbers are assigned to each mnemonic op-code by **assembler**, and are used to translate instructions to binary, creating an **object code** that would be saved as a file on disk. Two or more object files (if any) are linked to create an executable program file (e.g., an .exe file in a Windows environment). The linker program links object codes of static (e.g., math) library functions if such functions are called in the program. In addition, not shown in Fig. 8.1, a program may include some dynamically linked codes (i.e., DLLs) that are linked during run time. The OS loader program loads an executable code (generally not in its entirety) into memory for execution.

8.1.3 Instruction Cycle

Figure 8.2 illustrates an instruction execution data path, also called an **instruction cycle**, with four circuit modules labeled fetch, decode, execute, and write back. The data path may be implemented as single-cycle, multicycle, or pipelined. Instructions are fetched from the **instruction memory** (IM) and data read from or written to the **data memory** (DM). Because the clock frequency used to operate a modern processor is higher than that used to operate synchronous dynamic random access memory (SDRAM), the IM and DM are organized to operate as fast cache memories built from static random access memory (SRAM) technology. Instructions and data are copied from SDRAMs to these cache memories during program execution. Therefore, the cache memories increase the overall performance of the system by keeping the frequently executed instructions (e.g., those forming a loop) or frequently accessed data inside the processor. Cache memory organization is discussed in Chap. 10.

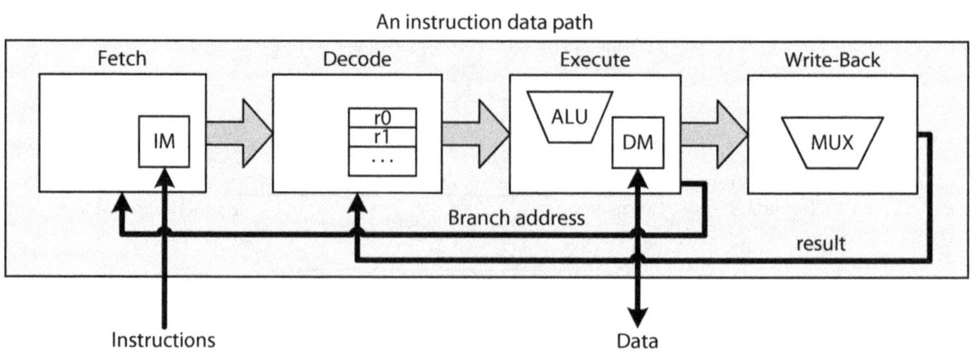

Figure 8.2 An instruction data path with instruction memory (IM) and data memory (DM).

8.2 Types of Instruction Set Architecture

Over the years, ISA developments are guided by advancements in integrated chip (IC) technologies and computer architecture concepts, such as pipelining. In addition to an op-code, an instruction includes a set of **operands** that are specified explicitly, implicitly, or both. The operands are grouped into input operands and, typically, one output operand. An input operand specifies either a constant value, also called an immediate data such as the number 9, or a register or memory content. An output operand is either a register number or a memory address. **Addressing modes** define the many ways input operands are interpreted to identify a target data used in the execution of the instruction.

8.2.1 Addressing Modes

Table 8.1 lists examples of how various addressing modes are declared. In the table, parentheses are used to distinguish a memory address from an immediate value. These notations are used by the assembler to translate an instruction to its equivalent binary, typically called a **machine instruction**. An immediate (I) operand is a 2's complement number and is immediately available to be used in the execution of the instruction. A direct (D) operand is a memory address, and the data must come from memory. An indexed (X) operand defines the address of the next array element in memory. There are other addressing mode examples, and their coverage is referred to elsewhere.

Table 8.2 illustrates several instruction examples with explicit and/or implicit operands; however, the instructions do not all belong to a single processor. Each instruction computes the sum of two data values and stores the sum in a register. The first instruction in the table has no explicitly listed operands. In this case, the source for the two data values and the destination to store the computed sum is a hardware stack within the data path. The second instruction includes only a single explicitly declared immediate data operand, 9. In this case, a register that implicitly is known to the instruction is used both as a data source register and as the destination register to store the computed sum.

The third instruction example includes two explicitly declared operands: the register R1 and the immediate value –9. In this case, R1 is also used to store the

Operand Notation	Addressing Mode
V	I, immediate: V is an immediate input operand, a 2's complement number.
(V)	D, direct: V is a memory address and (V) indicates the content of memory address V (i.e., M[V]).
R	R, register: Indicates an input data register source or a destination register or both
R, (V)	X, indexed: V is a memory address and R + V is the address of the next data item in memory (i.e., M[R + V]).

TABLE 8.1 Examples of Addressing Modes

Example	Instruction	Description
1	ADD	The source and destination operands are known to the instruction (e.g., a hardware stack).
2	ADD 9	Immediate: R ← R + 9; R is a known register to the instruction.
3	ADD R1, −9	Register and immediate: R1 ← R1 + −9; R1 is used for both as the source and also as the destination register number
4	ADD R1, (9)	Register and direct: R1 ← R1 + Memory[9]
5	ADD R1, R2, (9)	Register and indexed: R1 ← R1 + Memory[R2 + 9]
6	ADD R1, R2	Register and register (two operands): R1 ← R1 + R2
7	ADD R3, R1, R2	Register and register (three operands): R3 ← R1 + R3; the destination register can be different from the source registers

TABLE 8.2 A List of Instruction Examples with Implicit and/or Explicit Operands

computed sum. That is, the instruction performs R1 ← R1 + −9. In the fourth example, the second operand is a memory address, and the instruction performs R1 ← R1 + M[9] where M[9] indicates the content of memory address 9. In the fifth example, the second and third operands specify the address of the next data item in memory as R2 + 9, and the instruction performs R1 ← R1 + M[R2 + 9]. In the sixth example, the two input operands are both register contents, and the first register is also used as the destination register number; the instruction performs R1 ← R1 + R2. In the seventh example, a destination register is explicitly declared and may or may not be the same as one of the two input operand register numbers. The instruction performs R3 ← R1 + R2.

8.2.2 Instruction Format

An instruction format is used to convert a mnemonic assembly instruction to a machine instruction. The format indicates the number of bits required to specify an op-code, an addressing mode, a source register (if any), a destination register (if any), a n-bit immediate data value (if any), and a memory address (if any). Figure 8.3 illustrates the format used for each of the instructions listed in Table 8.2.

For example, with 8-bit op-codes, 16 registers, and a 16-bit immediate data, the instruction 3 (ADD, r1, −9) in Table 8.2 becomes a 4B instruction, determined as follows, assuming that the op-code for ADD is $(000000001)_2$ and the register-immediate (RI) addressing mode is encoded as $(1000)_2$:

```
0000 0001, 1000, 0001, 1111 1111 1111 0111 or
0181FFF7 in hex
```

The size and the number of different instruction formats depend on the type of ISA. In general, ISAs are classified as Stack-ISA, Accumulator-ISA (Acc-ISA), CISC-ISA, and RISC-ISA, discussed next.

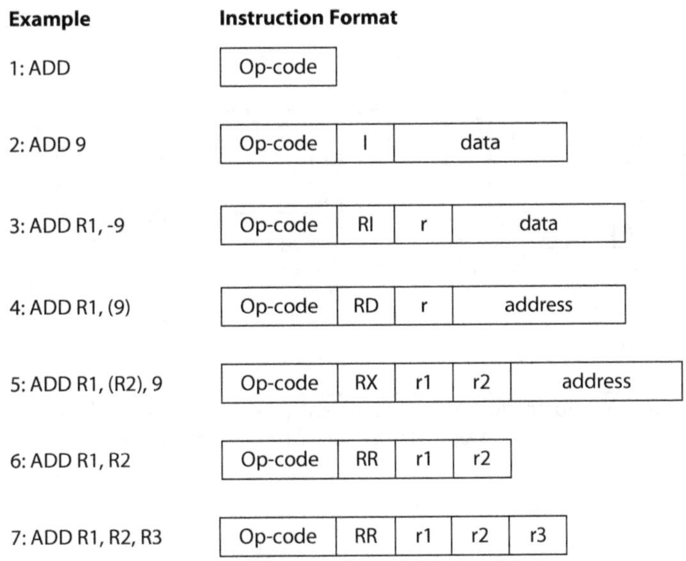

FIGURE 8.3 Examples of instruction formats for the eight instruction examples listed in Table 8.2.

8.2.3 Stack-ISA

A processor with a Stack-ISA uses a hardware stack that operates in a last-in and first-out (LIFO) order; the last value stored is also the first value retrieved. Data read from memory is pushed onto the stack. Data used in a computation or stored back into memory is popped from the stack. The result of a computation is always pushed on the stack.

A Stack-ISA has the advantage of using the shortest instructions for a majority of instructions. Therefore, a Stack-ISA is an ideal architecture to design a processor with limited I/O pins. The processor would operate like an Hewlett-Packard (HP) calculator, requiring a math statement to be first ordered into **reverse polish notation**. For example, to calculate 2(3 + 4) in an HP calculator, one must enter 3, then 4, then +, then 2, and then *. Consider the following high-level language program statement:

```
A = B * (C + D);
```

In order to execute the statement in a Stack-ISA processor, the compiler must first translate the statement to the reverse polish natation C D + B * = A. The compiler then converts the notation to the following Stack-ISA example assembly program. Note that the arithmetic instructions ADD and MULT require no operands, making each a very short instruction. The PUSH and POP, each a data movement instruction, require only one operand. In general, program control-flow instructions, such as branch and subroutine calls, also require a single operand.

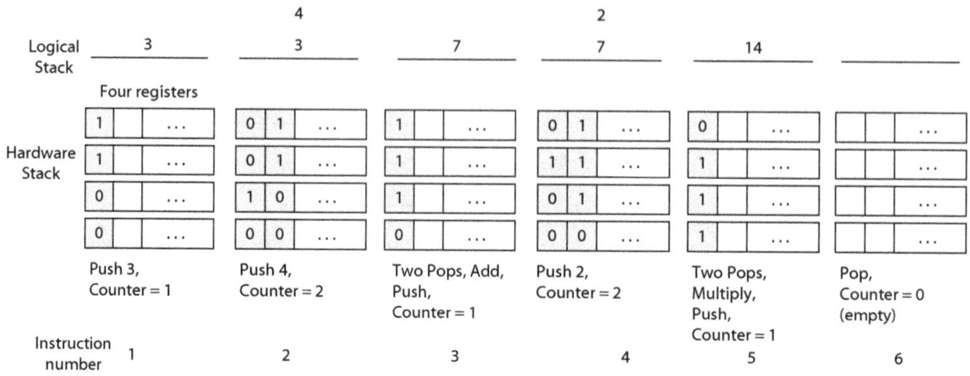

FIGURE 8.4 An illustration of stack content when computing the reverse polish notation C D + B ∗ = A; it is assumed that (B) = 2, (C) = 3, and (D) = 4.

```
Instruction
number
  1:        PUSH (C) //stack←M[C]
  2:        PUSH (D) //stack←M[D]
  3:        ADD      //stack← (C) + (D), values popped, added,
                     //result pushed
  4:        PUSH (B) //stack←M[B]
  5:        MUL //stack←((C) + (D)) * (B), values popped, added,
                //result pushed
  6:        POP (A)  //M[A]←(((C) + (D)) * (B)), value is popped
                     //and stored in memory
```

Figure 8.4 illustrates the execution of the previous program by a Stack-ISA processor. In the illustration, it is assumed that $(B) = 2$, $(C) = 3$, and $(D) = 4$. The figure shows the content of both a logical stack and a hardware stack as the instructions execute.

A Stack-ISA has the disadvantage of not being able to reuse memory content in a future computation. Once memory content is popped from the stack, it is no longer available inside the processor. For example, consider the program statement "A = (B + C) ∗ (B + D);" that requires the variable *B* to be pushed twice on the stack, thus increasing memory traffic.

8.2.4 Accumulator-ISA

An Acc-ISA is the simplest architecture and requires less hardware. The data path contains a special source and destination register called **accumulator** (ACC). The register is used as both an implicit input operand and an implicit output register name. For example, again consider the high-level program statement "A = B ∗ (C + D);" and its Acc-ISA example assembly program given next. As opposed to a Stack-ISA, arithmetic instructions in the Acc-ISA can directly operate on memory data; thus, when compared to a Stack-ISA, an Acc-ISA reduces the total number of instructions required to translate a high-level language program into an assembly program.

```
1:    LD     (C)     //ACC ← M[C]
2:    ADD    (D)     //ACC ← ACC + M[D]
3:    MUL    (B)     //ACC ← ACC * M[B]
4:    ST     (A)     //M[A] ← ACC
```

However, the disadvantage of an Acc-ISA is that the register ACC, like a hardware stack, may become a bottleneck. The content of ACC may need to be stored in memory so ACC becomes available to be used in the next computation. For example, consider the computation of A = (C + D) ∗ (E − F), where C + D and E − F must be computed first before their results can be multiplied. In this case, once C + D is computed, the sum in the ACC must be stored in a temporary memory location before E − F can be computed. An Acc-ISA may also include other registers, such as an index register to access an array element in memory and a link register to save a subroutine return address.

8.2.5 CISC-ISA

A complex instruction set computer (CISC) ISA is an improvement over the Acc-ISA by increasing the number of working registers in the data path. Each arithmetic instruction can now include one or two explicitly declared register operands. Like an Acc-ISA, arithmetic instructions of a CISC-ISA can reference data in memory directly. In addition, a CISC-ISA typically implements many addressing modes, requiring many instruction formats of various sizes.

The input operands for arithmetic instructions could be an immediate data, register, or memory content. However, only one of the operands can be an immediate value or memory content. Typically, a CISC-ISA processor implements many simple and complex instructions. Therefore, a high-level language program translated into CISC instructions would require the least number of instructions. However, some complex instructions may require more time to execute. The following is a CISC-ISA example assembly program to compute the statement "A = B ∗ (C + D);":

```
1.    LD     R1,    (C)    //R1 ← M[C]
2.    ADD    R1,    (D)    //R1 ← R1 + M[D]
3.    MUL    R1,    (B)    //R1 ← R1 * M[B]
4.    ST     (A),   R1     //M[A] ← R1
```

Because there are more registers to choose from, the statement "A = (C + D) ∗ (E − F);" can easily be computed using a register (e.g., R1) to compute C + D and another register (e.g., R2) to compute E − F, and then the contents of the two registers would be multiplied to generate the final product. However, even with having more working registers in a CISC-ISA, the number of available registers is still very small when compared to the number of variables in a typical high-level language program. Also, depending on the size of each high-level program statement, there could be many intermediate results (e.g., C + D and E − F) that must be saved either in registers inside the processor or in memory. Therefore, similar to the Acc-ISA, an intermediate result in one of the registers may still need to be stored in memory to free up a register for the next computation.

[A CISC-ISA compiler, as opposed to the Stack-ISA or Acc-ISA compilers, needs to implement a register selection policy to minimize memory accesses when all the registers contain intermediate results. A register allocation policy, such as **least recently used** (LRU), is used to free up a register by storing its content, if intermediate, in memory. On the other hand, if the LRU register content is a memory data, it is discarded.]

8.2.6 RISC-ISA

As stated earlier, typically, the data path of a CSIC-ISA implements many different addressing modes. On the other hand, the designers of a RISC ISA believed (correctly) in having simpler instructions to implement a simpler and more efficient data path. Alpha, MIPS, and Sparc are examples of processors that were designed from the beginning as a RISC-ISA processor.

A RISC-ISA, also known as **load/store architecture**, uses only two instructions (e.g., LD and ST) to access memory. No arithmetic instructions can operate directly on data in memory. A memory data must be loaded into a register before it can be used in a computation. The following is an example of a RISC-ISA assembly program to compute the statement "A = B * (C + D);":

```
1.   LD    R1,    (C)        //R1 ← M[C]
2.   LD    R2,    (D)        //R2 ← M[D]
3.   ADD   R3,    R1, R2     //R3 ← R1 + R2
4.   LD    R4,    (B)        //R4 ← M[B]
5.   MUL   R5,    R3, R4     //R5 ← R3 * R4
6.   ST    (A),   R5         //M[A] ← R5
```

The assembly program uses five registers to keep three memory contents and two intermediate results (i.e., C * D and B * (C + D)) inside the processor. Usually, there are more such registers in a RISC-ISA processor than there are in a typical CISC-ISA. The reason for this is to keep more data in registers and thus increase the processor throughput.

8.3 Design Example

The design of a simple Acc-ISA CPU is presented. However, the objective here is not to create a complete set of instructions, but rather, to provide a top-down design methodology that starts from a simple example high-level language code and includes the following:

- Instruction set design (the list of instructions necessary to translate an example high-level language program code to its equivalent assembly language program)
- Assembly language program code listing

- Binary executable code (machine instructions)
- Data path design
- HDL model
- Simulation

8.3.1 Acc-ISA Instruction Set Design

Example 8.1 presents a high-level program code that sums the elements of an array using a for-loop. We would like to create a set of Acc-ISA instructions to compile the code and generate an equivalent assembly program.

Example 8.1. A program code listing that sums the elements of an array with size 8:

```
int array[8];
int i, sum;
sum = 0;
for (i = 0; i < 8; i++)
    sum = sum + array[i];
```

A close examination of the example code reveals that we would need to create arithmetic instructions (e.g., add and compare); data movement instructions, including array indexing; and program control-flow instructions (e.g., jump and jump greater than). For array indexing, the data path must contain another register (X) that would hold the next index for the array. Table 8.3 lists the

Op-Code	Instruction	Addressing Mode	Example	Action
0	NOP		NOP	Do nothing
1	ADD	Immediate	ADD data	ACC ← ACC + data
2		Direct	ADD (address)	ACC ← ACC + M[address]
3	CMP	Immediate	CMP data	if ACC == data then GTF = 1 else GTF = 0
4	JGT	Immediate	JGT address	PP ← address if GTF = 1
5	JMP	Immediate	JMP address	PP ← address
6	LD	Immediate	LD data	ACC ← data
7		Direct	LD (address)	ACC ← M[address]
8		Indexed	LD X(address)	ACC ← M[X + address]
9	MVX	Register	MVX	X ← ACC
10	ST	Direct	ST (address)	M[address] ← ACC
ACC: Accumulator; GTF: Greater than flag; PP: Program pointer; X: Index register				

Table 8.3 Example Acc-ISA Instruction Set That Translates a High-Level Program into an Equivalent Assembly Language Program

instruction set required to translate the example program code to an equivalent assembly language program. In the table, a **program pointer** (PP), also called a **program counter** (PC), holds the address of the next executing instruction. Here, integer numbers starting from 0 are arbitrarily assigned to the op-codes. The op-code 0 is named no-operation (NOP). A NOP instruction, while not used here, is necessary to design a high-performance RISC processor.

Acc-ISA Example Assembly Program

Because there are no compilers for the example Acc-ISA processor, the code in Example 8.1 must be manually converted to an assembly program—for example, the one shown in Example 8.2. Each line in the Acc-ISA assembly language program contains four fields: an optional label field, an op-code field, an operand field (if any), and an optional comment field. The label field may include a jump address, such as the L1 and L2 in Example 8.2.

The ".code" and ".data" are called **assembler directives** and are used to separate the instructions and data in the program, respectively. This is necessary if the processor data path requires that instructions and data be stored in separate memory sections during execution. For high performance, this allows instructions to be stored in an IM and data in DM inside the processor, as illustrated in Fig. 8.2. The "RB," which here stands for reserve byte, is an example of a **pseudo-instruction**. The pseudo-instruction "RB" is interpreted by the assembler for allocating memory addresses to each of the variables *sum* and *i* and data structure *array* in the program.

Example 8.2. The listing of an Acc-ISA assembly language program for the program in Example 8.1:

```
.code       //start program code
        LD      0           //Initialize, ACC←0
        ST      (sum)       //M[sum] ←ACC
        ST      (i)         //M[i] ←ACC
  L1:
        CMP     7           //is i > 7? (is ACC == 7?)
        JGT     L2          //exit for-loop if yes (PP←L2)
        MVX                 //get next index (X←ACC)
        LD      X(array)    //get next array element (ACC←M[array
                            //+ X])
        ADD     (sum)       //and add it to the partial sum (ACC←ACC
                            //+M[sum])
        ST      (sum)       //store the partial sum in memory (M[sum]
                            //← ACC)
        LD      (i)         //do i = i + 1: get i (ACC←M[i]),
        ADD     1           //increment i (ACC←ACC + 1), and
        ST      (i)         //save i (M[i] ←ACC).
        JMP     L1          //loop back
```

```
L2: …

.data  //start program data
array: RB   16       //reserve 16 bytes for array in memory
i:     RB   2        //reserve 2 bytes for i in memory
sum:   RB   2        //reserve 2 bytes for sum in memory
```

The program in Example 8.2 uses an arbitrary syntax. However, for an example of a specific syntax, refer to the Microsoft assembler (MASM) [2].

Code and Data Memory Spaces

Figure 8.5 illustrates how program's code and data are typically stored in **virtual memory** during execution. The range for a virtual address determines the maximum program size in bytes. For example, with 32-bit user virtual addresses, user programs can be a maximum of 4 GB.

The amount of the virtual memory address space that is allocated to run a program is typically divided into program code and program data regions, as illustrated in Fig. 8.5. Virtual memory systems are covered in Chap. 10. However, here we will assume Fig. 8.5 represents the way our sample program is stored in memory.

Pentium IV Example Assembly Program

Example 8.3 presents Intel Pentium IV assembly code for the program in Example 8.1. It is generated by using the "gcc" compiler in CygWin, a Linux utility for the Windows environment [3]. In the listing, integer data types

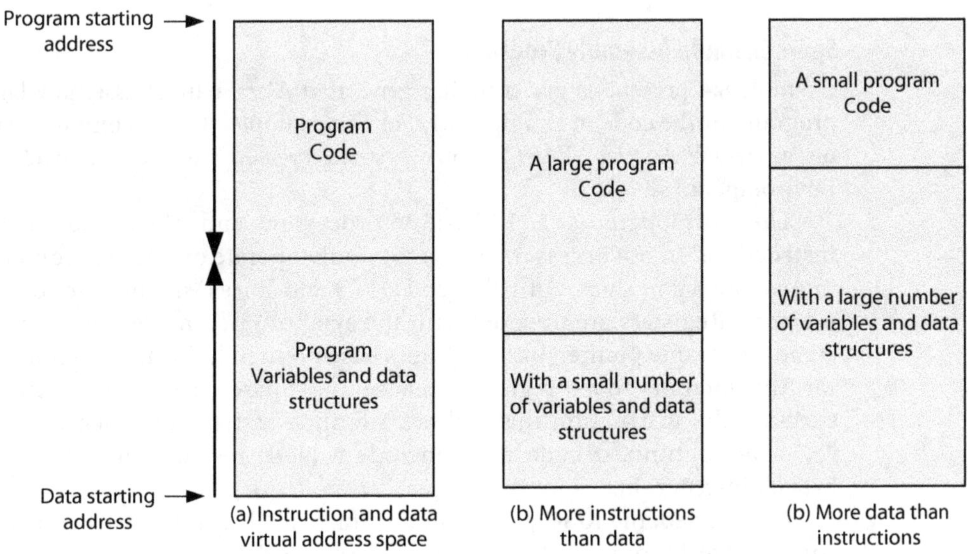

FIGURE 8.5 A program's code and data in virtual memory address space.

are 4B, and the rightmost listed operand in each instruction indicates destination. In the program, registers are prefixed with the symbol "%". The register %ebp is called a base pointer and holds the starting memory address for the program's data region. Also, in order to avoid cache and memory misalignments and achieve better performance [4], larger data structures are allocated first. However, with the availability of large main memory, data structures may be declared larger than necessary (i.e., padded) to achieve better cache performance [5]. In the code, the data address %ebp – 40 is assigned to the *array*, %ebp – 44 to the *i*, and %ebp – 48 to the *sum*. The load effective address ("leal") instruction loads a memory address (not memory content) into a register.

Example 8.3. A Pentium IV assembly code for the sample program given in Example 8.1:

```
    movl  $0, -48(%ebp)       //initialize sum, Memory[%ebp – 48] ← 0 (sum = 0)
    movl  $0, -44(%ebp)       //initialize i, Memory[%ebp – 44] ← 0 (I = 0)
L7:

    cmpl  $7, -44(%ebp)       //is i > 7?
    jg    L8                  //if yes, jump to L8
    movl  -44(%ebp), %eax     //get i, %eax←Memory[i]
    movl  -40(%ebp,%eax,4), %edx    //get array[i], %edx←Memory[array + i * 4]
    leal  -48(%ebp), %eax     //get address of sum, %eax←%ebp - 48
    addl  %edx, (%eax)        //Memory[sum] ←array[i] + Memory[sum]
    leal  -44(%ebp), %eax     //get address of i, %eax←%ebp – 44
    incl  (%eax)              //Memory[i] ←Memory[i] + 1
    jmp   L7                  //repeat
L8: ...
```

Sparc Example Assembly Program

Example 8.4 presents a gcc compiler generated AltraSparc II assembly language program for the code in Example 8.1. For this example, the gcc compiler was used on a virtual Aurara SPARC Linux system, created on the Virtutech Simics Environment [6].

The AltraSparc is a RISC-ISA processor, and thus, the arithmetic instructions do not access memory; they only operate on register contents and immediate data values. Only the load ("ld") and store ("st") instructions access memory. Registers are prefixed with the symbol "%," and register %fp, which stands for frame pointer, like the Intel's base pointer, holds the starting address for the memory data region. Because the Sparc processor automatically executes the instruction that follows a branch instruction, such as "bg" and "b," a nonoptimized code must include a NOP instruction following each branch instruction.

As expected, the RISC program in Example 8.4 has more instructions (19) as compared to 11 for the CISC program in Example 8.3.

Example 8.4. A Sparc assembly code for the sample program given in Example 8.1. The code, however, is not optimized.

```
     st   %g0, [%fp-48]    //Store, Memory[fp -48]←g0 (g0 always 0) (sum = 0)
     st   %g0, [%fp-44]    //Store, Memory[fp – 44]←g0 (i = 0)
.LL5:
     ld   [%fp-44], %g1    //Load, g1←Memory[i]
     cmp  %g1, 7           //Compare, is g1 > 7?
     bg   .LL6             //Branch if greater than 7
     nop
     ld   [%fp-44], %g1    //Load, g1←Memory[i]
     sll  %g1, 2, %g2      //Compute ptr to array[i]: Shift Left Logical (i * 2)
     add  %fp, -8, %g1     //g1←fp – 8 (get memory location of array)
     add  %g2, %g1, %g1    //g1←g2 + g1 (array location + next i * 2)
     ld   [%fp-48], %g2    //Load sum, g2←Memory[fp -48]
     ld   [%g1-32], %g1    //Load array]i], g1←Memory[g1 – 32]
     add  %g2, %g1, %g1    //Array]i] + sum, g1←g2 + g1
     st   %g1, [%fp-48]    //Store sum, Memory[fp – 48]←g1
     ld   [%fp-44], %g1    //Load i, g1←Memory[fp – 44]
     add  %g1, 1, %g1      //Increment, g1←g1 + 1
     st   %g1, [%fp-44]    //Store i, Memory[fp – 44]←g1
     b    .LL5             //Branch to instruction ay LL5
     nop
.LL6:
```

Executable Codes

An assembly language program is translated to its equivalent binary code by an assembler, typically, through a two-pass process. During the first pass, an assembler assigns a memory address to each label in the code and data sections. The instructions listed in Table 8.3 are represented in binary using one of the three instruction formats shown in Fig. 8.6. For simplicity, each instruction is made of an 8-bit op-code and an 8-bit data or address operand (if any). The "NOP" and "MVX" instructions do not have operands; thus, the operand field in these two instructions is set to 0.

Example 8.5 lists a typical output of an assembler using the program in Example 8.2. The listing uses an arbitrary syntax. For simplicity, the example Acc-ISA processor is assumed to be a 16-bit machine. It is also assumed that a program's code

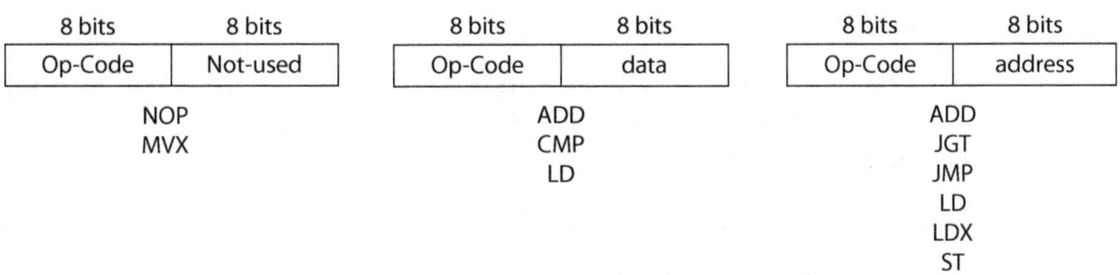

FIGURE 8.6 The instruction formats for the Acc-ISA example processor.

and data address space is 256B, with program codes starting at address 0 and data space starting at address 0xFF (a base address), as was shown in general in Fig. 8.5.

Example 8.5. The manually assembled output for the assembly program in Example 8.2. The *array* starts at data address 0xF0 for 16B, *i* at 0xEE for 2B, and *sum* at 0xEC for 2B.

```
Address     Instruction      Instruction in binary    Inhex
0:          LD    0          0000,0110;0000,0000      0600
2:          ST    0xEC       0000,1010;1110,1100      0AEC
4:          ST    0xEE       0000,1010;1110,1110      0AEE
6:          CMP   7          0000,0011;0000,0111      0307
8:          JGT   0x1A       0000,0100;0001,1010      041A
A:          MVX              0000,1001;0000,0000      0900
C:          LD    X(0xF0)    0000,1000;1111,0000      08F0
E:          ADD   (0xEC)     0000,0010;1110,1100      02EC
10:         ST    (0xEC)     0000,1010;1110,1100      0AEC
12:         LD    (0xEE)     0000,0111;1110,1110      07EE
14:         ADD   1          0000,0001;0000,0001      0101
16:         ST    (0xEE)     0000,1010;1110,1110      0AEE
18:         JMP   6          0000,0101;0000,0110      0506
1A:         ...
...
```

Example 8.6 is the assembler output for the Intel program in Example 8.3. The Intel's Pentium family of processors executes CISC instructions of different sizes. For example, "movl" is a 7B instruction, and "jg" is a 2B instruction.

Example 8.6. The output of the assembler for the Pentium IV program in Example 8.3. The numbers 0xffffffd0, 0xffffffd4, and 0xffffffd8 are the 2's complement representations of –48, –44, and –40, respectively.

```
401340:   c7 45 d0 00 00 00 00    movl   $0x0,0xffffffd0(%ebp)
401347:   c7 45 d4 00 00 00 00    movl   $0x0,0xffffffd4(%ebp)
40134e:   83 7d d4 07             cmpl   $0x7,0xffffffd4(%ebp)
401352:   7f 13                   jg     401367 <_main+0x77>
401354:   8b 45 d4                mov    0xffffffd4(%ebp),%eax
401357:   8b 54 85 d8             mov    0xffffffd8(%ebp,%eax,4),%edx
40135b:   8d 45 d0                lea    0xffffffd0(%ebp),%eax
40135e:   01 10                   add    %edx,(%eax)
401360:   8d 45 d4                lea    0xffffffd4(%ebp),%eax
401363:   ff 00                   incl   (%eax)
401365:   eb e7                   jmp    40134e <_main+0x5e>
401367:   b8 02 00 00 00          ...
```

Example 8.7 is the output of the assembler for the AltraSparc II program in Example 8.4. Because Sparc is a RISC-ISA, all the instructions are the same size for efficient implementation.

Example 8.7. The output of the assembler for the AltraSparc II program in Example 8.4:

```
...104f4:    c0 27 bf d0        clr   [ %fp + -48 ]
   104f8:    c0 27 bf d4        clr   [ %fp + -44 ]
   104fc:    c2 07 bf d4        ld    [ %fp + -44 ], %g1
```

```
10500:   80 a0 60 07    cmp   %g1, 7
10504:   14 80 00 0f    bg    10540 <main+0x90>
10508:   01 00 00 00    nop
1050c:   c2 07 bf d4    ld    [ %fp + -44 ], %g1
10510:   85 28 60 02    sll   %g1, 2, %g2
10514:   82 07 bf f8    add   %fp, -8, %g1
10518:   82 00 80 01    add   %g2, %g1, %g1
1051c:   c4 07 bf d0    ld    [ %fp + -48 ], %g2
10520:   c2 00 7f e0    ld    [ %g1 + -32 ], %g1
10524:   82 00 80 01    add   %g2, %g1, %g1
10528:   c2 27 bf d0    st    %g1, [ %fp + -48 ]
1052c:   c2 07 bf d4    ld    [ %fp + -44 ], %g1
10530:   82 00 60 01    inc   %g1
10534:   c2 27 bf d4    st    %g1, [ %fp + -44 ]
10538:   10 bf ff f1    b     104fc <main+0x4c>
1053c:   01 00 00 00    nop
10540:   ...
```

8.3.2 Acc-ISA Processor: Single-Cycle

Figure 8.7 illustrates a single-cycle data path for the example Acc-ISA processor. It consists of fetch, decode, execute, and write-back units. During each clock cycle, an instruction is fetched, decoded, and executed, and the result (if any) is written back to register ACC, X, or 1-bit status register (SR). The instructions are assumed already loaded into the IM (instruction cache memory) and data into the DM (data cache memory).

Also, during each clock cycle, the PP register either loads the output of the adder, if the next instruction is the next sequential instruction in the program, or the output of the multiplexer (MUX), if the next instruction is a result of a jump instruction, either "JGT" or "JMP." In the figure, it is assumed that PP retains byte addresses, but IM inputs a word address to output the next 16-bit instruction.

The decode unit contains a combinational circuit that inputs an op-code and generates all the control signals necessary to execute the current instruction. It also contains the ACC, X, and SR. In general, an SR is multibit register with a bit allocated for each condition, such as equal-to (E), less-than (L), not-equal-to (NE), arithmetic output is zero (Z), carry-out bit (C), arithmetic over-flow (O), and many other condition bits related to the state of the processor. The decode unit forwards all the control signals, the single immediate data operand (if any), and the content of the registers to the execute unit.

The execute unit contains all the components necessary to execute an instruction. MUXs are needed when there are multiple data sources for a module. For example, the adder/comparator (Add/Cmp) module either computes a sum or compares its inputs and outputs the greater-than-flag (*gtf*) signal. The module inputs both ACC and either an immediate or a direct memory operand. Therefore, a MUX is needed to choose either the operand when the operand is an immediate data or M[operand] when the operand is an address and indicates

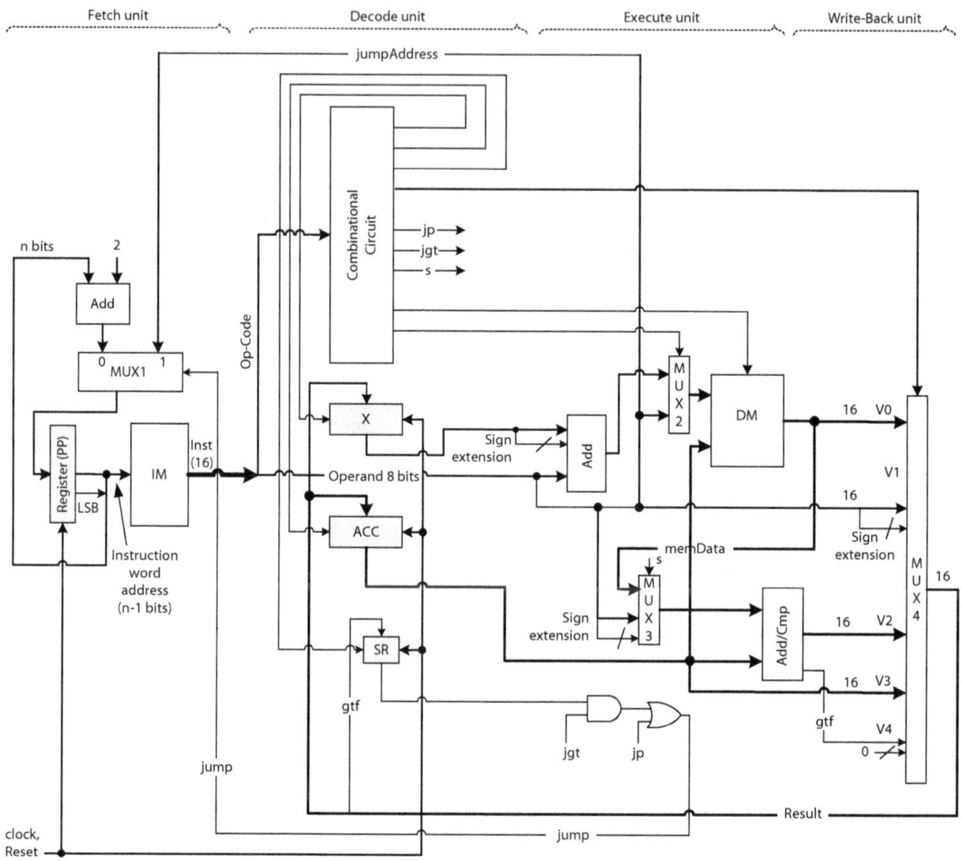

Figure 8.7 The Acc-ISA single-cycle data path to execute the program in Example 8.2.

memory content. Likewise, another MUX is used to select either a direct addressing or the quantity X + operand for index addressing. (Refer to the Exercises section in Chap. 3 for how to design a comparator.)

The write-back unit consists of a MUX. It is used to choose one of the several possible results generated by the execute unit. The result of executing "LD" and "ADD" instructions is stored (written back) in the ACC; the result of executing "MVX" instruction is stored in the X register, and the result of executing "CMP," which is a 1-bit greater than flag *gtf*, is stored in the 1-bit SR. Executing a "JMP" instruction causes the next instruction address to be loaded into PP.

Simulation

Example 8.8 presents the HDL model of the single-cycle Acc-ISA data path illustrated in Fig. 8.7. For simplicity, the IM and DM are assumed to be 64B each, organized as a 32 × 16 memory. In a real processor, both IM and DM are organized as cache memories (Chap. 10). Since the processor is a single-cycle data path, an instruction is executed within one clock cycle.

Example 8.8. An HDL behavioral description of the example single-cycle Acc-ISA data path includes two "initial" blocks to initialize IM with program instructions and DM with the *array* elements. The DM is initialized with eight values, 100 to 107. An extra "JMP 0x1A" instruction is inserted at the end of the program to create an infinite loop for simulation purposes. The description consists of two major code sections discussed next.

Decode and execute units—This combined behavioral description implicitly describes the control signals required to execute each instruction. The code also describes all the necessary combinational circuits required to create the data path. However, the HDL model does not explicitly describe any computing circuit module; their implementation details are left for the Altera's synthesis tool to determine.

Fetch and write-back units—These units update the state of the processor by changing the contents of registers PP, ACC, X, and SR. Thus, their descriptions are combined, but each could also be described separately. The combined unit includes the behavioral descriptions of the adder and the MUX used in the fetch unit and the MUX used in the write-back unit, as illustrated in Fig. 8.7.

```
module accISA(
  input clock, reset,
  output [15:0] inst,
  output reg [15:0] acc,
  output reg [4:0] x,
  output reg sr
 );

  reg [5:0] pp; //program pointer, points to the next instruction
  reg [7:0] opcode;
  reg [15:0] operand;
  reg [15:0] result;
  assign inst = {opcode, operand[7:0]};
  wire [4:0] imAddress = pp[5:1];      //IM is 32 X 16, addressing one instruction word
  wire [5:0] dmAddress = operand[5:1]; //DM is 32 X16, addressing one data word

  (* ramstyle = "M512" *) reg [15:0] IM[0:31];    //instruction memory

  (* ramstyle = "M512" *) reg [15:0] DM[0:31];    //data memory

  parameter       NOP =       0,
                  ADD_I = 1,
                  ADD_D = 2,
                  CMP_I = 3,
                  JGT = 4,
                  JMP = 5,
                  LD_I = 6,
                  LD_D = 7,
```

```
                LD_X = 8,
                MVX = 9,
                ST = 10;
   //Initialize IM with machine instructions
   initial begin //Note, memory is organized as 31 X 16
   IM[0] = 16'h0600;        //LD 0
   IM[1] = 16'h0AEC;        //ST (0xEC)     //(sum)
   IM[2] = 16'h0AEE;        //ST (0xEE) //(i)
   IM[3] = 16'h0307; //CMP 7
   IM[4] = 16'h041A; //JGT 0x1A
   IM[5] = 16'h0900; //MVX
   IM[6] = 16'h08F0; //LD X(0xF0) //(array)
   IM[7] = 16'h02EC; //ADD (0xEC) //(sum)
   IM[8] = 16'h0AEC; //ST  (0xEC) //(sum)
   IM[9] = 16'h07EE; //LD  (0xEE) //(i)
   IM[10] = 16'h0101;       //ADD 1
   IM[11] = 16'h0AEE;       //ST 0xEE //(i)
   IM[12] = 16'h0506; //JMP 6
   IM[13] = 16'h051A; //JMP 0x1A, extra instruction to cause looping end
   //Initialize the array
   initial begin //Note, memory is organzied 32 X 16
   //Therefore, for example, address 0xF0 = (1111 0000) is reduced to 0x30 = (11
0000)
   //and then to 0x18 = (1 1000), drppong the LSB to access a 16-bit word
   //DM[8'h16] reserve for "sum"
   //DM[8'h17] reserved for "i"
   DM[8'h18] = 100;        //array[0] = 100
   DM[8'h19] = 101;        //array[1] = 101
   DM[8'h1A] = 102;        //array[2] = 102
   DM[8'h1B] = 103;        //array[3] = 103
   DM[8'h1C] = 104;        //array[4] = 104
   DM[8'h1D] = 105;        //array[5] = 105
   DM[8'h1E] = 106;        //array[6] = 106
   DM[8'h1F] = 107;        //array[7] = 107
   end

   //---- Read instruction memory (IM) ------
   always@(*)
   begin
     opcode <= IM[imAddress][15:8]; //ImAddress is instruction word address
     operand[7:0] <= IM[imAddress][7:0];
     operand[15:8] <= {8{operand[7]}}; //sign extend 8-bit operand to create a 16-bit
                                        //operand
     end

     //------ Decode and Execute ---------
     always@(*)
     begin
              case (opcode)
              NOP: result = 16'hx;
```

```
        ADD_I: result = acc + operand;
        ADD_D: result = acc + DM[dmAddress];
        CMP_I: if (acc > operand)
                        result = {15'hx, 1'b1};
               else
                        result = {15'hx, 1'b0};
        JGT:   result = 16'hx;
        JMP:   result = 16'hx;
        LD_I:  result = operand;
        LD_D:  result =  DM[dmAddress];
        LD_X:  result = DM[dmAddress + x];
        MVX:   result = acc;
        ST:    result = acc;
        default: result = 16'hx;
endcase
end

//------- Fetch and Write Back --------
always@(posedge clock or posedge reset)
begin
 if (reset == 1)
 begin
        pp <= 0;
        acc <= 0;
        x <= 0;
        sr <= 0;
 end
 else

        case (opcode)
        NOP:     pp <= pp + 2; //incrementing in bytes not in words
        ADD_I: begin acc <= result; pp <= pp + 2; end
        ADD_D: begin acc <= result;  pp <= pp + 2; end
        CMP_I: begin sr <= result[0];  pp <= pp + 2; end
        JGT:     if (sr == 1)
                        pp <= operand[5:0];
                 else

                        pp <= pp + 2; //incrementing in bytes not in words
        JMP:     pp <= operand[5:0];
        LD_I: begin acc <= result; pp <= pp + 2; end
        LD_D:  begin acc <= result;     pp <= pp + 2; end
        LD_X:  begin acc <= result; pp <= pp + 2; end
        MVX:   begin x <= result[4:0]; pp <= pp + 2; end //indexing 16-bit
words
        ST:    begin DM[dmAddress] <= result;  pp <= pp + 2; end
        endcase
 end
endmodule
```

Figure 8.8 shows the simulation waveform illustrating the beginning of the program execution after an asynchronous reset. The waveform lists each

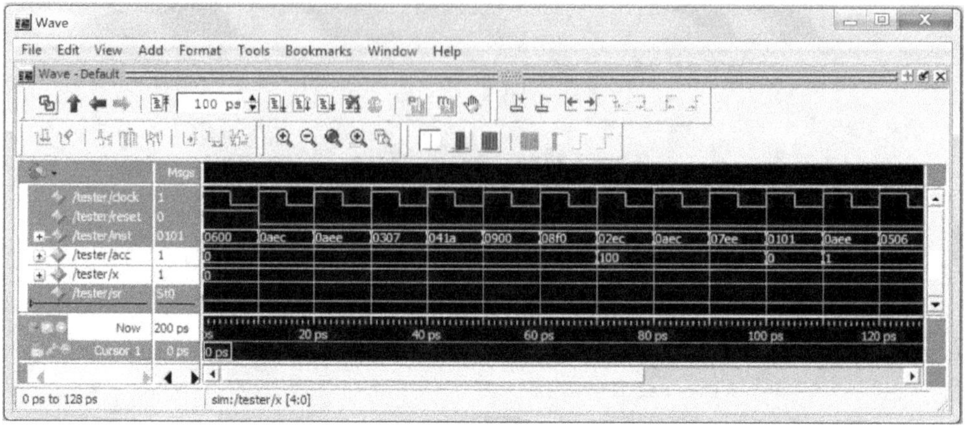

FIGURE 8.8 The example Acc-ISA single-cycle processor simulation waveform; illustrating the beginning section of the waveform.

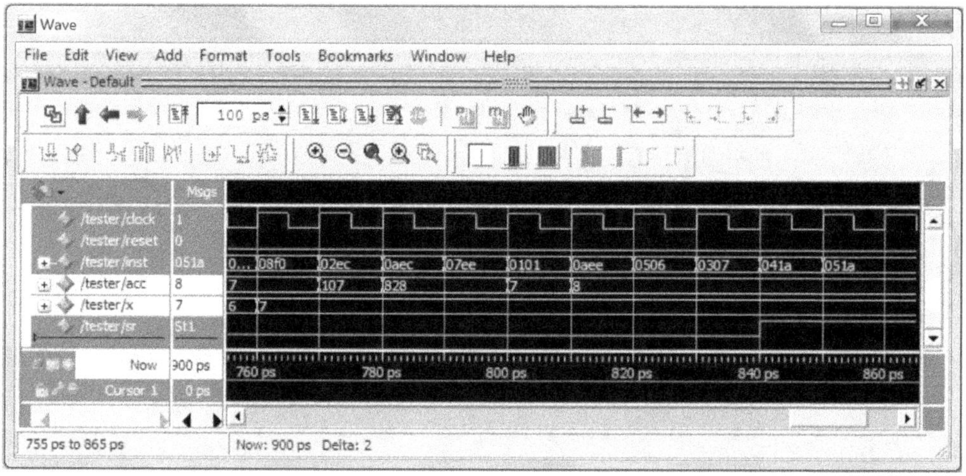

FIGURE 8.9 The example Acc-ISA single-cycle processor simulation waveform; illustrating the ending section of the waveform with the final sum = 828 ($\Sigma_{100}^{107} i$).

instruction in hex and the contents of ACC, X, and SR registers in decimal. Figure 8.9 shows the ending of the simulation waveform when SR bit becomes 1. As expected, the figure shows ACC = 828, the final sum of the *array* elements. The program execution continues in a loop with the "JMP 0x1A" (0x051A) instruction.

8.3.3 Acc-ISA Processor: Pipelined

Figure 8.10 illustrates a block diagram of a four-staged instruction pipeline for the example Acc-ISA, and Fig. 8.11 illustrates its detailed circuit. All the control signals to execute an instruction are generated in the decode stage and, along with

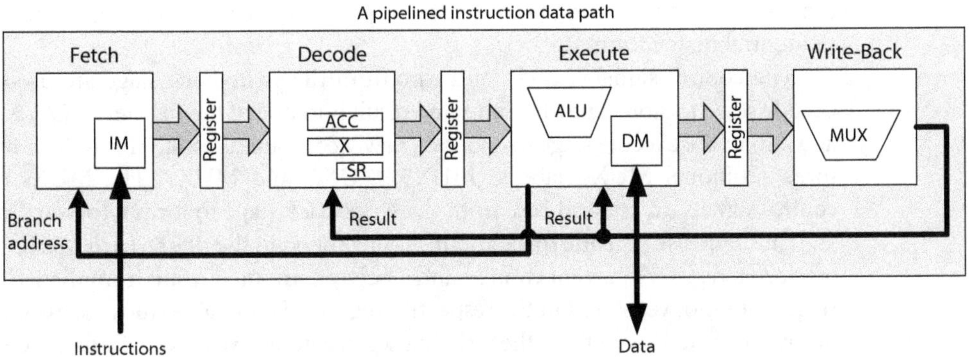

FIGURE 8.10 Illustrating a four-stage pipelined instruction data path.

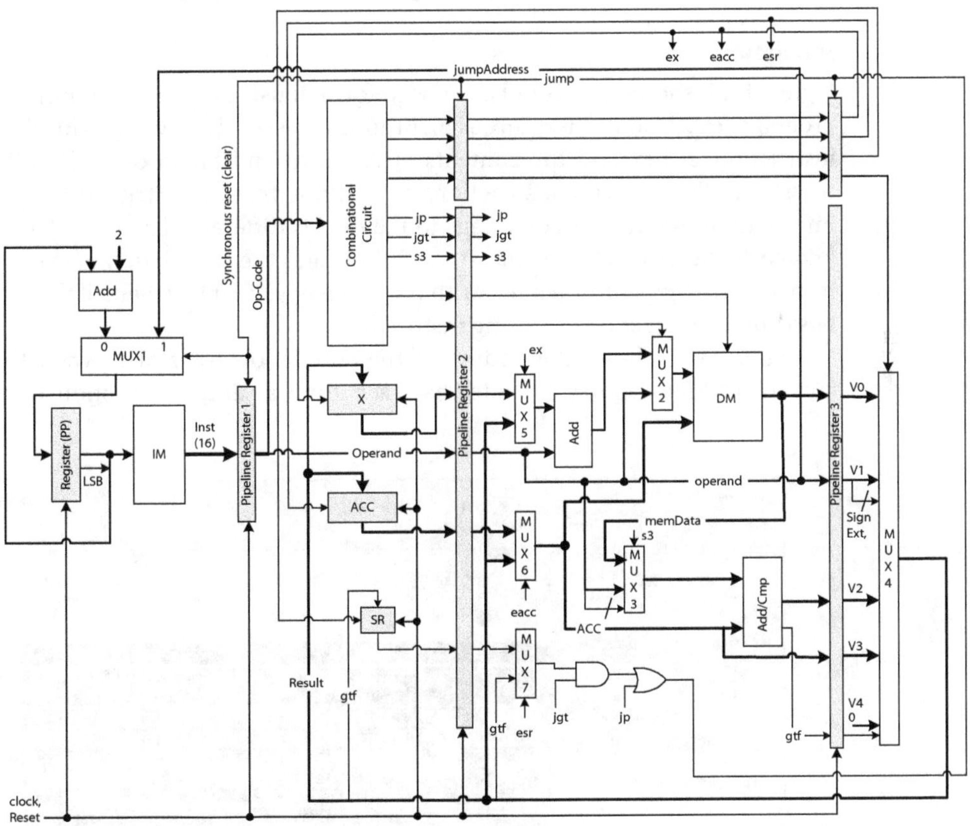

FIGURE 8.11 The pipelined data path of the example Acc-ISA processor.vsd.

the register contents, are fed to the execute stage. Those control signals that are needed in the write-back stage are forwarded from the execute stage, along with the computed results. The write-back stage then selects and passes one of the computed results (if any) to the decode stage. The write-back stage also forwards the selected result to the execute state for possible use in the execution of the next sequential instruction.

The control signals *ex*, *esr*, and *eacc* from the write-back stage are used in the decode stage to store a newly computed result into one of the registers ACC, X, or SR, and also in the execute stage to choose a newly computed result. This is done by using three additional MUXs, labeled MUX5, MUX6, and MUX7. The MUXs use the control signals *ex*, *esr*, and *eacc* from the write-back stage to form a **forwarding unit** and thus increase pipeline throughput. As illustrated in the figure, each MUX chooses between a register's current content and a newly computed result from the write-back stage that is not yet stored in the respective register. This enables the processor to use a newly computed result from the write-back stage in the execution of a data-dependent instruction that follows without waiting first for the result to be stored in a register. The complexity of a forwarding unit depends on the complexity of the specific ISA. The functions of a forwarding unit are discussed in more detail later in this section.

Simulation

Figure 8.12 shows the simulation waveform illustrating the beginning of the program execution after an asynchronous reset. The waveform lists each instruction in hex and the contents of ACC, X, and SR in decimal. While the waveform illustrates the instructions are moving from one stage to the next, the information passed from one stage to the next is different. The write-back is also known as the instruction **retire stage**. It indicates the completion of executing an instruction. A pipeline operates with the efficiency of 100% when all the stages are busy and concurrently executing instructions.

Figure 8.13 shows the ending of the simulation waveform when the 1-bit content of SR becomes 1. Note that each time a jump instruction ("JMP" or

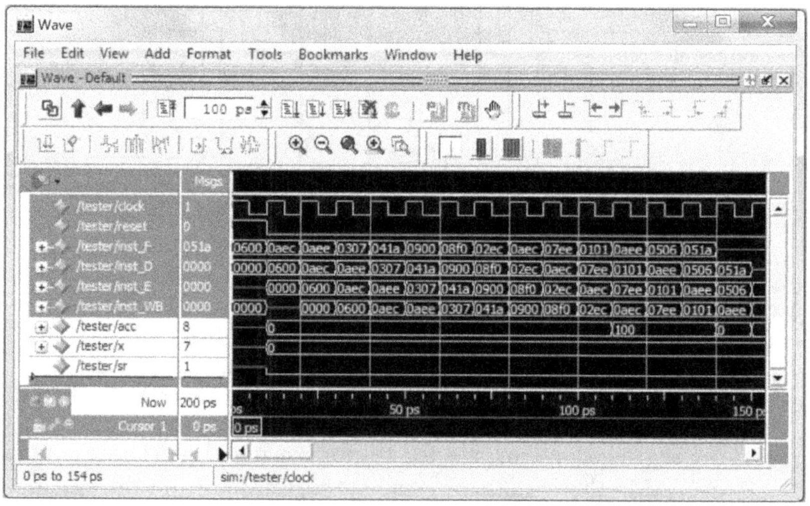

Figure 8.12 A simulation waveform for the example Acc-ISA pipelined data path; illustrating the beginning section of the waveform. The stages are displayed top to bottom.

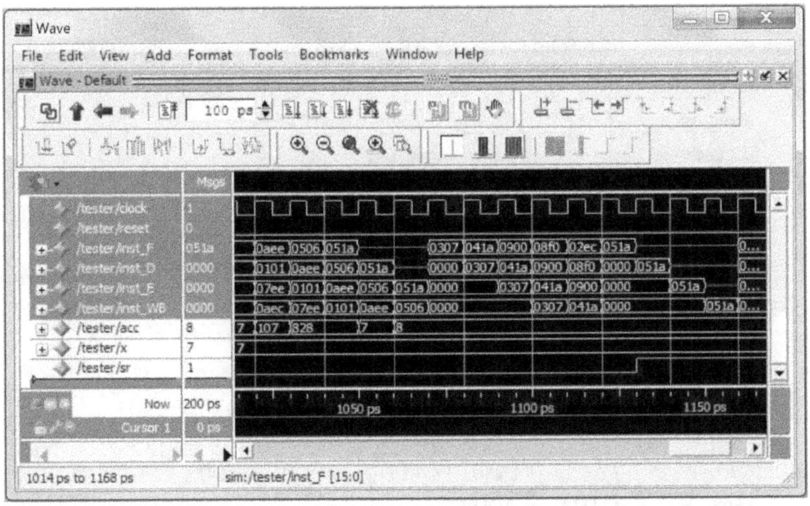

FIGURE 8.13 A simulation waveform for the example Acc-ISA pipelined data path; illustrating the ending section of the waveform with final sum = 828.

"JGT") executes, the pipeline starts over. This is called a **pipeline flush**. For example, as illustrated in Fig. 8.13, when the unconditional instruction 0x0506 ("JMP 6") executes or the execution of the conditional instruction 0x041A ("JGT 0x1A") results in a jump, it causes a pipeline flush and thus decreases the efficiency of the pipeline. In this case, when there is a pipeline flush, no instructions retire for three clock cycles as illustrated by the write-back stage (inst_WB) in the figure. The figure shows the content of ACC = 828 as the final sum of the *array* elements. The program execution continues in a loop with a "JMP 0x1A" instruction.

Forwarding and Hazard Units

A forwarding unit, which forwards a newly generated result from the write-back stage to the execute stage, was briefly discussed earlier for the example Acc-ISA pipelined processor. Here, the discussion is more general. Consider the following two data-dependent instructions from Example 8.2, the Acc-ISA example program:

```
LD   (i) //ACC←M[i]
ADD  1   //ACC←ACC + 1
```

The LD instruction loads a memory data into the ACC, and the ADD instruction increments the content of the ACC. Figure 8.14 illustrates two pipeline charts for executing the two instructions in sequence. In Fig. 8.14(*a*), the processor uses a forwarding unit to forward the newly read memory data (M[*i*]), not yet stored in the ACC, to the execute stage to be used in the execution of the ADD instruction.

On the other hand, the chart in Fig. 8.14(*b*) illustrates the execution of the same two instructions without a forwarding unit. In this case, when the LD instruction is executing, the ADD instruction is decoding. When the LD instruction moves to

No Hazard Unit is needed in this case

Arrow indicates Result forwarded from Write-Back stage to Execute stage

Write-Back				LD (i)	ADD 1
Execute			LD (i)	ADD 1	
Decode		LD (i)	ADD 1		
Fetch	LD (i)	ADD 1			
Clock Cycle:	1	2	3	4	5

(a) With a forwarding unit

Bubble is used to indicate a pipeline stall

Write-Back				LD (i)	O	O	ADD 1
Execute			LD (i)	O	O	ADD 1	
Decode		LD (i)	ADD 1	ADD 1	ADD 1		
Fetch	LD (i)	ADD 1					
Clock Cycle:	1	2	3	4	5	6	7

(b) Without a forwarding unit but with a hazard unit

Figure 8.14 Illustrating the effect of a forwarding unit; (a) a pipeline chart with a forwarding.vsd.

the write-back stage, the execution of the ADD instruction must be stalled until the ACC is updated with the M[*i*]. Therefore, the execution of the ADD instruction would be stalled for two clock cycles, as illustrated with bubbles in the figure.

Pipeline stalling is implemented by a **hazard unit** that delays and prevents a data-dependent next instruction (i.e., ADD) from moving into the execute stage by executing an implicit NOP instruction during that clock cycle. Once the ACC is updated with the new value, the dependent instruction (i.e., ADD) is allowed to move into the execute stage for execution.

Because the execution unit in this case consists of only one stage (i.e., "Execute," in Fig. 8.14), the processor either requires a forwarding unit or hazard unit, but not both. Figure 8.15 illustrates the Acc-ISA pipelined data path with a hazard unit, and Example 8.9 presents an HDL model for the hazard unit. The MUX4, MUX5, and MUX6 that were used to implement a forwarding unit in Fig. 8.11 are now removed from the data path in Fig. 8.15.

Example 8.9. The HDL code section describes the hazard unit for the Acc-ISA pipelined processor in Fig. 8.15 using explicitly declared register control signal names. The hazard unit compares the register control signals *ex*, *eacc*, and *esr* from the write-back stage with those in the execute stage. If each pair of control signals (e.g., *eacc* in the execute stage with the *eacc* in the write-back stage) are the same, the hazard unit located in the decode stage delays the execution of the current instruction. Specifically, the hazard unit asserts the *hazard* signal and synchronously resets (clears) the pipeline register that passes the register control signals for the current instruction from the decode stage to the execute stage, as shown in the figure.

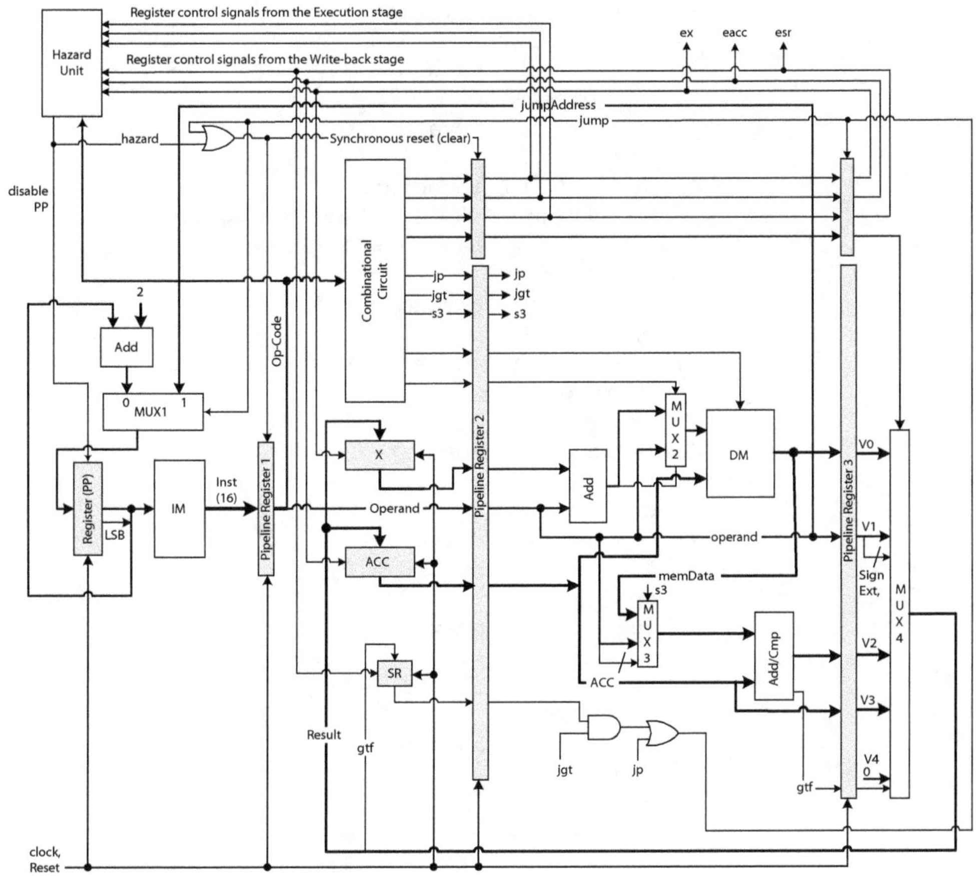

Figure 8.15 The example ACC-ISA pipelined data path with a hazard unit.vsd.

```
always@(*) //Hazard Unit
begin
        case(opcode)
        0: begin //NOP
           hazard = 0;
           end
        1: begin //ADD immediate
               if(eaccWB == 1 || eaccExe == 1)
                       hazard = 1;
               else
                       hazard = 0;
           end
        2: begin //ADD direct
               if(eaccWB == 1 || eaccExe == 1)
                       hazard = 1;
               else
                       hazard = 0;
           end
```

```
 3: begin //CMP immediate
        if(eaccWB == 1 || eaccExe == 1)
                hazard = 1;
        else
                hazard = 0;
    end
 4: begin //JGT immediate
        if(esrWB == 1 || esrExe == 1)
                hazard = 1;
        else
                hazard = 0;
    end
 5: begin //JMP address: PC <- address
     hazard = 0;
    end
 6: begin //LD immediate
                hazard = 0;
    end
 7: begin  //LD direct
                hazard = 0;
    end
 8: begin //LD indexed
        if(exWB == 1 || exExe == 1)
                hazard = 1;
        else
                hazard = 0;
    end
 9: begin //MVX
        if(exWB == 1 || exExe == 1)
                hazard = 1;
        else
                hazard = 0;
    end
10: begin //ST
        if(eaccWB == 1 || eaccExe == 1)
                hazard = 1;
        else
                hazard = 0;
    end
 default: begin
        hazard = 0;
        end
 endcase
```

In general, an execution unit is made of two or more stages to better distribute the required hardware into several stages and minimize the clock period of the pipeline. In such cases, a pipelined data path requires both a forward unit and a hazard unit, as will be discussed in Sec. 8.3.4 for a RISC processor. In addition, because a typical CISC or RISC ISA uses not one but several general-purpose

	i	ii	iii	iv	v	vi	1	2	3	4
Write-Back				LD 0	ST (sum)	ST (i)	CMP 7	JGT L2	MVX	LD X(array)
Execute			LD 0	ST (sum)	ST (i)	CMP 7	JGT L2	MVX	LD X(array)	ADD (sum)
Decode		LD 0	ST (sum)	ST (i)	CMP 7	JGT L2	MVX	LD X(array)	ADD (sum)	ST (sum)
Fetch	LD 0	ST (sum)	ST (i)	CMP 7	JGT L2	MVX	LD X(array)	ADD (sum)	ST (sum)	LD (i)
Clock Cycles	1	2	3	4	5	6	7	8	9	10

	5	6	7	7	8	10	11	12	1	2
Write-Back	ADD (sum)	ST (sum)	LD (i)	ADD 1	ST (i)	JMP L1	O	O	CMP 7	...
Execute	ST (sum)	LD (i)	ADD 1	ST (i)	JMP L1	O	O	CMP 7	JGT L2	...
Decode	LD (i)	ADD 1	ST (i)	JMP L1	?	O	CMP 7	JGT L2	MVX	...
Fetch	ADD 1	ST (i)	JMP L1	?	?	CMP 7	JGT L2	MVX	LD X(array)	...
Clock Cycles	11	12	13	14	15	16	17	18	19	...

	...	10	11	12	I	II		
Write-Back	...	JMP L1	O	O	CMP 7	JGT L2	O	O
Execute	...	O	O	CMP 7	JGT L2	MVX	O	
Decode	...	O	CMP 7	JGT L2	MVX	LD X(array)		
Fetch	...	CMP 7	JGT L2	MVX	LD X(array)	?		
Clock Cycles	...	100	101	102	103	104		

FIGURE 8.16 A pipeline chart illustrating the execution of the Acc-ISA program in Example 8.2.

registers, forwarding and hazard units must be able to check data dependency among several different registers.

Performance Analysis

The performance of a processor is often measured in terms of **cycles per instruction** (CPI). It is calculated as the total number of clock cycles used to execute a program divided by the number of instructions in the program (see Eq. (8.1)).

$$\text{CPI} = \frac{\text{Number of clock cycles used } (N)}{\text{Number of instructions executed } (n)} \tag{8.1}$$

Figure 8.16 illustrates a pipeline chart for the example Acc-ISA program. As illustrated in the chart and summarized in Table 8.4, it takes 6 clock cycles (labeled i through vi in the chart) to execute the instructions before the for-loop and 12 clock cycles to execute each iteration of the for-loop. Finally, two clock cycles (labeled I and II) are required to exit the for-loop. The program has three instructions before the for-loop, 10 instructions in the body of the for-loop, and two instructions ("CMP" and "JGT") to exit the for-loop.

Number of clock cycles to execute the instructions before the for-loop:	6
Number of clock cycles to execute an iteration of the for-loop:	12
Number of clock cycles to end the for-loop:	2
Number of instructions before the for-loop:	3
Number of instructions in the body of the for-loop:	10
Number of instructions to exit the for-loop:	2

TABLE 8.4 Data from the Pipeline Chart in Fig. 8.16

Equation (8.2) presents N, the total number of pipeline clock cycles, and n, the total number of instructions using m for-loop iterations.

$$\begin{aligned} N &= 6 + m(12) + 2 \\ &= 12m + 8 \qquad \text{clock cycles} \\ n &= 3 + m(10) + 2 \\ &= 10m + 5 \qquad \text{instructions} \end{aligned} \tag{8.2}$$

For $m = 8$ iterations, the CPI of the example program is calculated as follows:

$$\begin{aligned} \text{CPI} &= \frac{12m+8}{10m+5} \qquad \text{for } m \text{ iterations} \\ &= \frac{12(8)+8}{10(8)+5} = \frac{104}{85} = 1.22 \qquad \text{for } m = 8 \end{aligned} \tag{8.3}$$

Equation (8.4) presents the CPI for the example program as m approaches infinity (∞). The CPI is larger than 1 due to executing m "JMP" instructions and one "JGT" instruction that results in a jump, where each causes a pipeline flush. However, even though the CPI of a single-cycle processor is always 1, a single-cycle data path requires a much longer clock period than an equivalent pipeline data path, as was discussed in Chap. 6.

$$\text{CPI} = \frac{12m+8}{10m+5} = 1.2 \qquad \text{as } m \to \infty \tag{8.4}$$

In Eq. (8.4), the CPI is a **lower-bound** (the lowest value). If CPI is 1.0 for a program, the corresponding pipeline chart will have no bubbles, indicating 100% efficiency. The CPI limit of 1.2, which is greater than 1.0, indicates the efficiency of the pipeline is less than 100% and, at best, it takes, on average, the duration of 1.2 clock cycles to execute each instruction. A CPI value can be used to estimate the execution time of n instructions, as given in Eq. (8.5), where τ is the period of the clock signal in seconds.

$$\text{Execution time} = n * \text{CPI} * \tau \tag{8.5}$$

Achieving CPI = 1.0 is a difficult task due to presence of data-dependency relationships among the instructions, branch instructions in the program, and latency to access instructions or data from a cache or memory. However, while a forewarning unit can resolve some data dependencies among instructions, there are other techniques to improve CPI that will be discussed later.

8.3.4 RISC-ISA Processor

Figure 8.17 illustrates a block diagram of a five-stage RISC-ISA instruction pipeline. The DM (as data cache) is now placed in its own separate stage. This is similar to the five-stage data path used in an initial MIPS processor. The execute stage is now responsible for all the arithmetic operations and memory address calculations, but

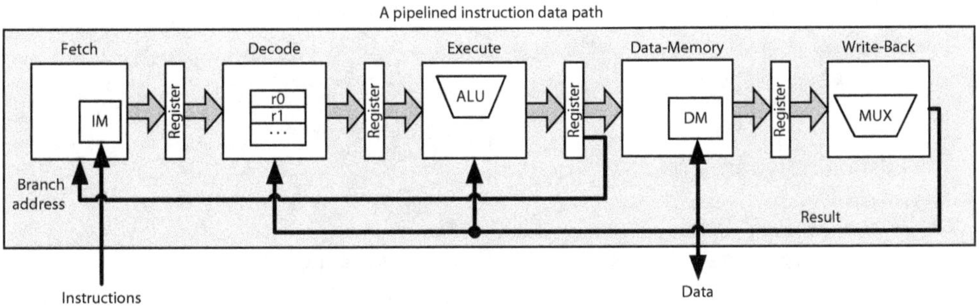

FIGURE 8.17 A five-stage RISC-ISA pipelined data path.

not for accessing data from cache that is performed in the DM stage. An arithmetic result is now passed unchanged from the DM to the write-back stage. The five-stage data path is more suitable for a RISC-ISA because data items must first be loaded into registers before they can be used by an arithmetic instruction in the execute stage. However, all the arithmetic direct (D) and indexed (X) instructions in the example Acc-ISA must now be converted into register-register arithmetic instructions. The data path has the advantage of simplifying the complexity of each stage and achieving a more uniform propagation delays among the five stages.

Program Example

Example 8.10 presents a RISC-ISA assembly program corresponding to the high-level language program code in Example 8.1. In the program, it is assumed that register R0 is always 0, and registers R1 through R4 are general-purpose registers. However, no code optimizations are performed.

Example 8.10. A RISC-ISA assembly program for the example program in Example 8.1; no compiler optimization is performed.

```
.code  //start program code section

        ST          (sum), R0     //Initialize: M[sum] ← R0
        ST          (i), R0       //M[i] ← R0

L1:

        LD          R1, (i)       //R1 ← M[i]
        CMP         R1, 7         //is i > 7?
        JGT         L2            //exit for-loop if greater (PP ← L2)
        LD          R2, (sum)     //load sum, R2 ← M[sum]
        LD          R3, R1, (array)      //load next array element, R3 ←
                                         //M[array + R1]
        ADD         R4, R2, R3    //compute sum + array[i]
        ST          (sum), R4     //store sum, M[sum] ← R4
        ADD         R1, R1, 1     //increment i
```

```
        ST          (i), R1          //store i (M[i]←R1).
        JMP         L1               //loop back
L2:     Ix                           //some instructions x, y, and z
        Iy
        Iz

.data  //start program data section
array:   RB   8                      //reserve 8 bytes
i:       RB   1                      //reserve 1 byte
sum:     RB   1                      //reserve 1 byte
```

Compiler Optimization

Because the DM cache resides in a separate stage, the RISC-ISA data path must implement a forwarding unit as well as a hazard unit. This is because the "LD R3, R1, (array)" and "ADD R4, R2, R3" instructions are data dependent, and therefore there is a one-cycle delay for the new content of R3, read from DM, to be forwarded to the execute stage, as illustrated in Fig. 8.18. At the time when instruction "LD R3, R1, (array)" is in the execute stage, instruction "ADD R4, R2, R3" is in the decode stage. When the "LD" instruction moves to the DM stage, the hazard unit must prevent the "ADD" instruction from moving to the execute stage by inserting a bubble (an implicit NOP instruction), as shown in the figure.

However, it is occasionally possible for the compiler to optimize programs and eliminate bubbles due to memory load instructions, such as the one shown in Fig. 8.18. In this case, the compiler may be able to rearrange and delay the execution of some or all instructions that depend on LD instructions. Example 8.11 presents an optimized code where instruction "ADD R1, R1, 1" is moved between instructions "LD R3, R1, (array)" and "ADD R4, R2, R3" in Example 8.10; thus, the execution of instruction "ADD R4, R2, R3" is delayed by one cycle, eliminating one bubble.

Example 8.11. Program in Example 8.10 optimized by compiler to delay the execution of instruction "ADD R4, R2, R3" by one clock cycle:

```
.code  //start program code section

        ST    (sum), R0        //Initialize: M[sum]←R0
        ST    (i), R0          //M[i]←R0

L1:

        LD    R1, (i)          //R1 ← M[i]
        CMP   R1, 7            //is i > 7?
        JGT   L2               //exit for-loop if greater (PP←L2)
        LD    R2, (sum)        // load sum, R2←M[sum]
        LD    R3, R1, (array)         //load next array element, R3←
                                      //M[array + R1]
```

```
        ADD   R1, R1, 1      // increment i; instruction moved
here
        ADD   R4, R2, R3     // compute sum + array[i]
        ST    (sum), R4      // store sum, M[sum] ← R4
        ST    (i), R1        // store i (M[i] ← R1).
        JMP   L1             // loop back
L2:       Ix                 // some instructions x, y, and z
          Iy
        ...Iz

.data    //start program data section
array:    RB   8                  // reserve 8 bytes
i:        RB   1                  // reserve 1 byte
sum:      RB   1                  // reserve 1
```

The compiler optimization must be done with care so that the program is not modified incorrectly. The program may generate invalid outputs, or may even fail to execute due to an out-of-bounds data access. For instance, in Example 8.10, moving both the "LD R2, (sum)" and "LD R3, R1, (array)" instructions to before the "JGT" instruction would not be a correct compiler optimization step in order to eliminate the one-cycle delay required to execute the "ADD R4, R2, R3" instruction. The reason for this is that after the last iteration, the "LD R3, R1, (array)" instruction, if placed before the "JGT," will try to access *array[8]*, which refers to an element outside the array boundary. The *array* in Example 8.1 has eight elements *array[0]* to *array[7]*.

Performance Analysis

The RSIC data path in Fig. 8.17 contains five pipeline stages versus four in its equivalent pipelined Acc-ISA data path shown in Fig. 8.10. The execute stage in the Acc-ISA data path, which contains the data cache and thus has the longest propagation delay, is divided into two stages (execute and DM) in the RISC data path, each with a smaller propagation delay. This reduction in the propagation delay of the longest stage enables RISC processors to use a faster clock, thus increasing the processor throughput.

In addition, as illustrated by Example 8.11, with the compiler optimization, it is often possible to further improve a RISC's CPI for an arbitrary program by overlapping a one-cycle delay cache access with the execution of another

Write-Back	...	...	...	...	LD R2, ...	LD R3, ...	O
Data Memory	...	...	...	LD R2, ...	LD R3, ...	O	ADD R4, R2, R3
Execute	...	...	LD R2, ...	LD R3, ...	O	ADD R4, R2, R3	...
Decode	...	LD R2, ...	LD R3, ...	ADD R4, R2, R3	ADD R4, R2, R3	...	...
Fetch	LD R2, ...	LD R3, ...	ADD R4, R2, R3	...	...	...	...

FIGURE 8.18 Illustrating RISC data dependency between an LD and ADD instruction.

instruction (i.e., “ADD R1, R1, 1”). Instructions “ADD R1, R1, 1” and “LD R3, R1, (array)” are data independent, and thus moving the “ADD R1, R1, 1” instruction from where it was in Example 8.10 to before the “ADD R4, R2, R3” instruction in Example 8.11 does not alter the correctness of the original program—only the order in which the instructions execute changes. “ADD R1, R1, 1” is executed earlier, and the execution of “ADD R4, R2, R3” is delayed by one cycle to allow time for the “LD R3, R1, (array)” instruction to load data from DM. The newly accessed data is then forwarded to the execute stage, where “ADD R4, R2, R3” is executed next.

As will be discussed in Chap. 10, if the target data is not in DM (a cache), it must be copied from a lower-level cache or the main memory requiring multiple CPU clock cycles. However, in the following section, we will further discuss multithreading to improve the efficiency of a pipeline when data is not in cache.

8.4 Advanced Processor Architectures

While the RISC pipeline improves instruction throughput, additional performance increases would come from reducing the pipeline clock period. One way to do this is to physically divide the data path of one or more pipeline stages into smaller sub-data paths, creating a **deeper pipeline**, one with many stages. Because each of the smaller stages will have a shorter propagation delay, a deep pipeline would operate with a faster clock, thus increasing instruction throughput.

If the data path of a pipeline stage is physically undividable (e.g., contains a memory), the stage is modified to include two or more copies of the undividable hardware. All the copies would operate in parallel but overlap, creating parallelism within the stage called **superpipelining**. While the propagation delay of a superpipelined stage remains about the same as that of the original stage, the stage, now with duplicate copies, is able to generate outputs more frequently. However, deep pipelining (including superpipelining) increases power consumption.

While deep pipelining improves instruction throughput, program flow control (i.e., conditional and unconditional) instructions reduce pipeline efficiency, increasing CPI and thus reducing throughput. Modern processors also implement **branch prediction** mechanisms to minimize pipeline flush and therefore increase efficiency.

In addition, modern processors increase the instruction throughput by using instruction level parallelism (ILP). The processor, in this case, is often referred to as **superscalar** because each stage contains additional resources and is able to execute multiple independent instructions in parallel. For a single program, the efficiency of an ILP pipeline, even with perfect branch prediction, is typically less than 100%. That is, a pipeline that implements k-instruction ILP would sometimes execute one, two, etc. up to k instructions due to data-dependency relationships among instructions, and sometimes would execute no instructions at all due to time lost for accessing data from caches or the main memory. Therefore, some or

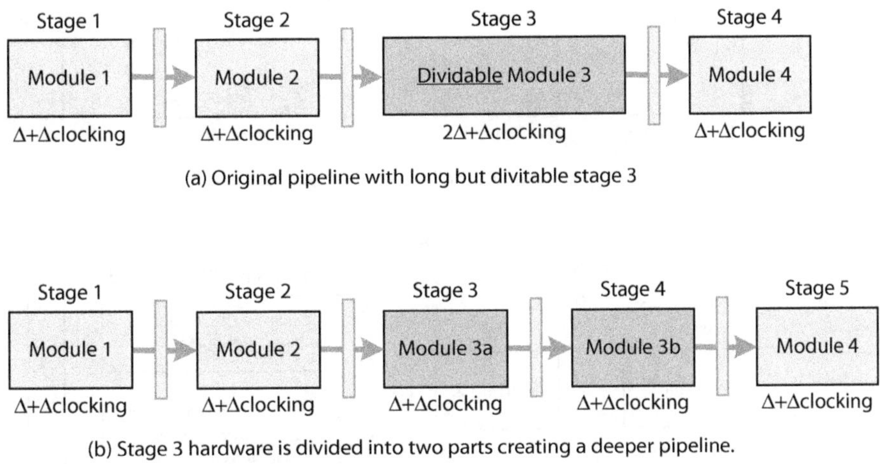

FIGURE 8.19 Deeper pipeline design: (*a*) original four-stage pipeline with dividable stage 3; (*b*) original pipeline converted into five stages, reducing the clock period.

all of the data path resources will remain idle for one or more clock cycles, unless the pipeline is equipped to execute multiple programs (threads) simultaneously. In this case, the pipeline is said to implement multithreading.

In the following sections, we will further explore deep pipelining, branch prediction, statically and dynamically scheduled ILP, and multithreading.

8.4.1 Deep Pipelining

Figure 8.19(*a*) displays an original four-stage pipeline where stage 3 has the longest propagation delay, indicated by $2\Delta + \Delta_{\text{clocking}}$, where $\Delta_{\text{clocking}} = \tau_{st} + \tau_{cq} + \tau_{cs}$; the τ_{st}, τ_{cq}, and τ_{cs} stand for setup time, clock-to-*q* time, and clock skew, respectively. In Fig. 8.19(*b*), the hardware of stage 3 (e.g., a multilevel MUX, CLA adder, combinational divider, etc.) is shown divided into two smaller modules, each with a smaller propagation delay.

For example, for simplicity, consider stage 3 to include one 2-level 4-to-1 MUX, designed using three 2-to-1 MUXs (Chap. 3). The 4-to-1 can be divided into two parts and organized to operate as two separate stages. The propagation delay of each of the two stages will be approximately one-half the propagation delay of the original stage 3 in Fig. 8.19(*a*). The new pipeline will have five stages and will operate with a shorter clock period $\tau_{\text{new}} = \Delta + \Delta_{\text{clocking}}$. The only extra hardware used in this case is one more pipeline register.

On the other hand, Fig. 8.20(*a*) illustrates a four-stage pipeline where the data path of stage 3 cannot be divided into smaller stages. For example, suppose stage 3 in the figure is a DM stage. In this case, the memory can be organized to operate as a two-way low-order (fine) interleaving (Chap. 7) that keeps the content of even memory addresses in one memory module and that of odd addresses in another module. As long as consecutive memory accesses are from alternating even and odd addresses, the operations of the two interleaved memory accesses can be

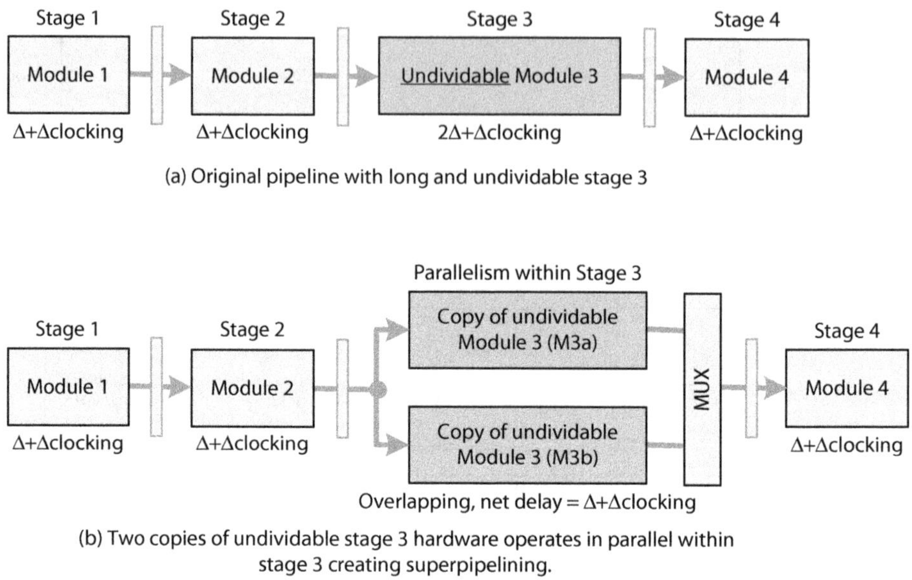

FIGURE 8.20 Superpipelining design: (*a*) original four-stage pipeline with undividable stage 3; (*b*) superpipeline organization of stage 3.

overlapped, creating superpipelining. This is illustrated with two modules in Fig. 8.20(*b*). In general, any pipeline stage can be superpipelined.

Table 8.5 illustrates a superpipelining chart using the data path shown in Fig. 8.20(*b*). The two identical undividable modules in stage 3 are labeled M3a and M3b. Using a clock cycle with period $\tau = \Delta + \Delta_{clocking}$, at time 3τ, instruction I1 enters stage 3 and is executed using M3a, which will take two clock cycles, or about 2τ, to complete. Note Δ also includes the delay of the MUX.

Because the pipeline controller takes turns and uses M3a and M3b alternately, one clock cycle later and at time 4τ when I2 enters stage 3, the pipeline controller

Stage 4 (Write-back)				O[&]	I1	I2	I3
Stage 3 (Execute)			I1 (M3a, 2τ)	I2 (M3b, 2τ)	I3 (M3a, 2τ)	I4 (M3b, 2τ)	...
Stage 2 (Decode)		I1	I2	I3	I4	I5	...
Stage 1: (Fetch)	I1	I2	I3	I4	I5	I6	...
Time	1τ	2τ	3τ	4τ	5τ	6τ	...

&: Bubble due to superpipelining.

TABLE 8.5 Illustrating Superpipelining of Stage 3 in Fig. 8.20(*b*) with Two Identical Modules Labeled M3a and M3b

selects M3b, which also takes 2τ to complete. Therefore, one result is available from stage 3 every clock cycle even though M3a and M3b each requires two clock cycles to generate a result.

At time 5τ, the output from M3a is fed to stage 4, allowing I1 to continue execution. At time 6τ (the next cycle), the output of M3b is fed to stage 4, allowing I2 to continue execution. This enables the pipeline to execute one instruction per τ, as illustrated in the table.

Deep pipelining enables the use of a faster clock to improve performance. In Fig. 8.19(*a*), the original clock period $\tau_{old} = 2\Delta + \Delta_{clocking}$ is reduced to $\tau_{new} = \Delta + \Delta_{clocking}$ in Fig. 8.19(*b*), approximately doubling the clock frequency. In Fig. 8.20(*b*), the original clock period, $2\Delta + \Delta_{clocking}$ in Fig. 8.20(*a*), is also effectively reduced approximately to $\Delta + \Delta_{clocking}$.

A deep pipelining (including superpipelining), however, increases the number of bubbles each time the pipeline restarts. In Fig. 8.20(*b*), one more bubble is introduced at time 4τ due to superpipelining as if the pipeline has five stages instead of four stages in the original pipeline in Fig. 8.20(*a*). As a deep instruction pipeline increases concurrency by operating on many more instructions at the same time, there are certain limitations, as follows, on how deep an instruction pipeline should be:

- Deep pipelining (including superpipelining) not only increases the amount of hardware, but also increases clock frequency. When the amount of hardware and/or clock frequency of a pipeline increase, so will the amount of power consumption (Chap. 6). Therefore, there is a limit for how deep a pipeline can be.
- Deep pipelining (including superpipelining) beyond a certain limit can also be counterproductive as it executes jump/branch instructions. Any change in program flow would cause a pipeline flush, which would, in this case, introduce more pipeline bubbles and therefore reduce pipeline efficiency and its performance.

8.4.2 Branch Prediction

Branch prediction means determining the direction of program flow in advance of executing a conditional or unconditional instruction. The target jump/branch address is determined early in the pipeline so that the fetch stage can start fetching instructions starting at a target address, thus minimizing the number of pipeline bubbles if there is a change in program flow.

[Note that for simplicity, a jump and a branch instruction is treated the same here. However, in general, they are implemented differently. A branch address is typically computed as a short distance (a displacement) from the current content of the PP, whereas a jump address is determined as an absolute address (not relative to PP). In general, a branch instruction is used when a jump distance is short—for example, in the implementation of a for-loop—and a jump instruction is used for a long jump—for example, a subroutine call. In the following sections, the terms "jump" and "branch" are used interchangeably, both causing a change in program flow.]

When the current instruction is a jump (e.g., "JGT" or "JMP"), the earliest time that the direction of program flow can change in Fig. 8.11 or in Fig. 8.17 is when the instruction is in the write-back stage. For example, in the Acc-ISA data path, when the "JMP" instruction is in the execute stage (e.g., clock cycle 15 in Fig. 8.15), on the next clock cycle, as "JMP" moves to the write-back stage, the content of PP also changes to the target address L1, causing the fetch stage to fetch the "CMP 7" instruction. Branching flushes the pipeline, discarding all the instructions that were partially executed (processed), and the program execution continues from the branch address. However, with additional hardware, branch directions can be predicted to reduce pipeline bubbles and improve efficiency.

Static Branch Prediction

Static or default branch prediction can be used by itself or in conjunction with dynamic branch prediction. Consider a branch instruction, such as "JGT L2" in Example 8.11, where L2 > PP is a **forward branching** address. In this case, using "Not Taken" as the default branch decision for this instruction would be correct as long as the for-loop is executing. The processor will mispredict the branch direction only once when the for-loop exits.

Likewise, when the conditional instruction causes **backward branching**, for example, during the execution of a "do-while" where the condition statement is at the bottom of the loop, using "Taken" as the default branch decision in this case would also be correct as long as do-while is executing. The processor will mispredict the branch direction again only once when do-while exits.

Typically, the rules of static predictions are taken into account during program compilation for optimal execution. For example, the condition "$i < 8$" in Example 8.1 would be compiled to "CMP 7" and "BGT L2" as they were listed in Examples 8.2 and 8.11. Likewise, the condition is compiled to similar instructions in Example 8.3 (Pentium processor) and in Example 8.4 (Sparc processor).

The earliest time that static branch prediction can be performed is when a branch instruction is in the decode stage and its op-code is known. However, this will result in a one-cycle delay necessary to perform static branch prediction. For example, if the pipeline implements only the static branch prediction described above (i.e., no dynamic branch predictor), the one-cycle delay may be overlapped with the execution of another instruction selected by the compiler.

Consider the unconditional jump instructions "JMP L1" in Example 8.11. It is possible to utilize this one-cycle delay by yet another compiler optimization step, as illustrated in Example 8.12. In order to utilize the one-cycle delay and execute a useful instruction during that clock cycle, the compiler can move an instruction from the body of the for-loop (e.g., "ST (i), R1") to after the "JMP L1" instruction. The pipeline, however, must be modified to always execute the instruction that follows an unconditional jump/branch instruction.

Example 8.12. A RISC-ISA assembly program for Example 8.1 optimized for static branch prediction for the "JMP" instruction. Instruction "ST (i), R0, R1" in Example 8.11 is moved to after the "JMP L1" instruction in order to eliminate the one-cycle delay that would be necessary to statically predict the branch direction for the "JMP L1" instruction.

```
.code   //start program code section

        ST    (sum), R0        //Initialize: M[sum] ← R0
        ST    (i), R0          //M[i] ← R0

 L1:

        LD    R1, (i)          //R1 ← M[i]
        CMP   R1, 7            //is i > 7?
        JGT   L2               //exit for-loop if greater (PP←
 L2)
        LD    R2, (sum)        //load sum, R2 ← M[sum]
        LD    R3, R1, (array)  //load next array element, R3 ← M[array
 + R1]
        ADD   R1, R1, 1        //increment i; instruction moved here
        ADD   R4, R2, R3       //compute sum + array[i]
        ST    (sum), R4        //store sum, M[sum] ← R4
        JMP   L1               //loop back
        ST    (i), R1          //store i, M[i] ← R1;
                               //instruction moved here

 L2:    Ix                     //some instruction x

.data   //start program data section
array:    RB   8      //reserve 8 bytes
i:        RB   1      //reserve 1 byte
sum:      RB   1      //reserve 1 byte
```

In those cases where the compiler is unable to move an instruction to after an unconditional jump/branch instruction, a no-op (NOP) instruction must be inserted, as was illustrated in the unoptimized Example 8.4 AltraSparc II program. Note that the Sparc program also includes a NOP instruction after the conditional instruction "gt." This is because the Sparc processor also executes the instruction that follows a conditional branch instruction and therefore provides compilers the option to further optimize the code when possible.

Dynamic Branch Prediction

It is simple to predict a branch decision as "Taken" for an unconditional jump/branch (e.g., "JMP") instruction when the instruction always branches and changes program flow. On the other hand, the prediction of "Taken" or "Not Taken" for a conditional branch instruction (e.g., "JGT") is data dependent, and it can be harder to always predict it correctly, especially when the conditional branch instruction controls an "if-else" statement within a loop. That is, depending on the state, true or false, the "if" condition evaluates, either the "then" or the "else" section of the code will execute. The static branch predictor described earlier works well with backward branching used in the execution of "for-loop" and "do-while" statements. However, good forward branch prediction is also necessary to improve pipeline efficiency.

Modern processors typically implement dynamic branch prediction mechanisms that collect and use branch history data for each conditional/unconditional branch instruction that executes. Initially, when the execution of a program starts and there is no branch history data, the rules of static branch prediction may be used the first time that a conditional/unconditional branch instruction executes. For best performance, the branch history data and prediction logic are kept in the fetch stage.

Before we discuss dynamic branch prediction techniques, Table 8.6 illustrates the execution of two iterations of the "for-loop" in Example 8.11. However, the following assumption is made:

Cycle	Fetch	Decode	Execute	Data Memory	Write-Back
1	ST (sum), R0				
2	ST (i), R0	ST (sum), R0			
3: L1	LD R1, (i)	ST (i), R0	ST (sum), R0		
4	CMP R1, 7	LD R1, (i)	ST (i), R0	ST (sum), R0	
5	JGT L2	CMP R1, 7	LD R1, (i)	ST (i), R0	ST (sum), R0
6	JGT L2	CMP R1, 7	**O**	LD R1, (i)	ST (i), R0
7	LD R2, (sum)	JGT L2 **(Static: Not Taken)**	CMP R1, 7	**O**	LD R1, (i)
8	LD R3, R1, (array)	LD R2, (sum)	JGT L2	CMP R1, 7	**O**
9	ADD R1, R1, 1	LD R3, R1, (array)	LD R2, (sum)	JGT L2 **(Decision: Not Taken, start history)**	CMP R1, 7
10	ADD R4, R2, R3	ADD R1, R1, 1	LD R3, R1, (array)	LD R2, (sum)	JGT L2
11	ST (sum), R4	ADD R4, R2, R3	ADD R1, R1, 1	LD R3, R1, (array)	LD R2, (sum)
12	ST (i), R1	ST (sum), R4	ADD R4, R2, R3	ADD R1, R1, 1	LD R3, R1, (array)
13	JMP L1	ST (i), R1	ST (sum), R4	ADD R4, R2, R3	ADD R1, R1, 1
14	Ix	JMP L1 **(Static: Taken)**	ST (i), R1	ST (sum), R4	ADD R4, R2, R3
15: L1	LD R1, (i)	Ix	JMP L1	ST (i), R1	ST (sum), R4
16	CMP R1, 7	LD R1, (i)	**O**	JMP L1 **(Decision: Taken, start history)**	ST (sum), R4
17	JGT L2	CMP R1, 7	LD R1, (i)	**O**	JMP L1

(Continued)

Table 8.6 Program Execution for Two Iterations Using a RISC Pipeline

Cycle	Fetch	Decode	Execute	Data Memory	Write-Back
18	JGT L2	CMP R1, 7	O	LD R1, (i)	O
19	LD R2, (sum)	JGT L2	CMP R1, 7	O	LD R1, (i)
20	LD R3, R1, (array)	LD R2, (sum)	JGT L2	CMP R1, 7	O
21	ADD R1, R1, 1	LD R3, R1, (array)	LD R2, (sum)	JGT L2	CMP R1, 7
22	ADD R4, R2, R3	ADD R1, R1, 1	LD R3, R1, (array)	LD R2, (sum)	JGT L2
23	ST (sum), R4	ADD R4, R2, R3	ADD R1, R1, 1	LD R3, R1, (array)	LD R2, (sum)
24	ST (i), R1	ST (sum), R4	ADD R4, R2, R3	ADD R1, R1, 1	LD R3, R1, (array)
25	JMP L1 **(Dynamic: Taken)**	ST (i), R1	ST (sum), R4	ADD R4, R2, R3	ADD R1, R1, 1
26: L1	**LD R1, (i)**	JMP L1	ST (i), R1	ST (sum), R4	ADD R4, R2, R3
27	CMP R1, 7	LD R1, (i)	JMP L1	ST (i), R1	ST (sum), R4
28	JGT L2 **(Dynamic: Not Taken)**	CMP R1, 7	LD R1, (i)	JMP L1 **(Decision: Taken, update history)**	ST (i), R1
29	JGT L2 **(Dynamic: Not Taken)**	CMP R1, 7	O	LD R1, (i)	JMP L1
30	LD R2, (sum)	JGT L2	CMP R1, 7	O	LD R1, (i)
31	LD R3, R1, (array)	LD R2, (sum)	JGT L2	CMP R1, 7	O
32	ADD R1, R1, 1	LD R3, R1, (array)	LD R2, (sum)	JGT L2 **(Decision: Taken, update history)**	CMP R1, 7
33: L2	Ix	ADD R1, R1, 1	LD R3, R1, (array)	O	JGT L2
34	?	Ix	ADD R1, R1, 1	O	O
35	?	?	Ix	O	O
36	?	?	?	Ix	O
37	?	?	?	?	Ix

Table 8.6 Program Execution for Two Iterations Using a RISC Pipeline (*Continued*)

The RISC pipeline in Fig. 8.17 implements both static and dynamic branch prediction logic. The static branch predictor discussed earlier is implemented in the decode stage, and a "perfect" dynamic branch predictor is implemented in the fetch stage. Note that in this case, there is no need for the compiler to optimize the code to utilize the one-cycle delay required to complete a static branch prediction.

For the conditional jump instruction "JGT L2" at clock cycle 6, the static prediction decision is "Not Taken" (no jump), and the program execution continuous with the next sequential instruction "LD R2, (sum)". When the

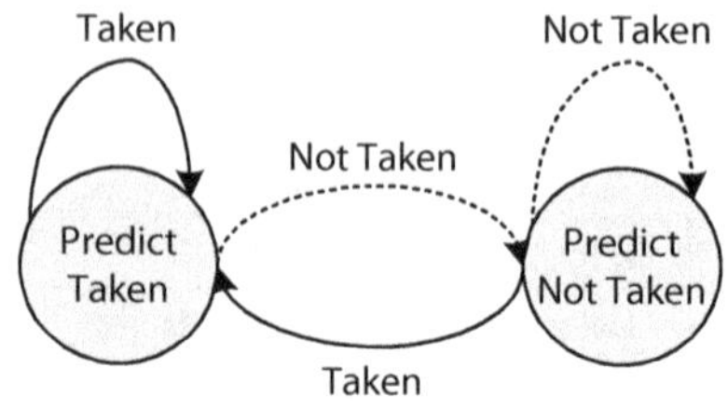

Figure 8.21 FSD for a 1-bit dynamic predictor.

execution of "JGT L2" instruction completes at clock cycle 9, the dynamic predictor is also updated with the branch decision "Not Taken." For the unconditional jump instruction "JMP L1" at clock cycle 13, the static prediction decision is "Taken" (jump), and the execution starts at the jump address L1 with the instruction "LD R1, (i)" resulting in one pipeline bubble at cycle 14. At clock cycle 16, when instruction "JMP L1" retires, the dynamic predictor is also updated with the branch decision "Taken." For the "JMP L1" instruction at cycle 23, the dynamic branch predictor predicts "Taken" and the program execution continues with instruction "LD R1, (i)" with no additional pipeline bubbles at cycle 24.

The program executes without additional branch-related stalls until after the last iteration, when the execution of the "JGT L2" instruction for the last time results in a misprediction at cycle 29. The jump causes a pipeline flush, introducing three pipeline bubbles. The program execution continues with the "Ix" instruction after the for-loop.

Figure 8.21 illustrates a finite state diagram (FSD) for implementing a 1-bit dynamic branch predictor. As illustrated in Table 8.6, when the "JGT L2" instruction executes for the first time, the static branch predictor in the decode stage would correctly predict "Not Taken" because L2 > PP indicates a forward branch address. The "Not Taken" decision at clock cycle 8 will be used to initialize the 1-bit predictor to the "Predict Not Taken" state during cycle 9. For the rest of the iterations, the "Not Taken" would be predicted correctly by the 1-bit predictor. After the last iteration, the execution of "JGT L2" instruction for the last time (e.g., at cycle 29 for two iterations of the for-loop) will result in a misprediction and the state of the 1-bit predictor will change to "Predict Taken."

If the for-loop executes only once, the 1-bit dynamic predictor would predict correctly as expected, illustrated in Table 8.6. However, suppose the for-loop in Example 8.11 executes as an inner loop and therefore would execute multiple times. The second time that the for-loop executes, the 1-bit predictor, which is now in the "Predict Taken" state, will cause a misprediction when instruction "JGT L2" executes at the start of the for-loop, and the state of the predictor will change to "Predict Not Taken." The 1-bit predictor will cause another misprediction at the end of the for-loop, resulting in two mispredictions for each time that the for-loop executes. Therefore, the 1-bit predictor will perform worse than the static predictor alone.

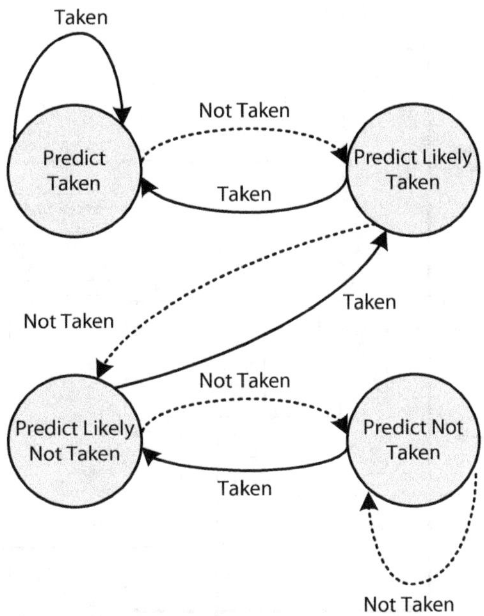

Figure 8.22 Illustrating two 2-bit dynamic branch prediction algorithms [7].

2-Bit Dynamic Predictor

Figure 8.22 illustrates the FSD used for a 2-bit dynamic predictor [7]. Again, consider the for-loop in Example 8.11 executing as an inner loop. During the first time that the for-loop executes, the state of the 2-bit predictor FSD for the "JGT L2" instruction would be initialized to "Predict Not Taken." The predictor will remain in this state until after the last iteration when the for-loop exits, and the 2-bit predictor will mispredict, causing the state of the predictor to change to "Predict Likely Not Taken."

The next time that the for-loop executes, the dynamic predictor, being in the "Predict Likely Not Taken" state, will correctly predict "Not Taken" for the "JGT L2" instruction in the first iteration. When "JGT L2" in the first iteration completes execution and it is determined that the "Not Taken" prediction was correct, the state of the 2-bit FSD correctly changes to the "Not Taken" state, with no misprediction, as if the for-loop was executing for the first time. The 2-bit predictor will cause one misprediction at the end of the for-loop, resulting, as expected, in only one misprediction each time the "for-loop" executes.

Figure 8.23 illustrates the data path of a 2-bit dynamic branch predictor implemented in the fetch stage. It includes a branch history table (BHT) that for each branch instruction it holds a branch instruction address (*BIA*), a branch target address (*BTA*), and a 2-bit current state (*CS*) of a 2-bit dynamic predictor. The 2-bit storage space used for each CS in the table represents the two flip-flops that would be required otherwise to implement a finite state machine (FSM). The table holds the history for the branch instructions that are executed recently depending on the size of the table. During the execution of a program, the lower bits of PP are

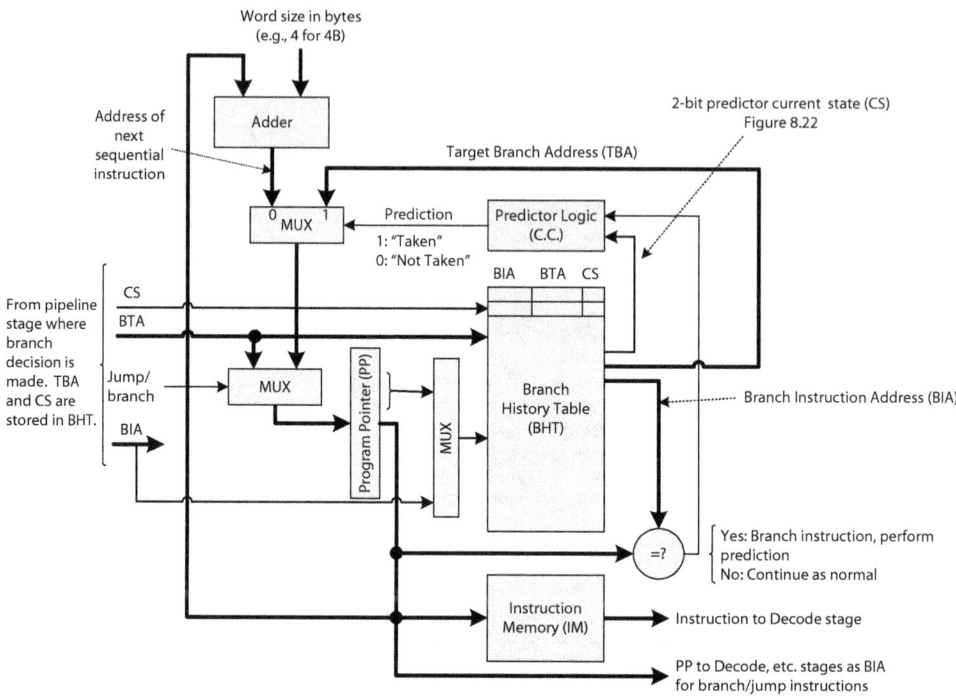

FIGURE 8.23 Partial illustration of 2-bit dynamic branch prediction logic placed in the fetch stage.

used to access the table. For example, with 10-bits, the table can hold 1K entries; however, not all entries will be branch instructions.

Each time that the content of the PP matches with a *BIA* in the BHT, it identifies a jump/branch instruction. The 2-bit *CS* read from the table is used to predict the direction of the branch. If, for example, the *CS* indicates "Predict Taken" or "Predict Likely Taken," then "Taken" is predicted and the content of the PP is replaced with the *BTA* saved in the table. The next instruction that is fetched is the one at address *BTA* stored in PP. On the other hand, if the *CS* indicates "Predict Not Taken" or "Predict Likely Not Taken," then "Not Taken" is predicted and the program execution continues from the next instruction.

If the content of PP does not match a *BIA* stored in the table, it indicates that there is no history for the current branch instruction in the table. In this case, on the next clock cycle, the static prediction is applied. Once the branch instruction executes and the decision "Taken" or "Not Taken" is determined as the initial value of the 2-bit *CS*, the *BIA*, *BTA*, and *CS* for the instruction are entered at an index determined from the *BIA* in the table as the instruction retires. The *CS* field in the table entry is then updated each time the branch instruction executes and retires.

It has been shown that using a 2-bit predictor (a four-state FSD) is as good as, or even better, than using a predictor with more states [8]. Because a 2-bit branch predictor works independent of previously executed branch instructions, it is called a **local predictor**.

It has also been shown that, for some programs, branch decisions are not always independent, and in some cases, there are correlations between the decisions made by recently executed branch instructions and the decision that will be made when a current branch instruction executes.

Branch Correlation-Based Prediction

For "if-else" statements that execute repeatedly (e.g., within a loop), it has been shown that a branch prediction algorithm works better when one uses a separate 2-bit predictor for each possible execution path [8, 9]. For instance, consider the following for-loop with three if-else statements. Also, assume the compiler translates the conditional statement of each "if-else" statement into a conditional branch instruction that the branch decision "Not Taken" refers to the "then" part of the "if-else" code.

Program code [8]

```
for(...)

{

    ...

            If(x == 2)
               x = 0;
               If(y == 2)
                  y = 0;
                  If(x != y)

            ...

    ...
}
```

Compiler output for the "if-else" statements

```
for(...)
{//compiler-generated instructions
   ...
   BNE ... // when x = 2, BNE does not branch
   ...
   BNE   // when y = 2, BNE does not branch
   ...
   BEQ   // when x ≠ y, BEQ does not branch
   ...
{
```

There are four possible execution paths to reach the branch instruction "BEQ," as shown in Fig. 8.24. Note that for the "BEQ" instruction, the branch

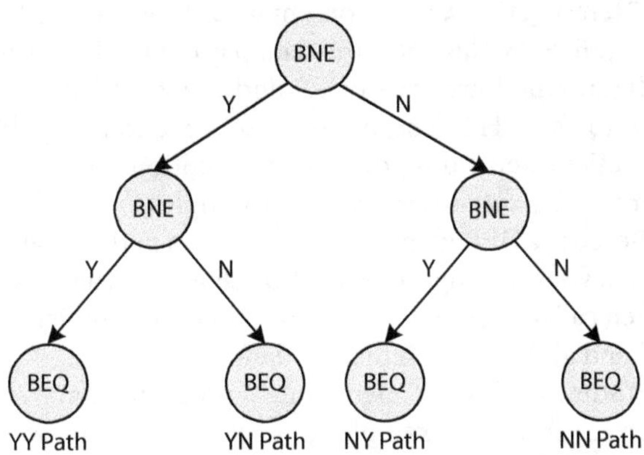

Figure 8.24 All possible execution paths to reach instruction "BEQ" within the for-loop noted earlier.

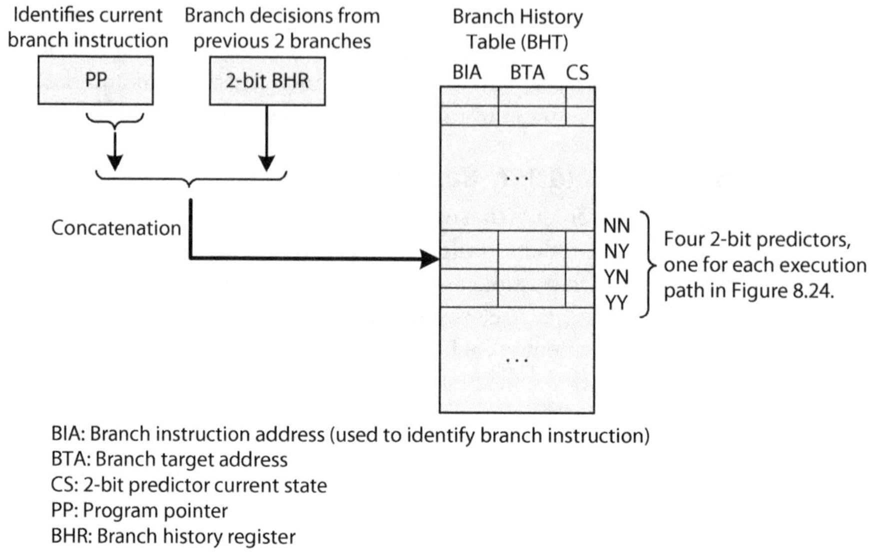

FIGURE 8.25 A correlation predictor using a global view of the branch decisions.

decision depends on an execution path. If x and y are both 2, then for sure the condition "$x != y$" is false. Therefore, in some programs, there are often strong correlations among the branch decisions of recently executed branch instructions. This information can be used to implement a better dynamic predictor.

In order to identify an execution path, a branch prediction register (BPR) is used to encode the recently executed branch decisions. For the previous for-loop, a 2-bit BPR register can encode the four execution paths in Fig. 8.24 as $(11)_2$, $(10)_2$, $(01)_2$, and $(00)_2$, where 1 implies yes (Y) and 0 no (N). If the branch decisions for the two instructions "BNE" and "BNE" were "Not Taken" (i.e., $x = 2$) and "Taken" ($y \neq 2$), respectively, then content of BPR becomes $(01)_2$.

Figure 8.25 illustrates the data path of a 2-bit correlation branch predictor. In the figure, the lower bits of PP and the 2-bit BPR are concatenated to create an index to the BHT. The index selects a different 2-bit predictor for the "BEQ" instruction, depending on which execution path the program took in Fig. 8.24. Refer to the Exercises section for examples.

Because a BPR identifies an execution path in the program, it is said to provide a global view of program execution. For this reason, a correlation-based predictor is often called a **global predictor**. Still other predictors use a combination of both local and global branch history data [9].

Studies of SPEC89 benchmark programs have shown that the frequency of misprediction—for example, for a "gcc" compiler—was 12% when a local predictor (e.g., Fig. 8.23) with a 4096-entry BHT was used versus 11% when a 2-bit global predictor (e.g., Fig. 8.25) with a 1024-entry but the same size ($1024 * 4 = 4096$) BHT was used. With the Spice circuit simulation program, the misprediction rates were 9% for local and 5% for global, and with Espresso logic minimization software it was 5% for local and 4% for global [10]. This shows that using multiple predictors often works better than using only one as discussed next.

Tournament Predictor

Modern processors often implement multiple predictors and dynamically select the best among them for each branch instruction. For example, consider implementing a local predictor and a global predictor and then using a 2-bit predictor (Fig. 8.22) to choose the best predictor. The mechanism is called a tournament predictor because for each branch instruction, the winner predictor (local or global) is the one that is selected more often.

For example, using a tournament predictor, 40% of the time the predictor selected a global predictor for the SPEC Integer benchmark programs versus only 15% for the SPEC floating-point (FP) benchmark programs [10]. Tournament predictors have been used in the AMD Opteron and Phenom processors.

8.4.3 Instruction-Level Parallelism

In order to execute two or more instructions in parallel on a superscalar processor, all those instructions must be data independent. With ILP, there are three possible dependencies, as defined with examples in Table 8.7. Data dependency exists among some instructions in all programs, no matter if ILP is used or not. An **antidependency** and/or **output dependency**, on the other hand, may exist between two instructions if the instructions execute in parallel (during the same pipeline cycle). An instruction pair with one or more of these dependencies, if executed in parallel, would result in hazards known as **read after write** (RAW), **write after read** (WAR), or **write after write** (WAW) hazards. Furthermore, these hazards cannot be resolved during execution.

Dependency Type	Instruction Pair Example	Definition

Data Dependence	LD **R1**, (sum)
	...
	ADD R3, R2, **R1**
		An instruction (e.g., "LD") changes (writes) the content of a register (i.e., R1) and another instruction (e.g., "ADD") uses (reads) the register content to perform additional computation. Executing these two instructions during the same pipeline cycle will result in a RAW hazard.

Antidependence	ADD R3, **R2**, R1
	...
	LD **R2**, (sum)
		An instruction (e.g., "ADD") uses (reads) the content of a register (i.e., R2) and another future instruction (e.g., "LD") changes (writes) the content of the same register. Executing these two instructions during the same pipeline cycle will result in a WAR hazard.

Output Dependence	ADD **R3**, R2, R1
	...
	LD **R3**, (sum)
		Two instructions (e.g., "ADD" and "LD") change (write) the content of the same register (i.e., R3). Executing these two instructions during the same pipeline cycle will result in a WAW hazard.

Table 8.7 Types of Dependencies in ILP

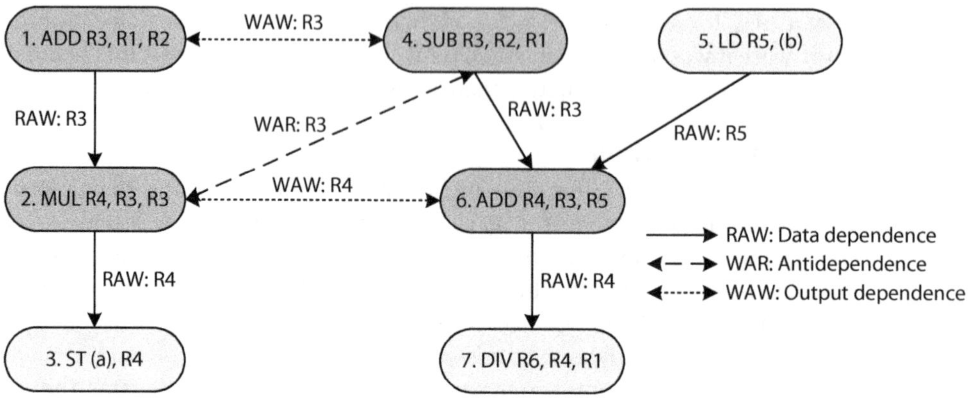

Figure 8.26 Using a graph to illustrate RAW, WAR, and WAW hazards among the instructions in Example 8.13.

For example, consider the execution of the program code given in Example 8.13 on a superscalar processor. There are data-dependency, antidependency, and output-dependency relationships among these instructions that, if executed in parallel, would cause RAW, WAR, or WAW hazards, as illustrated using a graph in Fig. 8.26.

Example 8.13. A program code example assuming that registers R1 and R2 are already initialized with values loaded from memory:

```
1:   ADD R3,    R1, R2    //R3 ← R1 + R2;
2:   MUL R4,    R3, R3    //R4 ← R3 * R3;
3:   ST (a),    R4        //M[a] ← R4;
4:   SUB R3,    R2, R1    //R3 ← R2 - R1;
5:   LD R5,     (b)       //R5 ← M[b];
6:   ADD R4,    R3, R5    //R4 ← R3 + R5;
7:   DIV R6,    R4, R1    //R6 ← R4 / R1;
```

There are five RAW, two WAW, and one WAR potential hazards in the example program. Any pair of instructions that are shown connected by a solid arrow (→) in the graph are data dependent and cannot be executed in parallel. The instructions "ADD R3, R1, R2" and "SUB R3, R2, R1" have an output-dependence relationship because they both update register R3. Their execution in parallel would cause a WAW hazard. Likewise, instructions "MUL R4, R3, R3" and "ADD R4, R3, R5" that both write to register R4 would cause a WAW hazard if they are executed in parallel. Instructions "SUB R3, R2, R1" and "MUL R4, R3, R3" have an antidependence relationship. There will be a WAR hazard if "SUB" executes before "MUL" that operates on the result produced by the "ADD R3, R1, R2" instruction and not on the result produced by the "SUB" instruction.

The requirement for ILP is that data-dependence, antidependence, or output-dependence relationships cannot exist among those instructions that are scheduled to execute during the same pipeline cycle. The instructions that can be executed in parallel are selected either statically by the compiler (i.e., a software solution) or dynamically by hardware during execution. Both the software and hardware solutions have advantages and disadvantages that will be discussed next.

Statically Scheduled ILP

For statically scheduled ILP, the compiler and not hardware decides which set of instructions should be issued for parallel execution during each pipeline cycle. The compiler for a superscalar processor must examine a group of instructions and divide them into sets. The instructions in each set must not have any dependency relationship among them that can cause a RAW, WAR, or WAW hazard. Consider the program code in Example 8.13. Suppose the processor is a two-issue superscalar—can issue maximum two instructions every clock cycle—with the following specifications:

- The compiler must organize the instructions into maximum two-issue ILP.
- The instruction pipeline contains enough resources (adder/subtractor, multiplier, and divider) to execute any two arithmetic instructions in parallel, if needed. For simplicity, it is also assumed that each arithmetic computation takes one clock cycle.
- "LD" instructions are scheduled as soon as possible to hide the single-cycle delay required to load data from cache (DM).

Example 8.14 lists the compiler generated two-issue organization of the instructions in Example 8.13. Table 8.8 illustrates the execution of the program code, with the assumption that both variables *a* and *b* are already loaded into cache. During each pipeline cycle, two instructions are fetched and executed in parallel.

Example 8.14. The instructions in Example 8.13 are organized for a statically scheduled two-issue ILP execution. The pipeline is capable of executing two arithmetic instructions in parallel.

```
ADD   R3,   R1,   R2              LD R5,  (b)
MUL   R4,   R3,   R3              NOP
ST    (a),   R4                   SUB R3,  R2,  R1
ADD   R4,   R3,   R5              NOP
DIV   R6,   R4,   R1              NOP
```

Because a superscalar processor typically executes multiple instructions during each clock cycle, a program's CPI is generally less than 1. For this reason, the preferred performance parameter is called instructions per cycle (IPC) and is calculated as the inverse of CPI, as defined by Eq. (8.6).

$$IPC = \frac{\text{Number of instructions executd } (n)}{\text{Number clock cycles used } (N)} \tag{8.6}$$

Cycle	Fetch	Decode	Execute	Data Memory	Write-Back
1	1: ADD R3, R1, R2 5: LD R5, (d)				
2	2: MUL R4, R3, R3 NOP	1: ADD R3, R1, R2 5: LD R5, (d)			
3	3: ST (c), R4 4: SUB R3, R2, R1	2: MUL R4, R3, R3 NOP	1: ADD R3, R1, R2 5: LD R5, (d)		
4	6: ADD R4, R3, R5 NOP	3: ST (c), R4 4: SUB R3, R2, R1	2: MUL R4, R3, R3 NOP	1: ADD R3, R1, R2 5: LD R5, (d)	
5	7: DIV R6, R4, R1 NOP	6: ADD R4, R3, R5 NOP	3: ST (c), R4 4: SUB R3, R2, R1	2: MUL R4, R3, R3 NOP	1: ADD R3, R1, R2 5: LD R5, (d)
6		7: DIV R6, R4, R1 NOP	6: ADD R4, R3, R5 NOP	3: ST (c), R4 4: SUB R3, R2, R1	2: MUL R4, R3, R3 NOP
7			7: DIV R6, R4, R1 NOP	6: ADD R4, R3, R5 NOP	3: ST (c), R4 4: SUB R3, R2, R1
8				7: DIV R6, R4, R1 NOP	6: ADD R4, R3, R5 NOP
9					7: DIV R6, R4, R1 NOP

TABLE 8.8 Execution Two-Issue Statically Scheduled Program Code Using a Five-Stage Superscalar Pipeline

Not counting the clock cycles required to fill the pipeline, the IPC for Example 8.14, as illustrated in Table 8.8, is determined as follows:

$$\begin{aligned} N &= 9 - 5 + 1 = 5 \text{ clock cycles (ignoring pipeline startup)} \\ n &= 7 \text{ instructions executed (from Example 8.13 program)} \\ \text{IPC} &= \frac{7}{5} = 1.4 \end{aligned} \tag{8.7}$$

An additional technique to further improve IPC is called compiler-based **speculative execution**. In this case, instructions independent of branch directions are selected for parallel execution. One compiler-assisted speculative execution method that is used with Intel Itanium architecture is to convert instructions of an "if-else" statement to conditional instructions called **predicated instructions**. For example, consider the following simple "if-else" statement:

```
if(a > 0)
   a = a - 1;
else
   a = a + 1;
```

The compiler would translate the if-else statement to the following assembly code by attaching the condition "$a > 0$" to all instructions (one in this case) in the "then" code section and the condition "! ($a > 0$)" to all the instructions (again, one in this case) in the "else" code section as shown next. This would also eliminate the branch instruction that would otherwise be needed to execute the instructions associated either with the "then" or the "else" code section.

```
LD R1, (a)
CMP R1, 0
SUB R1, R1, 1 (GTF = 1)     //if greater than flag (GTF) is set
ADD R1, R1, 1 (GTF = 0)     //if GTF is not set
ST (a), R1
```

Both the "SUB" and "ADD" instructions can now be scheduled to execute in parallel. The processor would compute both $a - 1$ and $a + 1$. However, only one of these results would be committed (written) to register R1, depending on the value of the greater than flag (GTF). If "$a > 0$," then R1 takes the result of $a - 1$; otherwise, R1 takes the result of $a + 1$. One of the computations ($a - 1$ or $a + 1$) is called speculative because its computed result may not be used.

An advantage of statically scheduled ILP is that the processor is more power efficient in that it doesn't use hardware to decide which set of instructions should execute in parallel during each pipeline cycle. For this reason, processors (e.g., ARM Cortex-A8) that implement statically schedule ILP are typically used in handheld devices such as smart phones. Another advantage of these processors as compared to those that implement dynamically scheduled ILP (discussed next) is that compiler can potentially examine a longer list of instructions for ILP than the list of instructions that must be dynamically and quickly examined during each pipeline cycle.

Dynamically Scheduled ILP

The pipeline of a superscalar processor that implements dynamically scheduled ILP is more complex and consumes more power. The fetch stage contains a queue of instructions that are not yet scheduled for execution. Dedicated hardware dynamically examines several instructions in the queue and decides which set of instructions to issue for parallel execution. When compared to a pipeline that is designed for statically scheduled ILP, the pipeline for a dynamically scheduled ILP provides certain advantages, such as

- The exact program flow is only known at run time. Therefore, there are more opportunities for the processor to select and issue more instructions for parallel execution.
- The processor could also implement a **register renaming** mechanism where antidependence and output-dependence relationships can be removed dynamically by changing register names in some instructions with temporary registers that are not visible to programmers. This enables the processor to schedule instructions **out of order** and achieve higher ILP.

- Programs are not required to be recompiled to take advantage of a next generation of the processor that implements dynamically scheduled ILP or when the dynamic scheduling in the next generation is implemented differently.

The pipeline of the Intel Core i7, for example, implements dynamically scheduled ILP. Its high-instruction throughput makes it suitable for high-end desktop or server computers.

Dynamic instruction scheduling requires that the pipeline implement certain techniques, known as **score boarding** or a more advanced version called **Tomasulo's algorithm** that also supports speculative execution. The pipeline can dynamically rename registers to avoid WAR and WAW hazards and allow out-of-order execution. For example, consider the program code in Example 8.13 that was discussed earlier with five RAW, two WAW, and one WAR potential hazards (Fig. 8.26).

Figure 8.27 illustrates the removal of two antidependence and one output-dependence relationships using register renaming. The output dependence between instructions "ADD R3, R1, R2" and "SUB R3, R1, R1" is resolved by changing the destination register R3 with temporary register T1 in the "ADD" instruction. This is because in program order, "MUL R4, R3, R3" that uses the result computed by the "ADD" instruction comes before the "ADD R4, R3, R5" that uses the output computed by the "SUB" instruction. R3 is also renamed to T1 in the "MUL" instruction. This removes the antidependence relationship that exists between the "MUL" and the SUB" instructions.

Likewise, the output-dependence relationship between the "MUL" and the "SUB" instructions is removed by renaming R4 in the "MUL" instruction with another temporary register T2. R4 is also renamed T2 in the "ST (a), R4" instruction. This leaves R4 in "ADD R4, R3, R5" unchanged, as it should, since this instruction computes the last value stored in register R4 as if the program is executed one instruction at a time.

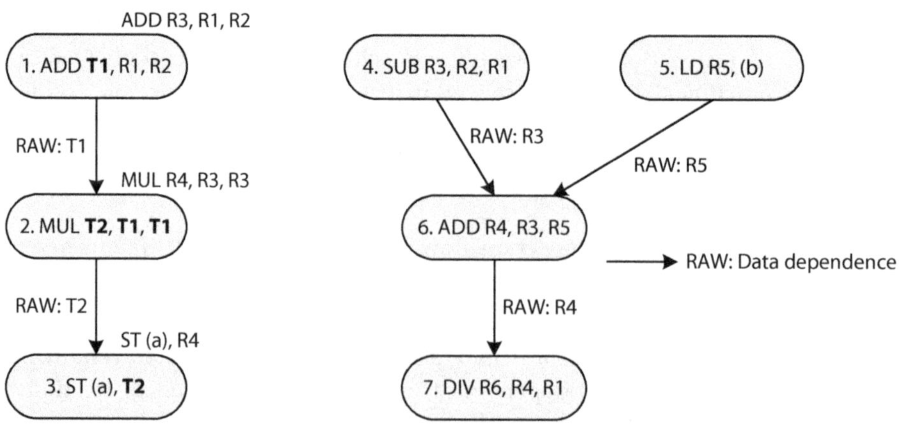

FIGURE 8.27 Illustrating register renaming to remove WAR and WAW dependencies in Fig. 8.26.

Cycle	Fetch	Decode	Execute	Data Memory	Write-Back
1	1: ADD T1, R1, R2 5: LD R5, (b)				
2	2: MUL T2, T1, T1 4: SUB R3, R2, R1	1: ADD T1, R1, R2 5: LD R5, (b)			
3	3: ST(a), T2 6: ADD R4, R3, R5	2: MUL T2, T1, T1 4: SUB R3, R2, R1	1: ADD T1, R1, R2 5: LD R5, (b)		
4	7: DIV R6, R4, R1 Nop	3: ST(a), T2 6: ADD R4, R3, R5	2: MUL T2, T1, T1 4: SUB R3, R2, R1	1: ADD T1, R1, R2 5: LD R5, (b)	
5		7: DIV R6, R4, R1 Nop	3: ST(a), T2 6: ADD R4, R3, R5	2: MUL T2, T1, T1 4: SUB R3, R2, R1	1: ADD T1, R1, R2 5: LD R5, (b)
6			7: DIV R6, R4, R1 Nop	3: ST(a), T2 6: ADD R4, R3, R5	2: MUL T2, T1, T1 4: SUB R3, R2, R1
7				7: DIV R6, R4, R1 Nop	3: ST(a), T2 6: ADD R4, R3, R5
8					7: DIV R6, R4, R1 Nop

TABLE 8.9 Dynamically Scheduled Two-Issue Execution of Program Code

Table 8.9 illustrates the execution of the program code in Example 8.13 on a two-issue dynamically scheduled superscalar processor using the modified instructions shown in Fig. 8.27. Again, the following list of requirements is assumed; they are the same as those we used with the two-issue statically scheduled superscalar processor.

- The instruction pipeline contains enough resources (adder/subtractor, multiplier, and divider) to execute any two arithmetic instructions in parallel, if needed. For simplicity, it is also assumed that each arithmetic computation takes one clock cycle.
- "LD" instructions are scheduled as soon as possible to hide the single-cycle delay required to load data from cache.

During the first pipeline cycle, from the three independent instructions "ADD T1, R1, R2," "SUB R3, R2, R1," and "LD R5, (b)" as shown in Fig. 8.27, two must be issued. "ADD" is selected because it comes before "SUB" in program order. "LD" is selected so data can be loaded early to hide the one-cycle cache latency required before the "ADD R4, R3, R5" instruction can execute.

During cycles 2 and 3, the scheduler issues the two independent instructions "MUL T2, T1, T1" and "SUB R3, R2, R1" followed by the two independent instructions "ST (a), T2" and "ADD R4, R3, R5." Finally, during cycle 4, "DIV R6, R4, R1" is issued. Normally, as instructions are scheduled, new instructions are fetched and stored in the instruction queue. However, in order to keep the illustrations simple, such steps are ignored here.

Assuming that the initial cycles to fill the pipeline are ignored, the IPC for the program code, as determined by Eq. (8.8), is 1.75. This makes the dynamically scheduled two-issue superscalar processor 25% (1.75/1.4 = 1.25) faster than its equivalent statically scheduled superscalar processor for the program example.

$$\begin{aligned} \text{IPC} &= \frac{7 \text{ instructions}}{(8-5+1)\text{ cycles}} \\ &= \frac{7 \text{ instructions}}{4 \text{ cycles}} = 1.75 \end{aligned} \tag{8.8}$$

However, because a typical RISC processor has many (e.g., 32) registers and compilers do avoid reusing recently used registers, compilers can assign different registers to instructions to reduce some of the antidependency and output dependency among instructions to achieve better statically scheduled ILP.

Speculative execution is also possible with dynamic scheduling. In this case, the processor executes instructions as soon as their operands are available, independent of branch directions. Some speculatively computed results may be discarded if there is a change in program flow, and the results that must be committed to registers and/or memory must be done in program order as if instructions were executed one at a time.

8.4.4 Multithreading

As was said earlier in this chapter and in Chap. 1, there is a limit to ILP. For a given program, there are only a few independent instructions, even with register naming, that can be executed at the same time during each pipeline cycle. Studies of some benchmark programs have shown that about 30% of the time three instructions and about 2% of the time six or more instructions can be executed in parallel, with the average being about 2.5 instructions [11]. Even with the availability of more transistors and allowable maximum power consumption, there is a limit on how quickly a single program can be executed. Therefore, the only way to perform a task faster is to divide the work into subtasks that can be performed concurrently.

Program Example

Consider the C program given next that has a for-loop with 100 million iterations. For simplicity, the array elements are initialized to 1.0.

```
main()
{
  double array[10000000] = {[0 ... 9999999] = 1.0};
  double sum;
  int i;
  sum = 0.0;
  for(i = 0; i < 100000000; i++)
      sum = sum + array[i];
  printf("sum = %e\n", sum);
}
```

One way to reduce the total time required to compute the sum of the array elements is to divide the array into two halves and compute the sum of each half using a different thread of code. For example, the following two-threaded program in C can be used to sum the array elements. The "main" program first creates a thread (Thread 0) to sum the first half of array elements, and then itself, as Thread 1, sums the second half of array elements. Once, the "main" (now Thread 1) is done with its half of the array, it must check to make sure Thread 0 is also done with its half of the array. The "main" then computes and prints the grand total sum. Note that both threads can access globally declared "array" and "sum."

```
/*compile as "gcc -pthread filename.c" */
#include <pthread.h>
#include <stdio.h>

//globally declared array and sum
double array[10000000] = {[0 ... 9999999] = 1.0};
double sum[2];  //to store two partial sums

typedef struct {
       int start;
       int end;
       int pid;
} range;

void *calculate_array_sum(void *arg)
{
 range *incoming = (range *) arg;
 int i;
 double partial_sum;
 int start, end, pid;

 start = incoming->start;
 end = incoming->end;
 pid = incoming -> pid;

 partial_sum = 0;
 for(i = start; i < end; i++)
        partial_sum = partial_sum + array[i];

 sum[pid] = partial_sum;

 return;
}

main ()
{

 pthread_t threadID;
 void *exit_status;
 range range_thread0;
 range range_thread1;
```

```
//define the array range for Thread 0
range_thread0.start = 0;
 range_thread0.end = 5000000;
 range_thread0.pid = 0;  //Thread 0

 //now create a thread to sum the first half of the array elements
 pthread_create(&threadID,   NULL,   calculate_array_sum,
&range_thread0);

 //define the array range for the main as Thread 1
 range_thread1.start = 5000000;
 range_thread1.end = 10000000;
 range_thread1.pid = 1;  //Thread 1
 calculate_array_sum(&range_thread1);

 pthread_join(threadID, &exit_status); //wait for Thread 0

 printf("Grand total = %e\n", sum[0] + sum[1]);

 }
```

The execution of the two-threaded C program will cause two sets of code, as follows, to execute concurrently if the CPU implements multithreading:

Thread 0

```
for(i = 0; i < 50000000;
i++)
     partial_sum = partial_
sum + array[i];
sum[0] = partial_sum;
```

"main" as Thread 1

```
for(i = 50000000; i <
100000000; i++)
     partial_sum = partial_
sum + array[i];
sum[1] = partial_sum;
```

The instruction pipeline must contain two copies of all the registers, including PP. The following sections describe three k-issue multithreading pipeline organizations.

Coarse-Grained

Each time the execution of an "LD" or "ST" instruction causes a long wait—for example, when main memory access is required—the pipeline switches and resumes execution of a second thread. For example, when executing Thread 0 and Thread 1 discussed above the pipeline continues and issues up to k instructions from Thread 0 until it determines that an array element is not in cache and it must be accessed from main memory, which has a long latency. At that time, while the data is being accessed from memory, the pipeline switches and continues issuing up to k instructions from Thread 1 until it encounters a long delay; the pipeline then switches back and issues k instructions from Thread 0.

It is called a coarse-grained multithreading architecture because the pipeline executes one thread at a time until it encounters a long delay. Because each thread uses its own private set of registers, the state of the switched thread is automatically saved. While this method is good when the delays are long, it is not efficient when delays are short—for example, when the access is from a lower-level cache with shorter latency. This is because each time that a switch to a different thread takes place, the pipeline must be flushed, wasting pipeline cycles. Furthermore, in some applications, thread switching may happen less frequently than necessary. For example, some real-time applications may require faster thread execution.

The pipeline organization, however, has the advantage of requiring the least amount of hardware as compared to other multithreading architectures discussed next because the pipeline resources at any time are used for executing only one thread. Another advantage of this architecture is that it can be implemented as a statically or dynamically scheduled ILP multithreading processor. Note that, in general, the threads may or may not belong to the same program. Threads from two different programs may be executed concurrently.

Fine-Grained

In this case, thread switching happens every clock cycle. For example, when executing the above Thread 0 and Thread 1 the pipeline issues up to k instructions from Thread 0 during the first pipeline cycle, then switches and issues up to k instructions from Thread 1 during the second pipeline cycle, and then it repeats. Again, in general, the threads may or may not belong to the same program.

It is called a fine-grained multithreading architecture because the pipeline switches threads every cycle. This pipeline organization is more efficient as compared to a coarse-grained multithreading pipeline, but the execution of each thread is still slow. As the pipeline executes instructions from only one thread during each cycle, other threads must wait their turn. The pipelines in the Sun Niagara processor and Nvidia GPUs, for example, implement this type of multithreading.

Because threads are switched every clock cycle, the architecture provides a higher thread-level concurrency as compared to the coarse-grained architecture. Likewise, the architecture can be implemented as a statically or dynamically scheduled ILP multithreading processor.

Simultaneous

In this case, the pipeline issues up to k instructions selected from all running threads during each clock cycle. For example, when executing the above Thread 0 and Thread 1, the pipeline issues $k0$ ($0 \le k0 \le k$) instructions from Thread 0 and $k1$ ($0 \le k1 \le k$) instructions from Thread 1 during each cycle, where $0 \le k0 + k1 \le k$. Because the pipeline selects independent instructions from several (e.g., two) threads and executes them simultaneously, this architecture is called simultaneous multithreading. The pipeline must implement dynamically scheduled ILP and, therefore, its implementation requires the most amount of hardware. Thread-level parallelism (TLP), where instructions from multiple threads execute at the same time, is the advantage of this architecture as compared to the others.

Example 8.15. Consider the execution of the following two program codes, labeled Thread A and Thread B, on a simultaneous multithreading four-issue superscalar processor with dynamically scheduled ILP. Thread A is the program code given in Example 8.13 with data-dependence, antidependence, and output-dependence relationships among some of the instructions. Thread B, for simplicity, contains only data-dependence relationships among its instructions.

Thread A (program code in Example 8.13)

```
ADD R3, R1, R2
MUL R4, R3, R3
ST (a), R4
SUB R3, R2, R1
LD R5, (b)
ADD R4, R3, R5
DIV R6, R4, R1
```

Thread B

```
ST (x), R3
ADD R4, R1, R2
MUL R6, R4, R5
DIV R7, R3, R6
ST (z), R7
LD R8, (y)
SUB R9, R7, R8
```

The following is also assumed:

- The execute stage has enough resources (e.g., adder/subtractor, multiplier, and divider) to execute any four arithmetic instructions in parallel, if needed. For simplicity, it is also assumed that each arithmetic operation takes one clock cycle.
- The organization of DM (cache) is two-way fine interleaved (Chap. 7), and the pipeline is designed to execute one "LD" instruction (if any) and one "ST" instruction (if any) simultaneously, if needed. It is assumed that there is less chance for a conflict in a fine interleaved memory when executing a load with a store at the same time than when executing two load or two store instructions at the same time.
- The pipeline will issue "LD" instructions as soon as possible to hide the single-cycle delay required to load data from cache.
- Variables *a*, *b*, and *x* to *z* reside in cache without conflicts.

Table 8.10 illustrates the execution of Thread A and Thread B in Example 8.15. During pipeline cycle 1, the pipeline issues two possible instructions, "ADD T1, R1, R2" and "LD R5, (b)" from Thread A, and two instructions, "ST (x), R3" and "ADD R4, R1, R2" from Thread B. During cycle 2, the pipeline again issues two instructions from Thread A and two instructions from Thread B. During cycle 3, the pipeline issues two instructions from Thread A and one instruction from Thread B. Finally, during cycle 4, the pipeline issues two final instructions, one from each thread. The instructions issued during each cycle are data independent and can execute in parallel. Because the pipeline can only issue one "LD" and one "ST" (if any) during a cycle, the instruction "LD R8, (y)" cannot be issued earlier and during cycle 1. Therefore, the execution of the "SUB R9, R6, R8" instruction is delayed until cycle 6.

Cycle	Fetch	Decode	Execute	Data Memory	Write-Back
1	A: ADD T1, R1, R2 A: LD R5, (b) B: ST (x), R3 B: ADD R4, R1, R2				
2	A: MUL T2, T1, T1 A: SUB R3, R2, R1 B: MUL R6, R4, R5 B: LD R8, (y)	A: ADD T1, R1, R2 A: LD R5, (b) B: ST (x), R3 B: ADD R4, R1, R2			
3	A: ST (a), T2 A: ADD R4, R3, R5 B: DIV R7, R3, R6 B: SUB R9, R6, R8	A: MUL T2, T1, T1 A: SUB R3, R2, R1 B: MUL R6, R4, R5 B: LD R8, (y)	A: ADD T1, R1, R2 A: LD R5, (b) B: ST (x), R3 B: ADD R4, R1, R2		
4	A: DIV R6, R4, R1 B: ST (z), R6 ? ?	A: ST (a), T2 A: ADD R4, R3, R5 B: DIV R7, R3, R6 B: SUB R9, R6, R8	A: MUL T2, T1, T1 A: SUB R3, R2, R1 B: MUL R6, R4, R5 B: **LD R8, (y)**	A: ADD T1, R1, R2 A: LD R5, (b) B: ST (x), R3 B: ADD R4, R1, R2	
5%		A: DIV R6, R4, R1 B: ST (z), R6 ? ?	A: ST (a), T2 A: ADD R4, R3, R5 B: DIV R7, R3, R6 **O**	A: MUL T2, T1, T1 A: SUB R3, R2, R1 B: MUL R6, R4, R5 B: LD R8, (y)	A: ADD T1, R1, R2 A: LD R5, (b) B: ST (x), R3 B: ADD R4, R1, R2
6			A: DIV R6, R4, R1 B: ST (z), R6 B: **SUB R9, R6, R8** ?	A: ST (a), T2 A: ADD R4, R3, R5 B: DIV R7, R3, R6 **O**	A: MUL T2, T1, T1 A: SUB R3, R2, R1 B: MUL R6, R4, R5 B: LD R8, (y)
7				A: DIV R6, R4, R1 B: ST (z), R6 B: SUB R9, R6, R8 ?	A: ST (a), T2 A: ADD R4, R3, R5 B: DIV R7, R3, R6 **O**
8					A: DIV R6, R4, R1 B: ST (z), R6 B: SUB R9, R6, R8 ?

% "SUB R9, R6, R8" is delayed one-cycle due to "LD R8, (y)"
A: Thread A instruction
B: Thread B instruction

TABLE 8.10 Dynamically Scheduled Two-Thread Simultaneous Multithreading of Program Codes on Four-Issue Superscalar Processor

Based on the four-issue ILP shown in Table 8.10, the IPC for Thread A or Thread B executed separately is 1.75 (7/4), where the number of clock cycles required to fill the pipeline is ignored. However, the IPC of simultaneous multithreading of Threads A and B—again with ignoring the cycles required to fill the pipeline—is 3.5, as determined by Eq. (8.9).

$$\begin{aligned} IPC &= \frac{(7+7)\,\text{instructions}}{(8-5+1)\,\text{cycles}} \\ &= \frac{14\,\text{instructions}}{4\,\text{cycles}} = 3.5 \end{aligned} \tag{8.9}$$

Multithreading, especially a simultaneous multithreading architecture, increases pipeline efficiency. In addition, a simultaneous multithreading pipeline implements small-scaled TLP. Higher levels of TLP require a multicore processor or multiprocessor system, which we will discuss in Chap. 10.

References

1. IEEE standard for microprocessor assembly language (IEEE Std. 694-1985), IEEE, 1985.
2. MASM, http://www.masm32.com/.
3. cygWin (GNU + Cygnus + Windows), http://www.cygwin.com/.
4. Intel Architecture Optimization Manual, 1997, www.intel.com.
5. Intel 64 and IA-32 Architectures Optimization Reference Manual, 2014.
6. Simics (system level instruction set simulator), http://www.virtutech.com/
7. J. E. Smith, A study of branch prediction strategies, *Proceedings of the 8th Annual International Symposium on Computer Architecture*, June 1981, pp. 135–147.
8. Shien-Tai Pan, Kimrning So, Joseph T. Rahmeh, Improving the accuracy of dynamic branch prediction. *Proceedings of the Fifth International Conference on Architectural Support for Programming Languages and Operating Systems* (ASPLOS V), Sept 1992, pp. 76–84.
9. M.-C. Chang and Y.-W. Chou, Branch prediction using both global and local branch history information, *IEE Proc-Comput Digit Tech*, Vol. 149, No. 2, 2002, 33–38.
10. J. Hennessy and D. Patterson, *Computer Architecture: A Quantitative Approach*, 5th ed. Morgan Kaufmann, Waltham, 2012.
11. David Culler, Jaswinder Pal Singh, and Anoop Gupta, *Parallel Computer Architecture: Hardware/Software Approach*, Morgan Kaufmann, San Francisco, 1999.

Exercises

8.1 Consider an Acc-ISA CPU that executes the pseudo-code shown. Do the following:

```
… //some code
while{(true)  //loop forever
{
      T =(-5 - A + B) ^ C;  //where ^ stands for bit-wise XOR
      … //some other code (not given)
}
```

a. Create a set of 8-bit Acc-ISA instructions with 3-bit op-codes and 5-bit operands. Any of the variables *A* through *C* may be a negative 2's complement number.

b. Write an assembly program using your instruction set. Also assume that the code starts from memory address 0 and increases by 1, and data starts from memory address 0x1F and decreases by 1.

c. Manually assemble your assembly program and write instructions in binary and in hex. Assign op-codes to the instructions in the order they were used in the assembly program, starting from address 0.
d. Draw the CPU data path with only the data paths necessary.

8.2 Consider the Acc-ISA assembly instructions "LD data" (ACC ← data), "LD (adrs)" (ACC ← Memory[adrs]), "ST (adrs)" (Memory[adrs] ← ACC), "ADD (adrs)" (ACC ← ACC + Memory[adrs]), "XOR (adrs)" (ACC ← ACC ⊕ M[adrs]). Do the following:

a. Write an assembly program for the following program:

```
X = -2;
Y = 6;
Z = 11;
T = X + Y - Z;
```

b. For the assembly instructions, draw a single-cycle instruction data path.

8.3 An Acc-ISA CPU executes the following instructions using 3-bit op-codes and 5-bit address or 2's complement data. Do the following:

```
LD (address)      //Acc←Memory [address], read from LM2
LD data           //Acc←data (a 2's complement number, sign
                  //extended)
ADD data          //Acc←Acc + data (data is a 2's complement
                  //number, sign extended)
SUB data          //Acc←Acc - data (data is a 2's complement
                  //number, sign extended)
ADD (address)     //Acc←Acc + Memory[address]
SUB (address)     //Acc←Acc - Memory[address]
ST (address)      //M[address]←Acc
JMP address       //PP←address
JZ address        //PP←address if ACC = 0
```

a. Draw a data path for the CPU, assuming the DM has separate input and output buses, as in the data path shown in Fig. 8.7. Do not include data paths not used by the instructions.
b. Draw a data path for the CPU, assuming the DM has a bidirectional data bus. Do not include data paths not used by the instructions.

8.4 For the high-level code segment shown next, create a set of instructions and write an equivalent assembly program for each of the following architectures:

```
int i, sum;
for (i=0; i> k, i++)
  if (i mod 2 == 0)
       sum = sum +i;
```

a. Stack-ISA
b. Acc-ISA
c. CISC-ISA
d. RISC-ISA

8.5 Design a simple 8-bit hardware stack with depth = 16 using registers. Also, include four signals *push*, *pop*, *ovf* (for stack overflow), and *udf* (for stack underflow). Assume active-high signals. The *ovf* becomes 1 (asserted) if the stack is full and *push* is asserted. The *udf* becomes 1 if stack is empty and *pop* is asserted.

8.6 Discuss how different CISC instruction formats versus the fixed RISC instruction formats can complicate the design of an instruction data path.

8.7 Given the single-cycle CPU data path in Fig. 8.7, estimate an upper bound for the clock frequency. Assume Δ_{IM} and Δ_{DM} are each 1.2 ns, $\Delta_{\text{add}} = 0.8$ ns, $\Delta_{\text{add/cmp}} = 0.9$ ns, $\Delta_{\text{MUX}} = 0.3$ ns, $\Delta_{\text{NAND}} = 0.1$ ns , and $\tau_{st} = \tau_{cq} = \tau_{cs} = 0.05$ ns.

8.8 Given the pipelined data path in Fig. 8.11 estimate an upper bound for the clock frequency. Assume Δ_{IM} and Δ_{DM} are each 1.2 ns, $\Delta_{\text{add}} = 0.8$ ns, $\Delta_{\text{add/cmp}} = 0.9$ ns, $\Delta_{\text{2-to-1 MUX}} =$ 0.3 ns, $\Delta_{\text{NAND}} = 0.1$ ns, and $\tau_{st} = \tau_{cq} = \tau_{cs} = 0.05$ ns.

8.9 Consider an instruction pipeline with four stages, fetch (F), decode (D), execute (E), and write-back (WB), where Δ_E is twice that of the other stages. Do the following as the pipeline executes *n* instructions.

a. Suppose the E stage can be divided into two stages, E1 and E2, each with $\Delta_{\text{E1}} = \Delta_{\text{E2}} = \frac{1}{2}\Delta_E$. Determine an expression for speed-up in terms of Δ_E and Δ_{clocking} as *n* approaches infinity (∞). Assume CPI = 1. Also, estimate the speed-up as $n \to \infty$ if $\Delta_E = 2$ ns and $\Delta_{\text{clocking}} = 0.1$ ns.
b. Suppose the E stage can be superpipelined using two copies of the stage hardware and an MUX. Determine an expression for speed-up in terms of Δ_E, Δ_{clocking}, and Δ_{MUX} as *n* approaches infinity (∞). Assume CPI = 1. Also, estimate the speed-up as $n \to \infty$ if $\Delta_E = 2$ ns, and $\Delta_{\text{clocking}} = 0.1$ ns, and $\Delta_{\text{MUX}} = 0.3$ ns.

8.10 Discuss the effect of switching the positions of the execute and DM stages in Fig. 8.17.

8.11 Given the assembly code in Example 8.10, do the following:

a. Show the pipeline chart for two iterations of the for-loop using the five-stage pipeline in Fig. 8.17. Do not include the time used to execute the instructions (i.e., *Ix, Iy, Iz*) that are after the for-loop.
b. Calculate the program's CPI for two iterations.
c. Determine an equation for CPI in terms of *k* iterations.
d. Determine the limit for the CPI as *k* approaches infinity.

8.12 Given the assembly code in Example 8.11, do the following:

a. Show the pipeline chart for two iterations of the for-loop using the five-stage pipeline in Fig. 8.17. Do not include the time used to execute the instructions (i.e., *Ix, Iy, Iz*) that are after the for-loop.
b. Calculate the program's CPI for two iterations.
c. Determine an equation for CPI in terms of *k* iterations.
d. Determine the limit for the CPI as *k* approaches infinity.

8.13 For the pipeline chart in Table 8.6, determine CPI for *k* = 10 iterations. Also, determine the lower bound for the CPI as *k* approaches infinity.

8.14 Consider the following for-loop with a single if-else statement and its compiler-generated branch instructions. Suppose the processor uses a correlation branch predictor. When the for-loop executes, the "then" section of the code executes when *i* is an even number (0, 2, 4, etc.) and the "else" section executes when *i* is an odd number. Do the following:

```
for (i = 0;…)
{
        if(..> …)
        …
        else
        …
}
```

```
L1:             …
                CMP …
                JGT endfor
                CMP …
                JLE else
                …
                JMP endif
else:           …
                …
endif:          JMP L1
endfor: …
```

a. Suppose the predictor uses a 2-bit BHR. Complete the following table for six iterations and determine the number of mispredictions, where "N" (predictor not taken), "T" (predict taken), "LN" (predict likely not taken), and "LT" (predict likely taken) are used to indicate the states of a 2-bit predictor where one 2-bit predictor is used for each of the execution paths BHR = NN, NY, YN, and YY. In the table, the state in a prior row is used to predict the branch direction as Y or N for the current instruction. For example, initially, when executing the "JMP endif" instruction where BHR = NN (no branching for "BGT" and no branching for "BLE"), the 2-bit predictor for BHR = NN is initialized to "T" because "JMP endif" branches. The next time that BHR = NN, the correlation predictor will predict "taken" for the current instruction. This is illustrated for BHR = NY in the table.

	Branch Instruction			2-Bit Predictor				
Iteration	**Previous**	**Current**	**BHT**	**Initial State in BHT**	**Prediction**		**Next State**	
1	JGT	JLE	NN	T				
	BLE	JMP endif	NY	T				"then"
	JMP endif	JMP L1	YY	N				
	IMP L1	JGT	YN	T				
2	JGT	JLE	NY		Taken		T	"else"
	JLE	JMP L1	YY		Not taken		N	
	JMP L1	JGT	YN		Taken	Miss	LT	

… …

b. Suppose the predictor uses a 3-bit BHR that identifies one of eight total execution paths as NNN, NNT, NTN, NTT, TNN, TNT, TTN, or TTT. However, which of the eight possible execution paths the program will follow depends on the data it processes.

	Branch Instruction				2-Bit Predictor				
Iteration	Previous Two		Current	BHR	Initial State in BHT	Prediction		Next State	
1	JGT	JLE	JMP endif	NNY	T				"then"
	JLE	JMP endif	JMP L1	NYY	N				
2	JMP endif	JMP L1	JGT	YYN	T				
	JMP L1	JGT	JLE	YNY	T				"else"
	JGT	JLE	JMP L1	NYY		Not taken		N	
				...	... ...				

8.15 Consider the program code shown next. Assume (1) DM is fine interleaved and any combination of two memory instructions can execute at the same time; and (2) each arithmetic instruction takes one clock cycle to execute. Do the following:

```
LD R1, (a)
LD R2, (b)
ADD R3, R1, R2
ST (c), R3
LD R4, (d)
MUL R5, R3, R4
ST (d), R5
SUB R3, R2, R1
DIV R6, R3, R5
ST (e), R6
```

a. Organize the program for execution on a two-issue (ILP) statistically scheduled superscalar processor.
b. Organize the program for execution on a two-issue (ILP) dynamically scheduled superscalar processor.

8.16 Consider the program code in Example 8.13. Suppose a dynamically scheduled two-issue (ILP) superscalar processor schedules arithmetic instead of "LD" instructions as soon as possible. Use a pipeline chart to show program execution and calculate IPC. Ignore the cycles required to fill the pipeline when calculating the IPC.

8.17 Consider a four-issue simultaneous multithreading superscalar processor. Also, consider a two-thread multithreaded program with approximately 10^{11} executing instructions per thread. Do the following:

a. Suppose the program's IPC = 3.5 when there is no memory access latency. How long will it take for the processor to execute the program, assuming a 1 GHz clock? Also, ignore delays due to OS overhead.
b. Suppose 20% of the instructions are "LD" and "ST" instructions and 10% of these instructions will cause data loading and storing memory latency where the IPC drops to 1.75. Determine the program execution time.

8.18 Computer security (secure co-processor): Exercise 11.27 (also see Secs. 11.4, 11.8, and 11.10).

8.19 Computer security (secure processor): Select Exercises 11.28 and/or 11.29 (also see Secs. 11.4, 11.9.2, and 11.11).

8.20 Computer security (spoofing, splicing, and replay attacks): Exercise 11.30 (also see Secs. 11.3 and 11.11).

8.21 Computer security (secure processor performance-related issue): Exercise 11.31 (also see Sec. 11.11).

CHAPTER 9

Computer Architecture: *Interconnection*

9.1 Introduction

A modern computer system is an interconnection of one or more processors, memory units, and input/output (I/O) devices. A personal computer (i.e., a microcomputer) additionally may include an optional special-purpose or custom processor used as an accelerator (e.g., GPU, FPGA). A keyboard, mouse, printer, network adapter, hard disk or flash drive, portable drive (e.g., memory stick), CD drive, microphone, etc. are examples of I/O devices used in a microcomputer.

While innovations in CPU and memory architectures have enabled instruction-level parallelism (ILP), multithreading and multicore processors and innovations in integrated chip (IC) technologies have increased the speed of CPUs from 16.7 MHz (Sun-4 Sparc) in 1986 to 3.33 GHz (Intel Nehalem Xeon) in 2010; it is the innovations in interconnection architectures that have increased the overall system performance. Today, a shared memory system runs multiple applications and allows programs to communicate with various I/O devices simultaneously. Furthermore, innovations in interconnection architectures have enabled "**plug and play**" I/O device interface where users of modern microcomputers are able to use an abundance of devices with ease.

All I/O devices do not operate the same; the frequency, speed, and amount of communicated data of each device are different. In addition, in some cases, the processor must be directly involved in communicating with a device, and in other cases, a device may be instructed to communicate directly with memory. And still in other cases, the processor may need to poll devices to provide service (i.e., send or receive data) when there are many devices in the system, or a device can inform the processor via interruption when it needs a service. And because various system components, from special-purpose processors, to memory units, to I/O devices, operate differently and with various speeds as compared to a processor, special hardware modules are needed to interconnect these components with one or more processors.

A **memory controller** controls the timing and responds to memory read/write requests. A device controller interface (DCI), introduced in Chap. 1, is a simple or complex embedded system and acts as a "middle man" between an I/O device and processor or both processor and memory. A **bridge** translates the communication protocol used by one component, such as processor, to a typically standard protocol used, for example, by a GPU or a disk DCI.

As a simple embedded system, a DCI is typically a **microcontroller**, which, as discussed later, is a small system with CPU, RAM, ROM, and other modules used for interfacing. Each I/O device additionally needs a **device controller** (DC), also typically a microcontroller, to control the actual hardware of the device. For example, a keyboard DC controls the hardware of the keyboard; a disk DC controls the hardware of a disk drive; etc. A DCI that connects via a cable or wirelessly to a DC performs the following two tasks:

- It controls the functions of a device by sending control data to the DC.
- It communicates with the DC to exchange data with the device. It can receive data from an input device (e.g., keyboard), send data to an output device (e.g., a printer), or both send and receive data from a disk drive or a network adapter.

Furthermore, in order to support a "plug and play" interface with devices, modern microcomputers use one or more general-purpose DCIs, such as the one commonly called a **host controller interface** used to interface with **Universal Serial Bus** (USB) devices. A USB host controller can interface and communicate with many different types of USB devices simultaneously.

In this chapter, we will discuss four generations of interconnection architectures, from single-bus to multibus, and integrated to linked-based point-to-point. The chapter then presents a list of I/O devices to highlight their communication needs and the I/O ports (initially introduced in Chap. 1) used in the design of a DCI and DC.

Interruption, interrupt structures, and the requirements for a direct device communication with memory are discussed, and circuit modules are presented. An example CPU data path with interrupt handling circuitry is used to familiarize readers with interrupt handling mechanisms and the steps the CPU must take to provide service to a device. Alternative interrupt structures to improve performance are also discussed and illustrated. Finally, for a better understanding of the tasks performed by a, host controller interface, the chapter presents both the internal organization and communication protocols of a USB host controller interface.

9.1.2 Interconnection Architectures

The architecture of single-bus system is simple. All system modules share a single bus to communicate. However, as the speed of microprocessors and memory started to diverge, single-bus architectures no longer worked efficiently. Processors were operating with higher-frequency clocks than the rest of the modules in the system. In addition, because processors typically use proprietary buses to communicate with memory and devices, manufacturers needed to use a set of standard I/O buses to interconnect DCIs with memory and/or processor. This, in turn, created several generations of multibus system architectures.

In order to simplify designs and bring the cost of personal computers down, many modules, including a memory controller and certain bridges and DCIs, were integrated into two ICs: one IC for interconnecting fast modules such as processors and an optional GPU to memory, and a second IC for interconnecting various DCIs designed to interface with standard I/O buses. However, while the integrated interconnections worked well with limited number of processing cores, this architecture created a memory bottleneck as the number of cores in each processor, as well as the number of processors, increased.

This integrated interconnection architecture was no longer scalable; you could not use more memory units to increase memory bandwidth. Therefore, there was a need for a scalable interconnection architecture that would allow memory bandwidth to increase without significant increase in memory latency.

Single Bus

Figure 9.1 illustrates a single-bus architecture with a CPU, a memory unit, and three DCIs that interface with three peripheral devices. The bus consists of an address bus (AB), a bidirectional data bus (DB), and a control bus (CB). The CPU would use the bus to access the memory or to communicate with a DCI. For instance, the CPU would send register content as data or a command to a DCI and receive, via the DCI, data and status information from the corresponding DC.

The bus would also be used to transfer a large amount of data (e.g., when reading a file) between a disk drive and the memory. However, in order for the computer to operate as expected, run a program, accept a user's input without delay, etc., the CPU would also use the bus to access memory while a disk transfer to/from memory is taking place. For this reason, another module, known as a **direct-memory-access** (DMA) controller would be used. Both the disk DCI and the DMA controller would receive commands from CPU to transfer large data blocks between a disk drive and memory without involving the CPU in the actual transfer of data. The CPU and DMA controller would share the bus, taking turns

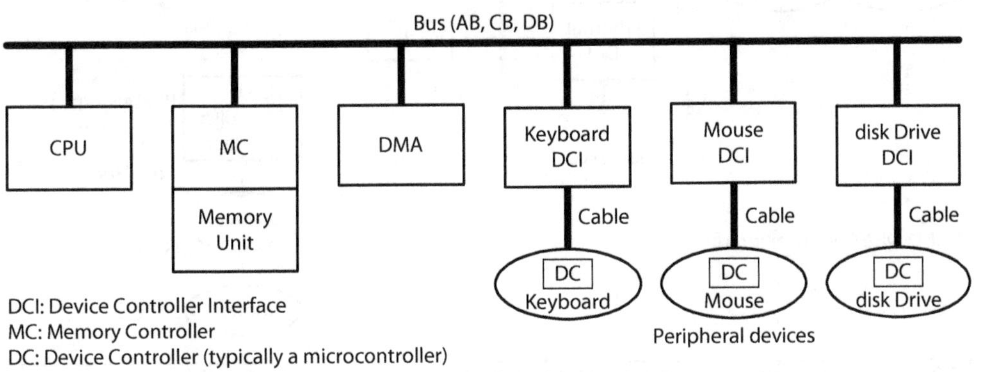

Figure 9.1 A simple microcomputer system architecture.

to access memory. Currently, however, only microcontrollers use single-bus architecture.

Multibus

Figure 9.2 illustrates an example of a multibus system architecture using a bus hierarchy known as mezzanine [1]. It uses a processor bus called a **front-side bus** (FSB) and a combination of standard I/O buses, such as those listed in Table 9.1. Peripheral devices are also not treated the same as in Fig. 9.1; they may be grouped into slow, medium, and fast devices. In the figure, a hierarchy of standard I/O buses is used to separate the fast and more frequently communicating devices from those that are slow and communicate less frequently.

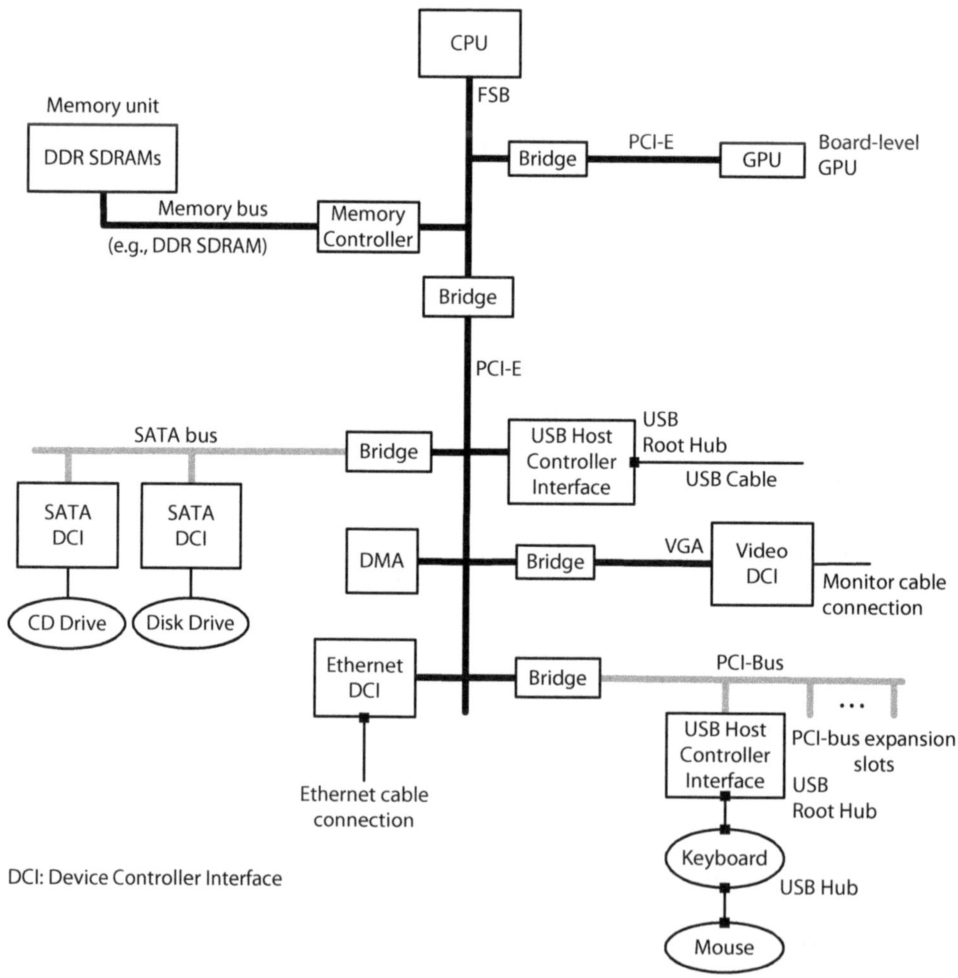

FIGURE 9.2 A multibus microcomputer architecture with front-side bus and I/O buses.

Name	Description
AHB	Advanced High performance Bus
AGP	Accelerated Graphic Port, including APG-2x, 4x, etc.
ATA	Advanced Technology Attachment, including ATA-1, 2, etc. also known as parallel ATA (PATA) or integrated drive electronics (IDE) and serial ATA (SATA)
DMI	Direct Media Interface bus, including HDMI
DVI	Digital Video Interface
FireWire	IEEE 1394
PCI	Peripheral Component Interconnect, including PCI-bus, PCI-X, PCI Express (PCI-E or PCIe), and compact PCI (CPCI)
SCSI	Small Computer System Interface, including SCSI-1, 2, etc. and Ultra SCSI
USB	Universal Serial Bus, including USB 1.x, 2.0, etc.
VGA	Video Graphics Adaptor

Table 9.1 A List of Contemporary Buses

A memory controller and two Peripheral Component Interconnect Express (PCI-E) bridges are shown interfacing with the FSB. The bridges are used to interface with a GPU and facilitate high-speed communications between the FSB and other I/O buses. In this case, the PCI-E bus that is interfacing I/O devices, connects an Ethernet DCI, a USB host controller interface, a serial advanced technology advancement (SATA) bridge, a video graphics array (VGA) bridge, a DMA controller, and a (slower) PCI-bus expansion bridge to the rest of the system, creating a **motherboard**.

The SATA bus supports chained connections of SATA hard and CD disk drives. A second USB host controller interface is also shown installed via one of the available PCI-bus expansion slots for communicating with a USB keyboard and a USB mouse. The mouse interfaces the system via a **USB hub** located in the keyboard. Each device connects to a **USB port** and communicates through one or more USB hubs and a **USB root hub** with a USB host controller interface.

Integrated Architecture

Interconnection chips, such as a memory controller hub (MCH) that was called a **north bridge** and an I/O controller hub (ICH) that was called a **south bridge** simplified the design of Intel and AMD next-generation computer system boards, as illustrated in Fig. 9.3. The modules that used to be scattered on the motherboard were integrated into the MCH and ICH chips. In the figure, an MCH includes a memory controller, a bridge connection to an optional accelerator (e.g., GPU), and a bridge to an ICH. The ICH, in turn, included a DMA controller, a high-definition audio (HDA) interface, a USB host controller interface, a network DCI, and a set of bridges to standard buses.

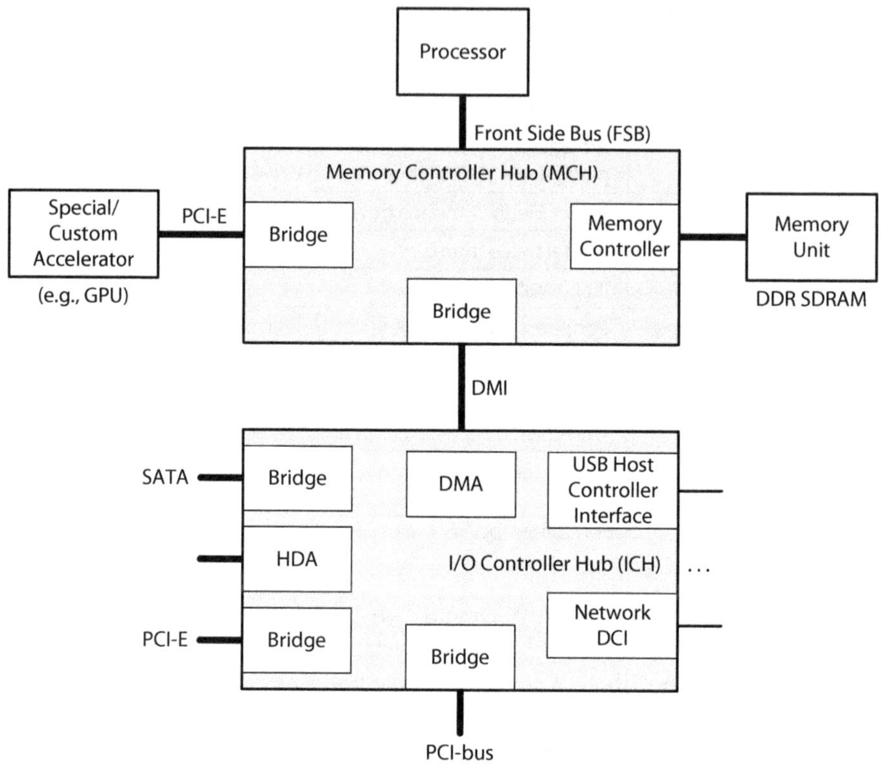

Figure 9.3 An integrated microcomputer system architecture using interconnection ICs.

The memory hub was later expanded to operate with multiple processors, thus creating a uniform memory access (UMA) architecture, which was only efficient when the total number of processing cores (CPUs) was small (e.g., four). However, as the number of cores increased, so did the number of requests to memory, resulting in long memory latency.

Point-to-Point Architecture

Figure 9.4 illustrates an example of a scalable nonuniform memory access (NUMA) architecture based on the Intel **QuickPath** links or AMD **HyperTransport** or tunnel architecture. The links are used to create one-to-one interconnections of processors for serial point-to-point communication. Instead of multiple processors sharing an FSB to access memory, each processor is directly interfaced with its own local memory, creating a NUMA architecture. As discussed in Chap. 7, average memory latency in a NUMA would be smaller than that of UMA architecture. Furthermore, the average memory latency in NUMA architecture would increase slowly as more processors are interconnected.

In addition, due to increased demand for single-chip embedded systems (system-on-chip, or SoC), more modules, such as the memory hub, migrated into the processor chip. For example, Intel's Sandy Bridge architecture includes a

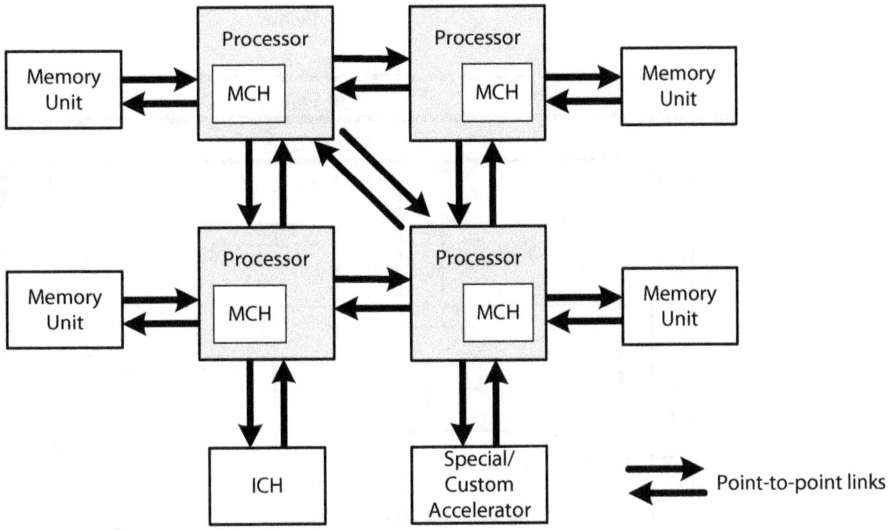

Figure 9.4 A modern microcomputer NUMA architecture.

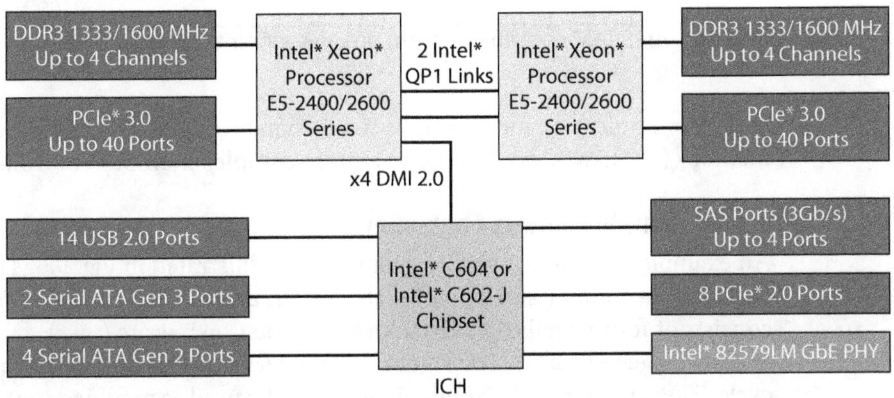

Figure 9.5 A two-node NUMA system. (Courtesy of Intel.)

memory controller, PCI-E, and other bridges all within a single processor chip. Figure 9.5 illustrates a two-node NUMA system.

9.2 Memory Controller

A memory controller is responsible for responding to requests made to memory. Its complexity depends on the communication protocols used by both the processor and memory unit. For example, a simple single-bus architecture with static random access memory (SRAM) requires a simple memory controller. On the other hand, a modern system that uses a complex bus, such as the FSB of an Intel processor,

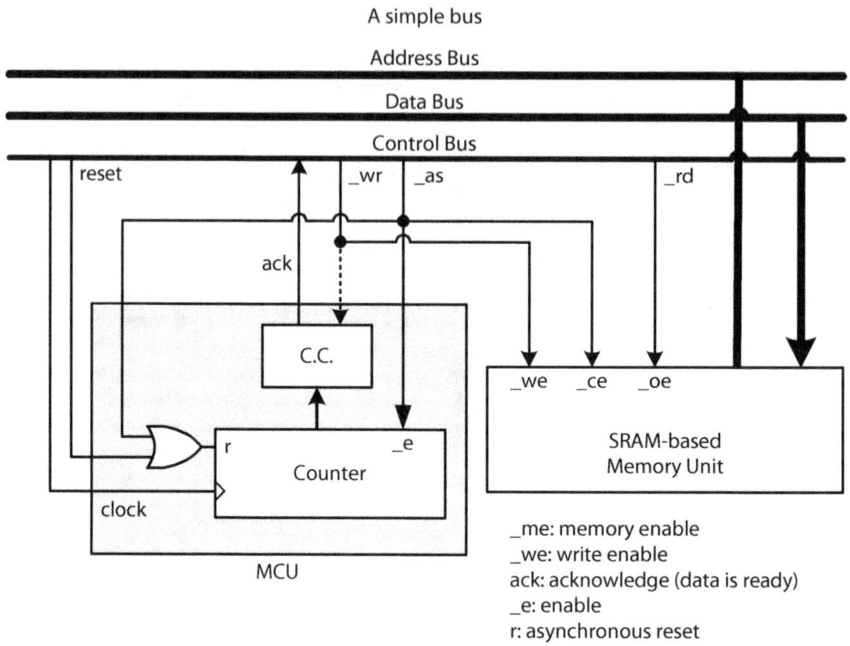

Figure 9.6 An SRAM control unit from processor point of view.

and communicates, for example, with synchronous dynamic random access memory (SDRAM), would require a more complex memory controller.

9.2.1 Simple Memory Controller

An example of a simple memory controller is illustrated in Fig. 9.6. The controller consists of a counter and a combinational circuit (CC). The processor control bus consists of four signals, labeled address strobe (*_as*), write (*_wr*), read (*_rd*), and acknowledge (e.g., *ack*). The signal *_as* is asserted to start a memory read or write cycle. The *_as*, *_wr*, and *_rd* signals are typically used as memory controller signals from processor points of view, whereas *_ce*, *_we*, and *_oe* (discussed in Chap. 7) are control signals used from memory point of view.

If there is more than one memory unit, *_as* and the target memory address are used to generate the *_ce* for a specific memory unit. In the figure, it is assumed there is only one memory unit in the system, hence *_as* directly connects to the *_ce* in the SRAM memory unit.

As shown in the figure, both the memory unit and counter are enabled when *_as* is asserted. Once enabled, the counter starts incrementing every clock cycle and asynchronously resets when *_as* is deasserted or the master *reset* is asserted. The counter is used to count the number of clock cycles, called **wait cycles**, required to access memory. The number of wait cycles is proportional to memory access time. The counter module asserts the *ack* signal and notifies the processor of the completion of a memory read or write access. The processor is said to be in a **wait state** (also called an **idle state**) while waiting for the *ack* to be asserted.

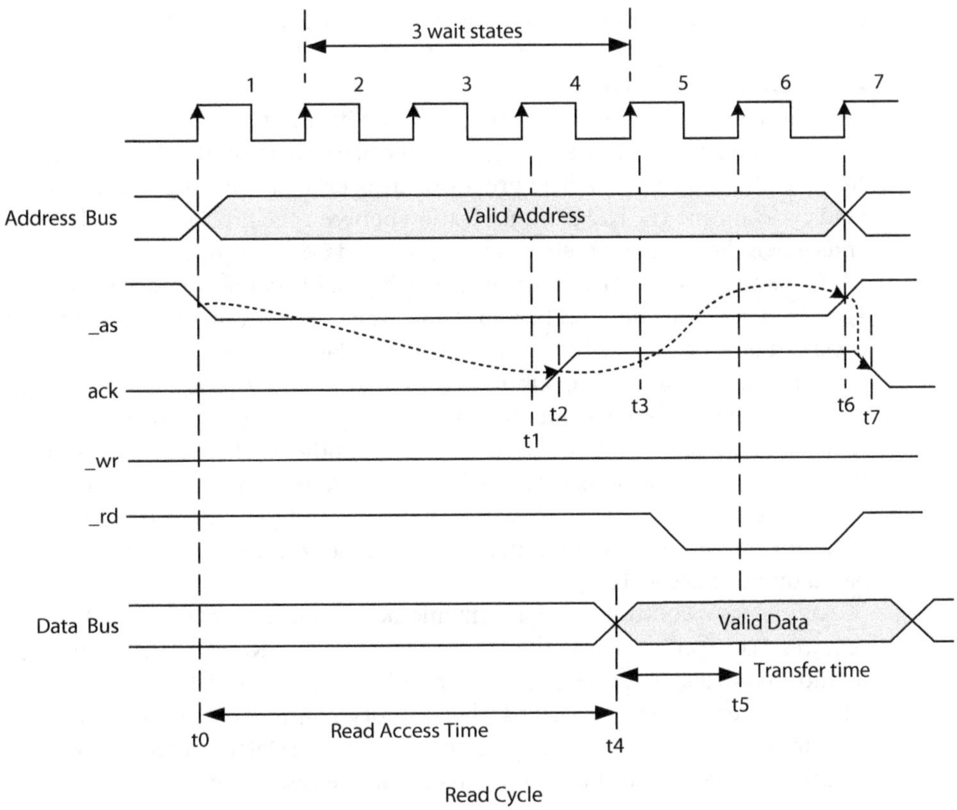

FIGURE 9.7 Illustration of a simple SRAM read cycle from processor point of view.

Figure 9.7 illustrates a memory read cycle from processor point of view. The number of wait cycles is determined using Eq. (9.1). The symbols $\lceil \ \rceil$ indicate the ceiling function, τ is the clock period of the bus, and m is the required number of clock cycles for the processor to detect *ack* = 1 and then end at he memory cycle. For example, suppose in the figure the clock period is 10 ns, memory read access time is 45 ns, and the processor requires one clock cycle to detect *ack* = 1 and another clock cycle to end the read cycle. From Eq. (9.1), the number of wait cycles is 3 (i.e., $\lceil 45\,\text{ns}/10\,\text{ns} \rceil - 2 = 3$).

$$\text{Wait cycles} = \left\lceil \frac{\text{Max(read access time, write access time)}}{\tau} \right\rceil - m \qquad (9.1)$$

In the figure, the three fine dotted arrows indicate signal dependencies commonly referred to as **signal handshaking**. A processor communication with the memory controller starts when the processor places a target address on the address bus and asserts *_as*, which marks the start of a memory cycle at time t0. The signal enables both the counter and the memory unit. The counter value becomes 3 at time t1 at the start of the fourth clock cycle since the start of memory cycle and causes the *ack* signal to become 1 at time t2 (Δ_{CC} = t2 − t1 in Fig. 9.6). This is indicated by an arrow from the time _as = 0 to when *ack* = 1.

When the processor detects *ack* = 1 at time t3 (one clock cycle later), it leaves the wait state and gets ready to load the data to its internal register on the next cycle at time t5. As shown in the figure, memory places data on the data bus starting at time t4 (45 ns = t4 − t0). The processor deasserts _as, making it 1, and ends the read cycle at time t6. The communication between the processor and the memory ends when the counter resets when _as = 1 and deasserts *ack*, making it 0 at time t7. These are illustrated by an arrow from when *ack* becomes 1 to when _as becomes 1, and again from when _as becomes 1 to when *ack* becomes 0. Another memory cycle can now begin starting at clock cycle 8 or later.

A memory write cycle is similar, except that when the processor detects *ack* = 1, it can remove the data from the data bus. Again, assuming that the processor requires one clock cycle to detect *ack* = 1 and another cycle to remove its data from the data bus, *ack* may be asserted earlier, as was done for a read cycle. In addition, if memory read access and write access times are different, the memory controller may implement two different wait cycles, one based on read access time and another based on write access time.

A memory controller for a dynamic access random memory (DRAM) also requires a counter, similar to the one discussed for an SRAM. In addition, a DRAM memory controller requires circuitry required to implement refresh cycles. However, when a refresh cycle is triggered, the memory controller typically would allow a current read or write cycle (if any) to complete before starting a refresh cycle. A new read/write cycle can start only after an ongoing refresh cycle is completed.

9.2.2 Modern Memory Controller

Figure 9.8 illustrates a modern memory controller interfacing with a complex FSB and a contemporary memory unit. The controller interprets memory requests it

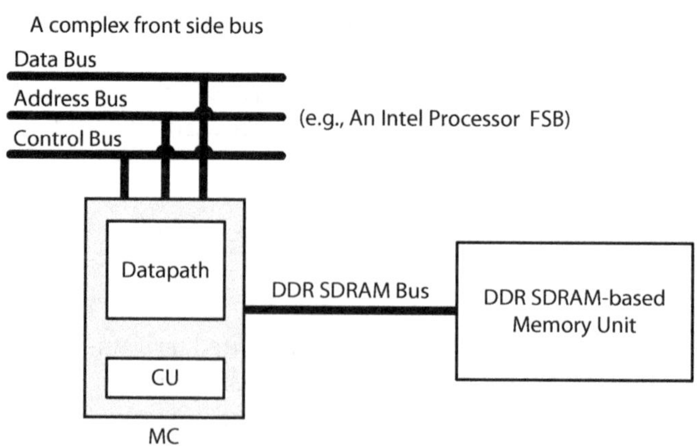

FIGURE 9.8 A complex memory controller.

receives over a complex bus and communicates, for example, with a double data rate (DDR) SDRAM over the standard DDR SDRAM bus. Table 9.2 is a partial signal listing of the 533 MHz FSB of the Mobile Intel Pentium 4 Processor. A FSB, or a complex serial link, for example, the Intel's QuickPath, typically implements **split transactions** to communicate with a memory controller. A split transaction consists of a **request transaction** and a **response transaction.**

Signal Names	Bus	Function
A[35:3]#	AB	Address bus (AB)
D[63:0]#	DB	Data bus (DB)
REQ[4:0]#	CB	Request bus command (defines a bus transaction type)
ADS#	CB	Address strobe indicating address on AB is valid
ADSTB[1:0]#	CB	Address strobe 0 to latch A[16:3] and REQ#[4:0] and address strobe 1 to latch A[35:17] by the destination module
AP[1:0]#	CB	Address parity bits
DBI[3:0]#	CB	Data bus inversion (shows polarity of DB signals, in groups of 16 bits)
DP[3:0]#	CB	Data parity bits
DRDY#	CB	Data ready signal
DSTBN[3:0]#	CB	Data strobes negative edge (used to load data from DB in groups of 16 bits)
DSTBP[3:0]#	CB	Data strobes positive edge
RS[2:0]#	CB	Response status (status of a completed transaction)
RSP#	CB	Response parity bit (parity protection for RS[2:0]#)
RESET#	CB	Resets the processor
BCLK[1:0]	CB	Differential pair FSB clocks (two clocks opposite of each other)
BINIT#	CB	Bus initialization signal (e.g., reset bus arbitration state machine)
BNR#	CB	Block next request (stalls bus)
BPRI#	CB	Bus priority request (used by others to request access to FSB)
BR0#	CB	Bus request signal used by the processor to request bus access
DBR#	CB	Data bus reset (used for debugging)
DBSY#	CB	Data bus busy (indicates DB is being used)
DEFER#	CB	Deferred transaction (when active indicates bus transaction cannot be completed)
HIT#	CB	Snoop hit (used for cache memory coherency)
HITM#	CB	Snoop hit modified (if used with HIT# indicates requiring a snoop stall)
LINT[1:0]	CB	Local APIC (advanced programmable interrupt controller) interrupt signals
LOCK#	CB	Indicates atomic bus transaction; when asserted no new request traction is allowed to the same memory location until a response transaction to an outstanding request transaction to the same location completes.

TABLE 9.2 Partial Signal Listing of the Mobile Intel Pentium 4 Processor 533 MHz FSB

A request read transaction includes an address and a command (e.g., read). A request write transaction also includes data. A response transaction includes one or more acknowledgement signals, as well as data if the request was a read transaction. A modern memory controller typically handles multiple outstanding split transactions.

In the table, the symbol # is used to indicate an active-low signal. The Mobile Intel Pentium 4 Processor with 533 MHz FSB [2] has 478 pins, with several pins reserved for power and ground connections, as well as power management, board-level design issues, performance measurement, bus error signaling, transaction flow control, and testing. It includes 36 pins for address bus (A[35:3]#) that can reference 2^{36} bytes of physical memory, 5 pins to issue a memory command (REQ[4:0]#), and 64 pins for data bus (D[63:0]#). A request transaction (i.e., address and command) is issued by an address strobe signal (ADS#), which is then latched by the memory controller in two parts using two additional strobe signals (ASTB[1:0]#). ADS# starts and ends a memory request transaction, and ASTB[1]# and ASTV[0]# are used as clock signals to operate two register buffers in the memory controller for loading the upper and lower bits of a target address.

A response transaction includes a data ready signal (DRDY#), a 3-bit response code (RS[2:0)]#), and data for a read command. If a response transaction includes data, it is latched by the processor in four negative (DSTBN[3:0]#) or positive (SDTBP[3:0]#) edge-triggered data strobe signals. The 36-bit address, 64-bit data, and 3-bit response code also include a parity error protection signal AP[1:0] for the address, DP[3:0] for the data, and RSP# for the response code. Some of the remaining control bus signals listed in the table deal with cache memory (discussed in Chap. 10), with request for service from peripheral devices, and with **atomic bus access** when two or more CPUs try to modify the content of a shared memory location. Only one CPU at a time is permitted to modify the content of a shared memory location.

A modern memory controller operates as a middleman between a processor by interfacing its FSB and memory unit using SDRAM modules. Using, for example, the A, D, REQ, ADS, and ADSTB bus lines in Table 9.2, the processor issues a memory transaction to the controller requesting to access the memory. The controller logs the transaction and then converts the communication protocol of the FSB to that of the SDRAM bus and vice versa. Refer to Chap. 7 for examples of SDRAM bus transactions.

9.3 I/O Peripheral Devices

Each peripheral device requires a specific communication protocol, operating speed, and data size. For example, a keyboard generates small data values (e.g., a key value) each time that a key is pressed. On the other hand, a magnetic hard disk drive is an electromechanical device designed to transfer a large amount of data fast. A hard drive has many recording circular surfaces (disks), each with many tracks (like the tracks on a CD). A track is divided into several equal-sized regions called **sectors**. Each recording surface has its own read/write head, and some hard drives can transfer data simultaneously from multiple sectors.

For example, consider a Samsung 260 GB hard drive (e.g., HD 642JJ). Its magnetic disks rotate at the speed of 7200 rotations per minute (RPM) and has 175 Mbps (megabits per second) peak transfer rate between its recording surfaces and its 16 MB internal (DRAM) memory. The internal memory, also called a buffer, is used as a temporary storage to store sector data to/from the main memory. Its peak data transfer rate between the buffer and the main memory is 300 Mbps. Its sector size is 512 B. Modern personal computers use disks with 4 KB sectors. Some modern disk drives operate at 10,000 RPM or higher.

The Samsung hard drive has 8.9 ms (millisecond) **average seek time**, the amount of time required to move a disk's read/write head to a specific track. The disk drive also has 4.17 ms **average latency**, the average amount of time required to rotate a disk 50% (half of its circumference). That is, at 7200 RPM, one-half (50%) rotation would require 4.17 ms (0.5/7200 RPM). The sum of the two averages is the approximate amount of time required to locate a target sector for read or write operation.

Example 9.1. Using the average and peak performance parameters of the Samsung disk drive, determine the approximate average required time to transfer 512 B (data from one sector) to memory.

Solution: Four sequential tasks are required to copy sector data to memory:

Task 1: Seek the target track with an average seek time of 8.9 ms.

Task 2: Locate the target sector with an average latency of 4.17 ms.

Task 3: Copy data from the sector to the internal buffer with the peak transfer rate of 175 Mbps.

Task 4: Copy data from the internal buffer to main memory with the peak transfer rate of 300 Mbps.

The average approximate required total time to transfer 512 B to memory is 13.075 ms, as calculated next:

$$T_{512\,\mathrm{B}} = 8.9\ \mathrm{ms} + 4.17\ \mathrm{ms} + \frac{512\ \mathrm{B}}{175\ \mathrm{Mbps}} + \frac{512\ \mathrm{B}}{300\ \mathrm{Mbps}}$$

$$= 13.07\ \mathrm{ms} + 0.0029\ \mathrm{ms} + 0.0017\ \mathrm{ms}$$

$$= 13.075\ \mathrm{ms}$$

However, transfers to/from multiple physically adjacent sectors would be faster. A transfer from memory to disk is similar, except that the data is first copied from memory to the internal buffer before it is written onto the target sector.

RAID (redundant array of independent disks) is designed to increase the bandwidth or both the bandwidth and the reliability of disks. For example, data interleaving, similar to memory interleaving (Chap. 7), is used to increase bandwidth by storing strips of data from a single record on several sectors, one strip on each independent disk. Interleaving can be done at a bit level (RAID-2 and RAID-3) or at a block-level (RAID-0 and RAID-4 to -6). Copies of files may be duplicated to create file storage redundancy (RAID-1). Error correction codes can be used to avoid the extra cost of storage redundancy by implementing a recovery mechanism if one or even two disks fail, such as in RAID-2 to -4, if only one disk fails, or RAID-6 and RAID DP (double parity) if two disks fail.

Other devices that have no mechanical parts can communicate at even faster rates. For example, an Ethernet network adapter is capable of transferring data between two communicating computers at the rate of 10 Mbps, 100 Mbps, or higher. In this case, data originates from memory and is communicated with a receiving computer via a DMA controller and the DCI of the adapter. The Ethernet adapter at the receiving computer receives the data and, via its DCI and a DMA controller, transfers the data to main memory.

9.4 Controlling and Interfacing I/O Devices

Peripheral devices are slow, medium, or fast; each requiring a clock with different operating frequency and data transfer rate. Some devices may contain electromechanical parts and operate differently from digital systems and may require data conversions to digital signal values. The data format and signal voltage levels in a device may also be different from those used by processors and memory. In addition, the processor must be able to communicate with each device without affecting the functions of other devices in the system.

A DCI communicates with the rest of the system via a processor bus (e.g., Fig. 9.1), I/O bus (e.g., Fig. 9.2), or ICH (e.g., Fig. 9.3 or Fig. 9.4). In the past, personal computers were designed with each device having its own dedicated DCI. Even the basic devices such as keyboard, mouse, and printer required its own DCI, as was illustrated in Fig. 9.1 for the keyboard and the mouse.

For example, older standards such as the parallel port (IEEE 1284) and RS-232 (recommended standard 232), including its smaller version DE-9, supported only a single point-to-point connection to a peripheral device and required a dedicated DCI. This meant they did not implement the "plug and play" device interface, and therefore increased the cost of a personal computer and, in addition, required system reboot each time that a device was installed. A typical system also supported only a limited number of slots for installing new devices, which created a restriction for personal computer users. Today, most, if not all, peripheral devices are "plug and play."

As was briefly discussed earlier, both DCIs and DCs use I/O ports. The ports are designed using tri-state buffers and registers, and each is identified by a unique address and accessed similar to memory. Isolated I/O or **port-mapped I/O** and **memory-mapped I/O** are two commonly used I/O port addressing schemes, as illustrated in Fig. 9.9.

In Fig. 9.9(*a*), port-mapped I/O requires two separate address spaces, one for memory (e.g., 512 B), and another with an equal size for I/O ports. An additional control bus signal, for example, *_m*, indicates whether a given address is a memory address if *_m* = 0 or an I/O port address if *_m* = 1.

On the other hand, memory-mapped I/O requires only one address space divided between memory and I/O ports, as shown in Fig. 9.9(*b*). In the figure, a 512 B address space is divided between memory and I/O ports. The same control signals (e.g., *_as*, *_rd*, *_wr*, and *_ack*) that are used to access memory are also used to access memory-mapped I/O ports. In this case, it is the responsibility of the

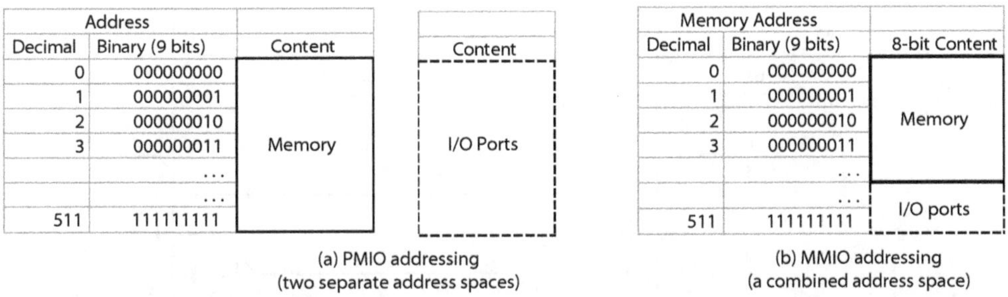

Figure 9.9 I/O port addressing: (*a*) port-mapped I/O, a 512-B address space for memory and another 512-B address space for I/O ports; (*b*) memory-mapped I/O, a single 512-B address space divided between memory and I/O ports.

memory controller and each of the DCIs to determine if an address is within the address range reserved for the memory unit or if the address belongs to an I/O port.

Any computer system may implement the memory-mapped I/O addressing scheme. Intel processors also support port-mapped I/O addressing, with two special I/O instructions: "IN" and "OUT." Typically, all reduced instruction set computer (RISC) processors support only memory-mapped I/O addressing, which has the advantage of using memory reference instructions to access I/O ports. On the other hand, the port-mapped I/O addressing has the advantage of not using any part of memory address space for I/O ports; however, this advantage has diminished as memory address space has increased over the years. For example, modern computer systems have main memory address space in GB, which is more than enough to easily support memory-mapped I/O port addressing to access many I/O ports.

9.4.1 I/O Ports

Figure 9.10 illustrates an example of a simple memory-mapped I/O port with an **input port** and an **output port**. The port refers to both the set of tri-state buffers

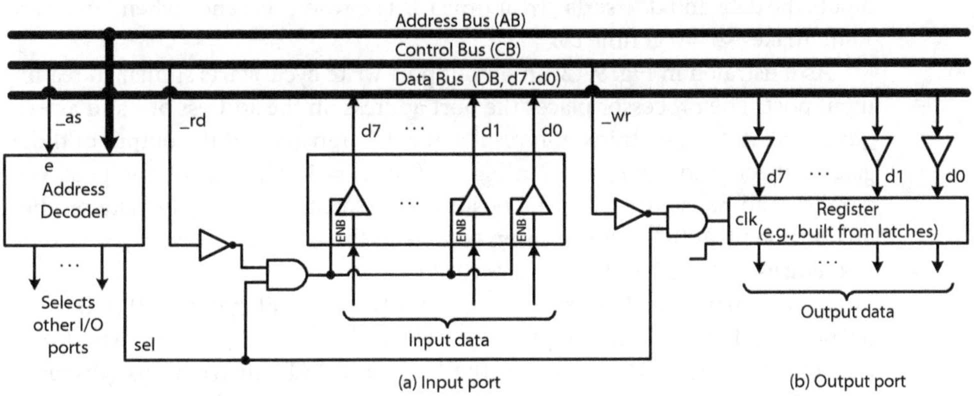

Figure 9.10 An illustration of an I/O port that includes an input port and an output port; the memory-mapped I/O port addressing is assumed.

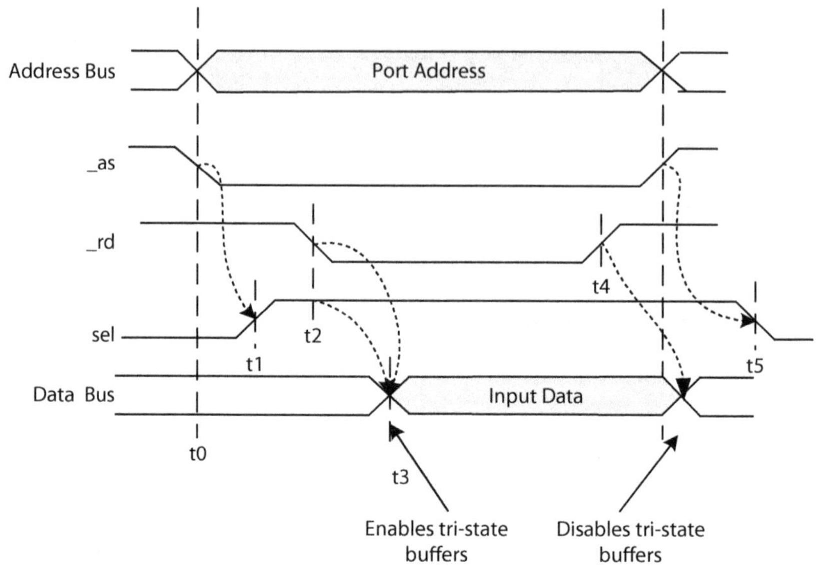

FIGURE 9.11 An illustration of an memory-mapped I/O input port read cycle from processor point of view; also may require an *ack* handshaking signal (not shown).

that isolates an input data from the data bus and to a parallel-load register that holds an output value. In the figure, the parallel-load register, also called a buffer, is built from positive-level latches.

For the input port, a read cycle starts when processor places the port's address on the address bus and asserts the *_as* (address strobe) signal, as illustrated in Fig. 9.11 at time t0. The *_as* signal, when asserted (*_as* = 0) by the processor, indicates the address currently on the address bus (AB) is valid. The *_as* = 0 enables the address decoder in Fig. 9.10, which in turn asserts the *sel* signal at time t1 if the address on the address bus is the target port address. A 1-0 *_rd* transition at time t2 enables the tri-state buffers, causing the input data to be placed on the data bus at time t3. The processor inputs the data and deasserts *_rd* at time t4. The read cycle ends when *_as* returns to 1 and makes *sel* = 0 at time t5.

As illustrated in Fig. 9.12, an output port write cycle starts similar to reading an input port. The processor places the port address on the address bus and asserts the *_as* signal. A 1-0 *_wr* transition results in a 0-1 transition at the output of the AND gate connected to the register in Fig. 9.10 if *sel* = 1. This causes the positive-level parallel-load register (built using positive-level latches) to load the data on the data bus at time t1. A 0-1 *_wr* transition completes the write. The write cycle ends when *_as* returns to 1 at t2 and makes *sel* = 0 at t3.

A port-mapped I/O port reading or writing is similar, except that the address decoder in Fig. 9.10 also inputs the control bus signal *_m* (discussed earlier). Assuming an Acc-ISA processor, the "LD" and "ST" instructions (discussed in Chap. 8) would also generate *_m* = 0 to access memory, and two new instructions, for example, "IN" and "OUT," would generate *_m* = 1 to access port-mapped I/O ports. With memory-mapped I/O ports, the *_m* signal is not needed, and the "LD" and "ST" instructions would be used to address memory as well as I/O ports.

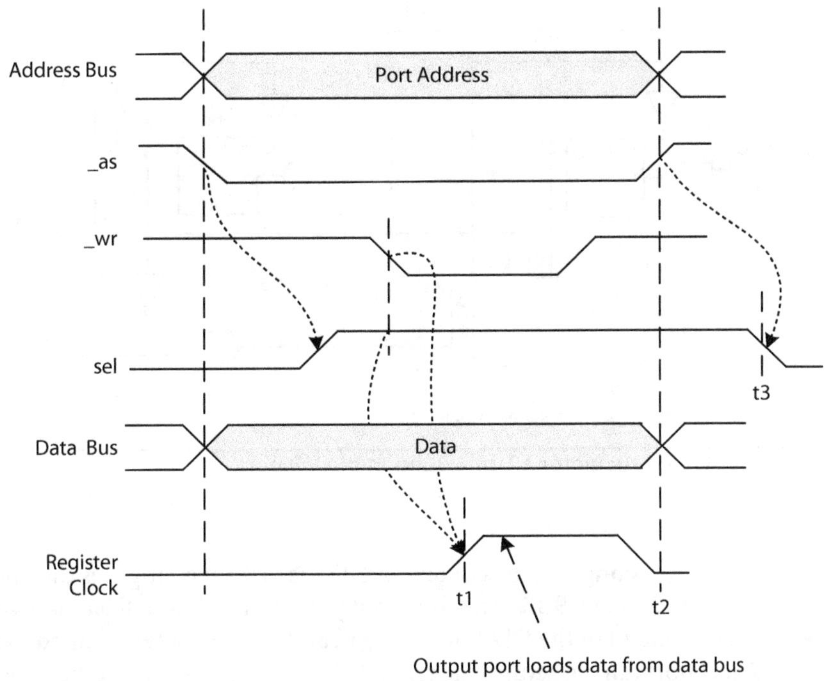

Figure 9.12 An illustration of an memory-mapped I/O output port write cycle from processor point of view; also may require an *ack* handshaking signal (not shown).

Configurable Ports

As illustrated in Fig. 9.13, a configurable I/O port also contains a **data direction register** (DDR) used for configuring each bit in the I/O port either as a 1-bit input port or a 1-bit output port. Figure 9.14 illustrates the design detail for 1-bit configurable I/O port. The DDR.$q_0 = 0$ configures the bit d_0 as a 1-bit input port by disabling the tri-state buffer 1. The tri-state buffer 2 is enabled during a read cycle. The DDR.$q_0 = 1$ configures d_0 as a 1-bit output port by enabling tri-state buffer 1 and disabling tri-state buffer 2.

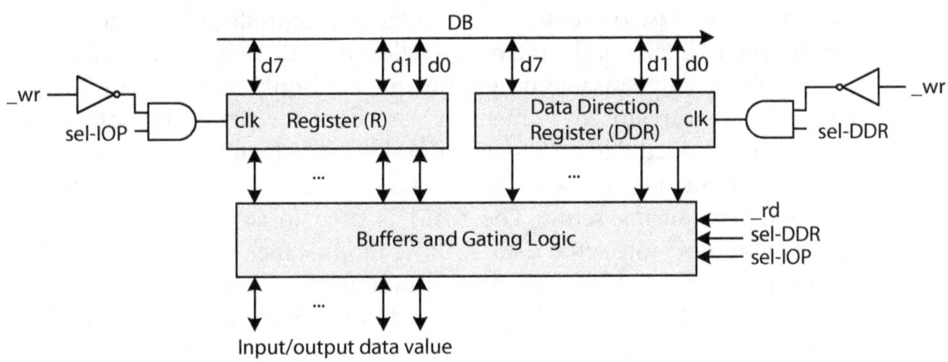

Figure 9.13 A block diagram for an 8-bit configurable port [3].

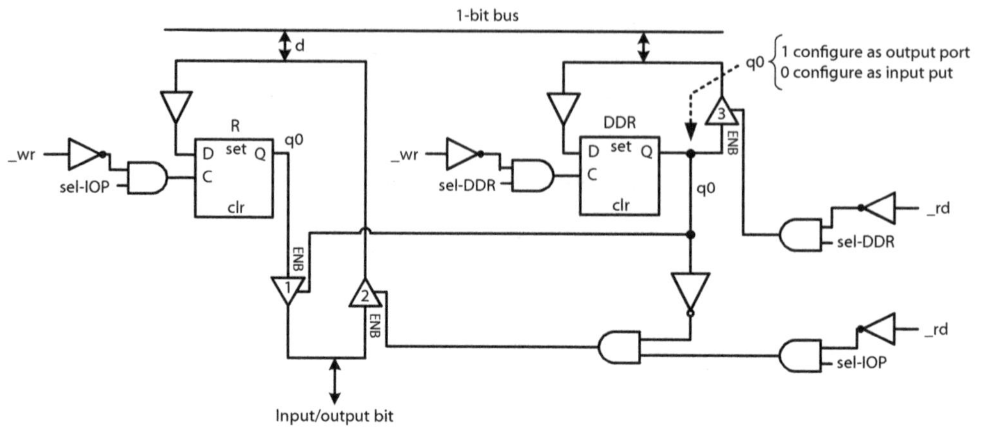

FIGURE 9.14 A detained circuit for a 1-bit configurable I/O port.

For example, if the content of DDR is 0x0F, it configures the upper 4-bits of the I/O port in Fig. 9.13 as a 4-bit input port, and its lower 4-bits as a 4-bit output port. The content of the DDR may be read and dynamically modified to reconfigure the I/O port during setup. The tri-state buffer 3 is used when the content of the DDR is read.

However, it is also possible to include, for example, a "command" port in a microcontroller that would be used to configure a set of I/O ports with limited configuration options but general enough to support the development of many DCIs and DCs.

Figure 9.15 shows an example of a microcontroller with three configurable I/O ports 0, 3, and 4 and two multipurpose I/O ports 1 and 2. Ports 1 and 2 are not only configurable, but also have dual use. In one application, the pins associated with ports 1 and 2 may be used as configurable I/O ports, and in another application, the same pins may serve as data or control signals.

A microcontroller, as an embedded system, includes CPU, RAM, ROMs, a set of I/O ports, one or more **timer modules**, an **interrupt controller**, one or more data communication modules, etc. A timer module is used when the microcontroller performs certain tasks periodically. The interrupt controller, which will be discussed later, is used to interrupt the microcontroller when there is an external event.

The electrically erasable programmable read-only memory (EEPROM) and flash memory (organized as memory) are used for **firmware**, software stored inside an embedded device. The EEPROM is loaded with a bootloader program, and the flash memory is used to store a DC or DCI firmware that can be updated by the system during setup. The RAM is used to store program data during execution and/or store setup data. A timer module includes a counter and is used to schedule events. For example, a timer module that uses a mod-12K counter and operates with a 12-MHz clock can be used to monitor the keyboard hardware for keystrokes once every 1 ms (12 K cycles / 12 M cycles/s $*$ 1000 ms/s = 1 ms).

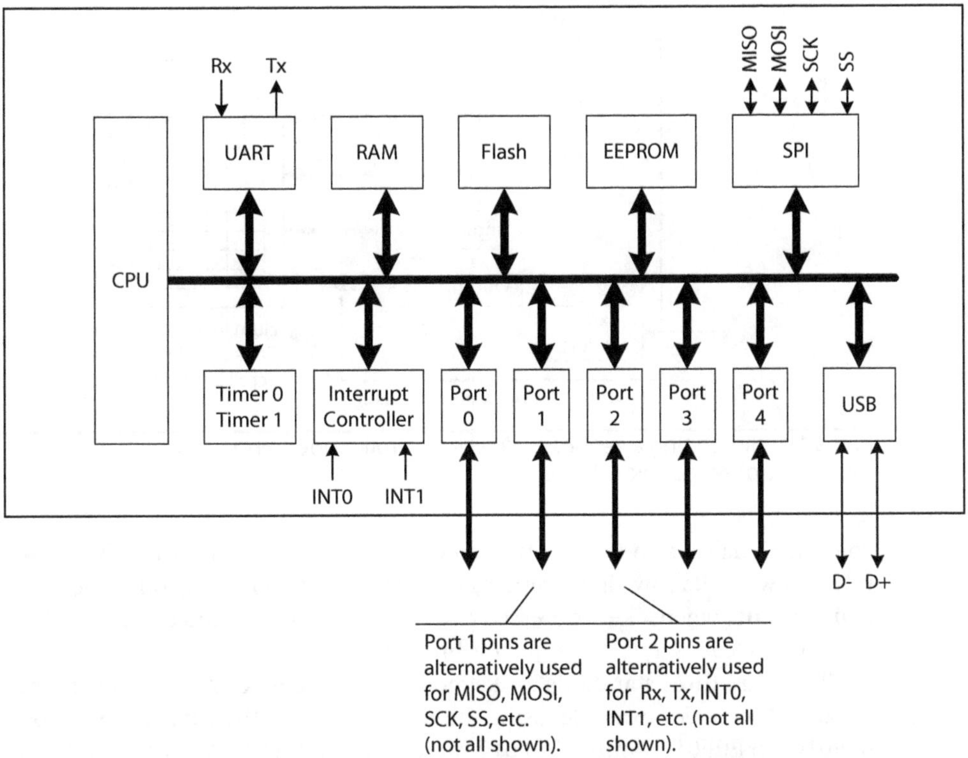

Figure 9.15 A microcontroller architecture: pins connected to ports 1 and 2 have dual use [3]; not all modules shown.

In the figure, the microcontroller also includes three different data communication modules to support various design applications. It includes a universal asynchronous receiver/transmitter (UART), serial peripheral interface (SPI), and a USB port, all within a single chip. UART, SPI, and USB are three different communication protocols used with I/O devices.

In general, not all the resources in a microcontroller are expected to be used in a single application, and thus some of the modules may share the same I/O pins. For example, in the figure, the pins from ports 1 and 2 could be used as the receive (Rx) and transmit (Tx) signals by UART module; as the master in slave out (MISO), master out slave in (MOSI), serial clock (SCK), and slave select (SS) signals by the SPI module; or the interrupt 0 (INT0) and interrupt 1 (INT1) signals by the interrupt controller.

Figure 9.16 illustrates a keyboard DC operating a **key matrix**. Each row-column intersection in the key matrix identifies a key. The DC scans the matrix by activating the 20 column signals, for example, setting each to 0 one at a time using Ports 0, 1, and 3, as illustrated in the figure, and reading the 8-bit row signals using Port 2. Each time a key is pressed, it creates a contact between a column wire and a row wire, making the corresponding row signal 0 if and only if its corresponding

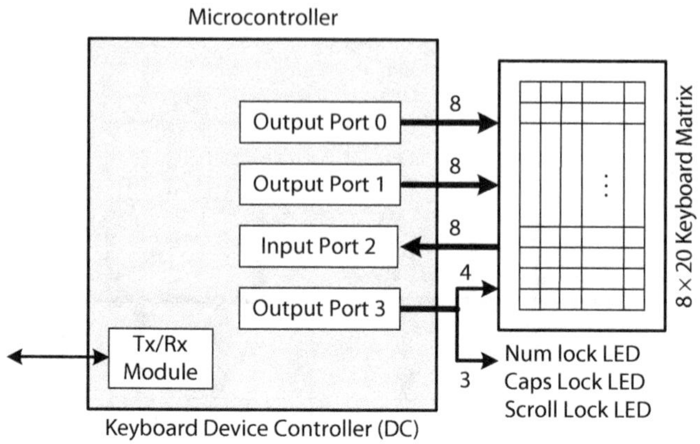

Figure 9.16 Illustrating a keyboard device controller operating a key matrix; not all microcontroller modules shown.

column signal is also 0; the other row signals remain at 1. A key is considered released when its row signal returns to 1. The DC generates a code, referred to as a **make code**, when a key is pressed and a contact is made, and a **break code** when the key is released and the contact is broken.

There are three standards for keyboard scan codes, known as **scan code** set 1, set 2, and set 3. For example, using the scan code set 2, the data sequence 0xE012, 0xE01C, 0xE0F01C, and 0xE0F012 indicate capital letter A. The code 0xE012 is the make code for the SHIFT key, 0xE0IC is the make code for the a key, 0xE0F012 is the break code for the a key, and finally 0xE0F012 is the break code for the SHIFT key. A key may also be held down to have its make code repeated every so often. The keyboard DC is initialized with a **key repetition rate** that determines how often a make code is generated if a key is held down.

The make and break codes for all the keys, including CTRL, ALT, etc., are transmitted via a DCI and stored in memory. The code is then converted, for example, to an ASCII code by a basic input/output system (BIOS) routine for use by application programs.

The keyboard DC also receives instructions through the keyboard DCI. For example, the DC receives a key repetition rate during initialization, or a command to turn on the Caps-On light-emitting diode (LED), if any, when the CAPS-LOCK key is pressed.

Ports with Status Bits

The I/O port shown in Fig. 9.10 is an example of an I/O port with no status bits. There is no way to know when input data is available to be read or when the output data has been read. I/O ports with no status bits are used in the design of a DC that directly controls the device hardware, such as those shown in Fig. 9.16 for controlling the key matrix.

On the other hand, I/O ports that are used in the design of a DCI require status bits to indicate when input data is available to be read or when output data has been read by the device. For instance, when communicating with a printer DCI, when the next print data should be sent to the printer DCI depends on how fast the printer is able to print the data it has already received. The next print data should be sent to the printer DCI when the DCI has already sent the previous data to the printer DC and the DCI indicates it is ready to receive the next print data.

Other examples are the keyboard DCI indicating the availability of a new scan code it has received from the keyboard DC, a DMA controller indicating the completion of a data transfer between main memory and a disk drive DCI or a network adapter DCI, etc.

An application example of I/O ports with status bits will be discussed in the next section, but first, there are three different mechanisms to transfer data to/from a DCI, one of which is DMA transfer, which was briefly discussed earlier.

9.5 Data Transfer Mechanisms

Data communication with a DCI is either done directly by a processor or directly by a DMA controller. Furthermore, even when a processor is directly communicating with a DCI to transfer data, the data is actually in main memory. The processor inputs data from a DCI and then writes it to memory, or the processor reads data from the memory and then outputs it to the DCI. **Interrupt-driven** and **programmed I/O** are two data transfer mechanisms that directly involve the processor.

9.5.1 Interrupt-Driven Transfer

An interrupt-driven transfer, also known as **interrupt-driven I/O**, is used when a device causes an interruption. The execution of a currently running program stops, and the processor invokes an interrupt handler (IH), also called an interrupt service routine. When the IH executes, it transfers data between the device and main memory. Because the processor is executing the IH, it is directly involved in the transfer of data between the device and memory.

An **interrupt structure** (discussed in Sec. 9.6) is used to assign each device an interrupt priority in case multiple devices wish to interrupt the processor. Some devices, such as the keyboard, have a lower interrupt priority than, say, a disk drive. If there are too many devices in the system, service for some devices could be delayed.

In order to illustrate interrupt-driven transfer and application of I/O ports with status bits, consider a legacy keyboard DCI shown in Fig. 9.17. In the figure, the DCI has two I/O ports, labeled port-0 and port-1. However, because both the processor and the DCI access these ports, an input port that the processor inputs data from is also an output port for the DCI. Likewise, an output port that the processor writes to is also an input port for the DCI. Two registers are associated with each port address, one as an input port and one as an output port.

In the figure, the registers are labeled “Scancode,” “Data,” “Status,” and “Command.” Table 9.3 lists the input and output ports from both the processor and the keyboard device DCI point of view. The ports, however, are labeled from the keyboard DCI point of view in the figure.

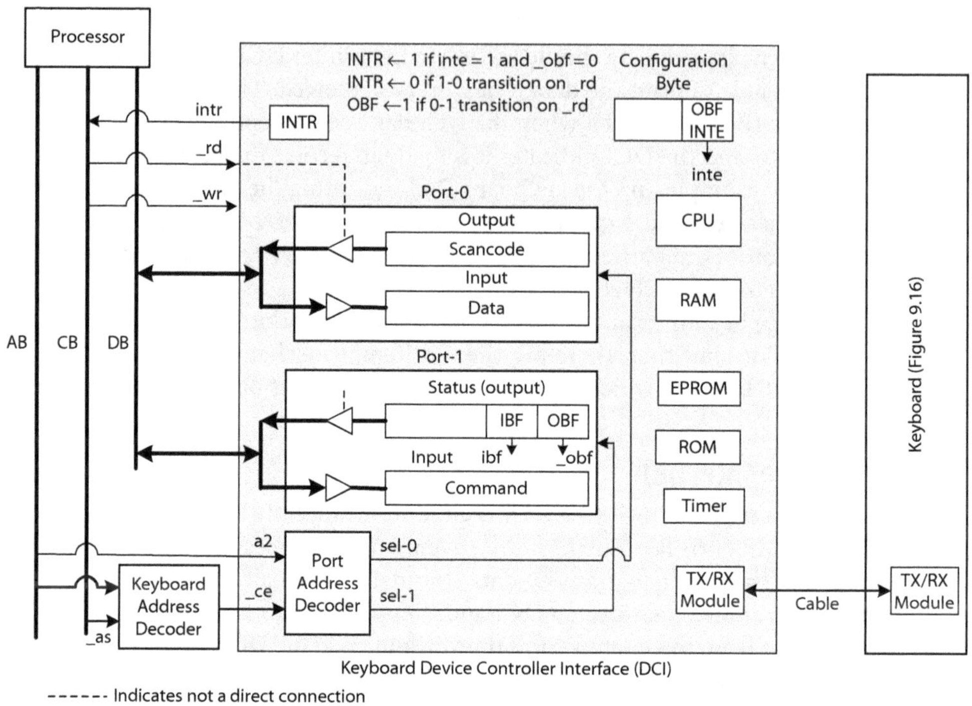

FIGURE 9.17 A keyboard DCI (device controller interface) to illustrate I/O port accessing and interrupt driven transfer; not all signal and data path modules are shown; ports are shown labeled from DCI point of view.

The processor can read the "Status" port, which contains the status bits for the remaining ports, at any time. The input buffer full (IBF) status bit (or active-high signal *ibf*) and output buffer full (OBF) status bit (or active-low signal *_obf*) indicate the status of the other three ports. When *_obf* = 0 (full), it indicates a scan code is loaded into the "Scancode" register for the processor to access. When *ibf* = 1 (full), it indicates that there is data from the processor available in the "Data" or in the "Command" port for the keyboard DCI to access.

A keyboard **driver routine** must first configure the DCI for either interrupt-driven or programmed data transfer. The driver routine executed by the processor

Port Name	From Processor Point of View	From Keyboard Device Controller Interface (DCI) Point of View
"Scancode"	Input port	Output port
"Data"	Output port	Input port
"Status"	Input port	Output port
"Command"	Output port	Input port

TABLE 9.3 How Ports Are Viewed by the Processor and Device Controller Interface in Fig. 9.17

writes a configuration command in the "Command" register initiating the configuration of the DCI. Upon receiving the command, the DCI starts the execution of a configuration program. The driver then writes configuration data in the "Data" register. The configuration program reads the data and stores it in a "configuration buffer" within the DCI.

For example, to illustrate how the I/O ports are accessed, the following program code is used to check the IBF bit, and if the bit is 0 (i.e., *ibf* = 0), a value as configuration command is stored in the "Command" register. The program code is written using instructions from an Acc-ISA, discussed in Chap. 8.

Keyboard Driver Routine Polls the "Status" Port and then Outputs to the "Command" Port:

```
         ...               ...
         ...               ...
LOOP:    IN (Port_1)       //input keyboard status
         AND 2             //select status bit 1, ibf; bit 0 is _obf.
         CMP 2             //is ibf = 1? 1 indicates full.
         JEQ LOOP          //wait for an empty input buffer
         LD ...            //get, configuration command
         OUT (Port_1)      //output to "Command" port, assuming
                           //port-mapped I/O port
         ...
         ...
```

For interrupt-driven transfer, configuration data must set the OBF interrupt enable (INTE) bit in the "configuration buffer." The interrupt request (INTR) bit (signal *intr*) is used to trigger an interruption if INTE is enabled. That is, for an OBF interruption, the *intr* becomes 1 (active) when we have both *_obf* = 0, which indicates the "Scancode" port is full, and *inte* = 1, which indicates interrupt-driven data transfer.

As soon as a key is pressed (Fig. 9.16), the keyboard DC sends a scan code to the keyboard DCI, which writes it to the "Scancode" register setting the OBF bit (*_obf* = 0). This asserts the interrupt request signal *intr* making it 1 if *inte* = 1. The *intr* = 1 notifies the processor, which invokes the keyboard IH, a routine. The handler then reads and stores the scan code in the main memory. Because this is an interrupt-driven transfer, the IH does not need to poll the OBF status bit; the *intr* = 1 is an indication that the "Scancode" buffer is full. When an input port read cycle starts (see Fig. 9.11), a 1-0 *_rd* transition clears the INTR bit, making *intr* = 0. When the processor reads the scan code in the "Scancode" buffer, a 0-1 *_rd* transition clears the OBF bit, making *_obf* = 1, which indicates the buffer is now empty. The DCI can now write another scan code in the "Scancode" buffer.

9.5.2 Programmed Transfer

The programmed transfer, also known as **programmed I/O**, is not interrupt driven; instead, the processor periodically executes a polling program, triggered by a timer module, and checks the I/O port status bits in each DCI to determine if any device requires a service. The polling is done in some priority order. The following program illustrates a programmed transfer where it checks the OBF and IBF flags of each device. The program either inputs from the port if *ibf* = 1 (input buffer full) or outputs to the port if *_obf* = 1 (output buffer is empty). In the program, this is shown for keyboard Port 0 in Fig. 9.17. Program code sections for other devices are labeled DCI_X, DCI_Y, etc.

A Polling Program For Implementing Programmed I/O:

```
POLL:       ...                //start polling programmed I/O DCIs
            ...
DCI_KB:     LD (Port_1)        //input keyboard status
            AND 1              //select status bit 0, _obf;
                               //bit 1 is ibf
            CMP 0              //is _obf = 0?0 indicates full.
            JNE KIBF           //if output buffer not full, check
                               //keyboard IBF
            LD (Port_0)        //else, output buffer full, read
                               //"Scancode" port, assuming
                               //memory-mapped I/O port
            ST ...             //store the scan code in memory
KIBF:       LD (Port_1)        //check keyboard IBF flag
            AND 2
            CMP 2
            JEQ DCI_X          //if input buffer not empty, check
                               //the DCI of another device X
            ...                //else, input buffer empty, read data
                               //from memory and write to "Data"
                               //port
DCI_X:      ...
            ...
DCI_Y:      ...
            ...
```

In this case, the processor not only is involved in the actual transfer of data between the devices and memory, but also is directly involved in the frequent polling of each device. A programmed transfer is only efficient when there are a large number of devices in the system. For example, when using a computer to monitor many sensors in a factory, if interrupt-driven I/O is used, this would cause frequent processor interruption. At the same time, a programmed transfer wastes valuable processor time. However, programmed transfer can be offloaded and performed by a host controller interface, such as a USB host controller interface, which would interrupt the processor when it needs services.

For example, consider a modern personal computer system with several different types of USB peripheral devices, such as digital speakers, digital phone, digital camera, digital fax machine, removable flash memory, etc., as illustrated in

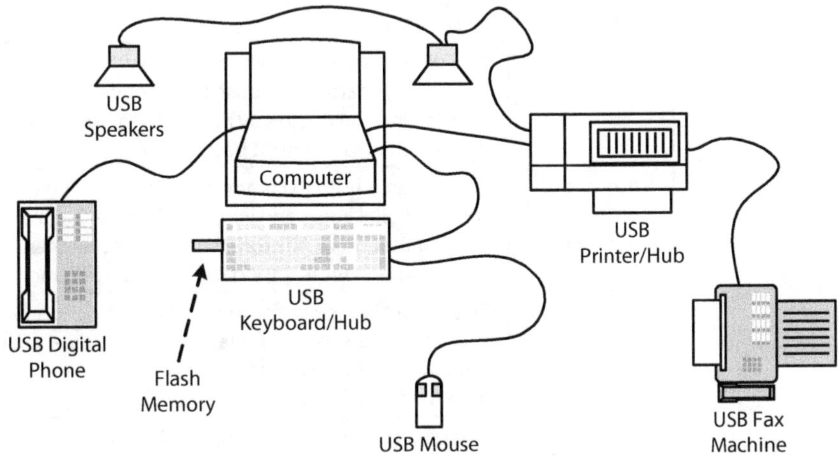

FIGURE 9.18 A modern personal computer with several connected USB devices [4].

Fig. 9.18. The list of USB interfaced devices is growing every day. Currently, a single USB host controller interface can poll 127 USB devices.

A USB host controller communicates with each device using **packets,** which is a collection of several data fields, each containing a piece of information. Figure 9.19 illustrates one way the USB devices in Fig. 9.18 may be interfaced with a single USB host controller interface that includes a root hub with three ports. Also, as illustrated, USB hubs in the printer are used to interface with USB speakers

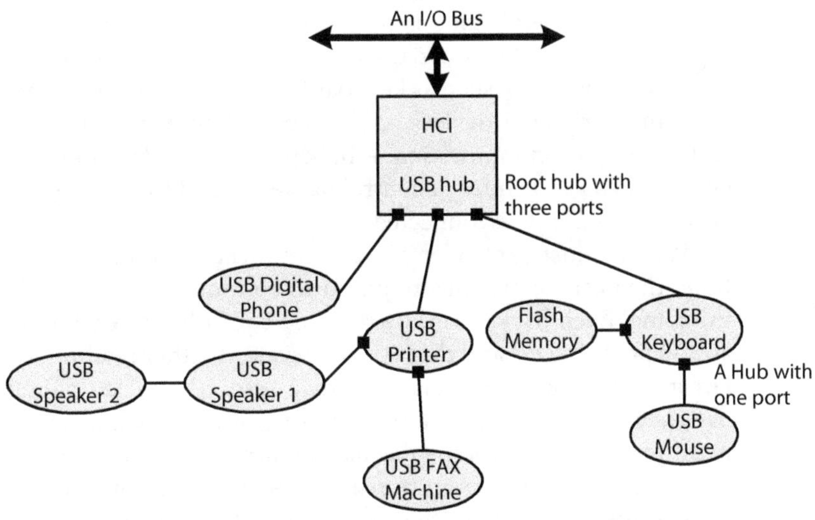

FIGURE 9.19 USB device connections to a single USB host controller interface with three root hubs.

Packet Class	Description
Interrupt	This packet is used to communicate with peripheral devices such as keyboard and mouse that require interrupt-driven transfer. Instead of directly interrupting the processor, the device sends as Interrupt packet to its HCI.
Isochronous	This packet is used to communicate with peripheral devices such as digital audio and video devices that require real-time communication with the system. In these devices the rate of data transfer is more important than data accuracy.
Bulk	This packet is used to communicate with peripheral devices such as printers to transfer large blocks of data such as a print job. In this case, data accuracy is more important than the rate of transfer.
Control	This packet is used to monitor the USB hub for setup as devices are connected or disconnected.

Table 9.4 Four Packet Classes Used with USB Devices

and a USB fax machine, and the hubs in the keyboard are used to interface a flash memory and the mouse.

The services required by all the connected USB devices are classified into four priority packet classes, as outlined in Table 9.4. Periodically, packets for all the connected devices are made into a frame and communicated serially from a USB host controller interface and through a USB root hub, USB hubs, and USB ports and USB cables to the USB devices; or vice versa from the USB ports and USB cables, through USB hubs and a USB root hub, to the USB host controller interface.

Frame communication is frequent enough (e.g., every few microseconds) to capture all the pending interrupt requests for timely service of devices that require interrupt driven I/O, such as keyboard and mouse, and also transfer data to/from real-time peripheral devices, such as digital speakers and phones, that require real-time data communication. A bulk packet is used to communicate with slower devices such as a printer. A control packet is used to poll the USB hubs as devices are connected and disconnected.

While a host controller can communicate with several devices at the same time, the basic operations required to communicate with each device are still the same. Each packet sent by the host controller to each device must contain an I/O port address and the type of service as either reading or writing the port. For a read packet, the device sends a response packet containing the device data (e.g., scan codes) back to the host controller, which stores the data in the main memory. For a write packet, the host controller must first access device data (e.g., print data) from the main memory and send the data to the device.

Because the size of the data communicated between a host controller and a device is in the order of many bytes at a time, both the host controller and the DCI of each device contain memory as buffer space to temporarily store the received or transmitted data. A USB host controller interface is described in more detail in Sec. 9.7.

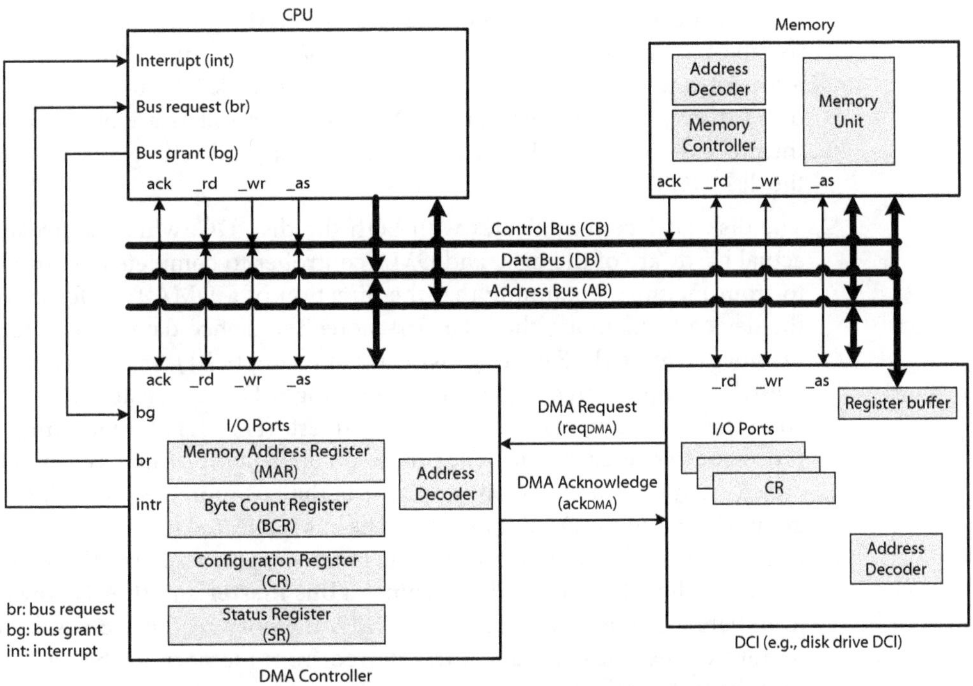

FIGURE 9.20 A simple DMA controller to illustrate DMA transfer; assumed memory-mapped I/O ports.

9.5.3 DMA Transfer

Interrupt-driven or programmed transfer is efficient if data transferred between devices and memory is small (a few bytes). For devices such as disk drives that require large (e.g., 4 KB) data transfers to/from memory, a DMA transfer, which does not involve the processor in the actual transfer of data, is more efficient. In order to illustrate a DMA transfer, consider the simple system organization shown in Fig. 9.20.

Both the processor and the DMA controller need to access the memory. In general, a module, known as an **arbitrator**, is needed to grant access to the shared bus. However, in a simple system organization like the one shown in the figure, often, the processor is also the arbitrator. The following steps explain a DMA transfer:

1. The operating system (OS), via the processor, initiates a DMA transfer by first writing the I/O ports of both the DMA controller and the DCI. Specifically, the processor (via a device driver) writes the memory address register (MAR) port with a starting memory address and the byte count register (BCR) port with a number of bytes to be transferred. The processor also writes the configuration register (CR) port to indicate both the direction of data transfer (to or from memory) and type of DMA as either **continuous**—for example, when starting up a computer—or **noncontinuous**—for

example, during normal system operation when the processor also needs to use the bus between each DMA transfer to/from memory. The processor also writes to the I/O ports in the, for example, disk DCI to configure the disk for a read or write operation. Once both controllers are programmed, the processor starts the DMA transfer by writing and setting a "start" bit in the disk DCI.

2. The disk DCI communicates with both the disk DC, which controls the actual hardware of the disk, and DMA controller to complete data transfer to/from memory. Assuming that the direction of a DMA transfer is from the device to memory, the DCI first stores the device data in its internal memory (refer to the Samsung disk drive in Example 9.1), and then it asserts the DMA request (req_{DMA}) signal requesting a DMA transfer. The DMA controller, in turn, asserts the bus request (*br*) signal requesting the processor to release the bus. The processor, upon completing its current bus cycle (if any), releases the bus and asserts the bus grant (*bg*) signal, which grants the bus to the DMA controller. The bus is released when the processor disables all its bus-connecting tri-state buffers. Upon receiving the asserted signal *bg*, the DMA controller becomes a **bus master** and thus is now able to initiate a memory write cycle. A DMA memory read cycle is similarly performed, except that data is transferred from memory to the DCI, and then from there the data is sent to the DC.
3. As a bus master, the DMA controller starts a memory write cycle using the content of MAR as the next memory address. It also asserts the DMA acknowledge (ack_{DMA}) signal for the DCI to place the data on the data bus. MAR is incremented and BCR is decremented after each bus transfer. For a memory read cycle, the ack_{DMA} is asserted when the memory data is on the bus (e.g., Fig. 9.7) and is ready to be transferred to disk DCI.
4. Once the DMA controller completes one transfer, it does one of two things: it either transfers the next data item (repeating step 3) if the transfer type is continuous, or it releases the bus if the transfer type is noncontinuous. The bus is released when the DMA controller deasserts *br*. Another DMA transfer can begin starting at step 2. Once all the data bytes are transferred and BCR becomes zero, the DMA controller interrupts the processor to initiate another DMA transfer if needed.

A modern DMA controller, however, may provide service to multiple devices by implementing multiple **DMA channels**, each equipped with its own I/O ports that can be configured to service a different device. A modern DMA controller may also include ports to minimize its communication with the processor and improve performance. For example, instead of requiring a processor to initiate a DMA transfer each time, the processor creates a **DMA transfer table** (or a linked list) in memory and passes its address to the DMA controller. Each of the table entries contains the information necessary to initiate a separate DMA transfer. When one transfer is completed, the DMA controller automatically fetches the information for the next DMA transfer from memory without interrupting the processor. The DMA controller only interrupts the processor when it has processed the entire table.

Alternatively, with some modern DMA controllers, the processor may also configure the controller to interrupt each time it has processed several DMA transfers. In this case, the processor is able to check the status and update the transfer table, for example, by adding more DMA transfers, if necessary.

A modern multichannel DMA controller may also need to operate differently if it is used to transfer data between two memory units (i.e., memory-memory DMA). A memory read cycle from one unit may be followed by a memory write cycle to another unit. Finally, each device may include its own dedicated DMA controller instead of sharing a multichannel DMA controller. In this case, several DMA controllers would compete to communicate with the main memory.

As we will see in Sec. 9.8, a USB host controller interface contains two DMA controllers. The controller uses one DMA controller to transfer data between the host interface and the main memory, and a second DMA controller is used to transfer data between the host interface and each of the connecting devices.

9.6 Interrupts

Interruptions are typically classified into **hardware interrupts**, caused by hardware modules internal or external to the processor, and **software interrupts**, caused by executing a special instruction, such as "INT" that invokes a system-level routine. Internal hardware interrupts, also sometimes called **synchronous interrupts**, **exceptions**, or **traps**, are due to an error, such as arithmetic overflow, divide by zero, or invalid op-code, that occurs within the CPU data path. They are called traps because they are instruction dependent. If the execution of a program results, for example, in an arithmetic overflow at a specific instruction—an "ADD" instruction at memory address X—the overflow will occur at exactly same instruction no matter how many times one runs the program (assuming that interruption on arithmetic overflow is enabled and input to the program is the same).

Other internal hardware interruptions, however, may not be synchronous. For example, consider a **multiprogramming execution environment** where programs are often too big to fit in the main memory. This creates a scenario in which a target instruction or data is not in main memory during the execution of a program. This results in an interruption, commonly known as a **page fault**. The program code or data page (e.g., 4 KB) must be copied on demand from a disk drive to main memory using a DMA transfer before the execution of the program can resume. The timing of the page fault may not be synchronous as pages from different programs move in and out of memory. Page faults will be discussed further in Chap. 10.

[In a multiprogramming execution environment, interruption of any kind stops the execution of the currently running program (called a process). In the case of a page fault interruption, the OS initiates a disk-to-memory DMA transfer and places the ID of the interrupted process (e.g., process-1) in a **wait queue**. While the page is being transferred, the OS starts or resumes the execution of another process. When the DMA transfer completes, the DMA controller interrupts the currently executing process (e.g., process-2), returning control to OS, which then moves the process-1 ID from the wait queue and places it in a ready-to-execute queue called a **ready queue**.

Processes (including process-1) that are in the ready queue take turns and execute for a fraction of processor time called a **time slice**. A timer module controls the duration of the time slice and causes an interruption when the current time slice expires. Each process may use one or more time slices to complete execution. Upon a time slice interruption, the OS places the ID of the interrupted process on the ready queue and assigns the processor to the process with its ID at the head of the queue. Process-1 (like other processes in the ready queue) resumes execution when its turn arrives and continues executing until there is another interruption, including another page fault, a time slice interruption, etc.]

Hardware interrupts, such as a DMA interrupt and a DCI interrupt, that are not internal to the processor, are called **asynchronous interrupts**. These interrupts can happen at random and at any time during the execution of an arbitrary program. For example, when a system uses interrupt-driven I/O to service the mouse or keyboard, the exact moment when mouse is moved or a key is pressed is not known.

The Intel "INT" and the ARM "SWI" are two examples of software interrupt instructions. Software interrupts are always synchronous, and they are used for many purposes, including allowing programs to take turn and access shared resources, such as a disk drive for reading and writing files. Here, we use the term interrupts to mean all types of interrupts.

9.6.1 Handling Interruptions

Figure 9.21 illustrates the difference between invoking a hardware interrupt handling (IH) routine and a subroutine call (i.e., function, procedure, or method). A subroutine is called by another routine when the routine executes a jump/branch subroutine instruction, such as "JSUB sort," that includes the starting address of subroutine sort. The starting and ending addresses of subroutine or an IH are labeled B and C in the figure. Because an instruction starts the execution of a subroutine, the call is illustrated, by solid arrows in Fig. 9.21(*a*), from A to B and subroutine return from C to D. After executing a subroutine, control is returned to the calling routine where its execution resumes starting at address D.

The execution of a hardware IH, on the other hand, is not started by an instruction; thus, this is illustrated using dashed arrows in Fig. 9.21(*b*). In this case, the CPU must first find out the cause of interruption and from that it determines which IH to invoke. For example, if the cause of the interruption is because the mouse was moved, the mouse IH is invoked.

After executing a current instruction, the CPU checks for pending requests for interruptions, such as checking whether or not the *intr* signal in Fig. 9.17 is asserted. For example, if the CPU finds the *intr* signal associated with the mouse is asserted, the CPU invokes the mouse IH. The IH then communicates with the

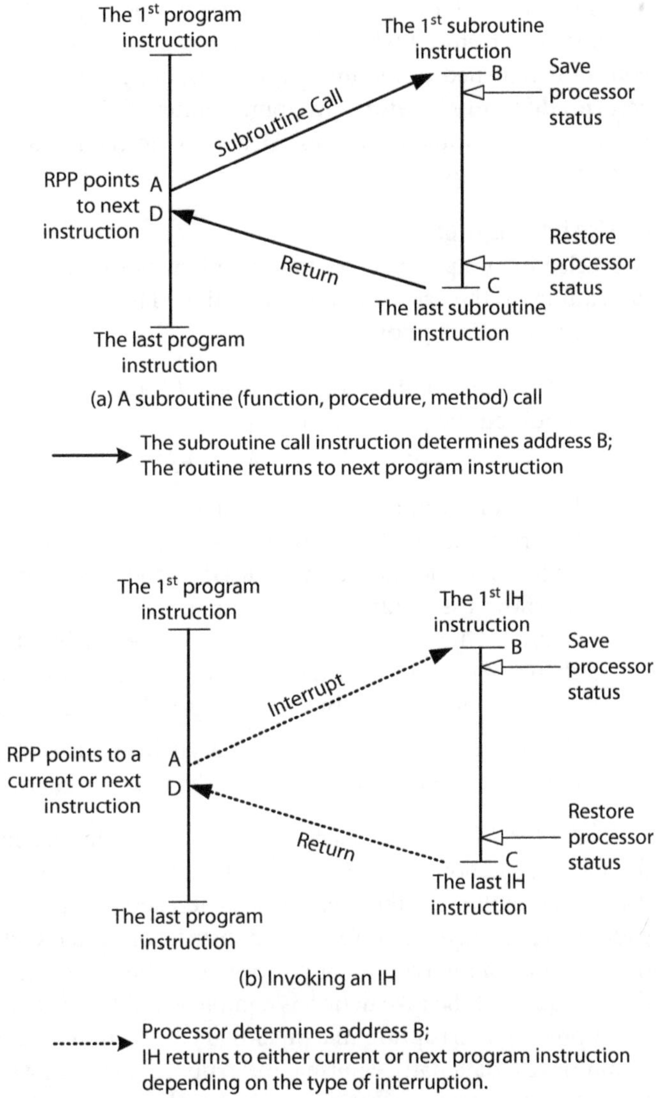

Figure 9.21 An illustration of subroutine call verses interrupt handler (IH); solid arrows verses dashed arrows.

mouse DCI, accessing its ports to input the mouse displacement information, which the IH uses to move the cursor on the screen.

The software interruption handlers are typically numbered with integer numbers. For example, Pentium instruction "INT 3," which stands for breakpoint exception, is used by debugger tools to create breakpoints in a program during testing [5]. When the program execution reaches a breakpoint, the IH is invoked that breaks the execution of the program under test.

When an IH is invoked or a subroutine call is made, the state of CPU, for example, the contents of ACC, X, SR, and program pointer (PP) in Fig. 8.7 (Chap. 8) are saved within CPU or typically in main memory by the IH or subroutine. The saved state is restored when the execution of the IH or subroutine completes and control is returned to the interrupted or calling routine. The saved PP is called a **return address** or a return program pointer (*RPP*). In the figures, the *RPP* for a subroutine call would be address D, and for an IH, it would be either address A or D, as discussed next.

Precise Interruption

A precise interruption refers to a set of rules that must be carried out by CPU, depending on the cause of an interruption. The following is a list of rules applied for a precise interruption:

1. All the instructions prior to the *RPP* must have been executed and have modified the state of CPU.
2. All the instructions starting at the *RPP* must not have been executed.
3. If the interruption is due to a synchronous event, except for software interrupts, the *RPP* must point to the instruction that caused the exception. In addition, if the exception is raised by the execution unit, such as arithmetic overflow, the result of the execution may not change the processor state. For example, the content of the ACC register in Fig. 8.7 (Chap. 8) should not change if an "ADD" instruction causes arithmetic overflow, so the content of ACC can be analyzed by the corresponding IH.
4. If the interruption is due to an asynchronous (external) event or software interrupt, the *RPP* must point to the next instruction.

If the CPU data path is single-cycle, the checking for interruption is performed during the execution of a current instruction, as will be illustrated in Sec. 9.7. If the CPU data path is multicycle, the checking for interruption is performed during the last data path operation required to retire each instruction. If the CPU data path is pipelined, the checking for interruption is performed in the write-back stage, where it indicates the execution of the instruction is complete and the instruction is about to retire.

A precise interruption in a pipelined CPU is more complex than a single-cycle or multicycle data path. Multiple interrupt requests may be generated at the same time as stages are operating concurrently in the pipeline. In this case, the instruction address (i.e., the content of PP) is also forwarded from the fetch stage to the decode stage along with the fetched instruction. The instruction address and any interrupt request generated in each stage are forwarded from one stage to the next until they reach the write-back stage.

Upon an interruption, the *RPP* is saved internally in a register as part of the CPU state that is saved in main memory, and the PP is changed to execute an IH. If the CPU data path is pipelined, this will also cause a pipeline flush. Note that the registers would not be cleared and would be left as is so they can be saved in main memory by the IH. As was discussed earlier, the state may also be saved internally within CPU (see Exercises section). The details of a precise interruption in a pipelined CPU are referred to elsewhere.

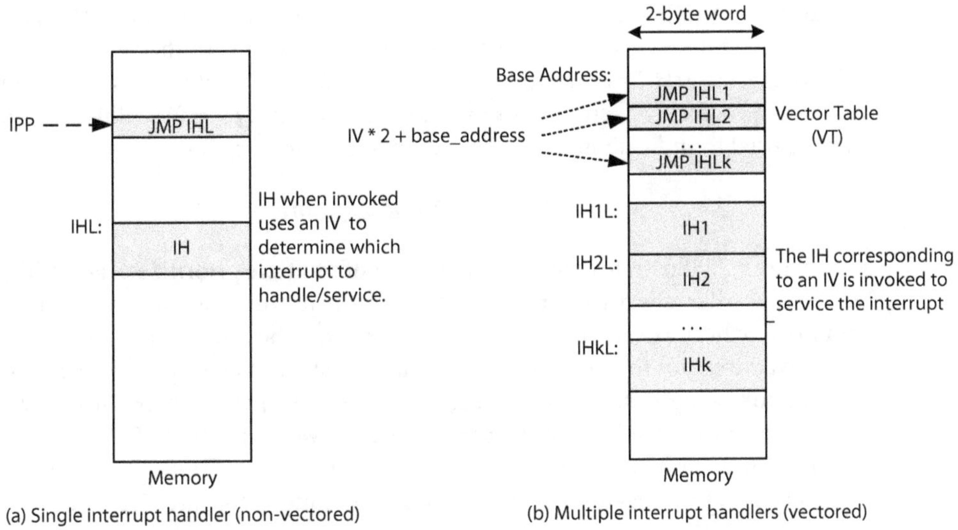

Figure 9.22 Handling of vectored interrupts; assuming 2 bytes per word memory.

Vectored Interrupts

Vectored interrupts refers to several IHs in the system. **Nonvectored interrupts**, on the other hand, means there is only one IH that handles all interruptions. Figure 9.22 illustrates the organization of a single IH and multiple IHs in memory. In the case of the nonvectored interrupts, the single IH is invoked each time there is an interruption. The handler then determines the highest-priority interrupting device to service. A nonvectored interrupt mechanism is simple to implement, but because all interruptions are handled by a single routine, interrupts are not handled quickly; thus, a nonvectored interrupt is not common in modern computer systems. However, as discussed in Sec. 9.7, a nonvectored interrupt mechanism is used for simplicity to illustrate the data path details of an interrupt handling CPU.

The following outlines the requirements to implement vectored interrupts:

- A small region in main memory must be reserved for a **vector table**, starting at a known address labeled "Base Address" as illustrated in Fig. 9.22(*b*).
- The interrupt requests from each DCI and DMA controllers must be prioritized in hardware to generate a unique number, called an interrupt vector (*IV*), associated with the highest-priority IH. The *IV* is used to invoke a corresponding handler.
- During system startup, the entries of vector table must be filled with jump instructions to different IHs, one for each *IV*.
- The processor data path must be extended by adding new registers and circuits to execute new instructions and features as necessary to implement vectored interruptions.

Equation (9.2) is an example of how an interrupt handler pointer (*IHP*) can be defined as a linear function of an *IV* and the *base_address* of the vector table. For *IV* = 1, 2, 3, etc., *IHP* = *base_address* + 2, *base_address* + 4, etc., is used as an index to access a vector table entry assuming each entry is 2B. *IV* = 0, as illustrated later, may be used to indicate no pending requests for interruption.

$$IHP = IV * 2 + base\_address \tag{9.2}$$

Figure 9.23 illustrates the execution order of multiple prioritized IHs. A higher-priority handler always interrupts a lower-priority one, but not the other way around. Furthermore, with each handler invocation, the processor status, including a return address for the interrupted program, must be saved (typically in memory).

The following describes the steps CPU takes to invoke an IH; it is assumed that IVs are prioritized so that IH1 (*IV* = 1) has lower priority to IH2 (*IV* = 2), and IH2 has lower priority to IH3 (*IV* = 3), etc., as shown in Fig. 9.23. Furthermore, *IV* = 0 is used to indicate there are no pending requests for interruption:

1. Processor ignores a receiving IV (labeled *IVr*) if the CPU's checking for interrupt requests feature is disabled or the IV of the currently executing IH (labeled *IVc*) is greater than *IVr*. That is, when *IVr* ≤ *IVc*, the CPU ignores *IVr*, keeping the request for an interruption pending until the execution of the current handler associated with *IVc* ends. Otherwise, if the CPU's checking for interrupt requests feature is enabled and *IVr* > *IVc*, CPU saves both the *RPP* for the currently executing IH and the *IVc* in special registers within the CPU and changes the content of PP with the *IHP* = *IV* * 2 + *base_address* and makes *IVc* = *IVr*. For example, if *IVc* = 1 and *IVr* = 2, then the CPU saves *RPP1* for IH1 and *IVc* = 1 and replaces the content of PP with the quantity *base_address+4* and makes *IVc* = *IV*2.
2. The CPU starts executing the jump instruction at memory address *IHP* and thus begins the execution of the IH corresponding to the current *IVc* (e.g., IH2 when *IVc* = 2).

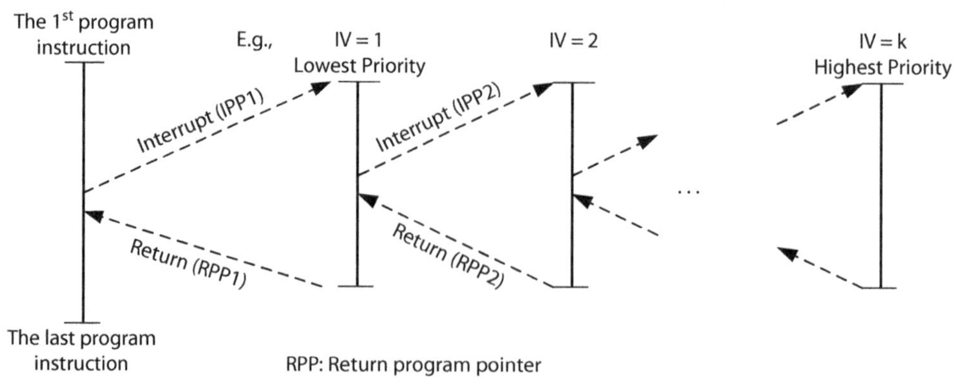

FIGURE 9.23 Invocation of a multiple interrupts handlers.vsd

3. The invoked handler must first save the processor state, including the saved *RPP* and saved *IVc* in memory (see Sec. 9.6.2). The CPU, however, continues monitoring the *IVr* for the next higher-priority IV.
4. The IH services the interrupting device—for example, a DCI or DMA controller. Upon completion, the handler restores the processor saved state, including the saved *RPP* and save *IVc*, and returns to the interrupted program so the execution of the program can resume, which could be a lower-priority IH (e.g., IH1) or a systems or application program.

9.6.2 Interrupt Structures

Figure 9.24 illustrates two hardware interrupt structures, **daisy-chained** and **independent request**, for prioritizing interrupt requests. They are used to determine the IV of the interrupting device. The IV, in turn, is used to determine which IH to invoke to service the device. The details of these structures are explained next.

Daisy-Chained

The structure in Fig. 9.24(*a*) orders all the interrupt requests into fixed priorities. When the signal *int* is asserted, the CPU asserts an interrupt acknowledge (*iack*) signal to identify the highest-priority IV.

A detailed circuit of a daisy-chain module is shown in Fig. 9.24(*b*). The *iack* signal is forwarded from one daisy-chain module to the next until the highest-priority IV (at that moment) as *IVr* is selected. A unique IV typically is assigned to each DCI during setup. With the USB devices, only the USB host controller interface interrupts the processor.

While the structure in Fig. 9.24(*a*) is scalable and more devices can be added to the system, it has the disadvantage of resulting in starvation of DCIs that are at the end of a long daisy-chain. A higher-priority DCI can prevent *iack* = 1 from reaching the lower-priority daisy-chain modules at the end of the chain.

Independent Request

The structure in Fig. 9.24(*c*) uses a priority encoder to quickly identify the highest-priority IV. In the figure, the 8-to-3 priority encoder is used as a 7-to-3 encoder by connecting its input-0 signal to ground. In this case, *IVr* = 0 when none of the seven interrupt request signals are active. Furthermore, *IVr* = 0 is used to indicate to CPU there are no pending requests for interruption. This would eliminate the *int* signal in Fig. 9.24(*a*), reducing one signal that the CPU inputs.

The independent request structure requires no *iack* signal, unless a hybrid—partly independent and partly daisy-chained—structure is used to implement **interrupt priority classes**. All the DCIs in a system would be grouped into different priority classes. For example, all the disk drives (e.g., hard disk and CD disk drives) may be considered part of a device class. In this case, an *IVr* identifies all the DCIs in the same class (i.e., all the disk drives), and an *iack* would then be used to identify the *IVr* of a specific device (e.g., hard disk drive) in that class.

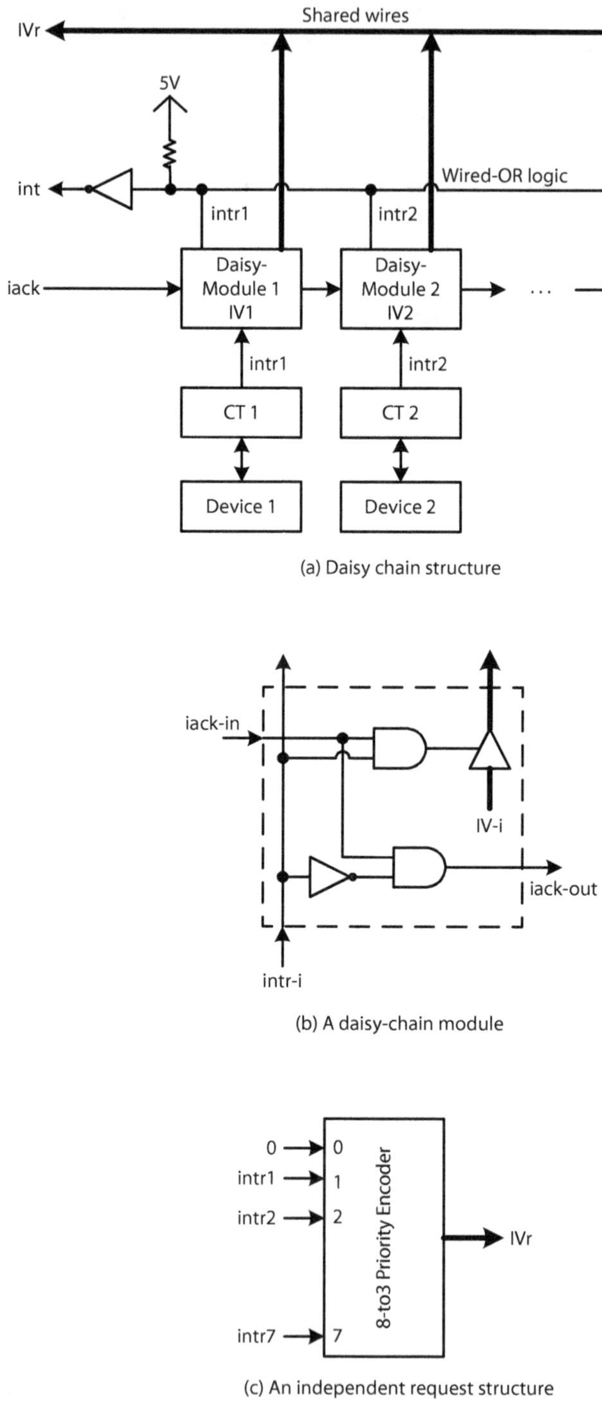

Figure 9.24 Interrupt structures: (*a*) daisy-chained structure; (*b*) daisy chain module; (*c*) independent request structure.

9.7 Design Example: Interrupt Handling CPU

This section presents the data path of an interrupt handling CPU with the following requirements and specifications:

- The data path of the single-cycle Acc-ISA CPU in Fig. 8.7 (Chap. 8) is extended to implement the invocation mechanism of a single IH that provides service to several devices.
- The *IHP* is hardcoded to memory address 0x40 within the CPU, assuming 8-bit addresses. Here, address 0x40 is arbitrarily selected.
- The IH can service a maximum of 16 devices, including a timer module, a DMA controller, and 1 to 14 I/O devices.
- A daisy-chained structure (Fig. 9.29(*a*)) is used to prioritize the interrupt requests from the 16 devices, where $IV = 15$ (the highest) is assigned to the timer module, $IV = 14$ to the DMA controller, and 0 (the lowest) to keyboard. Note that because we are using a daisy-chained interrupt structure, we may also use $IV = 0$ as a valid device IV.
- Upon an interruption, the state of the CPU, defined by the content of registers ACC, X, and SR, is saved in memory by the IH. The *the return address (RPP)* is saved internally within the CPU.

Figure 9.25 illustrates a 16-level daisy-chained interrupt structure that prioritizes a timer, DMA, and 1 to 14 I/O interrupt requests (only one, keyboard,

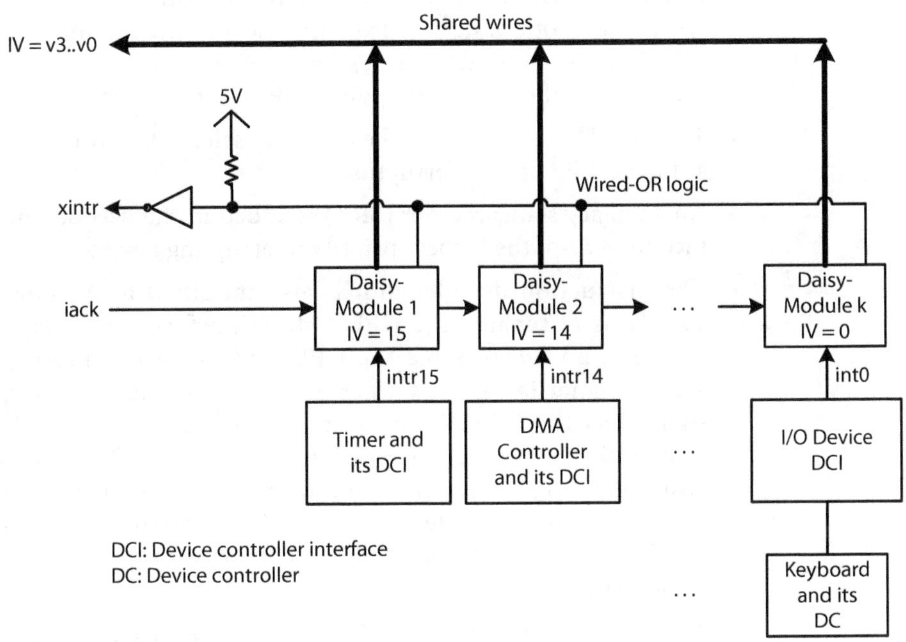

FIGURE 9.25 A 16-level daisy-chained interrupt structure that prioritizes interrupt requests from 16 devices from highest ($IV = 15$) to lowest ($IV = 0$).

is shown). The DCI of each device generates an interrupt request (*intr*) signal, which becomes 1 each time the device wants service from the CPU. For example, a key depressed on the keyboard would set its interrupt request signal as *intr0* (also see Fig. 9.17), causing the CPU to interrupt the currently executing program and start executing the IH to service the keyboard.

Likewise, a DMA controller would assert its respective interrupt request signal as *intr14* (also see Fig. 9.20) when it completes a DMA transfer. The timer module would interrupt, making *intr15* = 1, so the OS can start or resume the execution of another program. An I/O DCI can be device specific or universal, such as a USB host controller interface.

The *intr* signals from the timer module, DMA controller, and maximum of 14 DCIs are wired-ORed to create the external interrupt request signal *xint* as shown in Fig. 9.25. The *xint* signal, which enters the CPU, causes an interruption if asserted and if the CPU's interrupt handling feature is enabled. The data path of the interrupt handling CPU is presented in Fig. 9.26.

The new data path implements eight new instructions, listed in Table 9.5, and also includes added hardware as follows:

- A D flip-flop (FF) is used to enable or disable the CPU's checking for interrupts feature. The FF will be set to 1 (i.e., FF.q = 1) when the "EI" or "RTI" instruction executes.
- An auxiliary program pointer register (PPX) is added to save the *RPP*, which in this case, is the address of the next instruction in the currently running program. Upon returning from an interruption, the program resumes execution, starting from the address saved in the PPX. The added 2-to-1 MUX7 in the fetch unit selects *PPX* (the content of register PPX) as the next *PP* (the content of register PP) when instruction "RTI" executes.
- The added 2-to-1 MUX5 in the fetch unit selects the hard coded *IHP* = 0x40 as the next PP upon interruption.
- An auxiliary status register (SRX) is added in the decode unit to save the incoming *IV* of the highest-priority interrupting device.
- The instruction decoder unit is also modified to include three more signals as *ei* (enable the CPU's checking for interrupts feature), *rstr* (restore), and *di* (disable the CPU's checking for interrupts feature). The *ei* signal is asserted when the "EI" instruction executes. The *ei* signal sets the FF (FF.q = 1) on the next clock cycle, which in turn enables the CPU's checking for interrupts feature. The *rstr* signal, which also sets the FF to 1, is asserted when the "RTI" instruction executes. The *di* signal is asserted when the "DI" instruction executes, which synchronously resets the FF (FF.q = 0) and disables the CPU's checking for interrupts feature.
- The write-back multiplexer (MUX4) is also replaced with an 8-to-1 MUX to additionally implement ACC ← SRX, ACC ← SR, and ACC ← X, which are required for the IH to save the CPU state in memory.

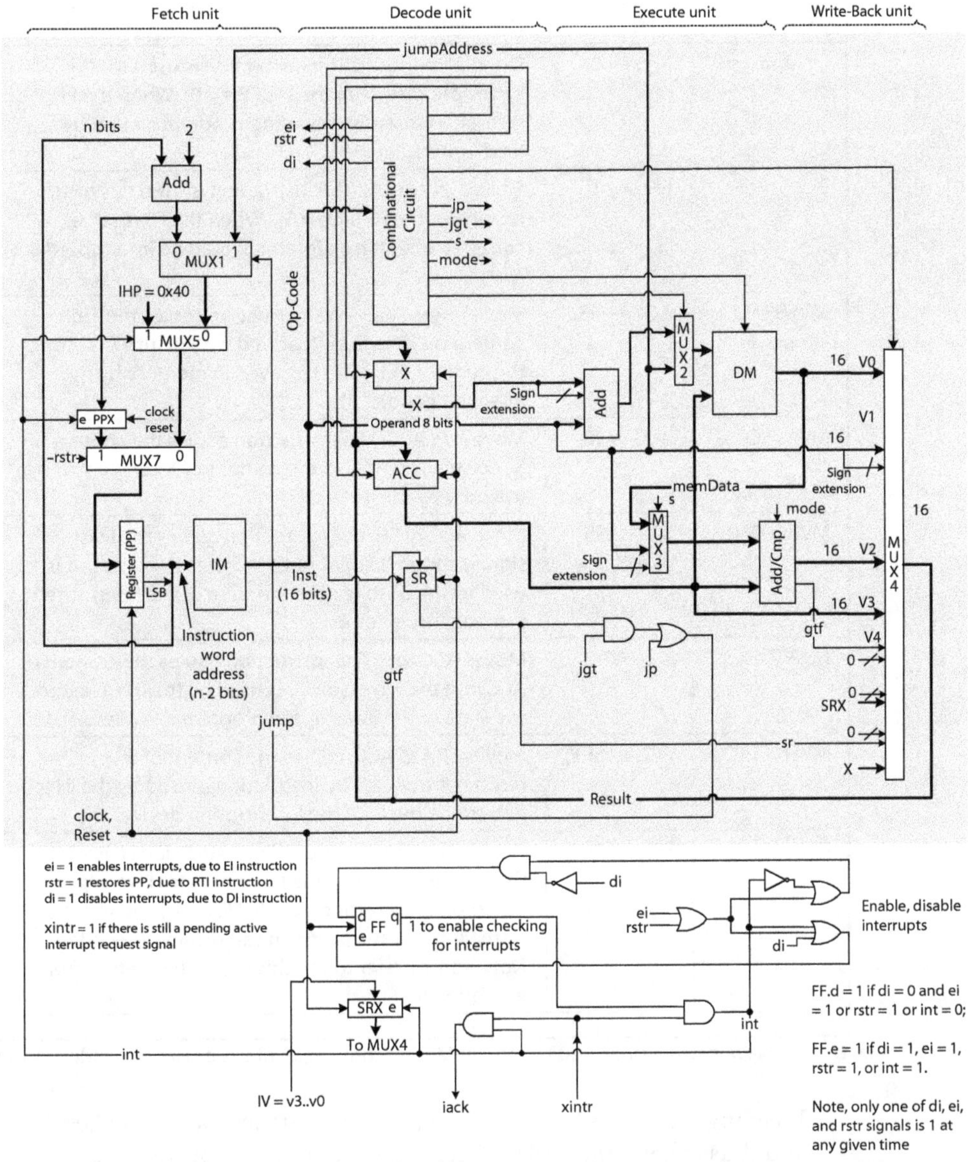

FIGURE 9.26 The data path of the interrupt handling CPU, which is an extension to the Acc-ISA discussed in Chap. 8.

Initially, upon reset, the FF would be set to 0 (FF.q = 0), disabling the CPU's checking for interrupts feature. FF.q remains 0 until the system starts and the operating system initializes the system by:

- Loading the single IH into memory starting, for example, at location 0x42
- Storing the instruction "JMP 0x42" at location *IHP* = 0x40

Instruction	RTN	Description
DI (optional)	FF.$q \leftarrow 0$;	Disable interrupts: The instruction is used to synchronously reset the FF (FF.$q = 0$). When the FF is reset, a currently executing program cannot be interrupted.
EI	FF.$q \leftarrow 1$;	Enable interrupts: The instruction is used to synchronously set the FF (FF.$q = 1$). When the FF is set, a currently executing program would be interrupted if $xint = 1$.
MVSR2ACC	ACC ← SR;	Moves SR to ACC: The instruction copies the 1-bit content of SR register (padded with 0) into ACC so the content of SR can be saved in memory by the interrupt handler (IH).
MVX2ACC	ACC ← X;	Move X to ACC: The instruction copies the content of X register into ACC so the content of X can be saved in memory by the IH.
MVACC2SR	SR ← LSB of ACC;	Move ACC to SR: The instruction copies the least significant bit (LSB) of ACC into the SR register. It is used by the IH to restore the state of the interrupted program upon return.
MVACC2X	X ← ACC;	Moves ACC to X: The instruction copies the content of ACC into the X register. It is used by the IH to restore the state of the interrupted program upon return.
MVSRX2ACC	ACC ← SRX;	Moves SRX to ACC, copies the content of SRX register into ACC. The instruction is used by the IH to determine the IV of the interrupting device.
RTI	FF.$q \leftarrow 1$, PP ← PPX;	Return from interruption: The instruction is used as the last instruction in the IH. It enables interrupts, synchronously setting the FF (FF.$q = 1$), and returns so the interrupted program can resume execution. Note, some prefer to call this instruction return from exception (RTE).

Table 9.5 New Instructions Required to Implement the Single Interrupt Handling Mechanism

- Enabling interrupts by executing an "EI" instruction, which makes FF.$q = 1$, and thus enables the CPU's checking for interrupts feature

During normal system operation, one or more interrupt request (*intr*) signals (Fig. 9.25) may be asserted. Each time that *xintr* becomes 1 and checking for interrupts is enabled (i.e., FF.$q = 1$), the *int* signal within the CPU becomes 1 and causes the following operations to take place during the next clock cycle:

- Via MUX7, $int = 1$ will cause data path to perform PPX ← PP + 2. This saves the quantity *PP* + 2 as the *RPP*. The interrupted program will resume execution (after returning from an interruption) starting from instruction address at *RPP* (see Fig. 9.21(b)).

- Via MUX5 and MUX7, *int* = 1 will cause data path to perform PP ← 0x40. This will cause the "JMP 0x42" instruction at memory location 0x40 to execute next, which will start the execution of the single IH.
- The *int* = 1 also causes SRX ← *IV* and thus saves the *IV* of the highest-priority interrupting device. Note that when both *int* = 1 and *xintr* = 1, *iack* becomes 1 and in turn selects the IV of the highest-priority device (Fig. 9.25) that is requesting service.
- The *int* = 1 also synchronously resets FF, making FF.*q* = 0. This disables the CPU's checking for interrupts feature. While FF.*q* = 0, *int* remains 0 and prevents a currently executing program, which could be an OS routine during system initialization or IH during normal operation, from interrupting.

The pseudo-code in Example 9.2 outlines the steps the single IH must take to provide service, one at a time, for an interrupting device among several devices. The handler performs four main tasks as follows:

1. The IH saves the state of the interrupted program, which is defined as the content of ACC, X, and SR registers, in memory. In the pseudo-code, this is shown by calling function `save_cpu_status()`, which can include the following code section:

```
STA Temp1;          // save ACC, for "STA" refer to Chap. 8
MVX2ACC
STA Temp2           // save X
MVSR2ACC
STA Temp3           // save SR
```

2. The IH must determine the IV of the interrupting device so it can call a driver routine to service the device and thus reset the device interrupt request signal (*intr*)—refer to Sec. 9.5.1 for the keyboard example. This is shown by "iv = get_iv()" and a "switch" statement in the pseudo-code. The "get_iv()" function would use the MVSRX2ACC instruction to copy the content of SRX into ACC, which will then be compared with 0, 1, 2, etc., to determine the IV of the interrupting device.
3. The interrupt handler must restore the state of the interrupted program before returning. This is shown by calling function `restore_cpu_status()`,which can include the following code section:

```
LDA Temp3;          // for "LDA" refer to Chap. 8
MVACC2SR            // restore SR
LDA Temp2
MVACC2X             // restore X
LDA Temp1           // restore ACC
```

4. The IH would then execute the "RTI" instruction, which will cause the data path to perform PP ← PPX and FF.*q* ← 1 on the next clock cycle. This

restores the return address, which resumes the execution of the interrupted program, and also enables the CPU's checking for interrupts feature.

Example 9.2. Describes a pseudo-code outlining the functions of a single IH that services an interrupting device, a timer module, DMA controller, or an I/O device one at a time:

```
interrupt_handler_routine() //invoked when PP is set to
                            //interrupt handler pointer (IHP =
                            //0x40 in Fig. 9.26)
{
save_cpu_status();          //save contents of ACC, X, and
                            //SR to memory

iv = get_iv();              //get IV from the SRX

switch(iv)

15: handle_timer_interrupt(); break;      //IV = 15, highest
14: handle_DMA_interrupt(); break;
. . .
0: handle_keyboard_interrupt();       //lowest
endswitch;
restore_cpu_status();       //restore contents of ACC, X,
                            //and SR from memory
return_from_intrrupt();     //RTI
}
```

If, upon returning from an interruption, the CPU finds *xintr* = 1, the handler will be invoked again to service another device that has its pending *intr* = 1.

Finally, the disable interrupts ("DI") instruction in Table 9.5 is included in case a "soft" reset—for example, using a "restart" menu option—is used. The disadvantage of a single IH mechanism is that if a higher-priority device requests a service while the CPU is executing the IH, the high-priority device must wait until the handler completes its task and returns. For this reason, modern computer systems implement a vectored interrupt mechanism so that the IH of a lower-priority device can be interrupted and the CPU can execute the IH of a higher-priority interrupting device (e.g., DMA) as was illustrated in Fig. 9.23.

9.8 USB Host Controller Interface

The need for a host controller interface was briefly discussed in Sec. 9.5.2. The USB host controller interface is designed to offload CPU from performing the task of directly providing services to potentially many I/O devices. In this case, the interrupt-driven transfer may not be a practical option because there could be many device interruptions. Programmed transfer would not be a viable option because it would waste valuable CPU time. In addition, the USB host controller interface implements "plug and play" device interface that can support numerous USB devices without requiring device installation and system restart.

9.8.1 Standards

The USB 1.x standard was designed according to the specification of either the universal host controller interface (UHCI) or the open host controller interface (OHCI) suitable for low-speed and full-speed devices requiring, respectively, 1.5 Mbps and 12 Mbps data transfer rates. The USB 2.0 standard, on the other hand, is based on the specification of the enhanced host controller interface (EHCI) and is designed for high-speed devices with a transfer rate of up to 480 Mbps. USB 3.0 is designed to support super-speed devices requiring transfer rates of up to 5 Gbps. Each generation of USB host controller interfaces also contains root hubs to service a slower device. For example, the USB 2.0 host interface contains root hubs to service low-speed and full-speed devices in addition to providing service to high-speed devices.

A USB cable consists of four wires, two of which are power and ground, and the other two are data signals labeled D+ and D–. A USB communication module uses the D+ and D– signals to send/receive packets with the non-return-to-zero inverted (NRZI) coding scheme (refer to the "Exercises" section in Chaps. 5 and 6 for information on NRZI). USB packets are grouped into **token**, **data**, and **handshake** as illustrated in Fig. 9.27. Each packet starts with a synchronization sequence and ends with an end-of-packet (EOP) marker. USB 1.x packets start with an 8-bit synchronization sequence (i.e., 00000001) and end with an EOP marker that keeps the D+ and D– signals at 0 for the duration of 2 bits. The USB 2.0 standard, on the other hand, uses a 32-bit (4-B) synchronization sequence and an 8-bit EOP marker.

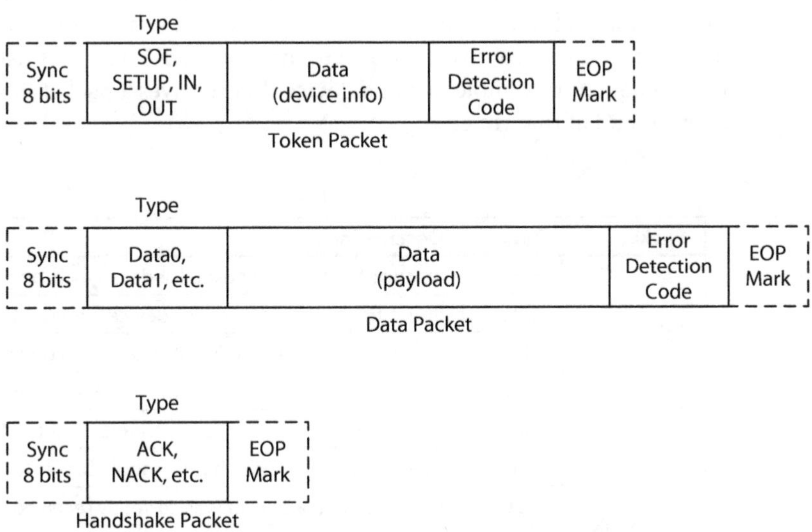

FIGURE 9.27 Three types of USB communication packets. Each packet starts with a synchronization sequence and ends with an EOP marker.

9.8.2 Transactions

Each USB transaction consists of one or more packets. For example, a transaction that sends print data to a printer, requires, in order, a token packet, a data packet, and a handshake packet.

As shown in Fig. 9.27, token and data packets each include a type field, a data field, and a field for an error detection code. Handshake packets have only a type field. A token packet identifies one of four possible transactions, called start-of-frame (SOF), setup (SETUP), input (IN), and output (OUT), as described in Table 9.6. A data packet contains a payload and is typed either as Data0 or Data1. The USB 2.0 has other data types, such as Data2, which is used to communicate the size of a large payload that would be transmitted via several transactions.

If a transaction contains an entire data payload from/to a device, its data packet would be typed as Data0. On the other hand, if two or more transactions are used to transfer a large payload, the consecutive data packets would be labeled alternating Data0 and Data1. Every source and destination module includes a toggle bit that is initially set to 0.

Each time that a Data0 or Data1 packet is sent by a source module and is received by a destination module, their toggle bits are toggled; if 0 (i.e., indicating Data0), it becomes 1 (i.e., expecting Data1 next), or if the tag was 1 (indicating Data1), it becomes 0 (expecting Data0 next). This ensures that a data packet at the destination is received only once in case of an error, such as when a destination module receives a packet correctly and toggles its toggle bit but the source thinks otherwise and keeps its toggle bit unchanged and retransmits the packet. In this case, since the two toggle bits are no longer the same, the receiver (correctly) rejects the retransmitted packet. Handshake packets serve multiple purposes. For example, an acknowledgement (ACK) packet is used to confirm the receipt of a data packet, or a negative acknowledgment (NACK) packet is used to control the flow of data, for example, when a device that receives a packet to send data has no data to send at that time.

Packet	Type	Description
Token	SOF	Start-of-frame packet used for data synchronization purposes
	SETUP	Setup transaction is used for device configuration
	IN	Input packet is used to input data from a device
	OUT	Output packet is used to output data to a device
Data	Data0	Used when the sender has its toggle bit set to 0
	Data1	Used when the sender has its toggle bit set to 1
Handshake	ACK	Indicates a data packet was correctly received
	NACK	Indicates device is currently unable to send or receive data

TABLE 9.6 Examples of USB Packet Identifiers

9.8.3 Transfers

The four types of data transfers, named **interrupt**, **isochronous**, **bulk**, and **control**, which were described in Table 9.4, are used to service all classes of USB devices. Also, as was discussed earlier, USB devices do not use interrupt-driven transfer. Instead, an interrupt packet is used to poll those devices, such as the keyboard and mouse, that normally use interrupt-driven transfer in legacy personal computers. In this case, a device, such as keyboard, will send a NACK handshake packet each time that it is polled but has no scan code to send.

Real-time USB devices, such as digital phones must be able to send or receive data and speakers must receive data frequently without interruptions when they are in use. These devices use isochronous transfers, and the corresponding transactions include token and data packets but no handshake packets; the data of these transactions is never retransmitted. Bulk transfers are used with devices such as printers, fax machines, scanners, and plotters that require accurate data communication but can tolerate interruptions in receiving data. Control transfers are used during a device or hub configuration.

Many devices may use a single host controller interface to send or receive packets. However, a single device cannot be allowed to communicate continuously and starve other devices, negatively affecting their normal operations. For example, consider the system shown in Fig. 9.18. Suppose a user wants to print a document while listening to music on USB speakers. Both the printer and speakers must share the host controller to receive print and music data, respectively. If the host controller is allowed to transfer print data continuously for a long time, the transfer of music data to the speakers will be interrupted, thus resulting in poor audio quality. Similarly, if the host controller is allowed to transfer music data continuously, the print job may never finish. In another scenario, a user may want to work on, say, a document at the same time that the printer is printing and the music is on. In this case, the keyboard and mouse must also be able to communicate with the system.

9.8.4 Descriptors

Each USB host controller interface implements one or more **access points**, each one called an **endpoint**. Each endpoint includes a set of I/O ports that are accessed by the host controller. Each endpoint contains a descriptor called an **endpoint descriptor** that identifies its transfer type—for example, control, interrupt, isochronous, or bulk—as well as other requirements, such as the maximum amount of data allowed in each data packet. The control, interrupt, and bulk transfers use a maximum data size of 64 B. Isochronous transfers, on the other hand, use a larger maximum data size (e.g., 1024 B in USB 2.0). A USB hub controller implements a **status change endpoint**, which is polled once every 255 ms to detect a possible port event, such as connecting or disconnecting a device from a USB port.

A device may contain multiple sets of endpoints to implement different types of interfaces. For example, a USB CD-ROM drive would implement three different interfaces: a mass storage interface for reading/writing files, an audio interface for handling music, and a video interface for handling video images. Each interface in a USB device includes an **interface descriptor** that includes a number of endpoint descriptors. The interface descriptors themselves are part of a **configuration descriptor**, which may consist of multiple different configurations, including, for example, a low-power configuration used with battery-powered systems. All these descriptors are organized in a hierarchy and included in a **device descriptor**, which also contains device information such as the manufacturer name and serial number. Each time that a newly plugged-in device is detected by polling the hub, the device's descriptor is accessed by a USB driver. The descriptor contains all the necessary device information, including a list of I/O port relative addresses within the device and the type of transfer used by the device. The driver assigns a unique address to the device and configures its endpoints.

9.8.5 Frames

A USB host controller interface is designed to service all the devices that are currently attached and active. It sends and receives packets at regular intervals called frames. For example, the USB 1.x uses 1 ms (millisecond) frame, and USB 2.0 uses 125 μs (microsecond) frames, each called a **microframe**. Each frame for full-speed USB 1.x devices is 12,000 bits (12 Mbits/s $*$ 1 ms $\approx$ 12,000 bits) long. A microframe is 60,000 bits (480 Mbits/s $*$ 125 μs $\approx$ 60,000 bits) long.

Assuming that the system shown in Fig. 9.18 is using a single USB host controller interface, Fig. 9.28 illustrates one possible frame content for the USB 1.x host controller. In the figure, each frame includes packets for several transactions and begins with an SOF transaction. For example, consider a scenario where a user is talking on a digital phone, working on a document, listening to music, printing a print job, sending/receiving a fax, and writing flash memory. In this case, a frame would contain packets for the following transactions as follows:

- Two isochronous audio IN and OUT transactions with the digital phone
- An isochronous stereo audio OUT transaction with the speakers
- An interrupt IN transaction with the keyboard
- An interrupt IN transaction with the mouse
- Zero or more control transactions with the hubs
- If possible, a bulk OUT transaction with the printer, fax machine, and flash memory

Packets that are associated with the interrupt, isochronous, and control transfers have higher priority and are included first in every frame. Packets that are associated with bulk transfers are included in the frame only if there is enough leftover bandwidth.

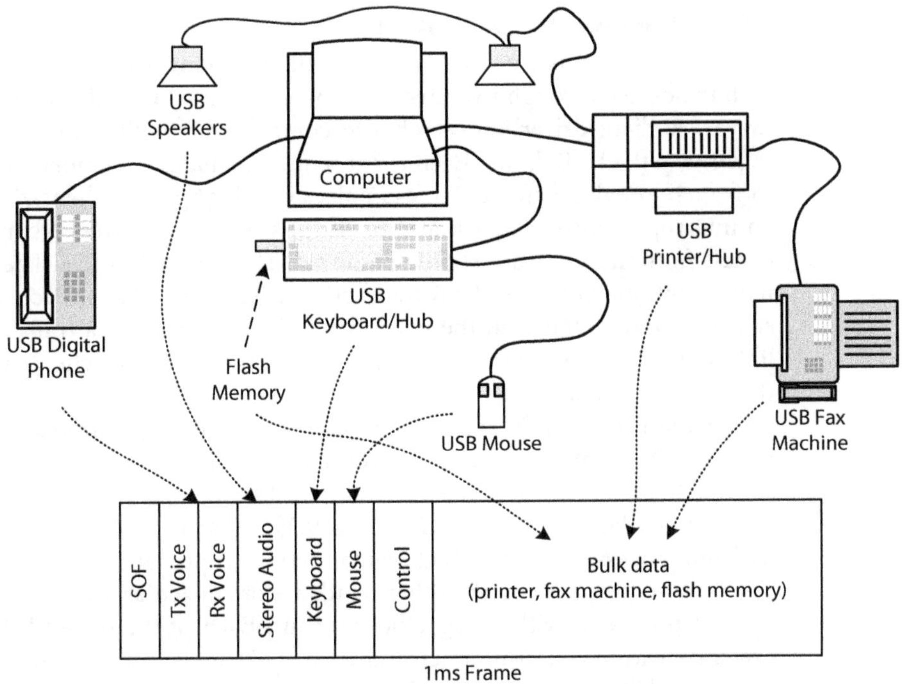

FIGURE 9.28 Illustrating USB transfers during a single frame; frame duration is 1 ms for USB 1.x [4].

In general, not all devices, except for those that require interrupt or isochronous transfers, need to have packets in every frame. Typically, 90% of the frame bandwidth is reserved for interrupt and isochronous transfers and 10% for control transfers, leaving up to 0% of frame bandwidth for bulk transfers. If at any time there are more interrupt or isochronous transfers, no bulk packets will be allowed, which may result in short interruptions in the operation of devices that use bulk transfers. Furthermore, a newly attached device would not be configured if its transfer type requires more bandwidth than is currently possible with the existing devices.

The USB 2.0 host controller uses split transactions to communicate with low- and full-speed devices that are interfaced through a high-speed hub. In this case, a low- or full-speed OUT transaction is split into several **start-split** (SS) microframe transactions. The data from these SS transactions are stored in a buffer in the high-speed hub and later are sent to the low- or full-speed devices. A USB 2.0 low- or full-speed IN transaction starts with a single SS microframe transaction. Once a target low- or full-speed device receives the IN request, it uses a lower packet rate of 1.5 Mbits/s or 12 Mbits/s to transmit its data to the high-speed hub, which receives it in an input buffer. The data in the buffer is then transmitted to a USB 2.0 host controller via several split transactions called **complete split** (CS) transactions.

9.8.6 Transaction Organization

Each USB device uses a driver routine called a **client driver** that communicates with the device through two pieces of host software: a **USB driver** and **USB host driver**. A client driver knows what to communicate with the device and issues requests to the USB driver. Each client request includes a memory space that data to/from the device is stored. For example, a USB keyboard (a client) driver initiates an interrupt transfer request and provides a memory address where data (scan codes) from the keyboard should be stored. The USB driver translates each client request into one or more USB transaction descriptors (TDs). Each TD is a data structure and includes all the necessary information about a transfer, as well as links to the next TD and memory space that either holds the client's OUT data or the client's IN data.

All the TDs from the active clients are grouped and organized as a set of linked lists of TDs in memory. The control and bulk TDs in USB 2.0 are organized separately than those for interrupt and isochronous packets. The interrupt and isochronous TDs, shown as iTDs in Fig. 9.29, are grouped into an array of linked lists and processed (executed) at regular periods (intervals). The control and bulk TDs, shown as qTDs, on the other hand, are organized as a queue of linked lists and are processed with no specific intervals. Each of the linked lists of iTDs is executed within a 125-μs microframe. A qTD is executed when there are no unexecuted iTDs left in the array.

A USB driver stores the starting addresses of the array and the head of the queue in a USB host controller interface via a USB host driver. A USB host controller

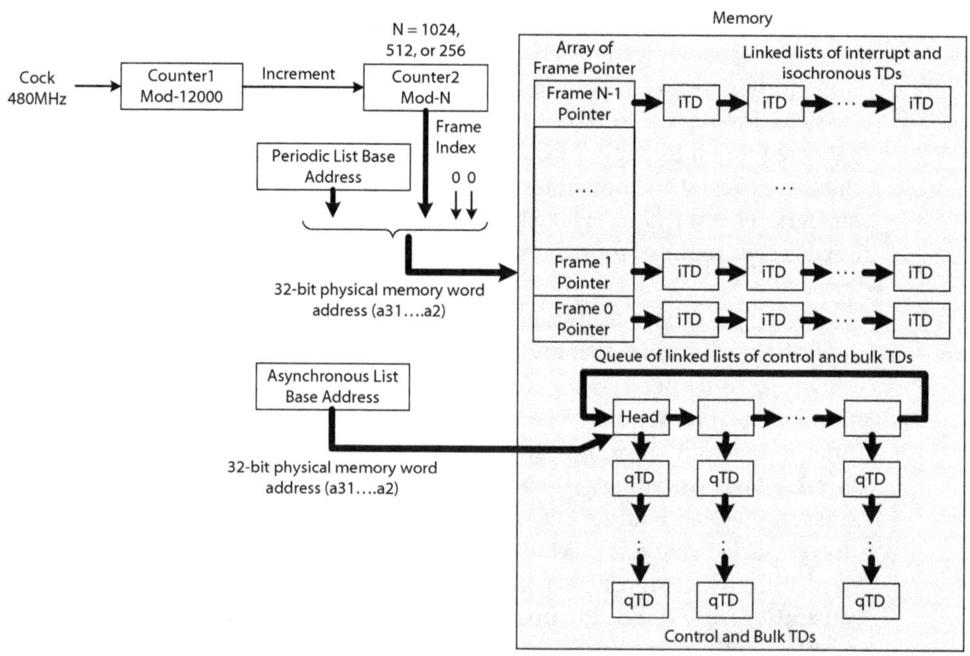

FIGURE 9.29 The organization of USB 2.0 transactions in memory [4].

uses two counters to execute the iTDs, as shown in the figure. The counter-1 is a mod-60000 counter and operates with a 480-MHz clock. The counter-2 holds a **current frame number** and is used as an index to the array. The counter-2 is incremented once every 125 μs. The array size is programmable and may be defined with 1024, 512, or 256 elements. The array base address is stored in a register called a **periodic list base address**, and is concatenated with the counter-2 to locate the next transaction in memory. The **asynchronous list base address** points to the head of the queue of linked lists in memory, as illustrated in the figure.

Each time that the counter-2 increments, an SOF transaction is executed, which informs all the devices that use isochronous transfers to synchronize their activities. If a TD (iTD or qTD) describes an OUT transaction, the host controller fetches its token and data packets from main memory and transmits them to the target endpoint. If the endpoint implements an interrupt, control, or bulk transfer, the controller receives a corresponding handshake packet from the endpoint. As stated earlier, isochronous transfers require no handshake packets.

If a TD is an IN transaction and indicates an interrupt, control, or bulk transfer, the host controller fetches the TD's token and handshake packets from memory and transmits the token packet to the target endpoint. After receiving a data packet from the endpoint, the controller transmits the handshake packet to the endpoint if the data was received error free.

9.8.7 Transaction Execution

Like any DCI (e.g., Fig. 9.17), a USB host controller interface includes a set of I/O ports that are accessed by both CPU and the host controller (an embedded system), as illustrated in Fig. 9.30 for the USB 2.0 host controller interface. No device data is directly communicated by the CPU using the I/O ports; instead, the ports are used only to configure and setup the host controller. Table 9.7 lists a few USB 2.0 host I/O ports called **USB operational registers**. They are used to set the size of the frame array with 1024, 512, or 256 entries; how often to interrupt CPU; etc. The interrupt frequency is specified in terms of the number of microframes. This is referred to as an **interrupt threshold** that is defined in terms of once every 1, 2, 4, 8, 16, 32, or 64 microframes.

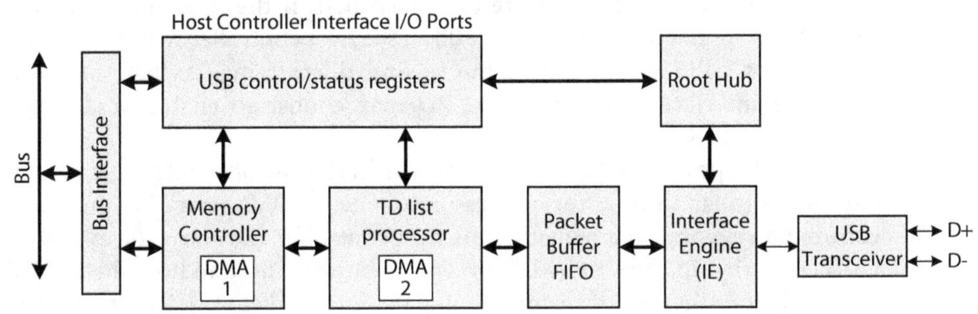

Figure 9.30 A single port USB 2.0 EHCI block diagram [6].

Register	Description
USB base address	Holds a base address for the memory mapped I/O ports in the host controller interface.
USB command	Used to configure the host controller interface. Examples of configurations are interrupt intervals (e.g., after every microframe, every two microframes, every 4 microframes, etc.), enabling/disabling periodic/asynchronous TD list processing, initializing the frame array size (1024, 512, or 256), Host controller interface reset, run/stop bit, etc.
USB status	Examples are periodic and asynchronous enable/disable bits, host controller interface halt/no-halt status bit; USB error interrupt bit; USB interrupt bit that becomes active if interrupt-on-complete (IOC) bit is set in a TD; interrupt on asynchronous advance status bit, which is set when the host controller interface increments the queue pointer; etc.
USB interrupt enable	Enables various interrupts; examples are enable interrupt on advancing asynchronous list so the USB driver would know the current status of the queue; enable interrupt on error; etc.
USB frame index	A relative address (index) to the frame array. Incremented once every 125 μs. It indicates the size of the frame array in Fig. 9.29.
Periodic list base address	A base address to the frame array. It is concatenated with the frame index to create a main memory address where the frame array is stored.
Asynchronous list address	A point to queue head (Fig. 9.29).
Port status and control	Used to configure USB ports. Examples are enabling wake-on-connect and wake-on-disconnect bits; line status (D+ and D–) used to detect low-, full-, and high-speed device connections; current port connection status (i.e., 1, device present and 0, device not present), etc.

Table 9.7 A List of the USB 2.0 Operation Registers (i.e., I/O Ports)

For example, if the interrupt threshold is set to once every two microframes, the host controller first will execute all the TDs (iTDs and qTDs) allowable within two microframes, where iTDs are processed first. If the interrupt-on-complete (IOC) bit is set in one or more of the TDs, the host controller will issue a request to interrupt CPU at the end of the second microframe. As part of the host controller interface configuration, CPU may choose to enable or disable the execution of TDs.

Once the processor fully configures the host controller interface, it sets the **run-bit**—similar to how the processor initiates a DMA transfer—in the **USB command register**, which enables the host controller and starts the execution of the TDs. A first in, first out (FIFO) memory buffer is used to hold the OUT data packets before they are transmitted and IN data packets received from devices before they are transmitted to the main memory. The memory controller in this case is responsible for generating memory addresses for fetching TDs, reading and writing endpoint data from/to main memory, and writing status data to main memory. For example, in Fig. 9.30, the DMA-1 is used to fetch TDs from the main memory and store them in a RAM within the list processor and to fetch endpoint OUT data and, via the DMA-2, store them in the FIFO buffer. The DMA-2 is also used to store the endpoint IN data from the FIFO buffer via DMA-1 in the main memory.

The DMA controllers are used for concurrent processing; they accelerate data transfer in and out of the host controller interface. The root hub, which also includes a set of configuration registers, is responsible for the management of the ports, such as port reset and resume, as well as port connections or disconnections. The **interface engine** (IE) is responsible for NRZI data encoding/decoding, token construction, etc.

References

1. William Stallings, *Computer Organization and Architecture*, Prentice Hall, 8th ed., 2010.
2. Mobile Intel Pentium Processor with 533MHz Front Side Bus, http://www.intel.com/Assets/PDF/datasheet/253028.pdf.
3. Microcontroller, http://www.atmel.com.products/, http://www.cypress.com/products/.
4. Don Anderson and Dave Dzatko, *Universal Serial Bus System Architecture*, 2nd ed., Addison Wesley, 2002.
5. Intel 64 and IA-32 Architectures Software Developer's Manual, Volume 3A: Systems programming Guide, Part 1, 2009.
6. USB 2.0 host controller core (inSilicon), http://www.synopsys.com/.

Exercises

9.1 Consider an FSB that interconnects four 1-GHz processors to a shared memory unit creating a UMA architecture. Suppose, on average, 10% of the instructions are "ST" that write to memory. Do the following:

a. Assuming that CPI = 1 and each write is 4 B, determine the memory-write bandwidth required by the four processors.

b. Considering that memory also needs to provide instructions and data to the processors, determine the required clock frequency of the FSB for twice the bandwidth calculated in part (a), assuming the FSB operates similar to an SDRAM bus with burst size = 1 and has a 32-bit data bus.

c. Suppose the memory controller operates a 32-bit 400 MHz SDRAM memory unit, determine a memory organization for peak performance; that is, there is 4 B transfers every FSB clock cycle.

9.2 Research and write a short paper on Intel's QuickPath link.

9.3 Research and write a short paper on AMD's HyperTransport tunnel.

9.4 Draw a six-node NUMA architecture with only nine serial links, where each processor is no more than two links away from another processor.

9.5 Consider a four-processor NUMA architecture with serial interconnection links versus a bus-based four-processor UMA system. Answer the following:

a. Compare serial versus bus interconnections.

b. Compare NUMA versus UMA architectures as the number of processors increases.

9.6 Consider the memory controller shown in Fig. 9.6. Suppose the SRAM has a read/write access time of 4 ns, the CPU bus clock frequency is 0.5 ns and requires one clock cycle to detect *ack* = 1 and one clock cycle to complete the memory cycle, either to load the data sitting on the bus to an internal register and end a read cycle or to remove the data from the bus and end a write cycle. Assuming that memory read and write access times are the same, determine the size of the counter.

9.7 Suppose the RPM of a new Samsung disk drive is twice the one in Example 9.1. Determine how much faster a 512-B transfer between memory and the disk drive will be.

9.8 Draw a circuit for a port-mapped I/O port access as in Fig. 9.10 that illustrates the circuit for a memory-mapped I/O port.

9.9 Consider the timing diagram in Fig. 9.12. Do the following:

a. Briefly explain why, when *sel* = 1 and *_wr* = 0, the output port will load the data on the bus.

b. Suppose the output port is designed using flip-flops. What changes would be required to interface and operate the port? Also, draw the output port and illustrate/explain how and when it will load the data from the bus. Note: the port flip-flops are not operated with a continuously changing clock signal.

9.10 Suppose port addresses 0x60 and 0x64 are, respectively, assigned to memory-mapped I/O ports 0 and 1 (each a 4-B port) in Fig. 9.17. Design an address decoding circuit for the two ports. For simplicity, assume an 8-bit address bus.

9.11 Explain in what way a modern DMA controller that processes a data structure containing DMA transfer information is better than a simple DMA controller shown in Fig. 9.20 and how a modern controller may affect the performance of a system.

9.12 Explain why when a receiving IV (*IVr*) is less than or equal to the IV of the currently executing interrupt handler (*IVc*), the CPU will ignore the *IVr* until *IVr* > *IVc*.

9.13 Assuming that the delay for a NAND gate is 0.1 ns and for a tri-state buffer is 0.2 ns, estimate the worst-case delay before an IV from a 16-node daisy-chained interrupt structure is placed on the bus.

9.14 The Sparc CPU implements eight-window "register windows" where the CPU data path includes eight copies of all the user-accessible registers. For example, ACC, X, and SR in Fig. 9.26 are the user-accessible registers. With a register window, the state of a currently executing program, upon subroutine call or interruption, would be saved within the CPU instead of on memory. Do the following:

a. Suppose the CPU in Fig. 9.26 has four register windows; explain how having register windows improves service to the waiting devices.

b. Suppose four copies of ACC, X, and SR registers are used in Fig. 9.26 to create register windows of size 4. Also, assume the registers are now labeled ACC0, X0, and SR0 for window 0, ACC1, X1, and SR1 for window 1, etc. Upon an interruption, the next instruction that executes and requires a register uses, for example, one of the registers in window 1. That is, for example, the instruction "MVSRX2ACC" in the IH when it executes performs ACC1 ← SRX, and the state of the interrupted program will remain saved as the content of the registers ACC0, X0, and SR0 in window 0. RTI will also switch back to window 0 so the interrupted program resumes execution using the registers in window 0. Design the four-window register windows circuit and briefly describe how it would be used to switch windows.

c. Discuss how to extend the register windows in part (b) to also support subroutine calls with the ability to pass a single parameter to a subroutine via ACC. Also, what should be done when there are several levels of subroutine calls.

d. A register window provides certain advantages. Suppose a CPU has eight register windows. What can programmers do to take advantage of register windows?

e. Research and write a short paper on how a Sprac processor implements overlapping register windows to pass parameters.

9.15 Briefly state why a USB host controller combines packets into frames that are periodically transmitted to/from devices.

9.16 Briefly state why the iTDs are processed before the qTDs in Fig. 9.29.

9.17 Briefly state why DMA-1 and DMA-2 in Fig. 9.30 are necessary.

9.18 Briefly state the purpose for the interrupt on asynchronous list advance in a USB host controller interface. (Hint: How modern modern DMA controllers work.)

9.19 Computer security (secure interruption): See Exercise 11.36 for detecting register spoofing, splicing, or replay attack upon returning from an interruption (also see Sec. 11.11.9).

CHAPTER 10

Memory System

10.1 Introduction

The performance of both uniform memory access (UMA) and nonuniform memory access (NUMA) systems depends on latency. Any improvement in latency, including memory latency, would increase the bandwidth of the system. The longer a CPU (processing core) stays idle, the number of clock cycles required to execute programs increase, reducing instruction throughput. In addition, the capacity of both nonvolatile storage and main memory has to be large enough for a system to store many programs and data so it can run systems and application programs concurrently. This means memory must be cost effective too. However, at the moment, there is no single technology available that can be used to build a low-latency, large-capacity, and low-cost memory.

Commonly used memory technologies today are static random access memory (SRAM), synchronous dynamic random access memory (SDRAM), and magnetic and flash memory. SRAM technologies are the fastest, but because SRAMs are used as cache memory inside the processor, they are also the most expensive. SDRAM technologies cost less, but are slower, requiring access time in the order of 100 CPU cycles. Magnetic and flash memories are nonvolatile and cost the least, but they are the slowest and require access time in the order of milliseconds; they are about 1,000,000 times slower than CPU.

However, programs contain loops and data is typically accessed from memory sequentially. Thus, only by using a combination of different memory technologies, organized in a hierarchy, as illustrated in Fig. 10.1, with the slowest memory at the bottom and the fastest memory on top, can one create a memory system that reduces average latency, minimizes cost, and has large storage capacity. Note that the instruction memory (IM) and data memory (DM) that were introduced in Chap. 8 as part of CPU data path are relabeled as instruction cache (Ic) and data cache (Dc) in the figure.

Other memory system design objectives may include low power usage, high reliability, and small physical size. The low power and small size requirements are especially important for handheld devices.

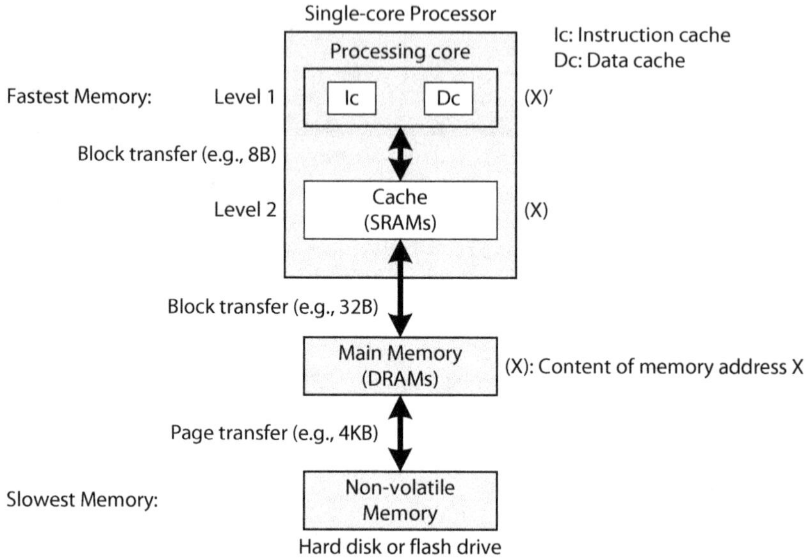

Figure 10.1 A memory system with four levels of memory hierarchy.

Data is copied from the lowest level of memory hierarchy to the next higher level until it is at the highest level before they are accessed by the CPU. In addition, multiple data items (as a block or page) are copied between levels of the hierarchy. Furthermore, the CPU will wait to receive data as long as data is in main memory. A program's execution will be stopped, as we see in Sec. 10.4, if the instruction or data the CPU requested is not in the main memory.

While a program is executing, modified data is copied from a higher level to the next lower level of memory hierarchy to create space or to inform the lower-level memory of changes when necessary. This creates scenarios where two copies of the same data in two different memories may not be the same. For example, consider a data item at memory address X that is copied from main memory to Level 2 (L2) cache in Fig. 10.1. The data is then copied from the L2 cache to Level 1 (L1) data cache before it is accessed by CPU. Now suppose the CPU, after operating on the data item, writes the new value in the L1 cache, modifying the copy, indicated as "(X)′" in the cache. The copy of the data item in the L1 cache is now different from the one still stored in the L2 cache and main memory.

Now, at this time, suppose the operating system (OS) instructs a direct memory access (DMA) controller to transfer data, including the data at address X, from main memory to the hard disk. Clearly, the data item that should be copied to the disk must be the modified one sitting in the L1 cache. Likewise, in another scenario, the OS may instruct the DMA controller to update the main memory, including the content of address X, with the copy on the hard disk. This time, the copy in the main memory would be the latest and valid and the ones in the two caches would be stale and not valid. Therefore, in order to prevent stale data from ever being saved on disk or ever being accessed by the CPU, certain **data coherency** protocols must be implemented between levels in memory hierarchy.

In addition, because volatile memory access time is very small (in the order of 100 CPU cycles maximum) as compared to the speed of a nonvolatile memory (disk or flash drive), the tasks of copying and maintaining coherency between memory levels, except between the lowest two levels in the hierarchy, are performed in hardware. Copying and maintaining data coherency between the nonvolatile memory (e.g., hard disk), known as **virtual memory**, and the main memory, known as **physical memory**, require certain functions be performed in software and certain operations in hardware. A virtual memory management system decides where in physical memory to store the codes and data of one or more programs stored on the hard disk.

This chapter presents the architecture and organization of memory hierarchy and provides alternative cache organizations and coherency protocols. Depending on the requirements, a cache can be designed to minimize hardware and access time or traffic. Power consumption and the relationship between a cache organization and system performance are also discussed.

The chapter also presents and illustrates virtual-to-physical address mapping schemes, and where exactly within the processor chip this mapping scheme must be implemented is discussed and alternative solutions are presented.

10.1.1 Memory Hierarchy

In Fig. 10.1, less costly nonvolatile memory is used to build potentially unlimited memory storage. DRAM technologies, typically SDRAMs, are used to build a large main (physical) memory, and SRAM technologies are used to build small but fast cache memories inside the processor.

Program code and data are copied from the nonvolatile storage to the main memory and from there to the L2 cache and then to the two L1 caches during program execution. Only a small fraction of each lower-level memory content can be retained in the next higher-level memory. In addition, because DMA transfers (Chap. 9) and burst memory transfers (Chap. 7) are more efficient, pages (e.g., each 4 KB) are transferred between a nonvolatile storage and main memory. Blocks (e.g., 32 B or 64 B), also called a cache block or **cache line**, are transferred between main memory and the L2 cache, and even smaller cache lines are transferred between the L2 cache and each of the L1 caches. Instruction cache lines are transferred from the L2 cache to Ic and data cache lines between the L2 and Dc.

A request from the CPU to access memory always goes first to the Ic if the address indicates an instruction or to the Dc if the address indicates data (also see Fig. 8.5 in Chap. 8). If the copy of the target memory block is in Ic or Dc, the access is called a **cache hit**. Otherwise, the access is called a **cache miss**, and the request is forwarded by the Ic or Dc to the unified L2 cache, which contains both instructions and data. Again, the access is called a cache hit if the copy of the block is in the L2 cache and a cache miss otherwise. If it is a cache hit, the L2 cache transfers a copy of the block, if instructions, to Ic or, if data, to the Dc. On the other hand, if the access is a cache miss, the L2 cache forwards the request to main memory.

In modern computer systems, a program is typically not loaded into main memory in its entirety. Instead, program and data pages are copied from the nonvolatile storage to main memory as needed. The modified data pages are copied back to nonvolatile storage to free up space in main memory. The instruction pages, however, are discarded if they are not referenced for some time and memory space is needed. This process is called **paging**, and it involves the OS and requires hardware to implement a **virtual memory organization**, discussed in Sec. 10.4.

Specifically, if a page in main memory contains a target instruction or data block on L2 cache miss, a copy of the block is transferred to the L2 cache. Otherwise, if the page is not in main memory, the request causes a page fault interruption (Chap. 9), and the execution of the program is suspended and control is returned to the OS. The execution of the program resumes once the page is loaded into main memory.

As was discussed in Chap. 8, during the entire time that a cache miss is being resolved, the processing core (CPU) is idle and does not execute instructions. Cores that implement multithreading can switch to execute instructions from a different thread. Modern processors are typically designed with three levels of cache memories, as illustrated in Fig. 10.2. Each of the processing cores communicates with its own L2 cache, and all the L2 caches communicate with a shared third-level L3 cache, thus creating a UMA system within the processor chip. A shared cache has the advantage of sharing cache lines used by two or more threads, but also has the disadvantage of one thread causing the removal of blocks used by another thread.

For example, consider Thread 0 and Thread 1 from the program example in Sec. 8.4.4 (Chap. 8). Recall that each thread operates on different array elements, computing their sum as $sum[0]$ by Thread 0 and $sum[1]$ by Thread 1. Thread 1 then outputs the quantity $sum[0] + sum[1]$. With a shared L3 cache, Thread 1 would be able to access $sum[1]$ from L3, saving one main memory access. In general, with the L3 shared cache, not only would the execution of two communicating threads be faster, but also the execution would generate less main memory traffic. An Intel's Xeon-E7-4870 (Nehalem architecture) processor has 10 cores, each attached to a 256KB L2 cache and then to a 30-MB shared L3 cache. The processor can access a maximum of 32-GB main memory space.

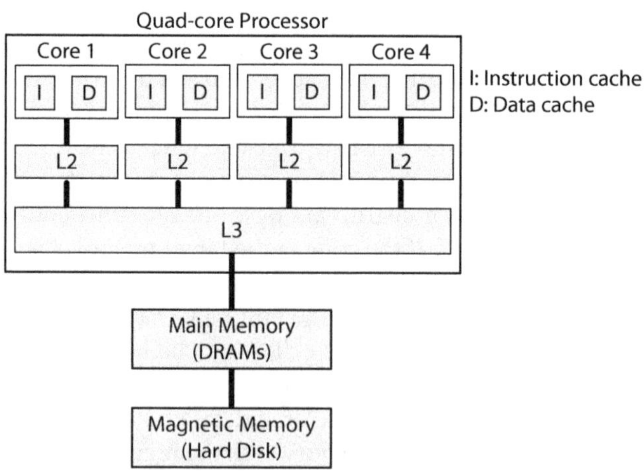

Figure 10.2 Example of a five-level memory hierarchy.

Memory Latency

The memory hierarchy reduces average latency because programs contain loops, and data accesses are typically sequential. Consider the memory hierarchy shown in Fig. 10.1. Suppose a block in main memory contains the instructions of a small for-loop. The first time that the for-loop executes, the request to fetch the first instruction will result in a cache miss. The block would then be copied from main memory to the L2 cache and from there to Ic. Therefore, the latency to transfer a copy of a block from main memory to Ic can be long.

However, once the block is loaded into Ic (a smaller SRAM), the subsequent instructions from that block are fetched more quickly (e.g., within 1 CPU clock cycle) from Ic. Therefore, this reduces the average memory latency for executing the for-loop. The same is true when data is accessed sequentially from main memory. It takes a longer time to access the first data item in each data block that causes a miss, but the subsequent accesses from the block in Dc (also a smaller SRAM) would be quicker.

In general, how soon each instruction in a program is executed again gives an indication of how much **temporal locality** exists in the program. Because the instructions in a small for-loop are frequently executed and once an instruction executes it will soon execute again (due to a small loop), high temporal locality is said to exist in the for-loop.

Likewise, in what order a program's data structure elements are accessed during execution indicates how much **spatial locality** exists in the program. For example, consider a for-loop that processes an array. During the execution of the for-loop, if the array elements are accessed from sequential memory addresses, then there is a high spatial locality among the elements accessed from main memory.

Note that temporal locality may also apply to program data—for instance, if the same set of data elements is accessed within a loop. Likewise, spatial locality would apply to program instructions.

Cache hit is determined in terms of probability of an access resulting in a hit. For example, while executing a program, if 95% of the time instructions are found in Ic, then the instruction **hit ratio** (also called the **hit rate**) is 0.95, resulting in an instruction **miss ratio** (also called a **miss rate**) of 0.05 (1.0 – 0.95). How often data is found in Dc determines a program's data hit and miss ratios. L2 and L3 caches are unified (both contain instructions and data). Thus, they would provide hit and miss ratios for the entire program. A high or low hit ratio depends on how much temporal and spatial localities exist in the program.

While a good programmer is expected to be mindful of temporal and spatial locality properties when writing programs, compilers do rearrange codes (e.g., switching nested loops) when possible to improve data spatial locality of programs during execution.

Example 10.1. Consider the memory hierarchy shown in Fig. 10.1. Given the following information, estimate its average memory latency. Also assume peak main memory bandwidth.

Ic and Dc:	Latency = 1 ns, hit ratio = 0.95 (assume the same for both caches)
L2 cache:	Cache line = 32 B, latency = 3 ns, hit ratio = 0.9
Main memory:	400 MHz SDRAM, 32-bit (4B) data bus

Solution: For peak performance, the time required to activate a row and issue a column address is ignored. This can happen when SDRAM memory operations to access multiple blocks are overlapped. For example, consider the SDRAM timing diagram shown in Fig. 7.19 (Chap. 7), where two separate burst accesses (each forming a block) of size 4 are read from memory in sequence. In the figure, the first data items in the two blocks are labeled x and y, respectively. Note that memory operations are overlapped. After the first data item x is accessed (i.e., appears on the data bus), one new data item is accessed every data bus clock cycle. The total time to access data items y, $y + 1$, $y + 2$, and $y + 3$ is only proportional to four data bus clock cycles. If memory is capable of supplying one data item every clock cycle, the memory is said to be operating with peak bandwidth. As calculated next, the SDRAM can deliver 1.6 GB peak transfer. Therefore, it would take 20 ns to transfer a 32-B block from main memory to L2 cache.

$$\begin{aligned}\text{Main memory peak bandwidth} &= 400\text{M cycles/sec} * 4\ \text{B/cycle}\\ &= 1.6\ \text{GB/s}\end{aligned}$$

$$\begin{aligned}\text{Main memory latency} &= 32\ \text{B}/1.6\ \text{GB/s}\\ &= 20\ \text{ns (transfer time only)}\end{aligned}$$

There is a 95% chance that a target instruction block is in the Ic cache and a target data block is in the Dc cache. During the 5% (100 – 95) of the time that a target block is not in the L1 caches (Ic or Dc), there is a 90% chance that the block is in the L2 cache. Finally, during the 10% (100 – 90) of the time that the block is not in L2, the block is in main memory. As discussed earlier, if the block is not in main memory, the execution of the program would be stopped and the CPU would be assigned by the OS to execute a different program. Therefore, the estimated average memory latency calculated next does not (and should not) include paging delays; however, paging delays are included in the program's total execution time.

$$\begin{aligned}\text{Average latency} &= (0.95)(1\ \text{ns}) + (1 - 0.95)(0.90)(3\ \text{ns}) + (1 - 0.95)(1 - 0.90)(20\ \text{ns})\\ &= 0.95\ \text{ns} + 0.135\ \text{ns} + 0.1\ \text{ns}\\ &= 1.185\ \text{ns}\end{aligned}$$

Note the 1.185 ns average latency is almost the same as the 1 ns latency assumed for both L1 caches, which are the fastest memories in the memory hierarchy.

10.2 Cache Mapping

Each lower-level volatile memory in a hierarchy, starting from main memory, has more blocks than there is space in the next higher-level memory. Therefore, each cache memory must implement a method to quickly verify if the copy of the requested block is in cache, or if the copy must be transferred from the next lower-level memory. Because a cache is simply a fast temporary storage space, a main memory address, issued by the CPU, is partitioned into a **block address** and an **offset**, which identifies a specific byte/word within the block. The block address is then used to determine the location of the block in cache called a **slot address** (or slot number) or an index.

Example 10.2. Consider 64 KB main memory, 1 KB L2 cache, and 8 B blocks. Determine the number of memory blocks, the number of cache slots, the range for block addresses, and the range for slot addresses.

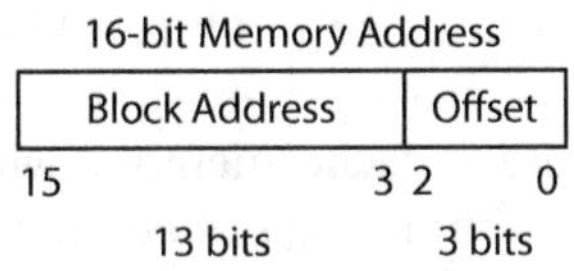

Figure 10.3 A memory address is made of a block address and an offset; each block is assumed to be 8 B.

Solution: The number of blocks is determined by dividing the main memory size by the block size as follows:

$$\begin{aligned}\text{Number of blocks in main memory, } N &= \frac{64\text{ KB}}{8\text{ B/block}}\\ &= \frac{2^{16}}{2^{3}}\\ &= 2^{13}\\ &= 8192\text{ blocks}\end{aligned}$$

The number of slots in the cache is similarly determined:

$$\begin{aligned}\text{Number of cache slots, } K &= \frac{1\text{ KB}}{8\text{ B/slot}}\\ &= \frac{2^{10}}{2^{3}}\\ &= 2^{7}\\ &= 128\text{ slots}\end{aligned}$$

The 16-bit (2^{16} B = 64 KB) memory address is partitioned into a 13-bit block address and a 3-bit offset, as illustrated in Fig. 10.3. The range for block addresses is 0 to 8191, and the range for slot addresses is 0 to 127. Note that the cache memory can only hold 1.56% (128/8192 ∗ 100) of total blocks in the main memory.

Direct mapping and **set-associative mapping** are two commonly used methods to map a block address to a slot address and are used between any two connecting memories in a hierarchy, starting with main memory. In a direct-mapped cache, a block copy can be stored in only one specific cache slot. In a set-associative mapped cache, a copy can be stored in a slot among a smaller set of slots. Direct-mapped caches are simpler, faster, and more power efficient because a block copy can only be in one slot.

Fully associative mapping is often used in some applications where a cache miss has a much longer latency. In this case, data can be stored in any cache slot, resulting in a higher hit ratio. A fully associative mapped cache, however, requires more hardware, as all the slots are searched in parallel (at the same time). In the following sections, direct mapped and set-associative mapped cache organizations are discussed in detail. The application of a fully associative mapped cache in the design of virtual memory system is presented in Sec. 10.4.

10.2.1 Direct Mapping

For a direct-mapped cache, we need two pieces of information to quickly determine whether or not the cache contains a copy of a target block. One is a slot address, which is determined using modular (Mod) arithmetic, and the other is called a **tag**, determined using integer division. For example, consider two block addresses, 129 and 1153, and a cache memory with K = 128 slots. For these two block addresses, the corresponding slot and tag values are calculated as follows:

129 Mod 128 => Slot 1

129/128 => Tag 1

1153 Mod 128 => Slot 1

1153/128 => Tag 9

However, if K is a power of 2 (i.e., $K = 2^m$), the mapping is simple and requires no hardware, as Eqs. (10.1) and (10.2) illustrate for an n-bit block address X.

$$\text{Slot address} = X \bmod 2^m \tag{10.1}$$
$$= (x_{n-1}x_{n-1} \cdots x_m x_{m-1} \cdots x_0)_2 \bmod 2^m$$
$$= (x_{m-1} \cdots x_0)_2 \Rightarrow \text{lower } m \text{ bits} = \text{Slot}$$

$$\text{Tag} = X/K \tag{10.2}$$
$$= (x_{n-1}x_{n-1} \cdots x_m x_{m-1} \cdots x_0)_2 / 2^m$$
$$= (x_{n-1}x_{n-1} \cdots x_m)_2 \Rightarrow \text{upper } n\text{-}m \text{ bits} = \text{Tag}$$

This is further illustrated as follows for block addresses X = 16 and 29 using $K = 8\ (2^3)$:

$16 \text{ Mod } 8 = (10\underline{000})_2 \text{ Mod } 2^3 = 0$ Slot = $(000)_2$ least 3-bits
$29 \text{ Mod } 8 = (11\underline{101})_2 \text{ Mod } 2^3 = 5$ Slot = $(101)_2$ least 3-bits

$16/8 = (\underline{10}000)_2/8 = 2$ Tag = $(10)_2$ upper 2 bits
$29 \text{ Mod } 8 = (\underline{11}101)_2/8 = 3$ Tag = $(11)_2$ upper 2 bits

Cache Organization

Table 10.1 shows two main memory addresses, 0x408 and 0x240B, that point to byte 0 in block 129 and byte 3 in block 1153, respectively. Only the copy of one of the blocks (shown by its address value) can be stored in slot 1, as illustrated in Fig. 10.4. The tag, which also is stored in cache, indicates the block address

Memory Address		Block Address (decimal)	Tag (dec.)	Slot Address (dec.)	Offset (dec.)	Target Byte
Address (hex)	Address Partitioned as Tag, Slot, and Offset					
408	000001,0000001,000	129	1	1	0	Slot 1: B0
240B	001001,0000001,011	1153	9	1	3	Slot 1: B3

Table 10.1 Direct-Mapped Cache Examples Using 64 KB Main Memory, 1 KB L2 Cache, and 8 B Blocks

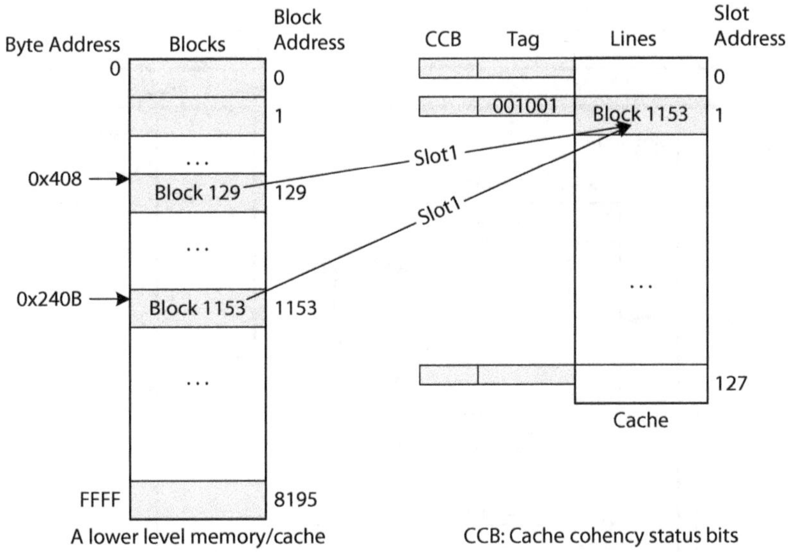

FIGURE 10.4 A logical view of a direct-mapped cache illustrating block 129 and 1153 mapping to slot 1; also shown is block 1153 copied to slot 1. Memory addresses 0x408 and 0x240B point to a byte in each of the blocks.

(129 or 1153) that the slot contains a copy of. In the figure, $tag = 9 = (1001)_2$ indicates that Slot 1 contains a copy of block 1153.

The cache also stores additional bits per slot that indicate the state of a block copy. The bits are used to implement a **cache coherency protocol** to make sure no stale data can be sent to the CPU or stored on disk. These bits are shown as cache coherency bits (CCB) in the figure.

Figure 10.5 illustrates the data path for the direct-mapped cache shown logically in Fig. 10.4. The data path consists of a 128-entry **tag memory** and a 128-entry **line memory**. The cache has the capacity to store a maximum of 128 blocks. Both the tag memory and the line memory are accessed at the same time using a slot address. The incoming tag is compared with the tag stored in the tag memory. If the two tags match, the cache access is a hit; otherwise, it is a miss. For performance reasons, while an incoming tag is compared with the tag stored in cache, the multiplexer (MUX) identifies the target byte (or word). If the cache access is a hit, the byte is sent to the CPU (assuming an L1 cache) on a read cycle, which is illustrated in the figure. If the access is a miss, the cache controller is triggered to access the block from its connecting lower memory. Until a cache miss is resolved, execution of the program is stalled.

10.2.2 Types of Cache Misses

In general, a cache miss is classified as a **cold miss, conflict miss, capacity miss, true sharing miss**, or **false sharing miss**. The following examples illustrate cold, conflict, and capacity misses. The true and false sharing misses relate to two or more threads accessing shared data blocks and are discussed in Sec. 10.3.

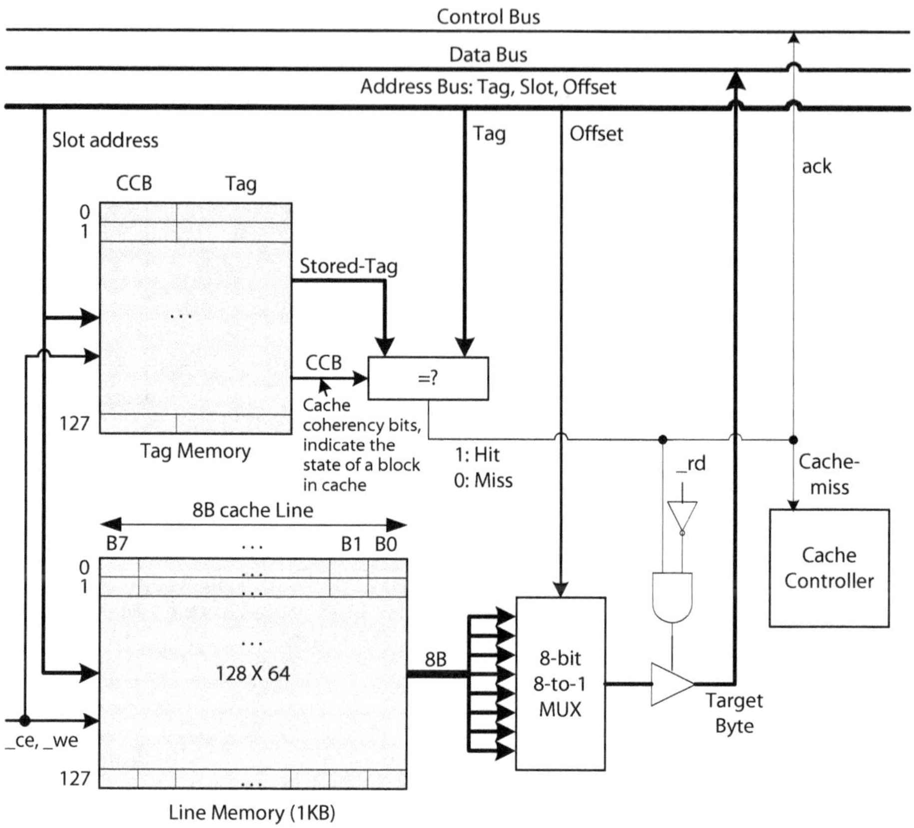

FIGURE 10.5 A direct-mapped cache data path with 128 slots and 8 B block illustrating cache read.

Example 10.3. Consider the direct-mapped cache shown in Fig. 10.4. Assuming an L1 cache, suppose the CPU accesses the following memory addresses in order 20 times. Determine the cache miss ratio for this sequence of memory accesses. Also, determine the number of cold, capacity, and conflict misses. The addresses are 16-bits, given in hex.

A.	0x3C10	assume address points to a byte in memory block Ba
B.	0x049C	assume address points to a byte in memory block Bb
C.	0x0410	assume address points to a byte in memory block Bc
D.	0x1C8D	assume address points to a byte in memory block Bd

Solution: We will first determine the number of cache misses and then use it to determine the miss ratio for this sequence of memory accesses. The mapping of the block addresses to slot addresses is given in Table 10.2. In the first round, the four memory addresses generate four cache misses. Those misses associated with blocks Ba, Bb, and Bd are cold misses because the blocks are copied into three initially empty slots in cache. The address C also maps to the same slot 2 as address A. However, because these two addresses have different tags, they are pointing to bytes in two different blocks in main memory; thus, a copy of Bc replaces the copy of Ba in slot 2, causing a conflict miss. It is called a conflict miss because there are still empty slots in the

First Time Accessed by CPU						
#	Address: Tag, Slot, Offset (binary)	Tag (dec.)	Slot Address (dec.)	Byte	Hit/Miss	Comment
A	001111,0000010,000	15	2	B0	M	Copy of Ba goes to slot 2
B	000001,0010011,100	1	19	B4	M	Copy of Bb goes to slot 19
C	000001,0000010,000	1	2	B0	M	Conflict in slot 2, copy of Bc replaces the copy of Ba
D	000111,0010001,101	7	17	B5	M	Copy of Bd goes to slot 17

#	Second Time		Third Time		Rounds 4 to 20	
A	M	Conflict in slot 2, copy of Ba replaces the copy of Bc	M	Same as the 2nd round	M	Same as the 2nd round
B	H	Copy of Bb in slot 19	H	Same as the 2nd round	H	Same as the 2nd round
C	M	Conflict in slot 2, copy of Bc replaces the copy of Ba	M	Same as the 2nd round	M	Same as the 2nd round
D	H	Copy of Bd in slot 17	H	Same as the 2nd round	H	Same as the 2nd round

TABLE 10.2 Direct Map of Four Addresses Given in Example 10.3

cache, but copies of Ba and Bc must still be stored in the same slot 2. In the first round, there are three cold misses and one conflict miss.

In the second round, the copies of blocks Bb and Bd are already in the cache and any read/write from these copies results in a cache hit. The copy of Ba, which was replaced with a copy of Bc in the first round, now replaces the copy of block Bc when address A is accessed again, causing another conflict miss. Accessing address C again causes yet another conflict miss and replaces the copy of Ba in slot 2 with a copy of Bc. In the second round, there are no cold misses but two conflict misses.

The hit/miss pattern in the second round repeats for the remaining 18 rounds, resulting in zero cold and two conflict misses per round. The four addresses generate zero capacity misses because there are still empty slots available in the cache. The four addresses, accessed 20 times, generate 42 total cache misses, resulting in miss ratio of 52.5%, as determined by Eq. (10.3).

$$\begin{aligned}\text{Total number of misses} &= 3 \text{ cold misses} \\ &\quad + 1 \text{ conflict miss} \\ &\quad + (20-1) \text{ rounds} * 2 \text{ conflict misses/round} \\ &= 42 \text{ total misses}\end{aligned} \tag{10.3}$$

$$\begin{aligned}\text{Miss ratio} &= \frac{\text{Number of total misses}}{\text{Number of memory accesses}} \\ &= \frac{42}{20*4} = \frac{42}{80} = 0.525\end{aligned}$$

Example 10.4. Consider the direct-mapped cache illustrated in Fig. 10.4. Suppose main-memory blocks 0 through 255 are accessed 10 times and in order. These blocks correspond to memory addresses 0x0 to 0x7FF. Determine the cache miss ratio for this sequence of memory accesses. Also determine the number of cold, capacity, and conflict misses.

Solution: Again, we will first determine the number of cache misses and then use it to determine the miss ratio for this sequence of memory accesses. The mapping of the block addresses to slot addresses is given in Table 10.3. In the first round, the copies of the first 128 blocks (0 to 127) fill the entire cache, each causing a cold miss, for a total of 128 cold misses because slots are initially empty. Each of the next 128 blocks (128 to 255) causes a miss and replaces the copy of one of the blocks 0 to 127 already in cache. For example, a copy of block 128 replaces the copy of block 0; a copy of block 129 replaces the copy of block 1; etc.

Therefore, in the first round, accessing blocks 128 to 255 would generate 128 capacity misses (the reason is forthcoming). In the second round, each of the blocks 0 to 127, whose copies in cache were replaced with copies of blocks 128 to 255 in the first round, will cause a capacity miss. A copy of block 0 will replace the copy of block 128 now in cache; a copy of block 1 will replace the copy of block 129 now in cache; etc. The access of blocks 128 to 255 in the second round will replace blocks 0 to 127 already in cache, causing a total of 256 capacity misses in the second round.

The capacity miss pattern in the second round will repeat in the remaining eight rounds. On odd-numbered rounds, copies of blocks 0 to 127 replace the copies of blocks 128 to 255 in cache, and on even-numbered rounds (starting at 2), copies of blocks 128 to 255 replace copies of blocks 0 to 127, resulting in 256 capacity misses in each round. The number of total cache misses for 10 rounds is 2560 and the miss ratio is 100%, determined by Eq. (10.4). There are no conflict misses in this case.

$$\begin{aligned}\text{Total number of misses} &= 128 \text{ cold misses} \\ &\quad + 128 \text{ capacity misses} \\ &\quad + (10-1) \text{ rounds} * 256 \text{ capacity misses/round} \\ &= 128 \text{ cold misses} + 2432 \text{ capacity misses} \\ &= 2560 \text{ total misses}\end{aligned} \tag{10.4}$$

Address (hex)	Address (binary)	Block Number (dec.)	Tag (dec.)	Slot (dec.)	Hit/Miss
0x0	000000,0000000,000	0	0	0	M
0x8	000000,0000001,000	1	0	1	M
0x10	000000,0000010,000	2	0	2	M
...	...		...	...	M
0x3F8	000000,1111111,000	127	0	127	M
0x400	000001,0000000,000	128	1	0	M
0x408	000001,0000001,000	129	1	1	M
0x410	000001,0000010,000	130	1	2	M
...	...	...	...	...	M
0x7F8	000001,1111111,000	255	1	127	M

Table 10.3 Direct Map of Main-Memory Addresses Given in Example 10.4

$$\text{Miss ratio} = \frac{2560 \text{ total misses}}{10 * 256 \text{ accesses}} = 1.0$$

Capacity miss is somewhat different and distinct from a conflict miss that does not depend on cache size. For example, if the cache size in Example 10.4 were doubled from 1 KB to 2 KB, the cache would have enough space to hold copies of all 256 blocks; thus, it would result in only 256 total cold misses in the first round and zero misses during the remaining nine rounds. This will improve the miss ratio from 100% to 10% (256/2560) and, therefore, the hit ratio from 0% to 90%. In general, a cache size is fixed and is determined during design. However, cache simulations can be used to choose a reasonable cache size that will not result in too many capacity misses. Conflict and capacity misses are sometimes grouped together.

In general, miss ratio also depends on block size. For example, as expected, with smaller blocks, there will be more cold misses. Using larger block sizes will reduce the number of cold misses. However, the relationship between block size and miss rate is program dependent, but in general, when blocks are too small or too large relative to cache size, miss rate increases.

10.2.3 Set-Associative Mapping

Direct mapping is simple and requires less hardware, but it has the disadvantage of being too restrictive. Two or more frequently referenced block addresses that map to the same slot address may generate frequent cache misses and cause delays, as was illustrated by Example 10.3. Figure 10.6 illustrates a program example with two such blocks located 1 KB apart in memory. The cache is assumed to be 1 KB

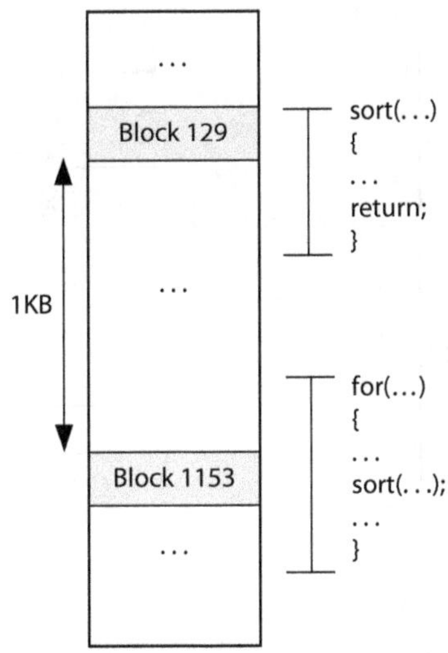

Figure 10.6 A program example illustrating the limitations of a direct-mapped cache; cache is 1 KB and block size is 8 B.

with an 8 B block size. As illustrated in the figure, block 129 contains instructions for a "sort" routine, and block 1153 contains instructions for a for-loop. As the for-loop executes and the sort routine is called repeatedly, blocks 129 and 1153, which both directly map to slot 1 (Table 10.1), will replace each other's copy in cache repeatedly. This will increase execution time because the Ic must frequently retrieve instructions in blocks 129 and 1153 from the L2 cache. This will also increase CPU idle time; however, multithreading (Chap. 8) could minimize this CPU idle time.

A set-associative mapping is a way to decrease misses due to conflicts, such as the memory access scenario shown in Fig. 10.6. In a set-associative mapping, the cache memory is organized in sets, each with a small (e.g., two, three, four, or eight) number of slots. A block address is directly mapped to a set address, but not to a slot address as is done in a direct-mapped cache. Within each set, a block can be stored in one of the slots determined by a **replacement algorithm**, such as least recently used (LRU), **cyclic** (in a circular fashion, first-in, first out), or **random**.

A replacement algorithm is implemented in hardware, and this could make set-associative caches even more complex and slower. LRU would require the most hardware as compared to the other two replacement algorithms, and random would require the least hardware. It has been shown that LRU generally is the best in terms of a lower miss ratio and random is the worst when cache size is small. The performance of LRU and random for large cache sizes is about the same and better than cyclic [1]. For further discussion on implementation complexities, refer to the Exercises section.

Figure 10.7 illustrates the mapping of block addresses to a **two-way** (two slots/set) 1 KB set-associative cache, assuming 8 B blocks. The 128 cache slots are

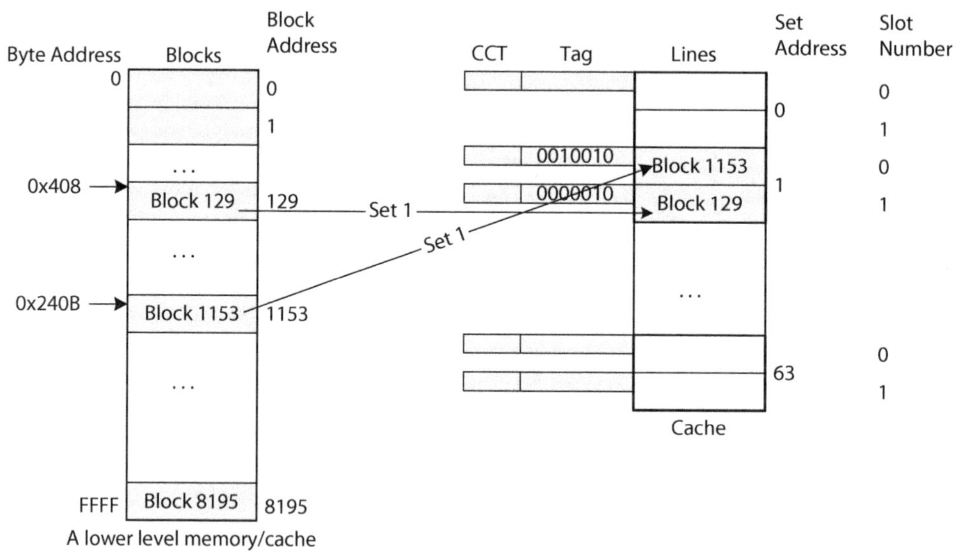

Figure 10.7 Logical view of a two-way set-associative cache with 128 slots grouped into sets of two slots each.

Target Byte Address		Block Address (decimal)	Tag (dec.)	Set (dec.)	Offset (dec.)	Target Byte
Address (hex)	Address: Tag, Set, Offset (binary)					
408	0000010,000001,000	129	2	1	0	Slot 1: B0
240B	0010010,000001,011	1153	18	1	3	Slot 0: B3

TABLE 10.4 The Mapping of Blocks 129 and 1153 to a Two-Way Set-Associative Cache with 64 Sets

grouped into $64 = 2^6$ sets, with two slots in each set. A block address is converted to a set address and a tag using Eqs. (10.1) and (10.2), with $m = 6$ instead. The mapping of two memory addresses, 0x408 and 0x240B, to a two-way set-associative cache with 64 sets is given in Table 10.4.

As illustrated in Fig. 10.7, block 1153 is copied to the cache first (assuming the memory access scenario shown in Fig. 10.6) and is stored in slot 0 of set 1. A copy of block 129, which also maps to set 1, is now stored in slot 1 of set 1. Therefore, the execution of the program in Fig. 10.6 will generate only two cache misses for accessing blocks 129 and 1153, as compared to many misses if the direct-mapped cache of Fig. 10.4 is used.

Cache Organization

Figure 10.8 illustrates the data path for the two-way set-associative cache logically shown in Fig. 10.7. The top tag and line memories are reserved for slot 0 in each of the 64 sets. The bottom tag and line memories are reserved for slot 1 in each of the 64 sets. During a cache read/write cycle, all four memories are accessed at the

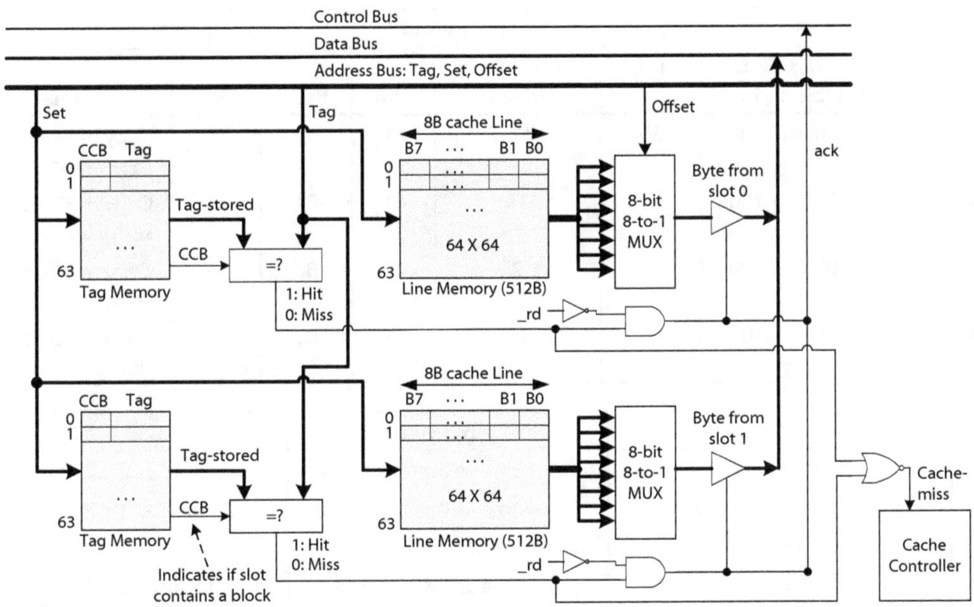

FIGURE 10.8 A data path for a two-way 64-set set-associative cache illustrating a cache read hit.

same time, and the incoming tag is compared with both tags stored in the two tag memories. If a copy of the target block is in the cache, one of the two tags read from the tag memories will match the incoming tag, resulting in a cache hit. Otherwise, the copy is not in the cache and the access will result in a cache miss.

Because all four memory modules in a two-way set-associate cache are accessed at the same time, set-associative caches consume more power. One way to reduce power consumption is to use **way-predicting** set-associative caches, where only one of the tag-line memory pairs is searched first. If this produces a cache miss, then all the tag-line memory pairs are searched next at the same time [2].

The number of ways in a set-associative cache needs not be multiples of two. For example, a three-way set-associative cache would require three pairs of tag and line memory modules; there are only two such pairs in Fig. 10.8. Intel's eight-core Xeon processor, includes a 24MB shared L3 cache organized as an eight-ported three-way (called 24-way) set-associative cache. Eight connecting L2 caches can access the L3 cache at the same time as long as accesses are from eight different sets.

Example 10.5. Consider the set-associative cache illustrated logically in Fig. 10.7. Suppose the CPU references the addresses given in Example 10.3 20 times and in order. Determine the cache miss ratio for this sequence of memory accesses. Also, determine the number of cold, capacity, and conflict misses.

Solution: Table 10.5 presents the calculations to map block addresses to set addresses. In the first round, the addresses generate four cold cache misses. Copies of blocks Bb and Bd are loaded into two empty slots in two different sets. The addresses for blocks Ba and Bc both map

	First Round						
#	**Address: Tag, Slot, Offset (binary)**	**Tag (dec.)**	**Set Address (dec.)**	**Slot**	**Byte**	**Hit/ Miss**	**Comment**
A	0011110,000010,000	30	2	0	B0	M	Copy of Ba goes to set 2/slot 0
B	0000010,010011,100	2	19	0	B4	M	Copy of Bb goes set 19/slot 0
C	0000010,000010,000	2	2	1	B0	M	Copy of Bc goes set 2/slot 1
D	0001110,010001,101	7	17	0	B5	M	Copy of Bd goes slot 17/slot 0

#	**Second round**		**Third round**		**Rounds 4 to 20**	
A	H	Copy of Ba in set 2/slot 0	H	Same as the 2nd round	H	Same as the 2nd round
B	H	Copy of Bb in set 19/slot 0	H	Same as the 2nd round	H	Same as the 2nd round
C	H	Copy of Bc in set 2/slot 1	H	Same as the 2nd round	H	Same as the 2nd round
D	H	Copy of Bd in set 17/slot 0	H	Same as the 2nd round	H	Same as the 2nd round

Table 10.5 Illustrating Two-Way Set-Associative Mapping of the Addresses Given in Example 10.3

to set 2, but this time, the copies of both Ba and Bc are stored in two separate slots in set 2. In the remaining 19 rounds, because all the copies of the four blocks are in the cache, none of the four addresses would cause a cache miss. There are also no capacity misses. The number of total misses is now four and miss ratio improves to 5% (hit ratio = 95%), as calculated next. This is compared to 42 total misses and miss ratio = 52.5% (hit ratio = 47.5%) using a direct-mapped cache (see Example 10.3).

$$\begin{aligned}\text{Total number of misses} &= 4 \text{ cold misses} \\ &\quad + (20-1)*0 \text{ misses/round} \\ &= 4 \text{ total misses} \\ \text{Miss ratio} &= \frac{4 \text{ total misses}}{20*4 \text{ accesses}} \\ &= 0.05\end{aligned} \tag{10.5}$$

10.3 Cache Coherency

Each cache memory must implement a coherency protocol to ensure that a read cycle always returns data from the latest copy of the block, no matter where the latest copy is (in a cache or main memory). For example, consider a two-processor system illustrated in Fig. 10.9. For simplicity, each processor is shown with one core and an L2 cache that connects to the memory bus. The system also contains a bridge between the memory bus and I/O bus. The DMA controller transfers pages

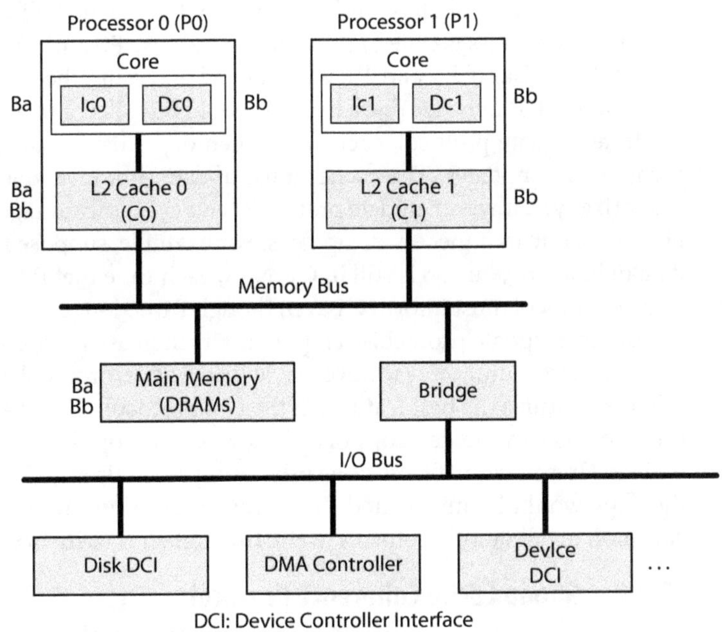

FIGURE 10.9 Two-processor UMA architecture.

between main memory and the disk drive. Also shown in the figure are the copies of two memory blocks, Ba and Bb, already copied to some of the cache memories.

In the figure, processor 1 (P1) may execute a store instruction to update a word, say, in the copy of block Bb in Dc1. Because the copies of Bb also exist in other caches and also in main memory, unless the other caches and the main memory are made aware of this update, these copies will be old and different from the copy in Dc1. Likewise, a DMA transfer may write a block, such as Ba, in main memory where copies of Ba are in one or more caches.

Because CPUs take turns to complete a write cycle, copies of blocks (if any) are either invalidated or are updated in the other caches. The two options are known as invalidation and update cache coherency protocols. A hybrid cache implements a combination of both invalidation and update coherency protocols.

10.3.1 Invalidation versus Update Protocols

In an invalidation protocol, when a block is updated in one cache, the other caches invalidate their copies and thus prevent stale data from ever being accessed. A cache memory with an invalidated copy must request to receive the updated copy when needed. Invalidation protocols have the disadvantage of invalidating an entire block even if only a single word is updated. This can increase the miss ratio especially when processors (or cores) access a shared block.

For example, suppose P0 in Fig. 10.9 accesses the first half of block Bb, and P1 the second half of Bb. Each time that P0 writes to its copy of block Bb, the copies of Bb in C1 and Dc1 are invalidated. If P1 accesses block Bb again, there will be a cache miss in Dc1. Dc1 must now request an updated copy of Bb, even though the update was made to the first half of the block and not the second half accessed by P1. In this case, the cache miss is called **false sharing** because P0 and P1 are not really sharing data in block Ba. If P0 and P1 were indeed accessing the same one or more data items in block Ba, the Dc1's cache miss would be called a **true sharing** miss.

In an update protocol, each cache memory must broadcast and inform other caches of any updates. The caches must update (not invalidate) their copies of the block (if any). However, update protocols have the disadvantage of increasing overall bus traffic due to unnecessary updates. For example, suppose P1 is done processing data in block Bb, but Bb is still in C1. Now, each time that P0 writes to block Bb, an update protocol must inform C1 even though P1 no longer needs to access block Bb.

Because update protocols can potentially increase protocol-related traffic, they are not very common. However, a hybrid cache protocol can use an adaptive scheme to utilize the best features of the two protocols. For example, a hybrid cache protocol may invalidate a copy of a block when the copy has been updated a number of times. In this case, if P1 is done processing block Ba and C1 still has a copy of Ba, the copy would be invalidated thus, preventing future unnecessary updates. Two common invalidation protocols, **write through** and **write back**, are discussed next.

10.3.2 Snoop Cache Coherence Protocol

Each cache memory requires two controllers: a **cache controller** and a **snoop controller**. A cache controller responds to requests it receives from its higher-level cache or requests from a processing core if the controller belongs to an instruction or data cache (Ic or Dc). An snoop controller responds to requests it receives from its lower-level cache or responds to read/write transactions that appear on the memory bus if the controller belongs to the lowest-level cache.

For example, in Fig. 10.9, the cache controller of Dc0 responds to requests it receives from P0. The snoop controller of Dc0 responds to requests it receives from cache C0. Likewise, the cache controller of C0 responds to requests it receives from either Ic0 or Dc0, and its snoop controller responds to memory bus transactions.

Assuming an invalidation protocol, if the DMA controller in Fig. 10.9 initiates a write transaction to block Ba, the C0's snoop controller, which continuously monitors (snoops) the memory bus, detects the write and invalidates its copy of Ba. C0, in turn, communicates with its higher-level cache Ic0 or Dc0 and invalidates the copy of Ba (if any). A memory write transaction initiated by one of the two L2 caches would be similarly detected by the snoop controller of the other L2 cache, and the copies of the block (if any) in that processor would be invalidated. The cache and snoop controllers that implement an update protocol operate similarly, except that, in this case, cache copies are updated with the new values (instead of being invalidated).

10.3.3 Write-Through Protocol

Write-through is an invalidation protocol. As its name implies, all write requests made by a processing core go "through" the cache to update main memory. On a **write hit** (i.e., cache hit is due to a write cycle), not only the copy of the block in cache is updated—the copy in main memory is also updated. However, if the copy of the block is not already in cache, the write cycle updates main memory, and typically, the copy of the block is not loaded into cache. The reason for this is that because a cache write transaction will always be forwarded to main memory, whether a copy is in the cache or not, there is no need to copy the block to the cache. This is known as a write through with **no allocation** cache protocol, as illustrated by a finite state diagram (FSD) in Fig. 10.10.

A write-through cache uses 1 bit per block that indicates the state of the block either as valid ("V") or invalid/not present ("I"), as shown in the FSD. The "V" and "I" states are encoded into a 1-bit CCB (Fig. 10.5) and stored in tag memory. For example, 1 may be used to indicate the copy in cache is valid, and 0 indicates the copy is either invalid or not yet loaded and therefore not present.

The FSD shows five possible transitions as $I_{WM} \rightarrow I$, $I_{RM} \rightarrow V$, $V_{WH} \rightarrow V$, $V_{RH} \rightarrow V$, and $V_{XWH} \rightarrow I$. The WM, RM, WH, RH, and XWH, respectively, stand for write miss, read miss, write hit, read hit, and external write hit (a write hit in another cache). For example, as stated earlier for a write hit, a write miss means the cache miss is due to a write cycle, read miss means the miss is due to a read cycle, etc.

The write-through protocol has the advantage of being simple; a copy of a block can only be in one of two states in each cache. However, this protocol has the disadvantage of potentially causing excessive bus and cache traffic, and thus, it is especially not recommended for multicore or multiprocessor systems.

For example, consider the multiprocessor system in Fig. 10.9 and assume the caches implement the write-through protocol shown in Fig. 10.10. Suppose the

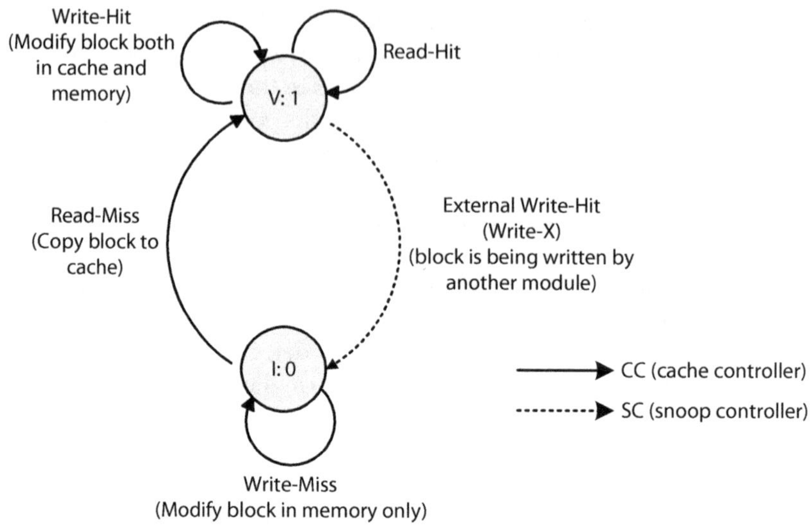

FIGURE 10.10 A finite state diagram illustrating write-through protocol with no allocation.

following code segment runs in P0. Because *sum* is declared global, the array elements will not be summed in register; instead, the block that contains *sum* (e.g., Ba in Fig. 10.9) will be updated in caches as well as in main memory each time that the next array element is added to the partial sum. Now because *sum* will be updated 100 times during the execution of the for-loop, Dc0 will need to issue 100 memory transactions to update the L2 cache, and the L2 cache in turn would need to issue 100 write transitions to update main memory, wasting valuable memory bandwidth.

```
int sum = 0;  //global variable sum
main()
{
    int array[100];  //locally declared array
    . . .
    for(i = 0; i< 100; i++)
    {
            sum = sum + array[i];
    }
    . . .
}
```

10.3.4 Write-Back Protocols

Write-back protocols are designed to reduce unnecessary bus and cache traffic and still keep caches coherent. A commonly used write-back invalidation protocol is known as the MESI (pronounced “messy”) protocol. Two other MESI-type protocols, called **MESIF** (used by Intel) and **MOESI** (used by AMD), are designed for more efficient cache-to-cache communication.

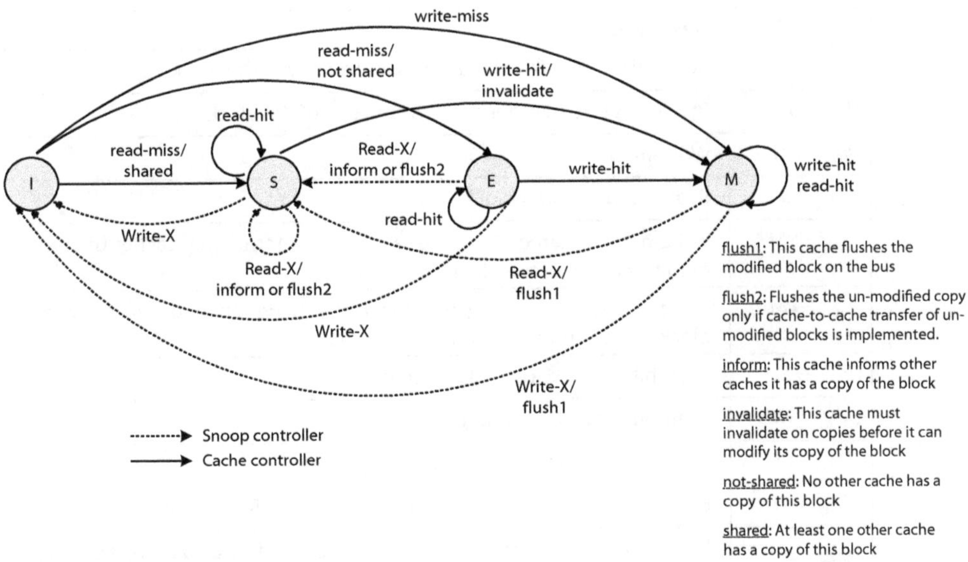

FIGURE 10.11 MESI protocol finite state diagram.

MESI

In the MESI protocol, a cache copy of a memory block can be in one of four states: modified (M), exclusively owned (E), shared (S), and invalid (I) or not present. The FSD of the MESI protocol is shown in Fig. 10.11. The "M" means the copy in the cache is modified ("dirty") and is no longer "clean"—that is, the same as the copy in main memory. The state "E" indicates the cache has the only copy outside main memory. The "S" indicates two or more caches contain a copy, and in addition, each copy is "clean."

Table 10.6 describes each of the MESI state transitions. The protocol greatly reduces traffic, but is more complex than the write-through protocol. For example, in order to update a shared ("S") copy, a MESI cache must perform the following tasks:

1. The cache (via its snoop controller) must inform all other caches before it can modify an "S" copy. This is to make sure the copies in other caches are invalidated first.
2. The cache then modifies its copy of the block and changes the state of the copy from "S" to "M." Its snoop controller is responsible for responding to a memory read/write request, which may be issued by another cache or by a DMA controller, for the block.
3. The cache must write the modified copy back to memory if the copy is replaced.

A write hit to an "E" copy does not, however, require the snoop controller to inform other caches; thus, it saves bus transactions and reduces cache traffic.

Transition	Description
$I_{RM} \rightarrow S$	Read-miss but another cache has also a copy of the block
$I_{RM} \rightarrow E$	Read-miss but no other cache has a copy of the block
$I_{WM} \rightarrow M$	Write-miss
$S_{RH} \rightarrow S$	Read-hit to a shared block
$S_{WH} \rightarrow M$	Write-hit to a shared block, the cache must inform other caches before it can modify its copy of the block.
$E_{WH} \rightarrow M$	Write-hit to an executively own block, the cache can modify its copy of the block without notifying other caches first.
$E_{RH} \rightarrow E$	Read-hit to an exclusively owned copy
$M_{RH} \rightarrow M$	Read-hit to an already modified copy
$M_{WH} \rightarrow M$	Write-hit to an already modified copy
$S_{XWH} \rightarrow I$	External write-hit, another cache wants to modify this block.
$M_{XRH} \rightarrow S$	External read-hit, another cache wants a copy of this block. This cache must "flush" (place) the block on the bus (flush1), so that another cache can load it.
$E_{XWH} \rightarrow I$	External write-hit, another cache is writing to this block.
$M_{XWH} \rightarrow I$	External write-hit, another cache wants to write to this already modified block. This cache must "flush" (flush1) the modified block and invalidate.
$S_{XRM} \rightarrow S$	External read-miss, another cache is loading a copy of the block from the main memory or another cache (flush2).
$E_{XRH} \rightarrow S$	External read-hit, another cache is loading a copy of this block from the main memory or another cache (flush2).

RM: read-miss, WM: write-miss, RH: read-hit, WH: write-hit, X: external

Table 10.6 State Transitions in the MESI Protocol

In many existing computers that implement the MESI protocol, it is common that main memory and not a cache is responsible for sending an "S" (clean) copy to a requesting cache. In addition, when a copy of a modified block in a cache is transferred from the cache to a requesting cache, the main memory is updated, and thus the state of the copy in both the caches changes to "S."

In a NUMA system the total memory space is divided among different nodes, such as the one shown in Fig. 10.12. Each node of a NUMA system includes a communication interface (CI) for internode communications. The CI of a requesting node routes a remote memory transaction to the CI of a destination node and therefore creates a virtual connection between the two nodes. For example, in the figure, CI0 and CI1 would make the memory buses in nodes 0 and 1 appear connected, providing seamless communication between nodes 0 and 1. However, this point-to-point communication can potentially increase latency when several caches request copies of a shared block. For example, consider block Ba in the figure. In this case, if the caches implement the MESI protocol without the "flush2" option (i.e., without the arcs Read-X/flush2 in Fig. 10.11), M3 would need to send a copy of Ba to any

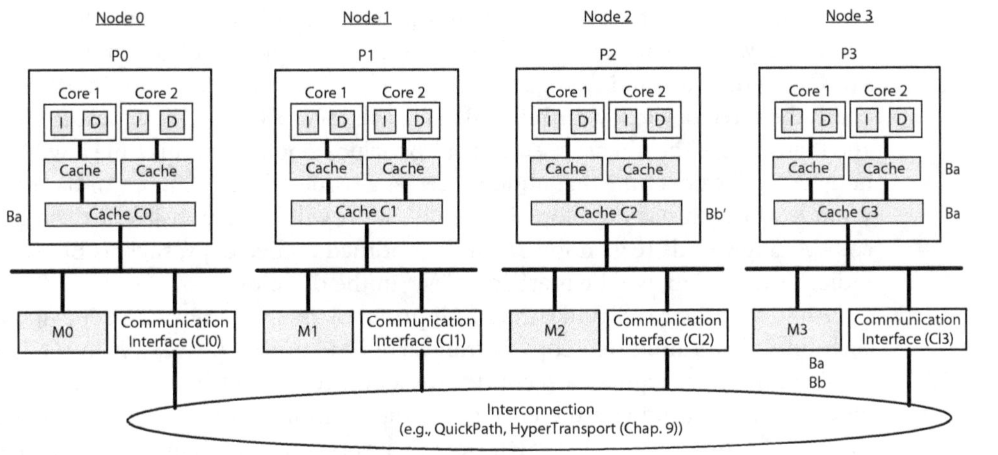

FIGURE 10.12 A three-node NUMA system block diagram.

cache that requests it. This could potentially make M3 a **hot spot** that would introduce delays and thus increase average latency. Likewise, if the caches implement the MESI protocol with the "flush2" option for cache-to-cache communication, the processor with a shared copy can become a hot spot. Two protocols, introduced earlier, that resolve this issue are MESIF and MOSEI, described next.

MESIF

This protocol is designed for cache-to-cache communication of shared ("S") copies. When two or more caches contain shared copies of a block, one of the cache copies (the first) is marked "F" (forward), and the copies of the other caches are marked "S." The cache with the copy in the "F" state is responsible for transferring (forwarding) a copy to another cache that is next in line to receive a copy. However, the transmitting cache changes the state of its copy from "F" to "S" after the transfer, and the receiving cache stores its copy in the "F" state.

Because only one cache can have the copy of a block in the "F" state, and the state of the copy changes to "S" after one transfer, the MESIF protocol prevents any cache from becoming a hot spot. The "F" state enables caches to take turns transferring a shared block copy to a requesting cache. Note that in the NUMA organization in Fig. 10.12, because each node contains only a single processor, MESIF would enable node-to-node communication of shared copies. If each node was a UMA multiprocessor system, the cache-to-cache communication would allow each cache within a node to receive a shared copy (if any) from a local cache instead of from another node with longer latency. The "M," "E," "S," and "I" states work the same as in the MESI protocol.

For example, suppose the four-node NUMA system in Fig. 10.12 implements the MESIF protocol. Also, suppose each node contains a directory (not shown in the figure) and for each memory block in a node, the directory holds a list of nodes that have cached copies of a block (if any). Furthermore, assume that block Ba in Fig. 10.12 was loaded in cache C3 (node 3) first and then a copy was sent to cache

C0 (node 0). The entry for block Ba in the directory of node 3 would list nodes 0 and 3 and would mark node 0 "F." A request for a copy of Ba from another cache, say, C1 (node 1), would first go to node 3. The CI3 communication interface will check the directory entry for block Ba and will forward the C1's request to node 0 and then change the directory entry to also include node 1, but this time it will mark node 1 "F." The next time that node 3 receives a request for an "S" copy of Ba, a cache in node 1, which has the block in the "F" state, will be responsible for sending a copy. A single node (e.g., node 2) with a modified cached copy, such as Bb′ where′ indicates modified, will be marked as "M" in the directory.

Also, with the MESIF, like the MESI protocol, main memory must be updated each time that a modified copy is transferred to another processor. In a NUMA architecture, this requires a second transaction to be sent by the cache with the modified copy to update the corresponding memory unit. Note that in a MESI cache UMA architecture, where the lowest-level caches and main memory share a common bus, main memory is updated (via its snoop controller) when a transfer of a modified block on the bus is detected.

MOESI

As opposed to MESIF, this protocol is designed to support cache-to-cache communications of modified copies. The "O" state stands for "owned" and is used to share a modified copy of a block without updating main memory. When a cache issues a request for a copy of a block that is modified, the cache that has the modified copy transfers a copy to the requesting cache and changes the state of its copy from "M" to "S." The receiving cache, however, stores the copy in the "O" state. The next time that the modified cache copy is requested, the cache with the copy in the "O" state would be responsible for transferring a copy to the requesting cache. After the transfer, the state of the copy in the source cache changes from "O" to "S." The destination cache, however, stores the copy in the "O" state. Therefore, systems that implement the MOESI protocol, like those implementing MESIF, prevent a cache from becoming a hot spot.

Note that in the MOESI protocol, a request for a "clean" copy has to come from the corresponding memory unit unless an "M" or "O" copy exists in another cache. In the MOESI protocol, the replacement of an "M" or "O" cache copy requires a transaction to update memory. Likewise, in a MOESI NUMA organization, a directory in each node logs the list of caches that have an "S" copy where only one is marked "O" (if any). Also, note that, in the MOESI protocol, an "S" copy may or may not be the same as the copy in the main memory. Directory logging of nodes with modified cached copies is the same as that in the MESIF protocol.

10.4 Virtual Memory

Modern single-core, multicore, and multiprocessor computer systems implement multiprogramming, where several single- and/or multithreaded programs (Chap. 8) execute concurrently. That is, the operating system (OS) takes turns and allocates a fraction of CPU time to each thread. On a multicore processor or multiprocessor system where there are several processing cores (CPUs), multiple threads would be executing in parallel. If each core also implements simultaneous multithreading (Chap. 8), an even greater number of threads would be executing concurrently.

In addition to the OS allocating CPU time to each thread, the OS would need to allocate main (physical) memory space for each running single- or multithreaded program, called a **process**. While individual processes may not share their allocated memory spaces, the threads of a multithreaded program do share the memory space allocated to the process (i.e., all the threads of a program can access the globally declared variables in the program). Therefore, a modern computer system must implement the following requirements involving the bottom two levels (nonvolatile memory and main memory) in memory hierarchy:

- To run a program too big for the physical memory. The program contains many instructions and large data structures that cannot be stored in their entirety in the main (physical) memory.
- To execute multiple processes, including those of the OS when there is not enough space in physical memory to store instructions and data structures for all the processes at the same time.
- To protect processes so that one process cannot access another process's memory space without permission.

A virtual memory system is a way to implement these three requirements of a modern computer system. As each program runs, its instructions and data must be copied from the nonvolatile memory (e.g., hard disk) to physical memory for execution. However, because the size of physical memory is smaller than the size of the hard disk, like a cache, only a small fraction of each process's instructions and data on the hard disk can be stored in the available space in the physical memory.

For example, a 32-bit CPU that has 32-bit address and 32-bit data buses can only read or write a maximum 4 GB (2^{32} B) memory space, organized as a $2^{30} \times 32$ memory unit. Even with this much physical memory space, it is not possible to fit all the instructions and data structures for all the processes, including those of operating system, in the memory unit.

Therefore, the total 4 GB memory space that a 32-bit CPU is able to access is interpreted as a virtual and not a physical space. Furthermore, in order to be able to run both OS as well as user processes, half of the 4 GB virtual space (i.e., 2 GB) may be reserved for a user process and the other half (2 GB) for a systems process. Alternatively, an additional bit in the CPU status register could indicate whether the CPU is operating in user mode and the address belongs to a user process or whether it is in supervisor mode and the address belongs to a systems process. The allocated virtual memory space to each process (e.g., 4 GB) is further divided into instruction and data regions (see Fig. 8.5 in Chap. 8). A very large program (>4 GB in size), however, would need to be compiled to run on a 64-bit computer system.

When the OS allocates a fraction of CPU time to each thread, the OS is said to be performing a **context switch**, which is invoked by a timer interruption (Chap. 9).

During a context switch, the state of a thread that was just executing for a fraction of CPU time is saved and the state of a thread that is waiting to execute is restored so that the execution of each thread is the same as if no context switching takes place. This gives users the illusion that there is one CPU per thread.

A context switch is typically called a **thread switch** if the switch does not change the current virtual address space the CPU is accessing. Otherwise, it is called a **process switch** where the CPU will begin to execute a thread from a new virtual space. In the rest of this section, we will focus on a process switch, which involves memory.

Like instruction and data blocks that are copied from the physical memory to cache memory, instruction and data blocks, each called a page, are copied as needed from virtual memory (e.g., hard disk) to physical memory (refer to DMA transfers in Chap. 9). The content of a physical page if dirty (i.e., modified) must be copied back on to the hard disk before it is replaced with the content of a new virtual page.

The size of each page in a modern PC system is typically 4 KB (a relatively large block), which makes DMA transfers between the hard disk and the physical memory more efficient. If memory space is divided into bigger size pages, for a given process, there would be fewer page misses (faults), much like larger block sizes that would cause fewer cache cold misses. Some systems may use variable size pages, called **segments**, but here we will focus on a page-based virtual memory management system.

10.4.1 Virtual Address Translation

Using a page-based virtual memory system, Fig. 10.13 illustrates, as an example, the state of a system with two processes. In the figure, Process 0 is shown with 12 and Process 1 with 9 **virtual pages**. The physical memory is also shown with eight **physical pages**. During the execution of a process, the memory system must map and store the content of process virtual pages in the physical memory as needed. The mapping is fully associative; therefore, the content of a virtual page can be stored in any page in the physical memory. However, unlike a memory-to-cache mapping, which is performed 100% in hardware, virtual-to-physical memory address mapping is performed partly in software (i.e., the OS) and partly by a hardware module, called a memory management unit (MMU).

In Fig. 10.13, the virtual pages are numbered 0 to 11 for Process 0 and 0 to 8 for Process 1. Also shown are, respectively, the mapping of Process 0 virtual page numbers 3, 5, and 8 to physical page numbers 5, 2, and 7, and Process 1 virtual page numbers 2 and 5 to physical page numbers 3 and 0, respectively. Physical page numbers 1, 4, and 6 are shown not occupied and therefore are free (available).

A virtual memory system uses a **page table** to keep records of virtual-to-physical page mapping information. For example, with 4KB pages, a 2-GB virtual memory would require a page table with 512K entries $\left(\frac{2\text{GB}}{4\text{KB}}\right)$. Because each process

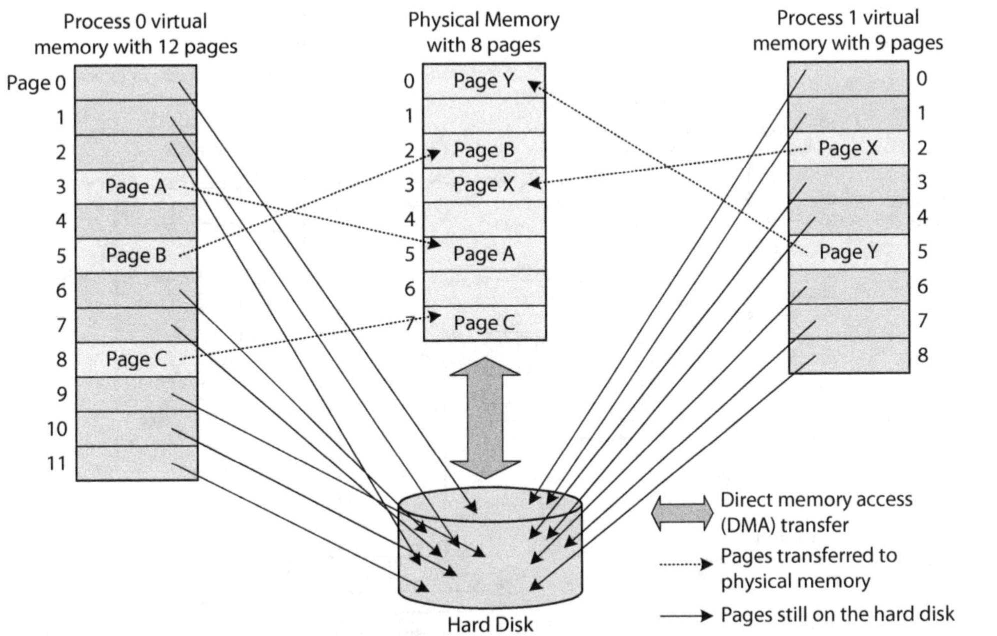

FIGURE 10.13 Illustrating virtual-to-physical memory page mapping.

has its own page table, as the number of processes increase in a system, so will the number of page tables. Therefore, in general, some least accessed table entries may be temporarily stored on the hard disk and would be copied back to physical memory as needed.

Figure 10.14 conceptually illustrates the mapping of Process 0 virtual page numbers 3, 5, and 8 to physical page numbers 5, 2, and 7, respectively. In the figure, it is assumed that each page is 256 B, each virtual memory space is 32 KB (2^{15} B) with 128 virtual pages $\left(\frac{32\text{KB}}{256\text{B}}\right)$, and the physical memory space is 8 KB with 32 physical pages $\left(\frac{8\text{KB}}{256\text{B}}\right)$. In this case, a 15-bit **virtual address** is viewed consisting of a 7-bit virtual page number (VPN, the upper 7-bits) and an 8-bit **page offset** (the lower 8-bits). A page offset, similar to a cache block offset, identifies the target byte or word within a virtual or physical page.

Suppose each 128-page table has 16-bit (2 B) entries $\left(\frac{256\text{B}}{2\text{B}}\right)$. Each entry would hold a 5-bit ($2^5 = 32$) physical page number (PPN, if any) and a set of status bits, such as a valid bit (v) and access control bits (e.g., read, write, dirty, user vs. supervisor). The valid bit, when 1 ($v = 1$) indicates the table entry contains a valid PPN.

Specifically, the execution of a program starts from virtual address 0 when a virtual memory system is implemented. Both an instruction address and a data address—for example, during the execution of an "LD" or "ST" instruction (Chap. 8)—represent virtual addresses. Using the illustration in Fig. 10.14, the following steps describe the operations performed by the MMU to translate the

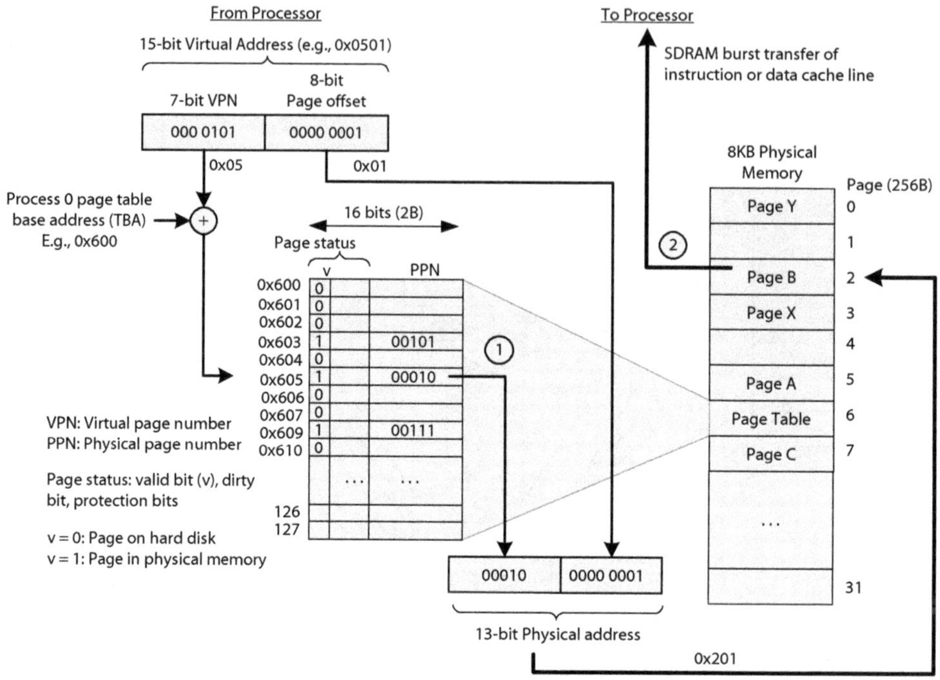

Figure 10.14 Illustrating MMU virtual-to-physical address translation steps shown for the virtual pages of Process 0 in Fig. 10.13.

15-bit virtual address 0x0501 to the 13-bit physical address 0x0201 and for the physical memory to transfer the corresponding cache line to the processor:

1. The MMU would view the 15-bit virtual address 0x0501 as a 7-bit $VPN =$ 5 (0x05) and an 8-bit page offset 0x01. Using the $VPN = 5$ as an index, the MMU would access the Process 0 page table, which is stored in the physical memory, starting at the page table base address $TBA = $ 0x600, as illustrated in the figure. The table entry that corresponds to $VPN = 5$ contains $PPN = 2 = (00010)_2$ and valid bit $v = 1$, which indicates physical memory page 2 is valid (contains up-to-date data). The $PPN = 2$ is then concatenated with page offset 0x01 to create the 13-bit valid physical memory address 0x0201.
2. Next, the physical memory would transfer the block containing the content of address 0x0201 as a cache line to processor, as illustrated in the figure.

Therefore, two separate physical memory accesses are required for the processor to receive the content of virtual address 0x0501. During the first access, the MMU accesses the physical memory directly to translate the virtual address 0x0501 to the physical address 0x0201. During the second access, the physical memory, responding to a cache miss, transfers the block containing the content of physical address 0x0201 to the processor.

In systems where virtual memory space is very large and thus requires a large page table, MMU may need to access the physical memory a number of times before completing a virtual-to-physical address translation. Multilevel is one way to organize the storage of a very large table. Each entry in a multilevel page table, except the lowest level, would contain a *TBA* for the next lower-level page table. The lowest-level page table would contain PPNs.

Suppose a system with a 256-B page size has 4 MB (2^{22} B) virtual memory space, or 16K virtual pages $\left(\frac{4\text{MB}}{256\text{B}}\right)$, much more than the one with only 128 virtual pages shown in Fig. 10.14. Figure 10.15 illustrates the organization of the 16K virtual pages as a two-level page table. This time, the MMU will view the larger 22-bit virtual address consisting of three parts as shown in Fig. 10.15: a 7-bit index to access the first-level page table (Table 1 in the figure), a 7-bit index to access the second-level page table (Table 2, again in the figure), and an 8-bit page offset (the lower 8-bits).

The MMU would use the first (highest) 7-bit index 0x00 to read *TBA* = 0x400 from Table 1 in order to access Table 2. The *TBA* to access Table 1 is 0x600 as shown in Fig. 10.15. The MMU would then use the second 7-bit index to access the target $PPN = 2 = (00010)_2$ and its $v = 1$ from Table 2, the lowest-level page table. The $PPN = 2$ is then concatenated with the 8-bit page offset to create the target 13-bit physical address. As it is illustrated, in this case, it would take two physical memory accesses, as compared to only one access in Fig. 10.14, for the MMU to translate a virtual address (e.g., 0x0501) to its corresponding physical address (i.e., 0x0201).

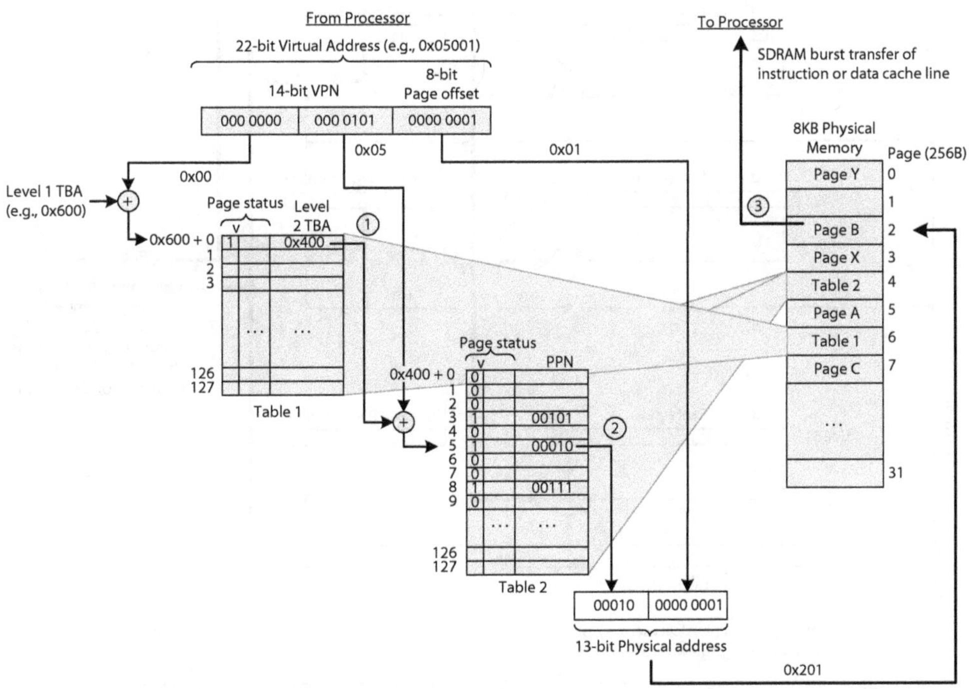

Figure 10.15 A 2-level page table organization.

10.4.2 Translation Lookaside Buffer

In order to reduce the long latency of a virtual-to-physical address translation, the most recently referenced PPNs are also kept in a special fully associative cache memory called a translation lookaside buffer (TLB). Figure 10.16 illustrates a fully associative TLB organization with 32 slots. Instead of tag and data memories that are used in direct and set-associative mapped caches, a fully associative TLB requires registers to store tags and PPNs. Without a slot or a set address, the search for a target tag is performed in parallel, as illustrated in the figure.

A TLB requires no cache coherency bits. Instead, as shown in the figure, it holds the status bits from the page table including a dirty (*d*) bit. Initially, the first time that a data page is accessed, its PPN in the TLB is marked not dirty ($d = 0$). If the processor writes (either write-through or write-back) and therefore modifies a block from that page, the *d* bit in the TLB is set to 1. The *d* bit when 1 indicates the page in the physical memory is modified, and that the page must be copied to the hard disk before it is replaced with the content of a newly translated virtual page. Also, a TLB, being fully associative, must implement a slot replacement algorithm.

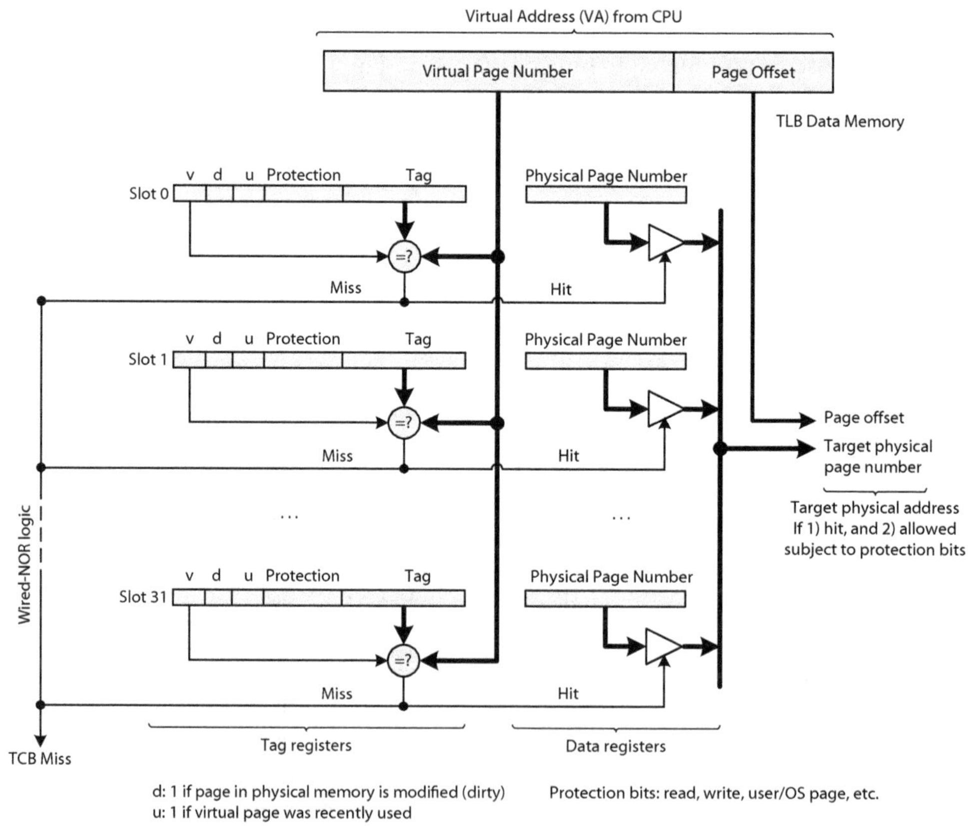

Figure 10.16 The data path of a fully associative TLB illustrating a TLB read (not all details are shown). A miss will cause the MMU to translate the virtual address to a physical address.

A replacement algorithm that requires less hardware, as compared to, say, LRU, is to use a single use (u) bit in each tag register as shown in the figure. Each time that a PPN is accessed from the TLB, its u bit is set to 1. All the u bits would then be periodically reset to 0 so that the list of pages that are not accessed recently is identified. Any time that a TLB access results in a miss, the MMU would start a new virtual-to-physical address translation, and would then select one of the slots with its $u = 0$. If the d bit in the selected slot is 1 (indicating a modified page in physical memory), the d bit in the page table for the modified page would be set to 1. The slot will then be updated with the newly determined PPN.

10.4.3 Processor Organization

Figure 10.17 illustrates three processor internal organizations. In Fig. 10.17(a), two TLBs quickly translate two virtual addresses, one for instruction and one for data (if any), to their corresponding physical addresses before each physical address is applied to its respective L1 cache. This organization has the advantage of using

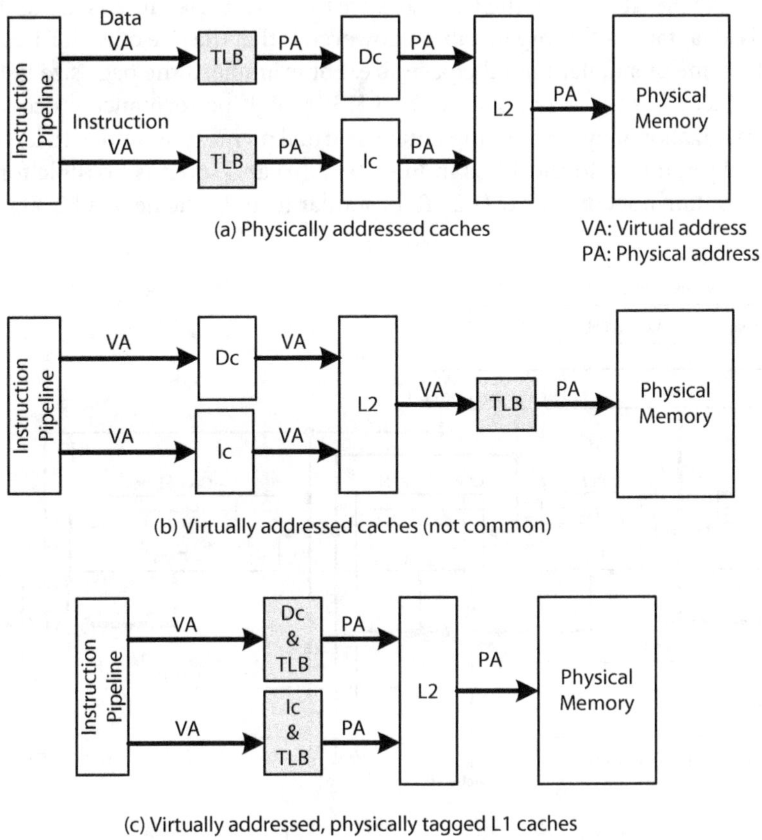

Figure 10.17 Alternative processor organizations: (a) long L1 cache latency; (b) short L1 cache latency, but OS must flush caches on each process switch; (c) short L1 cache latency but its size is tied to page size.

physically addressed caches as opposed to **virtually addressed caches** used in Fig. 10.17(*b*). Its disadvantage, however, is a relatively long L1 cache latency.

The organization in Fig. 10.17(*b*), which uses virtually addressed caches, requires a single TLB to quickly translate a virtual address from the lowest-level cache (L2 in this case) to its corresponding physical address before an access from the physical memory is made. This organization, however, is not very common, as it requires the OS to flush all the caches during a process switch. Without a flush, caches would not be able to differentiate between the content of, for example, $VPN = 5$ of Process 0 from $VPN = 5$ of Process 1 in Fig. 10.13. The advantage of this organization is that the design does not increase the latency of L1 caches.

In the organization of Fig. 10.17(*c*), a TLB is embedded with each of the L1 caches. As illustrated in Fig. 10.18, an L1 tag memory retains PPNs as tags. During a cache access, a hit/miss is determined by comparing a PPN produced by the TLB with the one read from the tag memory. Such a cache is said to be virtually addressed but physically tagged, such as the one used in the AMD Opteron processor.

This organization has two advantages: (1) it uses physically addressed caches; and (2) the latencies of the L1 caches are the lowest, the same as in Fig. 10.17(*b*). The disadvantage of this organization, however, is that the size of each L1 cache has to be the same or smaller than the page size. For example, if the page size is 4 KB, then an L1 cache has to be a maximum of 4 KB. A high-performance processor using this organization may need to implement a virtual memory system with larger page sizes.

In addition to the TLBs in Fig. 10.17 (*a*) and (*c*), it is possible for a processor to contain a second-level (L2) TLB, similar to an L2 cache. An L2 TLB would store

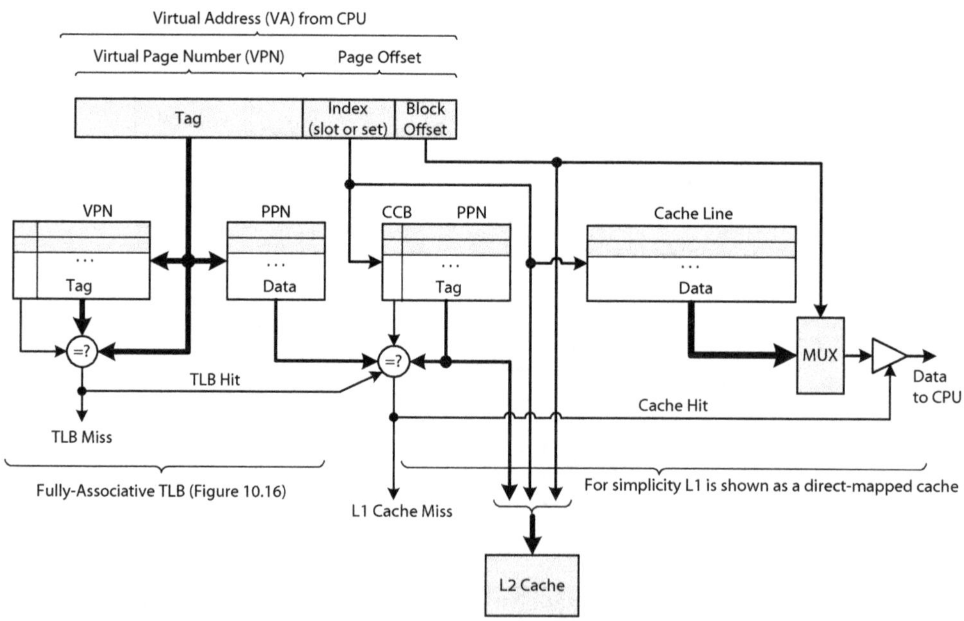

FIGURE 10.18 The organization of a TLB embedded with the cache.

a larger set of the recently referenced PPNs. If the search for a PPN in one of the L1 TLBs results in a miss, the search will continue with the L2 TLB. If the search still results in a miss, the MMU would be triggered to translate a target VPN to its corresponding PPN. An L2 TLB, typically being larger than an L1 TLB, would be implemented as a direct- or set-associative mapped cache.

References

1. John Hennessy and David Patterson, *Computer Architecture: A Quantitative Approach*, Morgan Kaufman, 5th ed., Waltham, 2012.
2. Inoue Koji, Ishihara Tohru, Murakami Kazuaki, Way-predicting set-associative cache for high performance and low energy consumption, *ACM*, 1999, 173-275.

Exercises

10.1. We would like to improve the estimated average memory latency in Example 10.1. Suppose, instead of SDRAMs, the memory unit is designed using DDR SDRAMs. Recalculate the average memory latency.

10.2. The estimated average memory latency in Example 10.1 is computed for peak performance. Suppose in the worst-case scenario none of the request and response transactions can be overlapped. Recalculate the estimated average memory latency, assuming that every SDRAM access (from the time a row address is issued until the first data item appears on the bus) takes five SDRAM clock cycles. Ignore the time required to deactivate a row.

10.3. Consider the following four memory locations accessed *N* times in a loop by CPU, and suppose a memory address is partitioned into tag, slot, and offset, as shown. Do the following:

0x3C1C (16-bit address)
0x0421
0x041F
0x0C88

Tag	Slot	Offset
6	6	4

a. Determine the number of misses in the first round, assuming the cache is initially empty.

b. Determine the total number of misses for *N* rounds, assuming the cache is initially empty.

c. Suppose the cache mapping is changed to a two-way set associative. Calculate the tag, set, and offset field sizes, and then determine the number of misses in the first round, assuming the cache is initially empty.

d. Determine the total number of misses for *N* rounds, assuming the cache is initially empty and a cyclic replacement policy (if needed) is used.

10.4. Repeat Exercise 10.3(a) through (d) for the following four addresses:

0x0C1C (16-bit address)
0x0521
0x041F
0x4D28

10.5. Repeat Exercise 10.3(a) through (d) for the following four addresses:

0x3C1F (16-bit address)
0x042C
0x0460
0x3C1D

10.6. Repeat Exercise 10.3(a) through (d) for the following four addresses but assume a four-way set-associative cache for parts (c) and (d):

0x3C17 (16-bit address)
0x3817
0x3917
0x1C17

10.7. Suppose main memory is 64KB, cache is 4KB, and block size is 16B. Determine the tag, slot, and offset field sizes for direct-mapped cache.

10.8. Consider a four-way set-associative cache. Discuss the complexity (e.g., hardware required) of implementing one of the following replacement algorithms.

a. Cyclic: Slots within a set are selected in cyclic, first-in, first out fashion (i.e., slot 0, 1, 2, 3, 0, 1, 2, etc.).

b. LRU: The least recently used slot is selected to be replaced. (Hint: Consider a four-element LRU algorithm using a 4 × 4 matrix with 1-bit entries. Suppose the four slots are numbered 0 to 3. Each time that one of the slots is referenced, the corresponding four row entries in the matrix are set to 1 and then the corresponding four column entries are set to 0. For example, suppose slot 0 is accessed first; then row 0 of the matrix will be $(0111)_2$, and the remaining rows will be $(0000)_2$. Suppose slot 2 is accessed next; then the rows of the matrix, in order, will be $(0101)_2$, $(0000)_2$, $(1101)_2$, and $(0000)_2$. The LRU slot is the one when its corresponding row is $0 = (0000)_2$.)

c. Random: A slot within a set is randomly selected to be replaced.

10.9. Consider the write-through protocol in Fig. 10.10. Determine which FSD transition will take place when the high-level language program statement "A = 1;" executes the first time.

10.10. Consider the MESI cache protocol in Fig. 10.11. Determine which FSD transitions will take place when the high-level language program statement "A = A + 1;" executes the first time.

10.11. Consider the MESI $I_{RM} \rightarrow E$ and $E_{WM} \rightarrow M$ transitions that take place when variables that are not shared are accessed. State how programmers can use this information when writing a multithreaded program so the program runs more efficiently on a multicore or multiprocessor system.

10.12. Consider a two-processor system with two MESI caches, C0 and C1. Suppose the processors execute two threads, T0 and T1, that share variable A. Outline a scenario when the MESI transition $E_{XWH} \rightarrow I$ will take place, assuming memory block B_A contains A.

10.13. Consider the following two threads, T0 and T1, and the system in Fig. 10.9. Assume initially $x = 0$ and $y = 0$. Suppose P0 executes T0, P1 executes T1, B_x contains x, and By contains y. Use the following table to indicate the state transitions of block B_x and B_y in the caches as T0 and T1 execute. In the assembly code listings, the execution order of memory reference instructions are shown as comments. For example, "STA (y)" of T0 executes first (//1), then "LDA (x)" of T1 executes next (//2), etc. Also, state the number of times memory is updated or will eventually be updated in each case.

Thread T0		Thread T1	
Program Code	**Assembly Code in Acc-ISA (Chap. 8)**	**Program Code**	**Assembly Code in Acc-ISA (Chap. 8)**

y = 1;
x = 1;
	LDA 0
	STA (y) //1
	LDA 1
	STA (x) //3
		white(x == 0)
		{} //wait
		y = y + 1;
			L1: LDA (x) //2, 4
			CMP 0
			JEQ L1
			LDA (y) //5
			ADD 1
			STA (y) //6

Thread: Instruction	Block State in C0	Block State in C1	Update Cache, Memory, or Both (if any)
T0: STA (y) //1			
T1: LDA (x) //2			
T0: STA (x) //3			
T1: LDA (x) //4			
T1: LDA (y) //5			
T1: STA (y) //6			

a. Write-through protocol
b. Assume MESI protocol
c. Assume MESIF protocol
d. Assume MOESI protocol

10.14. Repeat Exercise 10.13, except that this time, T0 and T1 execute the memory reference instructions in a different order, as given in the following table.

Thread T0	Thread T1
Assembly Code in Acc-ISA (Chap. 8)	**Assembly Code in Acc-ISA (Chap. 8)**

LDA 0
STA (y) //2
LDA 1
STA (x) //3
	L1: LDA (x) //1, 4
	CMP 0
	JEQ L1
	LDA (y) //5
	ADD 1
	STA (y) //6

10.15. Briefly explain how when a cache becomes a hotspot, it would increase average memory latency.

10.16. In each of the following architectures, state when will memory be updated:

a. Bus-based UMA with MESI protocol
b. NUMA architecture with MESIF protocol
c. NUMA architecture with MOSEI protocol

10.17. Suppose a system has 16 KB virtual memory space, 16 B page size, and 2 KB physical memory. Do the following:

a. Determine the number of virtual and physical pages.

b. Assuming that each page table entry is 2 B, what is the maximum size of a page table?

c. Design a page table organization to translate a 16-bit virtual address to an 11-bit physical address.

10.18. Consider a TLB; answer the following questions:

a. Briefly explain the purpose for a TLB (e.g., what if no TLB is used?).

b. Explain why a TLB should be designed as fully associative cache (e.g., what if it is implemented as a direct-mapped cache?).

10.19. Discuss the benefits of using a use (u) bit versus the implementation of LRU algorithm in hardware. Also see Exercise 10.8.

10.20. Computer security (secure virtual memory): See Exercise 11.32 (also see Sec. 11.11).

10.21. Computer security (virtual memory replay attack): See Exercise 11.33 for how to detect virtual memory replay attacks (also see Sec. 11.11).

10.22. Computer security (memory authentication task): See Exercise 11.34 (also see Sec. 11.9.2 and Sec. 11.11).

10.23. Computer security (preventing information leakage): See Exercise 11.35 for how to implement randomized encryption (also see Sec. 11.11).

10.24. Computer security (secure program execution): See Exercise 11.37 for how to set up a trusted program for secure execution (also see Sec. 11.11).

10.25. Computer security (preventing information leakage more efficiently): See Exercise 11.38 for how to prevent information leakage using less memory (also see Exercise 10.22).

10.26. Computer security (organization of virtual address space to support secure execution mode): See Exercise 11.39 for how to allocate multiple virtual address spaces (also see Sec. 11.11.8).

CHAPTER 11

Computer Architecture: *Security*

11.1 Introduction

Throughout the previous chapters, we focused on digital design techniques and computer architecture concepts to improve performance. Additional techniques and concepts are needed to design a secure computer. Today, more people and organizations are using computers, and thus, not only are they generating large amounts of data, but also creating new application software, some possibly with security holes. This creates opportunities and benefits for a range of attackers, from an individual hacker to a cyber-war army. Many organizations, such as government (e.g., military), financial institutions (e.g., banks), infrastructures (e.g., power grids), service industries (e.g., law firms), commercial businesses (e.g., e-commerce), industrial complexes (e.g., factory control systems), and social networking companies (e.g., Facebook) all have digital assets (programs, documents, data, etc.) that need protection.

Assets, especially those in electric power grids and industrial factory control systems, may have numerous security problems [1]. For example, they can become a cyber-war target for an enemy country that is willing to spend time and resources to develop a complex cyber-weapon. All security problems, however, can be characterized as having one of three **security properties** as follows:

Confidentiality. Concealing assets and/or preventing unauthorized access (e.g., eavesdropping). This includes programs, documents, data, etc., stored on the hard disk and, in some cases, instructions and data stored in memory.

Integrity. Ensuring that unauthorized modifications of assets are detectable and, if possible, preventable. For example, **code injection** would change the integrity of a program in memory, and illegal data modification would change the integrity of a database.

Availability. Preventing attacks that deny service to legitimate users. These attacks can have many forms and include **server overload**. A malicious attack can consume resources like memory and network bandwidth, and may cause overloading that slows down or stops the computer from carrying out its intended services.

How important each of the three security properties is for an organization depends on the types of assets and how they are used. For instance, the integrity of bank accounts, perhaps more so than their confidentiality, is critical for a banker; maintaining accurate balances is more important than, say, safeguarding knowledge of assets. Likewise, while it is necessary, for example, for students to have access to computers at university campuses, an interruption and thus unavailability of computers for a few hours may not be that crucial. On the other hand, all three properties would be essential for a government agency such as the military.

While the scope of computer security is large and covers many subject areas, this chapter presents an introduction to computer security topics related to computer architecture. In addition, even within the computer architecture domain, the scope is large and evolving; new concepts and methods are currently being researched. The chapter introduces computer security for computer architects and presents heuristic solutions. Others, including information-flow tracking methods, are deferred to the Further Readings section and elsewhere. The information flow-tracking techniques require every critical bit or word in memory to be marked as secure or not secure. For example, data entered via an I/O device would be tagged as not secure, and systems data would be tagged as secure. A marked data item would then be tracked as it enters the CPU. Controls would be implemented to prevent unauthorized modifications of CPU state (i.e., registers). These methods require additional memory space to store tags and may require altering hardware design practices (logic circuits, data path, memory organization) prevalent today.

Where a piece of software or hardware is developed and installed can lead to security problems. This is especially important today when many software and hardware companies rely on using third-party modules that may not be designed or implemented correctly or that may contain Trojans (illicit codes or HDL models). Software security policies and mechanisms are often based on some proven models, such as those used in military, and hardware security policies and mechanisms are based on a set of techniques to prevent an attack. The chapter presents examples of known software security models and their applications, as well as an introduction to security policy mechanisms applicable to hardware.

While there may be unlimited ways for attackers to exploit software security holes, software attacks are typically the result of inserting invalid data or copying data from one section to another in memory. For example, consider a poorly written C program that uses the "strcpy" (string copy) library function call to copy its command-line argument into a locally declared array (a buffer) within a subroutine. In this case, an attacker may use a specific argument value to "**spoof**" and cause a **buffer overflow attack** [2]. In general, attackers may use a statically or dynamically allocated buffer in a program to modify the memory stack area where subroutine return addresses are stored. They may then use the buffer to embed malicious codes and change a subroutine return address on the stack in order to, for example, run malicious software (malware) or invoke and use tools available on the system in privileged mode. Software attacks can be used, for example, to intentionally disrupt or overload a system, making it unavailable or too slow to conduct business as normal. An attacker may gain access to classified documents, modify a database, or trigger a **hardware Trojan** that could cause a hardware malfunction, leak secret information, etc.

Attackers may also use spoofing and other techniques to perform **physical attacks** [3–5] when they have exclusive access to a system. An attacker may be able to use sophisticated equipment to spoof or observe signal values for the purpose of gaining access to a portable device or performing reverse-engineering tasks on the device. The chapter also introduces spoofing and other techniques used to perform software/physical attacks.

Because, in general, there are too many security holes to fill, it is not possible to securely design, develop, and install every piece of hardware, firmware, and software used in a secure system. What is necessary, however, is to have a trusted computing base (TCB) to implement a secure system [6]. TCB, which must be designed to be secure and dependable (i.e., remain trustworthy), referrers to a minimum set of hardware and firmware and their secure implementation (design, develop, and install) requirements. In addition, depending on the security requirements of a system, TCB may include software that not only must be implemented securely, but the software must also execute securely. The following is a list of security application areas:

- For a handheld device to securely exchange data with a host computer and guard against physical attacks
- For system designers to implement security policy mechanisms to prevent unauthorized access to systems resources: password files, systems stack memory area, etc.
- For system designers to implement security policy mechanisms when the commodity operating system (OS) frequently is compromised [3, 7–9]
- For software companies to be able to develop and build secure application-dependent security policy mechanisms
- For a software company to securely distribute programs for remote installation
- For a user to be able to conceal information in the form of documents, data, pictures, etc., and store it securely on a local or remote disk drive; securely send and receive e-mails; perform secure remote login; etc.

- For a company to implement security policy mechanisms to prevent unauthorized access to its vital business resources: personnel data, customer data, intellectual properties, etc.
- For an entertainment company to target the delivery of its products to only authorized handheld devices
- For a cloud computing company to provide certified execution service to its customers, for example

Finally, the chapter introduces confidentiality and integrity techniques used in the implementation of software/physical security policy mechanisms, and provides architecture examples of coprocessor- and processor-based TCB and their application areas.

11.1.1 Security Engineering Methodology

The security engineering methodology (SEM) shown in Fig. 11.1 provides a step-by-step procedure for designers to identify potential threats; develop required security policies and mechanisms; and design, verify, and evaluate a computer architecture for security. The analysis of usage scenarios and potential security problems defines the scope of security risks. Usage scenarios are often application dependent and may involve a wide class of systems, such as embedded systems, real-time systems, and distributed systems, and may cover many industries, including IT, manufacturing, healthcare, business, etc. [10–11]. The analysis of threat model, security policy, and security mechanisms identifies a possible list of threats and required security policy and mechanisms for each threat.

To better understand the SEM, Table 11.1 presents the development of a paper-based security mechanism for changing students' grades as an example of university assets.

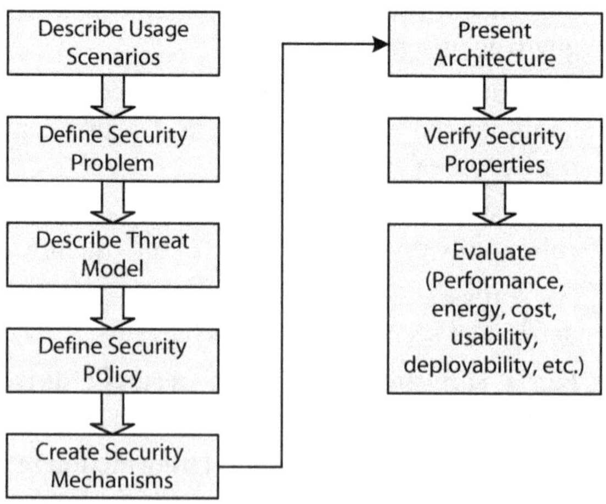

Figure 11.1 A security engineering methodology [12].

Methodology	Description
Usage scenario	How to change a student's grade
Security problem	Unauthorized grade change
Threat model	Inaccurate grades
Security policy	Only the instructor who taught the course can change a student's grade.
Security mechanisms	1. Instructor assigns a new grade for one of his or her student. 2. Department chairperson validates the grade change. 3. A fulltime staff (not a temporary helper) enters the new grade in the secure university database system.
Architecture	Create a grade-change form with spaces allocated for student name, student ID number, course number and semester taken, reason for changing the grade, and signature spaces for instructor and chairperson. Also, a completed form is either mailed using the secure on-campus mail system or hand-delivered to student records office.
Verification	Confidentiality: The grade-change forms are only handled by faculty and staff and thus remain confidential to other students. Integrity: The chairperson's signature authenticates a completed grade-change form. Availability: Depends on instructor's decision to change a student's grade.
Evaluation	The cost of grade-change forms, the time required to complete and process a grade-change form, etc.

Table 11.1 Developing a Paper-Based Security Mechanism to Change Students' Grades

In Fig. 11.1, the usage scenarios may be expressed as a set of **use cases**, and a threat model as a list of **threat vectors**. In general, a threat vector can be viewed as the attack path that could lead to stealing, damaging, or disabling a personal, business, IT, or other asset.

Table 11.2 lists data storage and remote server connection as two examples of computer use cases. Viruses and other types of malware, as well as the theft or other loss of a computer, can present certain security risks. A malware can delete, modify, and/or steal data stored on the hard disk. A lost or stolen computer (e.g., desktop, laptop, etc.) or a handheld device (e.g., smartphone) that contains valuable information can be a security risk; data stored on the hard disk or flash memory is susceptible to tampering. In addition, if the lost device belongs to a company, there is a chance that someone will wrongfully gain access to servers in the company and cause damage; important company files may be deleted or modified, or proprietary company business information may be stolen.

Each of the two use cases in the table includes potential security vulnerabilities, a threat model, a security policy, and a list of security mechanisms. However, the security mechanisms are described in layman's terms, and more precise solutions will be provided later.

Usage Scenario	Security Problem (Vulnerabilities)	Threat Model	Security Policy	Security Mechanisms*
Data storage on the hard disk	Storage of documents, user authentication data (e.g., username/ password), etc. on the hard disk is susceptible to tampering	Data theft	Keep data confidential	Keep data "locked" (concealed) on the hard disk requiring a "key"; without the "key" data remains confidential (stored in a meaningless way); do not store the "key" on the hard disk but hide it in the system's hardware.
Network authentication	Storage of network authentication data on the hard disk is susceptible to tampering	Unauthorized access to servers	Authenticate the platform (e.g., laptop), not the user	Use "dual-key locks" with each company laptop; use one "key" to "lock" and store the network authentication data on the hard disk and hide the "key" tied to the laptop's hardware; keep the second "key" with the company or with a trusted different company; use the second key to "unlock" the received but "locked" network authentication data; invalidate (trash) the second "key" if the laptop gets lost or stolen.
*For now the security mechanisms are stated in layman's terms.				

Table 11.2 Five Steps of SEM Applied to Two Computer Usage Scenarios

11.1.2 Threat Classes

Threats may be grouped into two broad categories: **operational** and **developmental**.

Operational Threats

Operational threats depend on asset types and usage scenarios, such as those listed in Table 11.2. Operational threats also depend on the tools used in the implementation of security mechanisms. While a security mechanism itself might be well designed, the tools (e.g., "locks") might be weak; for example, the "locks" themselves might be of poor quality and easily unlocked. Other examples of operational threats are briefly discussed next.

Devices, such as smart meters installed by utility companies in houses or portable devices used by the military or during an emergency by firefighters or emergency medical personnel, may face additional threats. These devices may require a secure communication channel with a host computer, but an attack may cause communication interference, for example, during an emergency.

Remotely installed devices that are physically accessible, as well as portable devices, are also subject to physical attacks. For example, an embedded device that is installed in an automobile and tracks the vehicle's odometer reading may be tampered with in order to change and reduce the odometer reading of an old car. Likewise, a high-tech portable device may be hacked to modify or reverse engineer its functions.

Developmental Threats

These threats depend on trust models used for software and hardware development, as well as delivery and installation of software. Incomplete specifications, incorrect implementations, and improper security policies and mechanisms are, in general, the three sources of software, hardware, and firmware vulnerabilities. Often, these vulnerabilities are unintentional, but they are sometimes intentionally caused by designers.

For instance, large integrated chip (IC) designs, especially processors, can include designer-injected malicious circuits, for example, by adding extra lines of hardware description language (HDL) code on purpose in the design [13]. In general, an intentionally caused vulnerability is harder to detect during validation. Many factors, including the growing use of third-party "soft" components (e.g., Verilog models) in hardware designs, increase vulnerability to attacks.

11.1.3 Access Control and Types

The data storage security mechanisms described in Table 11.2 are designed to protect data stored on the hard disk from malware or physical attacks. Other usage scenarios require mechanisms to restrict access. For example, in an organization, who should be able to access an employee's personal data (e.g., salary, Social Security number, etc.)? The answer might be that perhaps only an employee's supervisor should be allowed to examine employee's personal data, and the data should remain confidential to all other employees.

Likewise, in a computer system, only systems programs, not application programs, should be able to access systems data. Such security mechanisms are called **access control**. The **access control list** (ACL) used, for example, in Linux/Unix systems decides which files and directories (folders) each user can access. Users can use the command "chmod" to assign read (r), write (w), and/or execute (x) rights to each of their files and folders. For instance, for file foo, which initially had read/write (rw-) permission assigned to all users, the command "chmod 640 foo" assigns read/write (rw- indicated by 110 in binary) to the owner (i.e., Joe Smith), read-only (r-- indicated by 100 in binary) to the users in the group (e.g., system people), and no access (--- indicated by 000 in binary) to all other users. This is illustrated as shown:

```
>ls -l
-rw-rw-rw- 1 smithjoe faccsc 5 May 22 12:32 foo

>chmod 640 foo
>ls -l
-rw-r----- 1 smithjoe faccsc 5 May 22 12:33 foo
```

With the ACL, even if files and folders are not concealed, they will remain confidential and not accessible to some users. The Linux/Unix ACL is an example of a **discretionary access control** because each user decides what access rights he or she wants to assign to each of his or her files and folders. On the other hand, a **mandatory access control**, which may be rule based, enforces a set of confidentiality and integrity security rules to all subjects (people and programs) or all assets (objects) in an organization. For example, if only employees with the title of supervisor are permitted to examine employees' personal data, then the mandatory access control is called a **capability list** (CL) [14, 15]. Each subject is assigned a list of capabilities—a set of permitted actions that the subject is allowed to do with respect to all the objects in an organization. On the other hand, a mandatory access control that is organized by objects, not by subjects, is called a **mandatory ACL**, similar to the Linux/Unix **discretionary ACL** example discussed earlier. To illustrate an example, Table 11.3 presents an **access control matrix** created from a university's grading policy.

In the table, the rows are subjects (instructors, students, etc.), columns are objects (courses), and the matrix entries are a set of access rights to students' grades. The access rights are defined as read (R or r), write (W or w), both read and write, or neither read nor write. The capital letters R and W indicate read and write access for all grades in a single course. The lowercase letters r and w indicate a read and write access to a single grade, respectively. No-read and no-write accesses are shown as blanks in the table. The matrix is shown with one department chairperson *p*, two instructors *x* and *y*, two students *s1* and *s2*, one staff employee *e*, and three courses labeled *A*, *B*, and *C*. The matrix entries also show instructor *x* can assign (R/W) grades to students enrolled in courses *A* and *C*, instructor *y* can assign grades to students enrolled in *B*, student *s1* can read (r) her or his grades in courses *A* and *C*, and student *s2* can read his or her grades in courses *B* and *C*. The matrix contains no single write (w) access rights.

	Objects: Grades		
Subjects: People	**Course A**	**Course B**	**Course C**
Chairperson: p	R	R	R
Instructor: x	R, W		R, W
Instructor: y		R, W	
Staff employee: e	R	R	R
Student: s1	r		r
Student: s2		r	r
Student: ...	...	...	...
...	...	...	...

Table 11.3 An Access Control Matrix for Students' Grades

An access control matrix would produce a CL if the matrix is stored in terms of its rows. For example, chairperson *p*, instructor *x*, and student *s1* each has the following list of capabilities, shown as a relation:

```
 p: {(A, R), (B, R), (C, R)}
 x: {(A, R), (A, W), (C, R), (C, W)}
s1: {(A, r), (C, r)}
```

For example, the assigned capabilities for instructor *x* are read access and write access to all grades in courses *A* and *C*. The assigned capabilities for student *s1* are read access to his or her grade in each of the courses *A* and *C*.

On the other hand, as shown next for courses *A* and *B*, an access control matrix would produce a mandatory ACL if the matrix is stored in terms of its column data:

```
A: {(p, R), (x, R), (x, W), (e, R), (s1, r), ...}
B: {(p, R), (y, R), (y, W), (e, R), (s2, r), ...}
```

For example, course *A*'s access list indicates chairperson *p* and staff *e* have read access to all students' grades, *x* has both read and write access to all grades, and student *s1* has only read access to his or her grade. Because an ACL is a data-oriented mechanism, it is easier to change rights for an object—for instance, to add student *s3* to course *A*.

A program that wants to access an object (e.g., a file, data item, certain memory locations, network connection, USB port, etc.) must be listed in the object's access list; otherwise, access is denied. In contrast, a CL is subject oriented, and thus a subject can delegate (i.e., pass) its assigned capability list fully or partially to another subject. For example, instructor *x* can delegate his/her capability item "(*C*, W)" to chairperson *p*, who then becomes the responsible person (not instructor *x*) to assign (enter) grades to all students in course *C*.

While an ACL-based system is easier to implement, a CL-based system can provide better protection; a user or process can only access objects that are in its capability list. A CL-based system can also provide finer protection; a delagetee's responsibilities could be limited to a subset of data, memory locations, tasks, etc. For instance, instructor *x* can delegate only the responsibility of assigning a single grade, such as (*C*, w(*i*)), to chairperson *p*, where index *i* identifies a single student, such as *s2*, in course *C*. However, once a capability is delegated, there is no control; the capability could again be delegated to another subject. A system may use a hybrid approach to take advantage of both ACL and CL schemes. The discussions on other access control schemes, such as role based and originator based, are deferred to elsewhere.

11.1.4 Security Policy Models

Confidentiality and integrity security policy models must be able to create a complete security perimeter for a computer system that covers all its hardware and software components. As an example, consider the Flame virus that can activate a computer's audio system to listen in and transfer office chatter through the network, capture screenshots, log keystrokes, and even steal data from Bluetooth-enabled cell phones that are near the computer. Likewise, the Stuxnet cyber-weapon is able to enter an industrial control system through a universal serial bus (USB) port and change the operating specifications of the control system—for example, it can cause an industrial motor to spin too fast and actually cause physical damage.

A mandatory access control is typically defined based on some proven confidentiality and integrity security policy models, like those used in military and business environments. Security policy models are characterized as **multilevel** (hierarchical) or **multilateral** (compartmental). The following is a description of some well-known multilevel and multilateral security policy models.

Multilevel Models

A multilevel model is used in places where access to information is naturally hierarchical, like the military or a medical office. In the military, both subjects (people) and objects (e.g., documents) are assigned clearances and classifications, such as, "top secret," "secret," "confidential," and "unclassified." In a medical office, only doctors are permitted to access certain patient medical records. A security policy model is then used to control who can read or write each type of document or a medical record.

Bell-LaPadula (BLP) [16] is a multilevel security model that was designed to enforce confidentiality in the military. The BLP's "**no write down**" policy prevents an employee with a higher clearance from writing or appending a document with a lower classification level. In addition, the policy prevents the **flow of information** from a higher classified object to a lower classified object. For instance, a corrupted army general with top-secret clearance will be prevented from reading a classified document and then transferring the information to an unclassified document. This is known as the ***-property** of the BLP model. In a computer system, the *-property can prevent the general from copying a classified file to a USB flash memory, which would have a low classification in the system. (Also, all high-clearance personnel may be prevented from taking smart phones into their offices.)

The BLP's "**no read up**" policy prevents subjects that have low clearances from accessing objects that have high classifications. The policy can prevent an unclassified employee in the military from reading a top-secret document. The policy can also prevent malware that was downloaded from the Internet, and thus is assigned a low clearance in the system, from accessing a high-classified object, such as the password file or confidential user data.

The integrity **Biba** [17] model enforces "**no write up**" and "**no read down**" policies. Both subjects and objects are assigned integrity levels (or labels); for instance, system files are labeled high and network files low. The "no write up" policy can prevent a low-labeled malware downloaded from the Internet from modifying a high-labeled systems data, such as replacing a subroutine return address stored on the system's stack. The "no read down" policy can be used to lower the integrity label of a systems program as soon as the program receives data from the network. In this case, even if malware is somehow able to gain an administrator privilege (i.e., [forks a root shell]), its integrity label will still be low and thus it cannot write the password file, for example. However, the malware would still be able to read (up) the password file and therefore would be able to transfer (write down) the file through a network connection unless the BLP policies are also implemented.

With both BLP and Biba policies implemented, "writing up," "writing down," "reading up," and "reading down" will not be permitted. As a result, the combined policies produce a stronger security model. However, the combination can potentially create access restrictions in some applications, such as a database, when data sharing might be necessary. An example implementation of Biba is the LOMAC, a mandatory access control extension to the commercial off-the-shelf Linux OS [18].

Multilateral Models

A multilateral model is used in places where access to information is not hierarchical but compartmental, such as separation of duties in business transactions or activities that are deemed to create conflicts of interest.

The BLP model for confidentiality does not apply to nonhierarchical service-oriented businesses such as law firms, accounting firms, and advertising agencies that may have competing clientele. An employee of these service industries may receive sensitive client information that must be protected and not shared with other clients in the same industry. This type of information confidentiality is not multilevel, but rather multilateral. The **Chinese Wall** [19] multilateral model is designed to prevent conflict of interest.

For example, a law firm that has clientele from various industries (e.g., banks, oil companies, etc.) should not allow its employees to engage in activities that have a conflict of interest and may result in sharing one client's business information (e.g., Citibank's) with another in the same sector (e.g., Wells Fargo Bank).

Likewise, Biba, a multilevel integrity model, does not work in commercial environments. For example, a subject (an employee in a company or software) that enters an order to purchase merchandise should be different from the subject that receives the merchandise and pays for it. In this case, a multilateral integrity model, such as the **Clark-Wilson** [20] model that implements the principles of separation of duties in business transactions, is used. Table 11.4 presents a summary of these security policy models.

11.1.5 Attack Classes

Developmental threats, as discussed earlier, can produce unintentional and sometimes intentional vulnerabilities in the form of backdoors. A hardware backdoor attack is generally due to the presence of one or more malicious circuits (hardware Trojans) within hardware modules used to build a system. Given the large size of the modern ICs, malicious circuits in the design are unlikely to be discovered during validation. Moreover, almost all FPGAs now built elsewhere and some ICs may contain a remotely activated "kill switch" [21]. Thus, hardware backdoor attacks could present serious security risks. After the IC is fabricated and installed in a system, a malicious circuit may be triggered remotely, either by using malware or by having full access to the system and executing a triggering program.

On the other hand, malware such as viruses and spyware are examples of attacks due to generally unintentional backdoors in software, which are exploited

Security Policy Model		Description	Example
Multilevel	BLP: Confidentiality	"no read up": Subject *S* can read object *O* if and only if $CL(S) \geq CL(O)$	An LC-S (e.g., a downloaded software) cannot access an HC-O (e.g., systems data)
		"no write down": Subject *S* can write object *O* if and only if $CL(S) \leq CL(O)$	An HC-S (e.g., a systems program) cannot transfer data through an LC-O (e.g., a network connection)
	Biba: Integrity	"no read down": Subject *S* can read object *O* if and only if $IL(S) \leq IL(O)$	An HC-S (e.g., a systems program) cannot access an LC-O (e.g., data down-loaded from the Internet or data from a USB device)
		"no write up": Subject *S* can write object *O* if and only if $LI(S) \geq IL(O)$	An LC-S (e.g., a downloaded software) cannot modify an HC-O (e.g., systems memory stack)
Multilateral	Chinese Wall: Confidentiality	Prevents conflict of interest (COI); requires both discretionary and mandatory access controls; organizes the objects (Os) into sets of COI; one can choose (i.e., discretionary access) and access only (i.e., mandatory access) one O in each COI set; also introduces a time element.	An S who has recently accessed object $O_i \in COI_k$, then S cannot access $O_j \in COI_k$ for some period of time. Two divorce attorneys who work in the same law firm and each represents a member of a married couple should not be able to access the other member's divorce file.
	Clark-Wilson: Integrity	Integrity of data and transactions; enforces separation of duty; if a certified transaction CT_i is allowed with a certified data CD_j, then they forms a certified relation shown as (CT_i, CD_j).	If subject *S1* operates on (e.g., orders) object *O* and creates the valid relation (*S1*, CT = *"order"*, CD = *O*), then (*S1*, CT = *"receive"*, CD = *O*) is an invalid relation; (*S2*, CT = *"receive"*, CD = *O*) where $S2 \neq S1$ is however a valid relation; thus, *S1* and *S2* have separate duties.
S: subject (person or program); O: object (file, network connection, USB port, data, etc.);			
LC: low-clearance or low-classification; HC: high-clearance or high-classification;			
CL: confidentiality level; IL: integrity level; CT: certified transaction; CD: certified data			

Table 11.4 Multilevel and Multilateral Security Policy Models

by attackers. Physical attacks are possible when special equipment is interfaced to the hardware to alter its behavior during its normal operation. Both software and physical attacks use similar attack mechanisms.

There are still other types of attacks known as **side-channel** attacks that are not due to backdoors, but rather are based on some side information gathered while a program is executing. Examples of information used for side-channel attacks are program execution time, known as a **timing attack** (see Sec. 11.5.3), electromagnetic radiation and acoustic signals that naturally are released during execution [22, 23], and cache-based side-channel attacks [24].

11.2 Hardware Backdoor Attacks

Figure 11.2 illustrates three triggering mechanisms for hardware Trojans. In this case, an attacker uses a **triggering input** to cause the multiplexer (MUX) to select the result generated by the malicious circuit instead of that produced by the original circuit. A triggering input may be data, control (e.g., instruction), or time dependent. Also, it may consists of one data item or one control or both, a sequence of data or control inputs or both, or a counter (timer) to trigger an attack. The latter case is called a ticking time bomb.

In addition, an attack may be classified either as **noncomputational**, which targets memory, registers, MUXs, and other items that do not operate on data and simply store or route it, or **computational**, which targets arithmetic logic units (ALUs), decoders, finite state machines (FSMs), etc. that manipulate incoming data.

Furthermore, a hardware backdoor attack can potentially take many forms. For example, if the output of the MUX in Fig. 11.2 is a data item and the attack causes a change in data values, it is called a **corrupter attack**. On the other hand, if the output bits of the MUX represent control signals and potentially cause more events to follow, it is called an **emitter attack** [13].

A backdoor attack may alter and simplify cryptography algorithms implemented in hardware, generate more cache memory traffic, cause computational errors, consume more power, etc. The following sections provide examples of data, control, and timer backdoor attacks.

11.2.1 Data and Control Attacks

Figure 11.3 illustrates two hardware Trojan examples with a single instruction and a three-instruction sequence triggering mechanisms. In Fig. 11.3(*a*), the Trojan is triggered by an "ADD" instruction with a specific operand "11001100...", and in Fig. 11.3(*b*), a three-instruction sequence—"ADD 0," "ADD 0," and "ST 0"—triggers the Trojan. The instructions are just examples and assumed to be of type Acc-ISA (Chap. 8). The triggering inputs for both types of Trojans would normally be selected

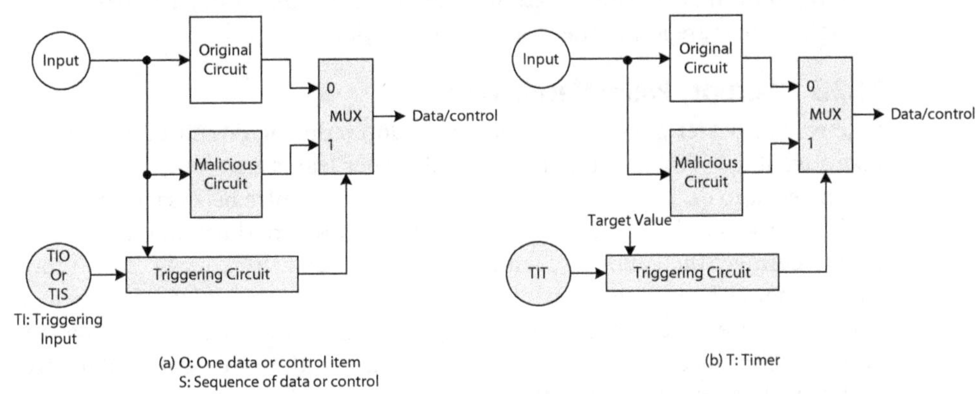

FIGURE 11.2 Examples of hardware Trojans [12]: (*a*) A one data or control triggered Trojan; A data or control sequence triggered Trojan; (*b*) A timer triggered Trojan (a ticking time bomb).

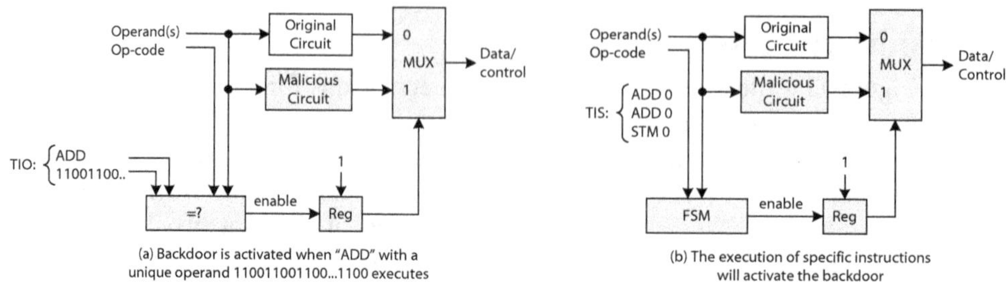

Figure 11.3 Examples of hardware Trojan triggering mechanisms: (*a*) a single instruction; (*b*) a sequence of instructions.

in such a way that it is unlikely for the Trojans to be detected during testing. For example, what are the chances of randomly selecting an "ADD" instruction with the specific operand "11001100...," or three instructions "ADD 0," "ADD 0," and "STM 0" in order during circuit testing?

Both the Trojans require an attacker to gain access to the hardware, either directly or via malware, and input the necessary triggering inputs. The attacker must be able to execute the "ADD" instruction for the circuit in Fig. 11.3(*a*) or the three instructions in sequence for the circuit in Fig. 11.3(*b*) to trigger an attack.

11.2.2 Timer Attack

A timer triggered Trojan, illustrated in Fig. 11.4, does not require the attacker to have access to hardware. As soon as the counter counts down to 0, the Trojan will be activated. Because most tests, especially those that are random, are not long and only require a few millions of cycles, the size of the counter would only need to be large enough to escape detection during testing.

In Fig. 11.4(*a*), an attack will switch the data generated by an original ALU with the data output by the malicious ALU, thus corrupting the result. In Fig. 11.4(*b*), when writing to certain memory locations, an attack could alter the signals of the cache controller to emit additional activities in the downstream memory and thus leak information. Many other examples of such Trojans can be thought of that are left to the reader's imagination.

11.2.3 Security Policy Mechanisms

One way to detect possible hardware backdoors during circuit design cycle is to take advantage of the fact that hardware designer teams are typically hierarchically organized, and each team does not design all the modules necessary for a complex IC (e.g., processor). Each team may design their own modules, and may also need to interconnect their modules with other in-house or third-party "soft" modules, such as some hardware description language (HDL) modules. However, while some in-house modules may have to go through code review and other quality control techniques to ensure their trustworthiness, others may still contain Trojans that can lead to security problems.

The following sections give examples of hardware backdoor security policy mechanisms.

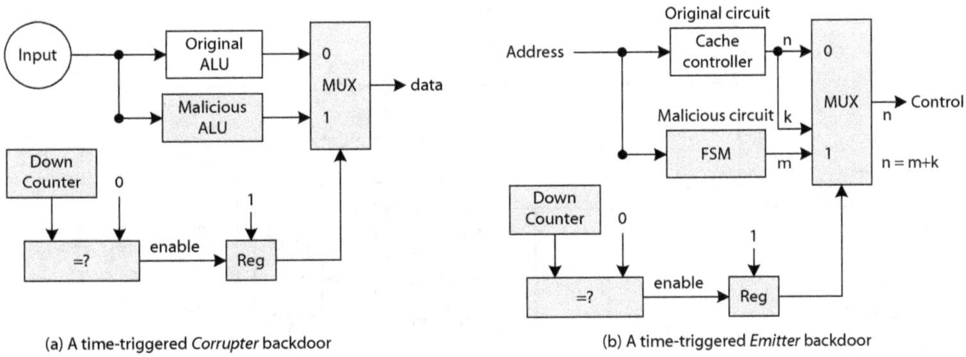

FIGURE 11.4 Examples of a timer-based corrupter and emitter backdoor: (*a*) a ticking time bomb *corrupter* backdoor; (*b*) a ticking time bomb *emitter* backdoor.

Data Obfuscation

A security policy that prevents a single input triggering attack is data confidentiality. However, depending on the type of the untrusted circuit module, different policy mechanisms are needed. If the untrusted module is noncomputational, such as the one shown in Fig. 11.5, only a simple data confidentiality technique known as data obfuscation [25] is needed to prevent an attack. In the figure, an output $(00110011)_2$ from the trusted module is XORed with a random number to obfuscate the data before it is input to the untrusted module. In this case, when an attacker generates the triggering input $(00110011)_2$ to activate the Trojan in the untrusted noncomputation module, the data actually presented to the triggering circuit changes to $(10010001)_2$, and, therefore, prevents an attack.

Homomorphic Encryption

Data obfuscation, however, is a bit difficult when it is used with a single input triggering Trojan in a computational module. For instance, if the untrusted computational module performs a square function, then a more complex data obfuscation technique known as homomorphic computation, also called homomorphic encryption, may be used [25–26]. Two functions, f and g, are said to have a homomorphic relationship if the equality in Eq. (11.1) holds.

$$f(g(x), g(y)) = g(f(x, y)) \tag{11.1}$$

For example, if f is the multiplication function and g is the square function, then f has a homomorphic relationship to g, as illustrated:

$$f(x, y) = xy \tag{11.2}$$

$$g(x) = x^2$$

$$g(y) = y^2$$

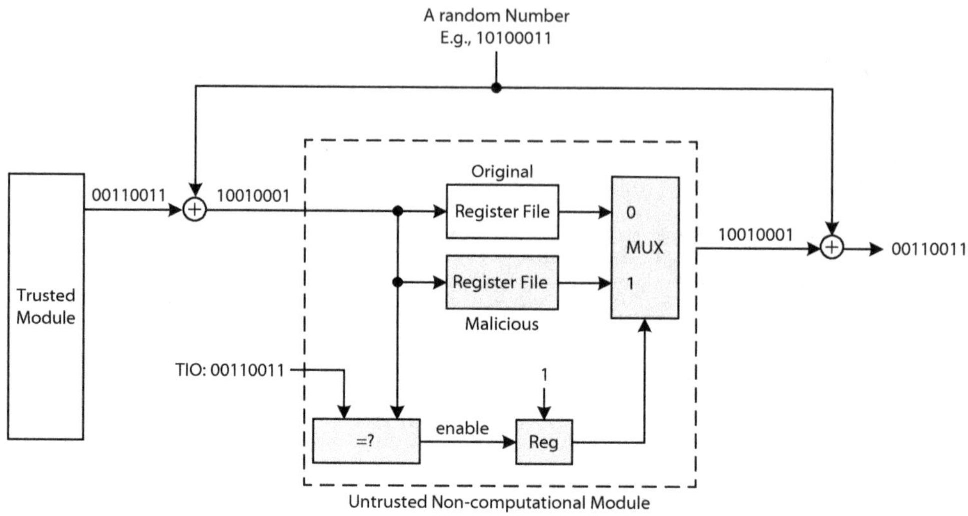

FIGURE 11.5 Untrusted noncomputational module interface with data obfuscation.

then,

$$f(g(x), g(y)) = f(x^2, y^2) = x^2y^2$$

is equal to

$$g(f(x, y)) = g(xy) = (xy)^2 = x^2y^2$$

If $x = 3$ and $y = 2$, then

$$3^2 * 2^2 = (3 * 2)^2$$
$$9 * 4 = 6 * 6 = 36$$

Suppose a design team decides to incorporate an untrusted HDL model of floating-point square function in the design of a processor. In this case, in order to prevent a single input triggering attack, either the design team must incorporate the homomorphic encryption discussed earlier in the hardware, or in systems that use the processor the homomorphic encryption is implemented in software, for example, as follows:

```
float homomorphic_encrypted_square(float x)
{
    float y;
    y = random();
    return (square(x * y) / (y * y));
}
```

In this case, because y is generated randomly and is used to conceal x, the quantity $x * y$ is also random, making it an unlikely triggering input. The square function computes $(x * y)^2$ and not x^2, which could trigger an attack. The quantity x^2 is then securely computed by dividing the quantity $(x * y)^2$ by the quantity $y * y$.

The advantage of homomorphic encryption is that operations for a computation are performed on encrypted (concealed) data and not on the original known values. For example, in order to illustrate another application and an advantage of homomorphic computation, suppose the "square" function in the previous code represents a unique remote function call for which a homomorphic relationship exists. Suppose the function is only available on a remote computer (e.g., a cloud) and the user does not want the value x (representing some secret input, for example, medical data) to be sent over the network or be accessible at the remote computer. In this case, the user is able to conceal (encrypt) inputs applied to the remote function without exposing the actual inputs to the outside world and still is able to securely use a remote function. Refer to the Exercises section for a potential problem with homomorphic computation.

In addition, for some functions such as the square root, while the homomorphic relationship $\sqrt{xy} = \sqrt{x}\,\sqrt{y}$ exits when both x and y are positive numbers, the relationship does not hold with negative numbers. For instance, $\sqrt{(-1)(-1)} \neq \sqrt{-1}\,\sqrt{-1}$.

In theory, in terms of circuit size, the cost of implementing a computational module that also implements a hardware homomorphic encryption/decryption algorithm can be unacceptably large.

Sequence Breaker

If a specific input sequence triggers an attack, a security policy to prevent such attacks is to alter the order in which a sequence of inputs is presented to an untrusted module. Random reordering of the inputs and inserting dummy inputs within the normal inputs are examples of policy mechanisms used to prevent input sequence attack. Many modern processors, such as a dynamically scheduled superscalar CPU (Chap. 8), already reorder instructions to improve performance; thus, it may be possible to introduce some randomness to this task. Random ordering of the inputs, however, may not work for memory read operations. In such cases, and for those where reordering due to data dependencies does not work, dummy inputs should be used. For instance, to alter the order of memory accesses, extra load instructions with pseudo-randomly generated memory locations can be inserted within a large sequence of load and store instructions [25].

Power Reset

A security policy for preventing a time bomb attack is to stop the counter from reaching the trigger value. A policy mechanism to do this is to frequently reset power to the untrusted module. How often an untrusted module must have its power reset would be determined based on how many clock cycles the module was tested. Because it is not known which type of backdoor may exist in an untrusted module, data obfuscation, input sequence reordering, and power reset all must be implemented to prevent an attack.

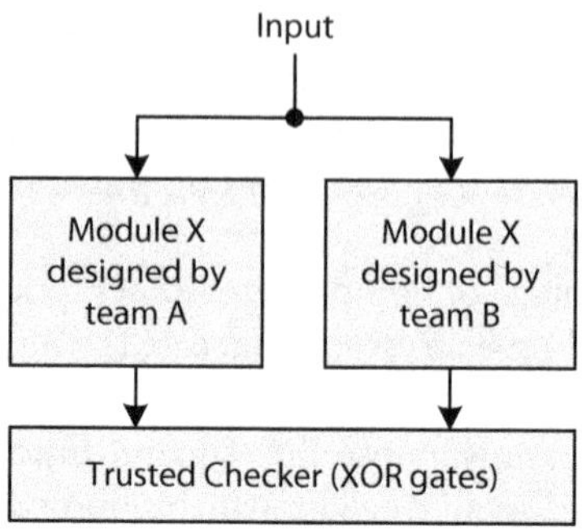

Figure 11.6 Module duplication to detect hardware backdoor attacks [12].

Duplication

An alternative but potentially costly solution is to use duplication in scenarios where attack prevention techniques may not work. Figure 11.6 illustrates an example of using a module duplication technique, where each copy of module X is designed by two teams, A and B. The outputs from the two copies are compared. As long as the outputs are the same, the copies are considered free from attacks. However, this requires that the modules be designed following the exact same specification; any small variation in the design could result in a false-positive backdoor attack.

Automatic HDL Code Analysis

Another technique to potentially detect a backdoor is to automatically analyze the HDL (e.g., Verilog) codes and tag the modules for possible backdoors. The tagged modules are then watched for malicious activities during run time [27].

Table 11.5 presents the summary of hardware backdoor attacks and a list of potential security policies and mechanisms to prevent attacks.

Triggering Type	Module Type	Security Policy	Security Mechanism
Single Input	Noncomputational	Data obfuscation	Use simple encryption techniques. For example, use a randomly generated number and a bitwise XOR to obfuscate the input to the untrusted module.
	Computational	Homomorphic computation	Use homomorphic functions.
Input sequence	Noncomputational	Alter the sequence	Reorder or insert dummy events.
	Computational		
Timer	Noncomputational	Prevent timer to reach the target value	Periodically reset power to the untrusted module.

Table 11.5 Summary of Security Policies and Mechanisms for Hardware Backdoors

11.3 Software/Physical Attacks

Spoofing, **splicing**, **replay**, and **man-in-the-middle** are four types of software and physical attack mechanisms. They can be used to physically monitor the behavior of, alter the functions of, or reverse engineer a computing device. Physical attacks are often called **hardware attacks**, but this should not be confused with hardware backdoor attacks discussed earlier.

These types of attack techniques, however, are detectable if appropriate confidentiality and integrity security policy mechanisms using cryptography are implemented.

11.3.1 Spoofing

A spoofing attack, as was discussed earlier, is caused by illegally inserting information (program code or data) into the system, as illustrated in Fig. 11.7. A virus can insert invalid data on the hard disk or memory. An attacker who gets physical hold of a computing device could use specialized tools to conduct a physical spoofing attack. For example, the attacker can intercept a memory transaction and spoof his or her data in place of the data in memory. Spoofing violates data integrity.

11.3.2 Splicing

A splicing attack is caused by illegally transposing information (instruction or data) with another one already in the system, as illustrated in Fig. 11.8. Malicious software that copies instructions from one memory section to another is an example of a splicing attack and so is a physical attack that intercepts a memory transaction and supplies the processor with previously accessed but different data from memory.

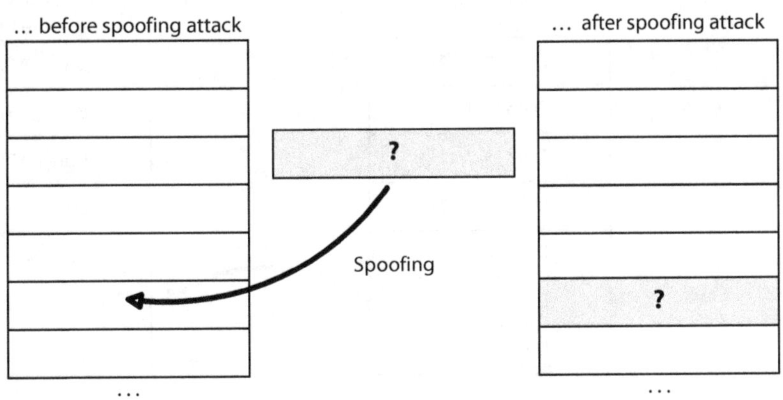

Figure 11.7 Illustrating a spoofing attack [28]: a different value is inserted in a target location.

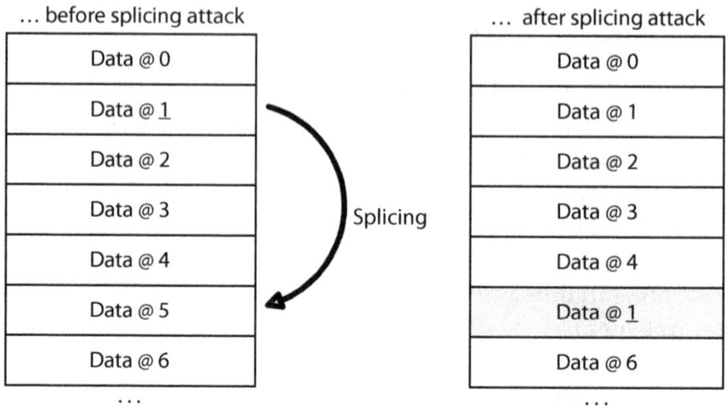

Figure 11.8 Illustrating a splicing attack [28]: data in location 5 is replaced with the data in location 1.

A splicing attack can occur unnoticed even if data is kept confidential; the attacker simply replaces one set of confidential data with another set. Splicing attacks also violate data integrity, but one must integrate data location information (e.g., memory address, register number, etc.) in the integrity security mechanism to detect such attacks. An attacker that changes an instruction physical page number *A* with *B* in a page table is an example of a splicing attack [29]. In this case, the attacker can force the OS or an application process to instead start executing instructions from physical page *B*.

11.3.3 Replay

A replay attack is caused by illegally replacing data with an earlier version in the system, as illustrated in Fig. 11.9. A replay attack is similar to a

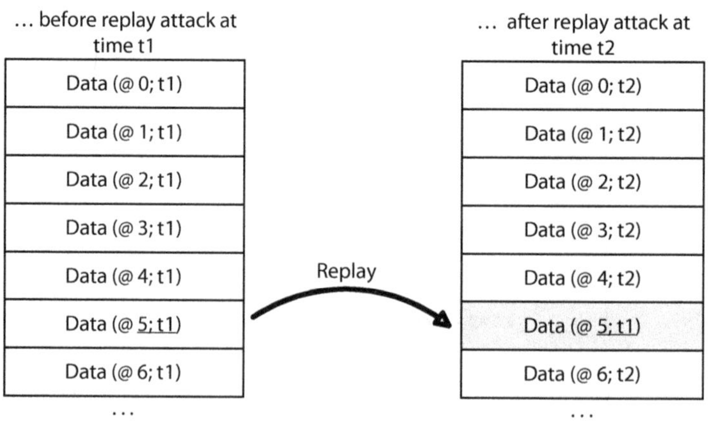

Figure 11.9 Illustrating a replay attack [28]: data in location 5 at time t2 is replaced with data in location 5 at time t1.

splicing attack, except that the replaced data is not relocated from another region in the system. For example, an attacker saves a specific data item at a memory location and uses it later when the same location is read again. Likewise, at the hardware level, an attacker can intercept a memory transaction for address *X* in order to save a copy of the memory content for later use [30]. The attacker then waits for a write-to-memory transaction for address *X* to complete. The next time that a read transaction for address *X* is detected, the attacker supplies the previously saved, but now old, content to the processor.

A replay attack may not need to happen in such a direct way. An attacker may enable both new and old data to remain in the system. For example, a replay attack may map a single virtual address to two different physical addresses and cause new data to be stored in one physical location but access older data from the second physical location.

A replay attack also violates data integrity, but in a different way—everything about the replaced data appears valid except that it is not current. For example, a replay attack can cause an older and lower reading of electricity usage to be communicated from a smart meter to the utility company. Replay attacks are much harder to detect; neither keeping information confidential nor implementing simple integrity mechanisms can help to detect them. A more complex security mechanism that integrates timing information (e.g., bus transaction number, communication session identifier, etc.) is necessary to detect replay attacks.

11.3.4 Man-in-the-Middle

A man-in-the-middle attack can happen in scenarios when two subjects (people, software, firmware) are communicating without using adequate confidentiality and integrity mechanisms. A malicious subject may intercept messages being exchanged between two legitimate subjects and then either copy the message (i.e., eavesdrop) or substitute a different message in place of the original message, as illustrated in Fig. 11.10. The A and B entities are unaware of entity C. Both A and B think they are communicating with each other.

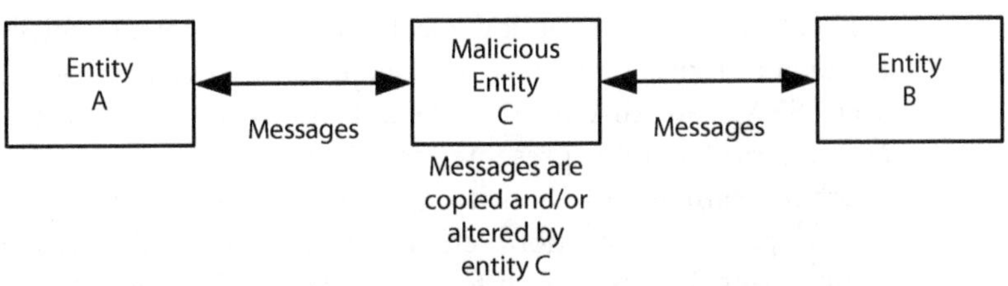

Figure 11.10 Illustrating a man-in-the-middle attack. The A and B entities are unaware of entity C.

11.4 Trusted Computing Base

A TCB encompasses secure design, development, installation, and functioning of hardware and firmware and possibly software modules responsible for maintaining security in a system. The hardware modules must be protected from backdoor attacks so they continue to operate correctly. We call this a trusted hardware module (THM). Firmware must also be securely designed, developed, and installed. We call this trusted firmware module (TFM), which would be embedded in THM, a tamperproof IC. Also, in some secure system application areas, one or more security-related software modules must also securely execute. In this case, the software modules must securely be designed, developed, and installed. Such a software module is called a trusted software module (TSM).

A THM-TFM would be organized as a secure coprocessor (SCP), for example, a **cryptoprocessor** that operates as an embedded system responsible for providing cryptography services to the OS as well as application software. The SCP would be responsible for generating secure **cryptography keys** (see Sec. 11.5) for protecting the confidentiality and integrity of files and data stored locally or remotely on a server and keys required for secure communication. The TFM, being embedded in the THM, is protected from software spoofing, splicing, and replay attacks, but not from physical attacks. The trusted platform module (TPM) as an example SCP is discussed in Sec. 11.10.

A THM-TFM-TSM, on the other hand, would be organized as a general-purpose secure processor (SP) for maximum applicability. An SP may support multiple **secure execution environments**, each implemented as a secure execution mode (SXM) to run an arbitrary TSM. Depending on the application area, an arbitrary TSM may require confidentiality or integrity or both of its instructions, as well as confidentiality or integrity or both of its data. An SP-based system can support all security-related application areas, including those supported by an SCP. An example of security application areas that would require an SP is the implementation of mandatory access controls (Sec. 11.1.4) when a commodity OS frequently is compromised or when a portable device must be protected from physical attacks [31–33]. Furthermore, handheld devices, such as a smart phone, may require a power-efficient SP. Consider, for example, the digital rights management [34] policies that would permit an encrypted media file to be decrypted only by a target handheld device.

Because a TSM may be subject to software attacks, an SXM must implement the necessary security policy mechanisms to protect a TSM from spoofing, splicing, and replay attacks, as well as physical attacks if the system is a portable device. SXMs are discussed in Sec. 11.11, and the architecture of an SP for maximum protection is presented in Sec. 11.12.

Other examples of security application areas that require a TCB include software piracy prevention, cloud computing, and certified execution. For instance, consider the Search for Extraterrestrial Intelligence (SETI) project [35] and a general-purpose distributed computing project at www.distributed.net [36] that utilize at-home computing power from thousands of volunteer users to conduct research in many areas important to public and academia. A user can download a free program that analyzes research data, for example, from the radio telescope for SETI. However, without certified execution, it is not possible to verify the correctness of the results.

11.5 Cryptography

Confidentiality is enforced by applying an encryption algorithm, also called a **cipher**, to scramble and conceal the information in a plaintext document, e-mail message, authentication data, memory content, etc. The output of a cipher is called a **ciphertext**. Likewise, a decryption algorithm, also called a cipher, unscrambles a ciphertext and generates the original plaintext, as illustrated in Fig. 11.11.

A cipher is called secure if its generated ciphertext does not contain any information that could be used to obtain the original plaintext input. A **symmetric-key cipher** uses a single cryptography key for both encryption and decryption. An **asymmetric-key cipher**, on the other hand, uses one key for encryption and a different key for decryption. In practice, only cryptography keys need to be secret; ciphers (the algorithms), however, can be known [37].

A cipher is called a **stream cipher** if it encrypts or decrypts its input one bit at a time. Otherwise, a cipher is called a **block cipher**, where it encrypts/decrypts its input one block (multiple bits) at a time. For inputs that are longer than one block, there are multiple ways, called **modes of operation**, in which a cipher is repeated to encrypt/decrypt the remaining input blocks. Examples of symmetric-key and asymmetric-key ciphers are discussed next.

11.5.1 Symmetric-Key Ciphers

Figure 11.12 illustrates an 8-bit linear feedback shift register (LFSR) as an example of a simple symmetric-key cipher, called a stream cipher. The cipher is designed using a parallel-load/right-shift register with four **tap** bits: 4, 5, 6, and 8. The tap bits are numbered starting with 1 and refer to the register bits from right to left. The four tap bits are XORed to generate the next left-input (LI) bit as the register is shifted right.

The register is initialized with a **secret key**, and as the register shifts right, the bits that are shifted out create a **key-stream** one bit at a time. For example, after

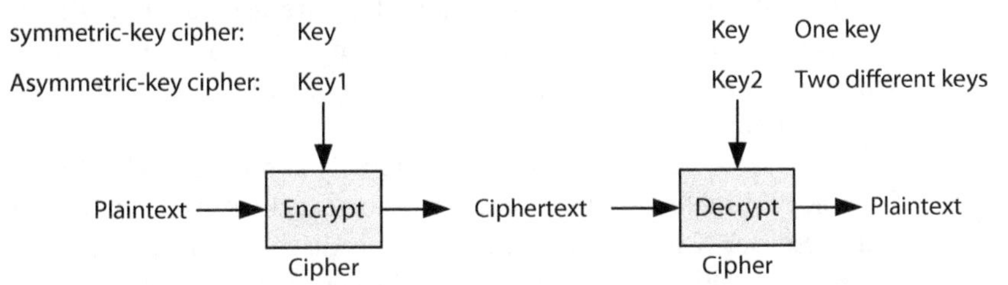

Figure 11.11 Illustration of encryption/decryption cryptography.

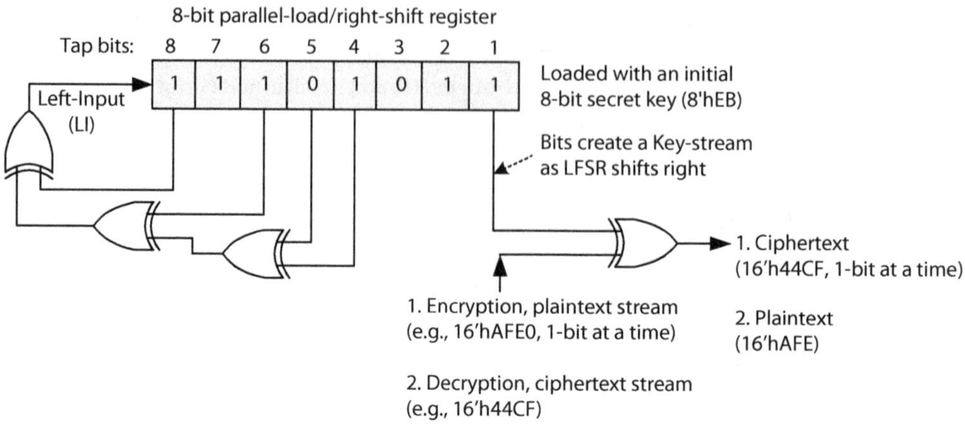

FIGURE 11.12 Illustration of an 8-bit LFSR cipher with tap bits 4, 5, 6, and 8.

16 shifts, the LFSR generates a 16-bit key-stream; after 32 shifts, it generates a 32-bit key-stream; etc. The **period** of a key-stream, which is determined by the selected tap bits, is the number of shifts before the key-stream repeats. A key-stream is bitwise XORed with an equal-size plaintext stream to generate an equal-size ciphertext stream one bit at a time. Likewise, the key-stream is bitwise XORed with a ciphertext stream to generate its corresponding original plaintext stream one bit at a time.

The initial secret key is carefully selected so that the register content does not become zero as it shifts right. In the figure, the LFSR cipher is initialized with an 8-bit secret *key* = 8'hEB (hex in Verilog) and requires 16 clock cycles to encrypt the 16-bit *plaintext* = 16'hAFE0 to its corresponding 16-bit *ciphertext* = 16'h44CF. Likewise, the register is initialized with the same secret *key* = 8'hEB before the 16-bit *ciphertext* = 16'h44CF is decrypted, in 16 clock cycles, to generate the original 16-bit *plaintext* = 16'hAFE0.

A5/1

A5/1 is a practical stream cipher and uses three 19-, 22-, and 23-bit LFSRs, as illustrated in Fig. 11.13. The three LFSRs are labeled X, Y, and Z, respectively. The symbol ⊕ indicates bitwise XOR. The LFSRs are not shifted during every clock cycle; instead, register bits x_8, y_{10}, and z_{10} are used as inputs to a majority circuit that outputs $m = 1$ if two or more of the inputs are 1, or $m = 0$ if two or more of the inputs are 0. For example, if $x_8 = 0$, $y_{10} = 0$, and $z_{10} = 1$, then $m = 0$, and if $x_8 = 1$, $y_{10} = 1$, and $z_{10} = 0$, then $m = 1$.

The x_8, y_{10}, and z_{10} bits are individually compared (XNORed) with m to either enable or disable each register during the next clock cycle. For example, if $\overline{x_8 \oplus m} = 1$, then register X is enabled; otherwise, X is disabled during the next clock cycle. This introduces additional randomness to the key-stream. The bits x_{18}, y_{21}, and z_{22} are used to generate the key-stream. The initial secret key is 64 (19 + 22 + 23) bits long.

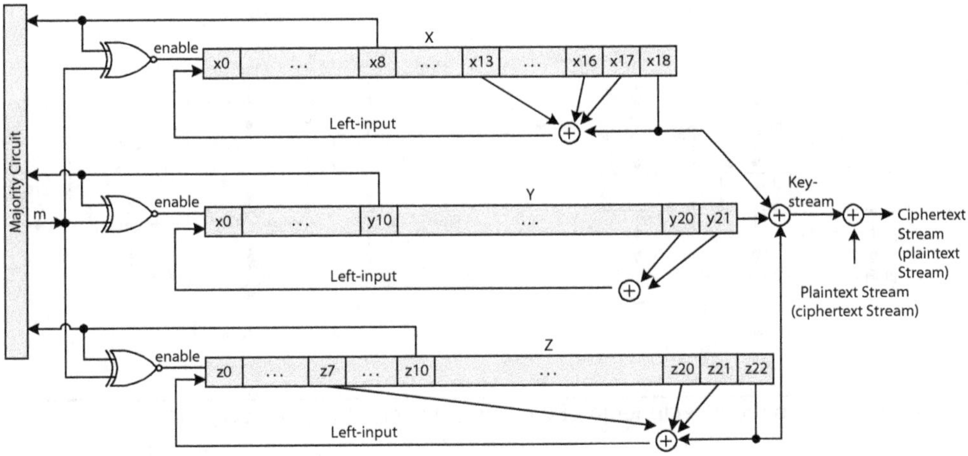

FIGURE 11.13 The A5/1 stream cipher.

Block Ciphers

Data Encryption Standard (DES) is one of the oldest block ciphers. It operates on 64-bit blocks and uses a 56-bit key. Advanced Encryption Standard (AES), the most common today, is recommended by the National Institute of Standards and Technology (NIST) [38].

Advanced Encryption Standard (AES)

The 128-bit AES cipher (the standard) requires a 128-, 192-, or 256-bit key to encrypt/decrypt blocks of 128-bits. If the number of bits in a plaintext input is not divisible by 128, the plaintext is padded with extra bits. AES organizes each input block into columns and rows. An encryption step requires 10, 12, or 14 rounds of operations, depending on the length of the key. Each round involves certain byte substitutions using a lookup table, row shifting, column mixing, and bitwise XOR operation to generate the input for the next round. The output of the last round is a 128-bit ciphertext. Depending on the mode of operations (discussed next), decryption operations may be performed in the reverse order or in the same order as encryption.

11.5.2 Modes of Operation

Figure 11.14 illustrates the application of the AES cipher in cipher block chaining (CBC) mode. Each encryption step starts with a bitwise XOR of the current 128-bit plaintext block and the preceding 128-bit ciphertext block; thus, the word "chaining" in the name. The first plaintext block, however, is XORed with a 128-bit randomly generated, but not necessarily secret, initialization vector (IV).

While the blocks of a plaintext are encrypted recursively in CBC mode, as illustrated in Fig. 11.14(*a*) and also outlined in Table 11.6, the blocks of a ciphertext can be decrypted in random order and even in parallel. In the table, the letter E

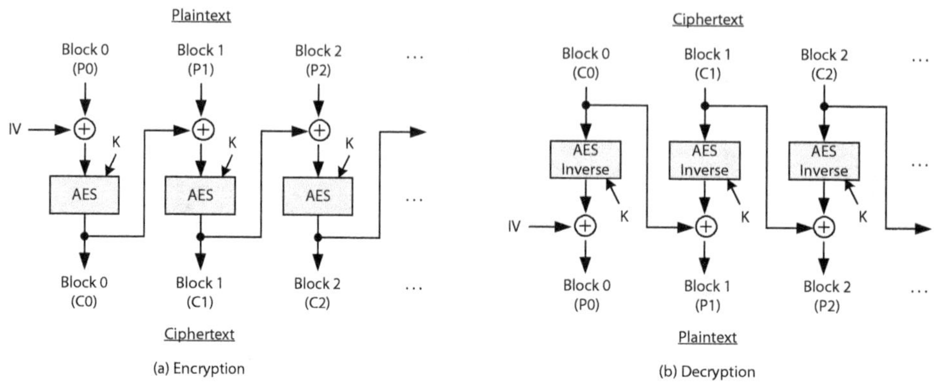

Figure 11.14 Illustrating the AES cipher in CBC mode: (*a*) encryption; (*b*) decryption.

Cipher Block Chaining (CBC)	
Encryption	Decryption
$C_0 = \mathrm{E}(IV \oplus P_0, K)$	$P_0 = IV \oplus \mathrm{D}(C_0, K)$
$C_1 = \mathrm{E}(C_0 \oplus \mathrm{P}_1, K)$	$P_1 = C_0 \oplus \mathrm{D}(C_1, K)$
$C_2 = \mathrm{E}(C_1 \oplus \mathrm{P}_2, K)$	$P_2 = C_1 \oplus \mathrm{D}(C_2, K)$
. . .	. . .
C: 128-bit ciphertext block; P: 128-bit plaintext block; E: encryption cipher (e.g., AES); IV: initialization vector; K: a secret key; D: decryption cipher.	

Table 11.6 Encryption/Decryption in CBC Mode

Counter Mode (CTR)	
Encryption	Decryption
$C_0 = P_0 \oplus \mathrm{E}(IV, K)$	$P_0 = C_0 \oplus \mathrm{E}(IV, K)$
$C_1 = P_1 \oplus \mathrm{E}(IV{+}1, K)$	$P_1 = C_1 \oplus \mathrm{E}(IV{+}1, K)$
$C_2 = P_2 \oplus \mathrm{E}(IV{+}2, K)$	$P_2 = C_2 \oplus \mathrm{E}(IV{+}2, K)$
. . .	. . .

Table 11.7 Encryption/Decryption in CTR Mode

stands for encryption; D for decryption; K for secret key; P_0, P_1, etc., for plaintext blocks 0, 1, etc.; and C_0, C_1, etc., for ciphertext blocks 0, 1, etc.

Another common mode of operation is the counter mode (CTR), as outlined in Table 11.7. It uses a sequence of IVs for concurrent processing. The function E(IV, K) indicates the encryption of an IV with K, which is then XORed with the first plaintext block P_0 to generate the first ciphertext block C_0. The IV is then incremented and is used to encrypt the next plaintext block (P_1) independent of C_0. This process continues until all the plaintext blocks are encrypted.

The CTR mode has the advantage of using a single cipher for both encryption and decryption. Note that, in the table, no decryption (D) cipher is used. In addition, blocks can be encrypted or decrypted concurrently. For example, a 16-block plaintext can be partitioned into two groups of eight blocks, such as, blocks P_0 to P_7 in one group and blocks P_8 to P_{15} in another group. The blocks of each group can then be processed concurrently, for example, using two threads. One thread operates on blocks P_0 to P_7 and uses *IV* to *IV*+7, and a second thread operates on blocks P_8 to P_{15} and uses *IV*+8 to *IV*+15.

Many processors, including those of Intel and AMD, have implemented the AES instruction set, where CBC, CTR, and other modes can be implemented in software [39].

11.5.3 Asymmetric-Key Ciphers

An asymmetric-key cipher, as stated earlier, requires two keys, one for encryption and another one for decryption. The primary application of asymmetric-key cryptography, as illustrated in Example 11.1, is for communication. In the example, Alice and Bob represent two subjects as people, programs, or hardware.

Example 11.1. Suppose Alice would like to send a secret message to Bob. Alice and Bob each have a nonsecret public key and a secret private key. Also, suppose, no one else will send a secret message to Bob and pretend to be Alice or alter Alice's message in any way.

Solution: Because Alice is only concerned with keeping the message confidential, Alice and Bob can use the following two steps:

1. Alice uses Bob's *public key* to encrypt her private message to Bob.
2. Upon receiving the message, Bob uses his *private key* to decrypt the message.

Since no one else knows Bob's private key, only Bob can decrypt Alice's encrypted message.

RSA

RSA, which stands for Ron Rivest, Adi Shamir and Leonard Adleman, the names of the three people who developed it, is an asymmetric-key cipher that requires one key for encryption and a different key for decryption. Each plaintext or ciphertext is viewed as an integer number. For example, using the ASCII coding scheme, the string message "HELLO" consists of five 8-bit ASCII codes; that is, 2'h48 or 72 for character H, 2'h45 or 69 for E, 2'h4C or 76 for L, 2'h4C or 76 for the second L, and 2'h4F or 79 for O.

The string may be partitioned and viewed as five separate ASCII codes with decimal numbers 72, 69, 76, 76, and 79, or viewed as single large 40-bit number: 40'h48454C4C4F (hex in Verilog). Other partitions of the string into integer numbers are also possible. For example, the string may be partitioned into three 16-bit numbers: 16'h4845 = 18501, 16'h4C4C = 19532, and 16'h4F00 = 20224. The last partition is padded with an 8-bit 0 to make it a 16-bit number.

For simplicity, it is assumed that the string "HELLO" is partitioned into five plaintext integer numbers as $P_0 = 72$, $P_1 = 69$, $P_2 = 76$, $P_3 = 76$, and $P_4 = 79$. The string can then be encrypted, one number at a time, using, for example, CBC or CTR mode of operation discussed earlier. However, asymmetric ciphers, as discussed next, are not designed to encrypt large inputs.

The RSA requires two relatively prime numbers as keys: a public key (e) used for encryption, and a private key (d) used for decryption. Equation (11.3) defines the relationships between an n-bit plaintext P and its n-bit ciphertext C.

$$C = P^e \, mod \, n \quad (11.3)$$
$$P = C^d \, mod \, n$$

Assuming that the public key $e = 5$ and $n = 91$, Eq. (11.4) illustrates the encryption of a plaintext $P = 72$ to its ciphertext $C = 11$, which is the remainder of 72^5 divided by 91. (Note that for 8-bit plaintext ASCII codes, n has to be 256).

$$C = P^5 \, mod \, 91 \quad (11.4)$$
$$C = 72^5 \, mod \, 91$$
$$C = 1{,}934{,}917{,}632 \, mod \, 91$$
$$C = 11$$

Equation (11.5) illustrates the calculations required to decrypt $C = 11$ and obtain its corresponding original plaintext number $P = 72$ using private key $d = 29$ and $n = 91$. Note that the integer quantity 11^{29} is too big to compute using a calculator or even some computers. It must be divided into smaller forms until each form is small and computable.

$$P = C^{29} \, mod \, 91 \quad (11.5)$$
$$P = 11^{29} \, mod \, 91; \text{ write } 11^{29} \text{ as } 11^{24} * 11^5$$
$$P = (11^{24} \, mod \, 91)(11^5 \, mod \, 91) \, mod \, 91; \text{ write } 11^{24} \text{ as } 11^{6*4}$$
$$P = (11^{6*4} \, mod \, 91)(11^5 \, mod \, 91) \, mod \, 91$$
$$P = ((11^6 \, mod \, 91)^4 \, mod \, 91)(11^5 \, mod \, 91) \, mod \, 91$$
$$P = (64^4 \, mod \, 91)(11^5 \, mod \, 91) \, mod \, 91$$
$$P = (1)(72) \, mod \, 91$$
$$P = 72$$

A very large integer quantity X^y may be written as X^{a+b}, X^{a*b}, or X^{a*b+c}. Equation (11.6) illustrates $X^{a*b+c} \, mod \, n$ divided into smaller forms.

$$X^{a*b+c} \, mod \, n = (X^{a*b} \, mod \, n)(X^c \, mod \, n) \, mod \, n \quad (11.6)$$
$$= ((X^a \, mod \, n)^b \, mod \, n)(X^c \, mod \, n) \, mod \, n$$

$$= (W^b \; mod \; n)(Y) \; mod \; n; \text{ suppose } W = X^a \; mod \; n, \text{ and } Y = X^c \; mod \; n$$

$$= (Z*Y) \; mod \; n; \text{ suppose } Z = W^b \; mod \; n$$

If, for example, quantity W^b is still too big a number in Eq. (11.6), the quantity is again successively divided into smaller forms until all such quantities are small numbers and the result of each mod function can be computed.

The following is an algorithm to determine an RSA encryption key *e* and decryption key *d* [40, 41]. Public keys are securely stored in a trusted center, such as the public key infrastructure (PKI).

1. Select two prime numbers *p* and *q*; for example, $p = 7$ and $q = 13$.
2. Determine $n = p * q$; that is, $n = 7 * 13 = 91$.
3. Determine $m = (p - 1)(q - 1)$; that is, $m = (7 - 1)(13 - 1) = 6 * 12 = 72$.
4. Find a prime number *e* such that *e* is relatively prime to *m* and $e < m$; that is, $\gcd(e, m) = 1$ and $e < m$, where "gcd" stands for greater common divisor; for example, select $e = 5$.
5. Find an integer $d = \frac{1 + k * m}{e}$, where *k* is an integer number and $d < m$; for example, for $k = 2$, $d = \frac{1 + 2 * 72}{5} = 29$. If there is no such integer *k*, select a different value for *e* in step 4 and repeat step 5. If there are no such *e* and *d* values, select two different prime numbers in step 1 and repeat.
6. Use the values of *e* and *n* as public (known) and *d* as private (unknown) information; that is, use *e* as a public key and *d* as a private key.

It is also possible to reverse the order and first select a value for *d* and then determine a value for *e* according to the requirements of steps 4 and 5.

The RSA cipher requires more computations when *P*, *C*, *e*, and *d* are very large numbers. For instance, a 128-character message may be interpreted as a 1024-bit integer (128 * 8) message, one of 256^{128} possible combinations made with 128 ASCII characters. This implies that *P* and *C* can be a very large number $<2^{1024}$. The encryption of a 1024-bit *P* would result in a 1024-bit ciphertext *C*, also containing 128 ASCII characters. Because typical processors do not have, for example, a 1024-bit arithmetic unit required for 1024-bit RSA cipher, large arithmetic functions must be either implemented in software or as a coprocessor in hardware [42].

The bigger the values of ciphertext *C* and the decryption key (*d*) are, the harder it is, requiring many days, months, or even years of computations, to break the RSA cipher. One possible technique is to use a brute-force approach and examine every possible decryption key value until the right key is identified and the resultant *P* makes sense. However, given a 2048-bit *C*, it could take a prohibitively long time to determine the corresponding 2048-bit *P*. A timing attack has been used to reduce the list of possible decryption key values based on the amount of time required to perform decryption, much like a thief guessing the numbers required to unlock a combinational lock based on how long it takes a person to turn the dial to unlock the lock.

Example 11.2. Suppose Alice would like to send a secret message to Bob, and she also wants to make sure no one else is able to send a secret message to Bob and pretend to be her.

Solution:

1. Alice uses her private key and encrypts her secret message. Since no one else knows Alice's private key, the encrypted message, in a way, has her signature.
2. Alice then uses Bob's public key and encrypts her already encrypted message; thus, only Bob can access the message and not anyone else who knows Alice's public key.
3. After receiving the message, Bob first uses his private key and decrypts the message.
4. Bob then uses Alice's public key and decrypts the secret message Alice sent.

These four steps require four applications of the RSA cipher: two for encryption by Alice and two for decryption by Bob. This is a disadvantage, considering the significant processing time of the RSA cipher. However, Alice and Bob could use the steps in Example 11.2 to share a symmetric key and use it instead to exchange large messages. Symmetric ciphers are faster and require fewer key bits to achieve the same level of security as compared to asymmetric ciphers; for example, 3072-bit RSA is comparable to 128-bit AES [43].

Example 11.3. Suppose Alice and Bob would like to exchange many secret messages. Therefore, they decide to be efficient and use a symmetric-key cipher instead of an asymmetric-key cipher, such as the RSA.

Solution:

1. Alice and Bob use the four steps outlined in Example 11.2 for Alice to send a message containing, for example, a 128-bit symmetric key to Bob.
2. Both Alice and Bob now know a common secret key that they can use to exchange secret messages using a symmetric-key cipher such as AES.

With both Examples 1.2 and 1.3, there is a chance for a man-in-the-middle attack. For instance, suppose Mary is able to access Alice's hard disk or performs a timing attack and discovers Alice's private key. Mary then is able to intercept and access Alice's secret symmetric key in Example 11.3. Once Mary knows the symmetric key, she can monitor secret messages exchanged between Alice and Bob. Therefore, additional security policy mechanisms are needed to detect man-in-the-middle attacks.

Example 11.4. Suppose Alice wants to send Bob a symmetric key and she wants to make sure there is no chance for any man-in-the-middle attacks. Also, assume the communication medium is secure and any data transmission errors (if any) can be resolved.

Solution: Both Alice and Bob need to include randomly generated numbers unknown to an adversary in their messages to make sure they are communicating with each other and not with a middle man [44]. In the following "ptxt" stands for plaintext, "ctxt" for ciphertext, "pr" for private key, and "pu" for a public key.

1. Alice uses Bob's public key (B_{pu}) and encrypts a random number (*r1*) plus her name or an ID number (ID_{Alice}); *r1* is generated by a trusted software. Alice sends the encrypted message to Bob and waits for a response.
2. Bob's receives the message and uses his private key (B_{pr}) to decrypt the message and discover *r1* (labeled $r1_{Bob\text{-}rcvd}$).
3. Bob uses Alice's public key (A_{pu}) to encrypt and send both $r1_{Bob\text{-}rcvd}$ and another random number (*r2*) that his trusted software generated to Alice. Bob then waits for a response from Alice.
4. Alice receives the message and uses her private key (A_{pr}) to decrypt the message and discover both *r1* ($r1_{Alice\text{-}rcvd}$) and *r2* ($r2_{Alice\text{-}rcvd}$). Alice compares $r1_{Alice\text{-}rcvd}$ with *r1*. If the two values are the same, Alice knows the message came from Bob and not from someone else.
5. Alice then uses Bob's public key (B_{pu}) and encrypts and sends $r2_{Alice\text{-}rcvd}$ back to Bob.
6. Alice also generates a plaintext symmetric key ($K_{sym\text{-}ptxt}$) and encrypts the key as $K_{sym\text{-}ctxt1}$ using her private key (A_{pr}). She then uses Bob's public key (B_{pu}) and encrypts $K_{sym\text{-}ctxt1}$ to generate $K_{sym\text{-}ctxt2}$ that she sends to Bob.
7. Bob uses his private key (B_{pr}) and discovers the *r2* he had sent to Alice ($r2_{Bob\text{-}rcvd}$). Bob compares $r2_{Bob\text{-}rcvd}$ and *r2*. If the two values match, Bob knows he is communicating with Alice.
8. Bob then first uses his private key to discover $K_{sym\text{-}ctxt1}$ from $K_{sym\text{-}ctxt2}$, and then he uses Alice's public key to discover Alice's $K_{sym\text{-}ptxt}$ from the $K_{sym\text{-}ctxt1}$.
9. Bob sends an acknowledgement message encrypted with $K_{sym\text{-}ptxt}$ to Alice.

These steps are summarized here using E for encryption, D for decryption, and symbols {} for concatenation:

```
Alice sends:     E(B_pu, {r1, ID_Alice}); r1 is a random number
Bob receives:    {r1, ID_Alice}_Bob-rcvd = D(B_pr, E(B_pu, {r1, ID_Alice});
Bob sends:       E(A_pu, {r1_Bob-rcvd, r2});
Alice receives:  {r1, r2}_Alice-rcvd = D(A_pr, E(A_pu, {r1_Bob-rcvd, r2});
Alice checks:    If r1_Alice-rcvd = r1 then continue; else stop,
                 insecure communication.

Alice sends:     E(B_pu, r2_Alice-rcvd);
Alice sends:     E(B_pu, E(A_pr, K_sym))
Bob receives:    r2_Bob-rcvd = D(B_pr, E(B_pu, r2_Alice-rcvd)
Bob checks:      If r2_Bob-rcvd = r2 then continue; else insecure
                 communication.

Bob receives:    K_sym = D(A_pu, D(B_pr, E(B_pu, E(A_pr, K_sym))));
Bob sends:       E(K_sym, message_ack);
```

Elliptic Curve Cryptography (ECC)

There are other asymmetric-key ciphers. The elliptic curve cryptography (ECC) requires smaller public/private key sizes to provide the same level of security as compared to RSA. For example, 2048-bit RSA provides the same level of security as 224-bit ECC, and 3072-bit RSA is comparable to 256-bit ECC [42, 43]. Furthermore, because both RSA and ECC ciphers require about the same amount of processing time with equal key sizes, ECC is more cost effective in terms of both required key storage space and processing time. This makes ECC advantageous when compared to RSA, especially in handheld devices where less power consumption is desirable. Mathematically, however, ECC is more complex. Its description is deferred to textbooks on cryptography.

11.6 Hashing

Hashing is like a fingerprint used for message authentication. A hash function converts the entire message into a unique **hash value**, also called a hash code, or simply just a hash, which is a number in the order of only a few bytes. A message is validated by verifying a received hash value with the one computed from the received message, as illustrated in Fig. 11.15. If the two hash values match, the message is considered valid; otherwise, the message, the hash, or both have been modified. Furthermore, a hash function is always one-way; it is impossible to retrieve the original message from a hash value of only a few bytes. A hashing function is called **standard** if no secret key is used to generate a hash.

A hash simplifies message authentication, which otherwise could be a very difficult task, especially if a message is, for example, a picture or binary file. Figure 11.16 illustrates a simple standard hash function using 8-bit bitwise XOR logic. In the figure, the ASCII string "HELLO" generates a *hash* = 8'h8E. It would be impossible to determine the original message "HELLO" from its hash value 8'h8E. A standard hash is also known as a **checksum**, which is typically used to detect data transmission errors.

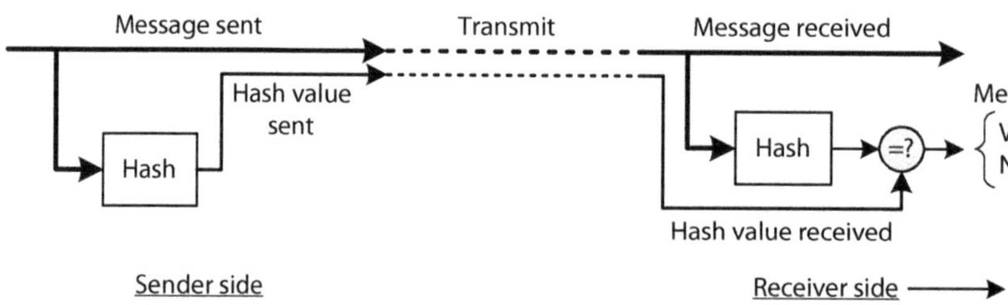

FIGURE 11.15 Message authentication using hashing.

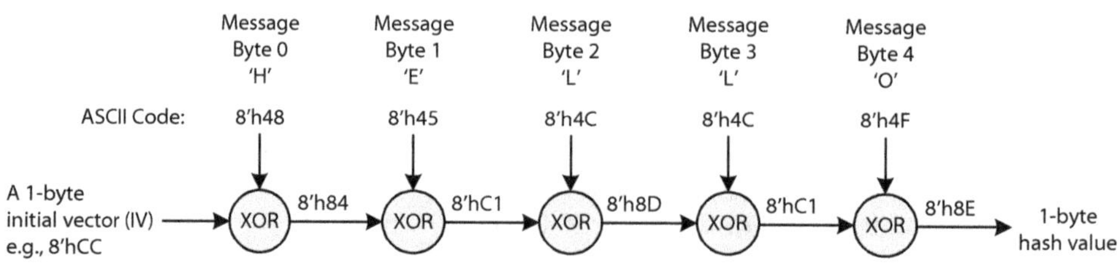

FIGURE 11.16 A simple standard hash function using 8-bit bitwise XOR; it generates a 1-byte hash for the string message "HELLO."

Example 11.5. Suppose Alice wishes to share an important personal message with her friends so she decides to digitally sign the message before sending it.

Solution: In the following, "H" indicates a hashing function, "M" a message, and "HV" a hash value:

Alice computes:	HV = H(M)
Alice signs:	$E(A_{pr}, HV)$
Alice sends:	$\{M, E(A_{pr}, HV)\}$; note M is not encrypted, anybody can read it
Friends receive:	$\{M, E(A_{pr}, HV)\}_{rcvd}$
Friends extract signature:	$HV_{rcvd} = D(A_{pu}, E(A_{pr}, HV)_{rcvd})$;
Friends check:	If $H(M_{rcvd}) = HV_{rcvd}$ then accept M_{rcvd}; else reject M_{rcvd}

Software companies can also use the technique illustrated in Example 11.5 to securely distribute nonconfidential software products.

A hash function must be **collision resistant**, never producing the same hash value for two different messages. For example, the simple hash function illustrated in Fig. 11.16 is not a collision-resistant hash function. The hash of string "WORLD" with the same *IV* = 8'hCC, as illustrated in Fig. 11.17, is the same as the hash of string "HELLO."

The Secure Hash Algorithm (SHA) that was developed by the NIST [38] now includes four hashing algorithms: the original SHA-1, SHA-256, SHA-384, and SHA-512. SHA-1 generates the following two hash values for the two strings "HELLO" and "WORLD" [45]. The two hash values are very different. However, SHA-1 has since shown to be not collision resistant [40].

Message: "HELLO"
SHA-1 hash value: `c65f99f8c5376adadddc46d5cbcf5762f9e55eb7`

Message: "WORLD"
SHA-1 hash value: `1a5db926797b9ae16ad56ec2c143e51a5172a862`

Each SHA algorithm successively processes a 512-bit (SHA-1 and SHA-256) or 1024-bit (SHA-384 and SHA-512) message block and generates a final 160-, 256-, 384-, or 512-bit hash value, respectively. A message must be padded (if necessary)

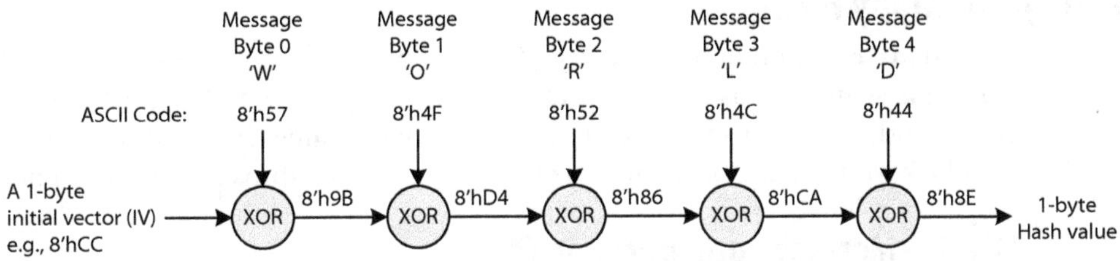

Figure 11.17 An example illustrating a noncollision-resistant hash function; two messages—"WORLD" and "HELLO"—generate the same hash value.

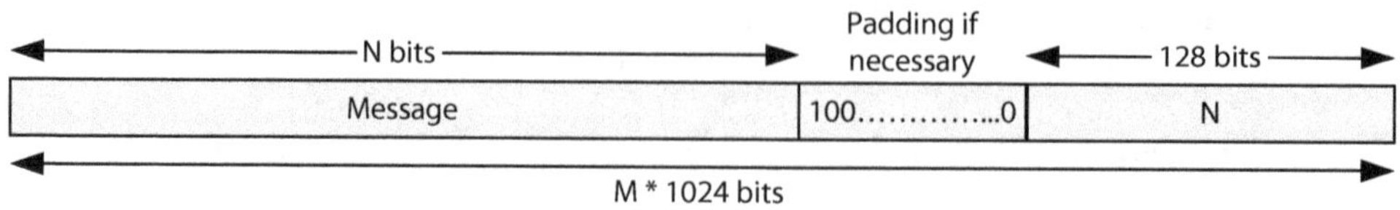

FIGURE 11.18 The message format used with the SHA-512 hash algorithm.

Properties	SHA-1	SHA-256	SHA-384	SHA-512
Size of hash value (bits)	160	256	384	512
Block size	512	512	1024	1024
Message size (plaintext or ciphertext)	$< 2^{64}$	$< 2^{64}$	$< 2^{128}$	$< 2^{128}$
Number of steps	80	64	80	80

TABLE 11.8 Properties of Secure Hash Algorithms

with extra bits and then concatenated with the size of the message before hashing. The padding and concatenation create an input message that is an integer multiple of block size in length. For example, Fig. 11.18 illustrates the format for the input message used with SHA-512. The message is first padded (as needed) with a 1 followed with zero or more 0's, and then it is concatenated with the size of the message (N) represented as a 128-bit unsigned number. The result becomes an $M * 1024$-bit message, with M being an integer number. Table 11.8 presents a list of properties for each of the SHA algorithms.

An SHA, like the simple hash function illustrated in Fig. 11.16, starts with a k-bit known IV, where k is typically less than the block size n ($k < n$), and performs several compression rounds to generate a hash value for the first block, which is then used as the IV to hash the next block. This process continues until the final hash, also called a **message digest**, is generated. Any change made to the message will result in a different hash value that will not match the one transmitted with the original message. The SHA algorithms could be used either with plaintext or ciphertext input.

11.7 Cryptography Hash

A standard hash cannot be protected. It is still possible for an adversary to alter both the message and its hash without detection. Cryptography hash algorithms, on the other hand, require a secret key to generate a hash value. In this case, the hash is called a **keyed hash**. Two examples of keyed hash algorithms are discussed next.

11.7.1 Message Authentication Code

A message authentication code (MAC), also called a cipher MAC (CMAC), requires a cipher to generate a secure hash. For example, Fig. 11.19 illustrates

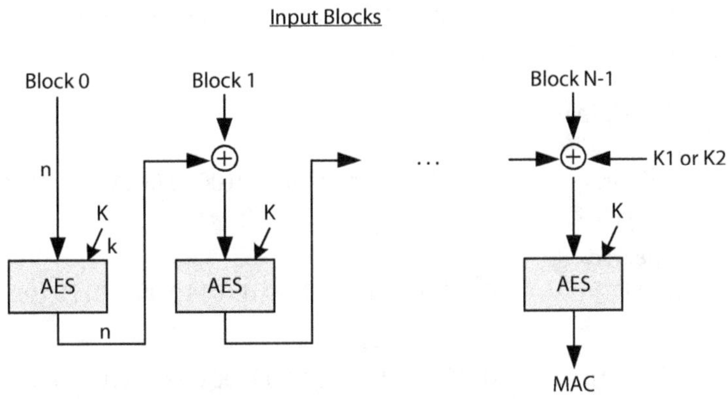

FIGURE 11.19 The AES-CBC-MAC; the MAC is m bits long where $m \leq n$.

the generation of a MAC using the AES-CBC cipher with n-bit blocks and a k-bit key (K). The n-bit keys $K1$ and $K2$ in the figure depend on the message size and are determined from K and a constant that depends on n [40]. If the message size is an integer divisible by n, then constant $K1$ is used; otherwise, the last block is padded with 1 followed by 0's to create an n-bit last block and the constant $K2$ is used. A few bytes from the last cipher block are selected as the MAC.

The combined abbreviations AES-CBC-MAC means an AES cipher in CBC mode is used to generate a MAC. AES-CBC-MAC has the advantage of performing both encryption and keyed hashing at the same time.

If an accidental or intentional modification is made to one or more of the input blocks, the algorithm produces a different MAC. Therefore, it is not possible for an adversary who doesn't know the secret key to alter the input and still generate the same MAC. However, both the sender and the receiver need to know the secret key and the technique used to generate a MAC. In addition, a MAC may be generated for a plaintext or ciphertext input.

11.7.2 Hash MAC

As opposed to a CMAC that requires a cipher, a hashed MAC (HMAC) requires a more efficient (less computationally intensive) standard hashing algorithm, such as SHA-256. An HMAC could be computed in two hashing cycles using two additional secret (S) codes *Si* (input S) and *So* (output S), derived from a secret key *K* and two integer constants called input pad (*iPAD*) and output pad (*oPAD*).

The *iPAD* and *oPAD* do not make the hashing algorithm more secure; instead, they are used to improve the quality of the secret key through a technique known as **whitening** [46]. For example, given an initial 16-bit *key* = 16'h1234 and *iPAD* = 8'h36, a 32-bit *Si* is generated as follows using bitwise XOR (⊕) as the whitening technique:

K: Original key padded with 0s:	00000000 00000000 00010010 00110100 (0x00001234)	(11.7)
iPAD repeated (*iPAD*$^+$):	00110110 00110110 00110110 00110110 (0x36363636)	
⊕	________________________________	
32-bit *Si*:	00110110 00110110 00100100 00110010 (0x36362432)	

Given a message *M*, a secret key *K*, the *iPAD* and *oPAD* constants, and a standard hash algorithm H, an HMAC is generated as follows:

The HMAC Algorithm:

1. Generate *Si* from *K* and *iPAD*, as illustrated in Eq. (11.7); that is, $Si = K \oplus iPAD^+$. The original secret key may need to be padded with 0's, and *iPAD* is repeated as needed to create *Si* of desired length.
2. Generate a hash value (*HV*) for *Si* concatenated with *M*; that is $HV = H(\{Si, M\})$, where {} indicates concatenation.
3. Generate *So* using *K* and *oPAD* (e.g., *oPAD* = *8h'5C*). *So* is generated in a similar way as *Si*; that is, $So = K \oplus oPAD^+$.
4. Generate the HMAC for *So* concatenated with *HV*; that is, $HMAC = H(\{So, HV\})$.

Equation (11.8) summarizes these four steps.

$$HMAC(K, M) = H(\{K \oplus oPAD^+, H(\{K \oplus iPAD^+, M\})\}) \tag{11.8}$$

11.8 Storing Cryptography Keys through Hardware

As was discussed at the start of the chapter, confidentiality and integrity are fundamental to computer security. However, storing cryptography keys on the hard disk creates a security risk and a threat to the system. Securely storing many secret keys, such as those used in an organization, requires creating a key structure tied to a secret key inside a tamperproof IC. This is known as **data storage through hardware** or **binding data to platform**.

The NIST lists a set of recommended key sizes and derivation techniques [38]. For example, 2048-, 3072-, or 4096-bit key sizes for the RSA, or 256- or 384-bit key sizes for the ECC are recommended for public key encryption. Likewise, 256-, 384-, and 512-bit SHA are recommended for hashing. AES-CBC-MAC is one of the recommendation for protecting both confidentiality and integrity. For integrity only, a 128-bit HMAC using SHA-256 is recommended. Readers are referred to the

NIST documents for application-specific recommendations. The secure storage of cryptography keys must be maintained by a trusted computing base (TCB).

11.8.1 Keychain Organization

A key structure, or keychain, is organized hierarchically as a tree with parent and leaf nodes. Figure 11.20 illustrates an example organization of one or more keychains, each with one or more nodes [47, 48]. In the figure, a parent node is shown by a square, and it refers to a **storage key** required to protect the content of its children nodes. A leaf node, shown by a diamond, refers to a **signature key**, also called a **signing key**. It is used, for example, to encrypt the hash of an e-mail message or the output of a program to certify results. A leaf node, shown as a circle, indicates a small amount of data, for example, a symmetric key used to encrypt a large data file. A key structure may also include other leaf nodes, such as an attestation identity key (AIK). It is an asymmetric key tied to the platform and used, for example, in server authentication applications.

The root parent node is protected by a secret root key (SRK) securely stored inside a tamperproof IC. Secret keys based on physical unclonable functions (PUFs) have been shown to be resistant to many types of attacks [49–51]. An arrow indicates a key derivation. Each parent and leaf node contains a set of **key material** that includes a unique number called **nonce** (number used once) and may also include one or more constants that identify a key type. The keys themselves are not saved; only the key material of each node is securely saved. In

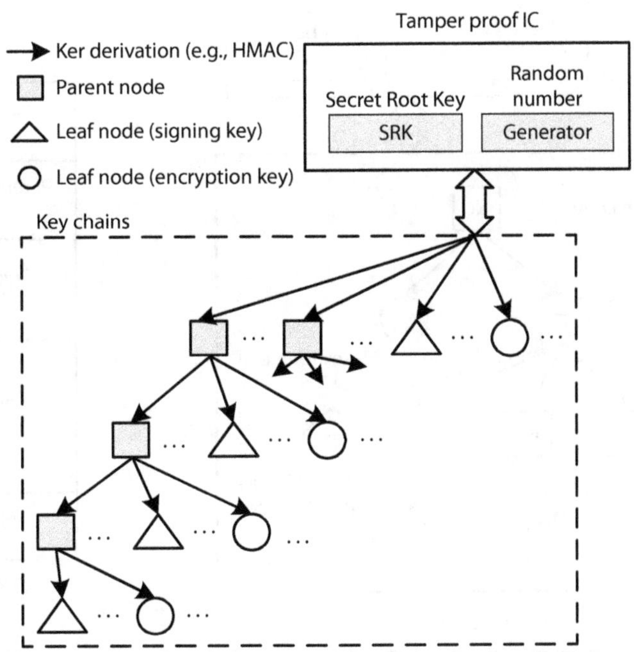

Figure 11.20 Keychains tied to an SRK inside a tamperproof IC.

addition, an authorized user typically may need to provide one or more correct passcodes, known as an **authdata**, before a child key can be used. A keychain may be organized in several ways [48, 52–55], such as the ones discussed next.

11.8.2 Storage and Access

Figure 11.21(*a*) illustrates a 10-node keychain with 4 parent nodes labeled 0, 2, 5, and 7, and 6 leaf-nodes labeled 1, 3, 4, 7, 8, and 9. The nonce N_0 is assigned to root parent Node 0, N_1 is assigned to data leaf Node 1, N_3 is assigned to signature leaf Node 3, etc. As shown in Fig. 11.21(*b*), a record is created for each node and securely stored either locally or remotely on a server. Each record contains a key identification number, for example, the node number; a parent identification number; the names of a cipher and a hashing algorithm, referred to as cryptography algorithm identifiers; the encrypted key material; and keyed hash for the entire record.

The SRK is a secure unique key embedded in each SCP or SP chip. The key may be programmed into the chip during setup through a set of secure I/O mechanisms without OS intervention [55]. In the figure, the key of root parent Node 0 is used to protect its two children nodes 1 and 2; the key of the data leaf Node 1 is used to protect, for example, a 128-bit AES key used for encrypting a

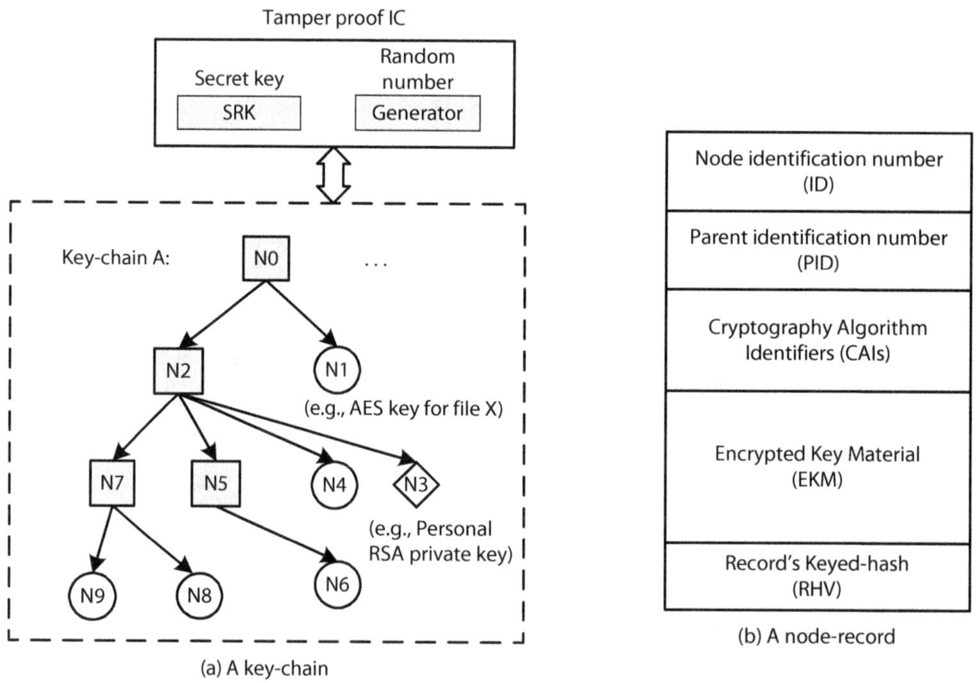

FIGURE 11.21 An example key structure: (*a*) a keychain; (*b*) a node as a record [53, 55].

large user file or a large application data; the key of parent Node 2 is used to protect its four children nodes 3, 4, 5, and 7; etc.

A keychain is maintained using trusted firmware module (TFM) in SCP or using a trusted software module (TSM) executed on the SP (also see Sec. 11.4). TFM/TSM (TFM or TSM) would include a set of application programming interfaces (APIs), such as "*Add2Keychain*" and "*Encrypt*" used by an OS or an application program to request secure cryptography services.

In the following example, it is assumed that each node may use different encryption and hashing algorithms. In addition, for simplicity, no *authdata* is required. All items marked plaintext ("ptxt") are considered secure within an SCP or an SP. The examples assume an application is requesting security services from SCP or SP.

Example 11.6. Application software requests TFM, if SCP, or TSM, if SP (indicated as TFM/TSM), to add parent Node 0 to key chain A in Fig. 11.21.

Solution:

Application task:

Using the following API, a request is sent to TFM/TSM to add root parent Node 0 and create a keychain; TFM/TSM will use the application-provided *cipher0* and hash algorithm *algHash0* to encrypt and perform a keyed hash of the key material. The parent of root Node 0 is Null.

$$Addkey2Keychain(keychain\_A, 0, Null, \text{cipher0}, \text{algHash0}, nodeType0, R0) \text{ //nodeType: parent}$$

TFM/TSM task:

Generates a nonce and then uses SRK to encrypt both the nonce and the Node 0's key material. The record is also keyed-hashed using SRK. The concealed record is returned to the application for storage. In the following, "nodeType" indicates parent or leaf (data) node, R indicates a reference to key record data structure in memory, "KID" indicates a key node ID (a number), "PID" indicates a parent node ID (a number), "EKM" indicates encrypted key material, N is a nonce, and HV is a hash value. SRK is the secure root key stored in the SCP chip if TFM implements the APIs, or in the SP chip if the TSM impalements the APIs. Specifically, TFM/TSM use SRK to perform the following tasks because $KID = 0$ and $PID = \text{null}$:

$$Addkey2Keychain(keychain, KID, PID, \text{cipher0}, \text{algHash0}, nodeType, R)$$

begin

$N_{ptxt} = nonce();$ //generate a nonce for root Node 0

$EKM_{ctxt} = E_{cipher0}(SRK, \{N_{ptxt}, nodeType\});$ //encrypt key material using SRK since PID=null

$HV = H_{algHash0}(SRK, \{KID, PID, \text{cipher0}, \text{algHash0}, EKM_{ctxt}\});$ //keyed hash of the record

$R = \{KID, PID, \text{cipher0}, \text{algHash0}, EKM_{ctxt}, HV\}$

end

Example 11.7. The application software requests TFM/TSM to add data leaf Node 1 to parent Node 0 in keychain A in Fig. 11.21.

Solution:

Application task:

Using the following API, the TFM/TSM is instructed to add an encryption key node to parent Node 0 in keychain A; the application provides *cipger1* and *algHash1* for encrypting and keyed hashing of Node 1's key material, where "K" stands for a key (also see Example 11.6).

$Addkey2Keychain(keychain\_A, 1, 0, \text{cipher1}, \text{algHash1}, nodeType1, R1);$ //node type: data

TFM/TSM task:

Performs the following operations because $PID = 0$:

$R0 = getParentRecord(keychain\_A, 0);$ //0, Null, $\text{EKM0}_{\text{ctxt}}$, cipher0, algHash0, HV0_{read}

$HV0' = H_{algHash0}(SRK, \{0, Null, \text{cipher0}, \text{algHash0}, EKM0_{ctxt}\});$ //re-compute hash

$HV0' = ?HV0_{read}$ if yes, continue; otherwise, raise an exception //authenticate Node 0

$EKM0_{ptxt} = D_{cipher0}(SRK, EKM0_{ctxt});$ //$EKM0$ contains N0_{ptxt}

$K_{0:1} = E_{cipher0}(SRK, N0_{ptxt}, 1, nodeType1);$ //generate key for Node 1;

$N1_{ptxt} = nonce();$ //generate a nonce

$EKM1_{ctxt} = E_{cipher1}(K_{0:1}, \{N1_{ptxt}, nodeType1\});$ //encrypt Node 1's key material

$HV1 = H_{algHash1}(K_{0:1}, \{1, 0, cipher1, algHash1, EKM1_{ctxt}\});$ //hash key material

$R1 = \{1, 0, cipher1, algHash1, EKM1_{ctxt}, HV1\}$ //Record is returned to application

Example 11.8. The application software requests TFM/TSM to encrypt application data using data-leaf key number 1 in Fig. 11.21.

Solution:

Application task:

Using the following API, the TFM/TSM is instructed to encrypt application data using encryption key 1 from keychain A, where $data_{ptxt}$ and $data_{ctxt}$ reference the application data structures in memory. (Also see Examples 11.6 and 11.7.)

$Encrypt(keychain\_A, 1, 0, data_{ptxt}, application\_cipher, data_{ctxt});$ //use application cipher

TFM/TSM task:

Generates encryption key 1 using the key material of data leaf Node 1, and then encrypts the application data using the application-provided cipher. Application data as ciphertext is returned to the application.

$R0 = getPrarentRecord(keychain\_A, 0);$ //0, Null, cipher0, algHash0, EKM0_{ctxt}, HV0_{read}

$HV0' = H_{algHash0}(SRK, \{0, Null, \text{cipher0}, \text{algHash0}, EKM0_{ctxt}\});$ //recompute hash

$HV0' = ?HV0_{read}$ if yes, continue; if no, raise an exception

$EKM0_{ptxt} = D_{cipher0}(SRK, EKM0_{ctxt});$ //will use N0_{ptxt}

$K_{0:1} = E_{cipher0}(SRK, \{N0_{ptxt}, 1, nodeType1\});$ //Key for Node 1;

$R1 = getRecord(chain\_A, 1);$ //1, 0, $cipher1$, $algHash1$, $EKM1_{ctxt}$, $HV1_{read}$

$HV1' = H_{algHash1}(K_{0:1}, \{1, 0, \text{cipher1}, \text{algHash1}, EKM1_{ctxt}\});$ //recompute hash

$HV1' = ?HV1_{read}$ if yes, continue; if no, raise an exception

$EKM1_{ptxt} = D_{cipher1}(K_{0:1}, EKM1_{ctxt});$ //will use $N1_{ptxt}$ and node$Type1$

$K_{data-leaf} = E_{cipher1}(K_{0:1}, \{N1_{ptxt}, nodeType1\});$ //generate the encryption key

$data_{ctxt} = E_{application\_cipher}(K_{data-leaf}, data_{ptxt});$ //encrypt data

Table 11.9 illustrates hypothetical records for Nodes 0 to 9 in Fig. 11.21(*a*); only four records are shown.The fact that in the figure, the keychain parent Node 0 is concealed by the SRK inside an IC; Nodes 1 and 2 are concealed by parent Node 0; Nodes 3, 4, 5, and 7 are concealed by Node 2; Node 6 is concealed by Node 5; and Nodes 8 and 9 are concealed by Node 7, the keychain is said to be sealed through hardware.

11.8.3 Application Example: Keychain as Access Control

The farther away a node is from the root of a keychain, the more computations are necessary to determine a key. Thus, a keychain can be used to implement a multilevel access control (Sec. 11.1.4) using a hierarchical authdata generation scheme. Figure 11.22 illustrates the data organization of a company that has three departments. In each department, the data is classified into several security levels. In the Project Department, for example, data is classified into four classifications as those accessed by engineers, by production engineers, by project leaders, and by the director. Anyone who has the role and privileges of an "engineer," for instance, would be able to access all the data that is classified as "engineering data." A production engineer would be able to access all the "production data" as well as all the "engineering data." The director, however, would be able to access the "director data" as well as all the other data in their department.

The keychain is organized using an SRK-protected RSA public/private key pair for each class of data (four in Fig. 11.22) and a prime nonce to each node [54]. An initial authdata is then computed for each data class $j \geq 0$ using all the public keys in the path from class j to class 0. To access a leaf node k in class j, an authdata is derived using the assigned initial authdata of class j, all the private keys in the path from class j to class 0, and the nonce assigned to node k.

Thus, the computation of an authdata—for example, for the lowest classified "engineering data"—would involve only one private key, while the computation of

Key Identification Number (KID)	Parent Identification Number (PID)	Cryptography Algorithm Identifiers (CAI)	Encrypted Key Material (EKM)[1]		Keyed-hash for the Entire Record (RHV)[2]
			Nonce[2]	Node Type	
0	null	cipher0, algHash0	6564...465	parent	3A5A...C20B
1	0	cipher1, algHash1	8815...489	data	1ECA...C360
2	0	cipher2, algHash2	6304...363	parent	45F6...2BEC
3	2	cipher3, algHash3	2467...427	signature	1F55...F8B9
...	...	...	...	...	...

1: Key material shown not encrypted; 2: Mock values

TABLE 11.9 Recorded Keychain in Fig. 11.21(*a*) Using the Format in Fig. 11.21(*b*)

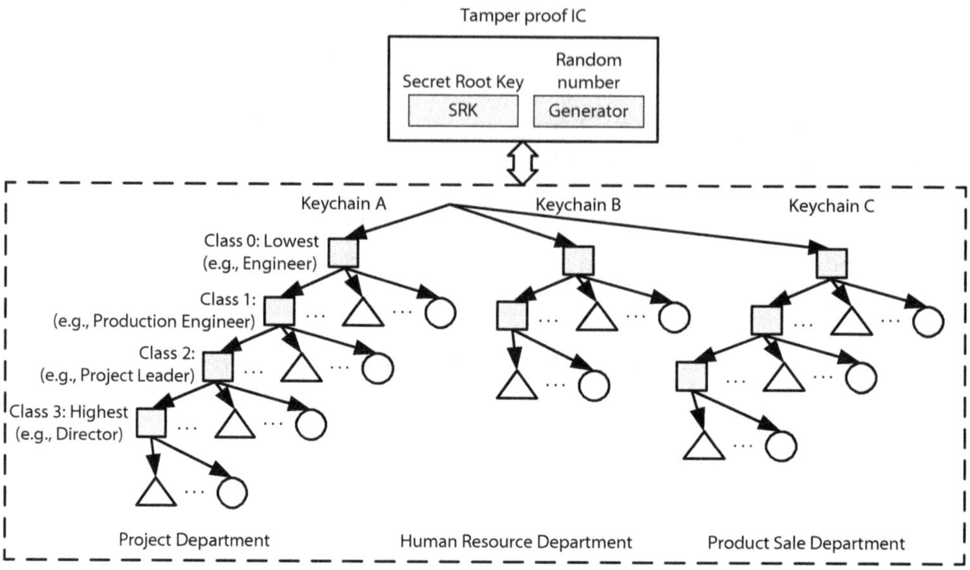

FIGURE 11.22 An access control keychain [54].

an authdata for a top classified "director data" would involve four private keys, making the "director data" more secure.

11.9 Hash Tree

While the keychain in Fig. 11.21 is concealed through hardware, it remains unprotected from replay attacks (Sec. 11.3.3). A malicious software can save the entire keychain, wait for the keychain to be updated, and then cause a replay attack. That is, it replaces the updated keychain with the one it has saved, and therefore, prevents access to some concealed data and potentially makes the system unavailable to the legitimate users. There is no way to detect a keychain replay attack unless a hash of the keychain is saved inside a tamperproof IC. The hash of the tree would need to be recomputed every time that the keychain is updated and every time that a key from the keychain is used. Given that a keychain could be large and may contain thousands of keys, this could be a costly task unless a hash tree, discussed next using examples, is used.

11.9.1 Application Example: Keychain Authentication

Figure 11.23(*a*) illustrates the keychain shown in Fig. 11.21(*a*) organized also as a **hash tree**, also known as a **Merkle hash tree** [56]. The arrows go in both directions—from a parent node to a child node when generating a key, and from a child node to a parent node when generating a hash. Instead of computing the hash of an entire keychain each time the keychain is updated or a key is used (a computationally intensive task), only the hash of children nodes of each parent node is computed and stored in the parent node. The hash of the root node, called the secure root hash (SRH), is stored inside an IC, as illustrated in the figure.

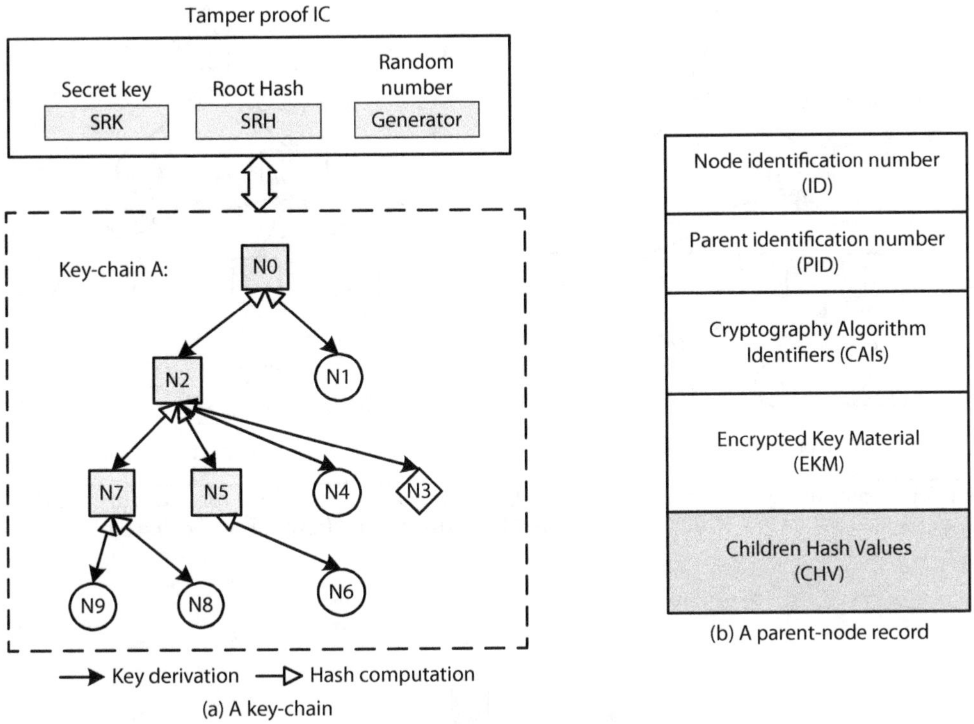

FIGURE 11.23 A keychain organized as a Merkle hash tree.

Figure 11.23(*b*) shows the structure of a record used for parent nodes. In this case, instead of the record hash value (RHV) (Fig. 11.21(*b*)), each parent record contains the hash of its children records. Figure 11.24 illustrates an example of a keychain hash tree using mock values as node contents. An 8-bit bitwise XOR, for illustration purposes, is used as the hashing function. Each parent node in the figure is assumed to contain a 16-bit content plus, as indicated in Fig. 11.23(*b*), a space for a hash computed from the content of its children nodes. A leaf node contains only a 16-bit data and no hash value.

Figure 11.24(*a*) illustrates the calculation of the root hash 8'hC0 (hex in Verilog) for the original hash tree. The hash would be stored as an *SRH* inside the chip. In Fig. 11.24(*b*), the content of a child node is shown changed from 16'h2345 to 16'h2355—a one-digit change. This results in a new root hash value of 8'hD0. If the change is a result of a normal update, 8'hD0 will replace *SRH* = 8'hC0. On the other hand, if the change is the result of an attack, the change would be detected because the new hash = 8'hD0 would not match *SRH* = 8'hC0 stored inside the IC.

11.9.2 Application Example: Memory Authentication

A Merkle hash tree has other applications, such as the *n*-ary hash tree shown in Fig. 11.25 used for memory authentication. The integrity of a TSM can be protected

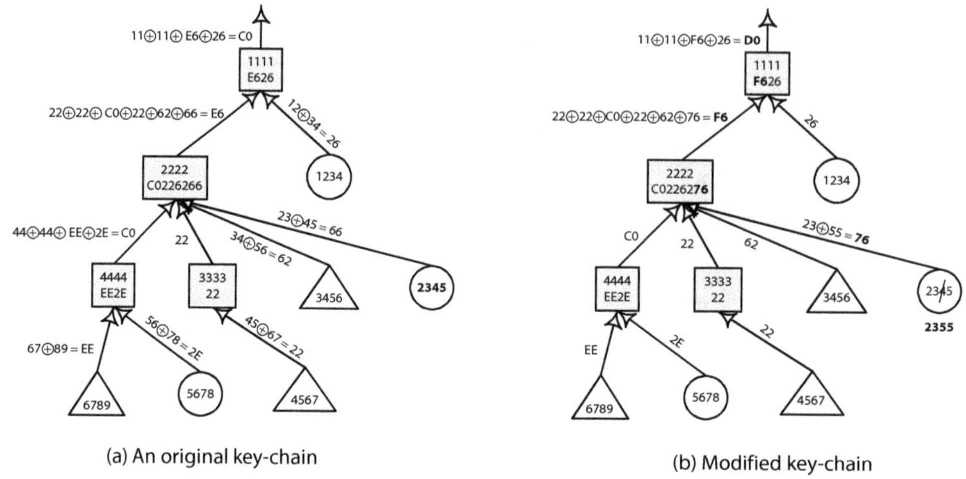

FIGURE 11.24 Computing a root hash using an 8-bit bitwise XOR as the hash function: (*a*) an original hash tree with root *hash* = 8'hC0; (*b*) a modified hash tree with different root *hash* = 8'hD0.

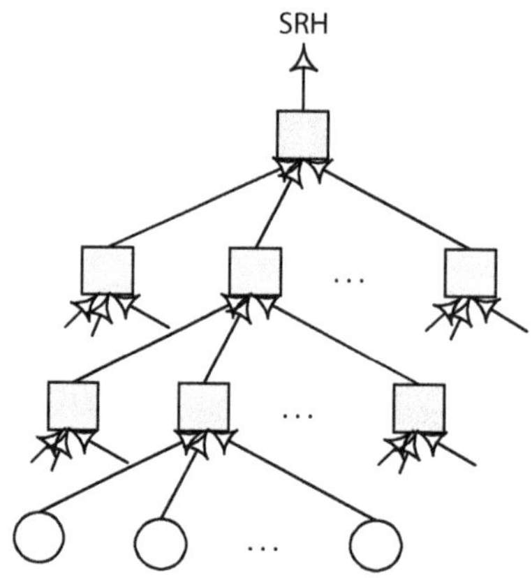

FIGURE 11.25 An *n*-ary hash tree with leaf nodes (circle) and data and parent nodes (rectangle) as hash values.

by authenticating its instructions and data in memory. Each node has exactly n children nodes. For $n = 2$, the tree is binary. An entire virtual or physical memory, or just a section, can be organized as an n-ary hash tree with a memory block at each node. A leaf block contains instructions or data, but a parent block contains only the hash computed from the contents of its children blocks.

Figure 11.26 illustrates a binary hash tree with four data blocks as leaf nodes and three hash blocks as parent nodes. In the figure, the size of each block is assumed to be two bytes (2B), and an 8-bit bitwise XOR is used as the hash function. Each parent block stores two 8-bit hash values, one from each of its two children blocks. The hash of the root block is stored as an *SRH* inside an IC. In the

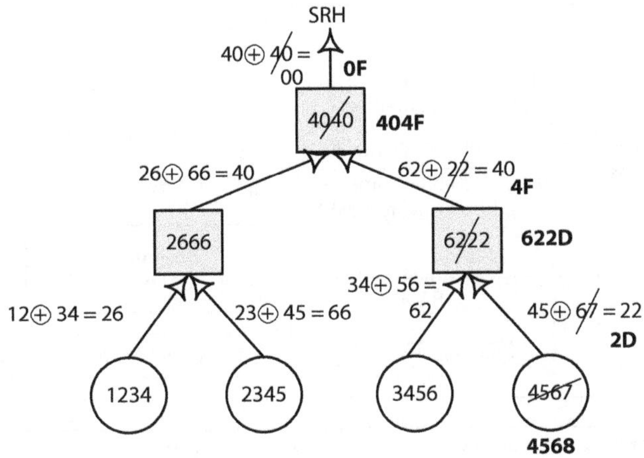

Figure 11.26 A binary memory authentication hash tree illustrating a change in one of the leaf memory blocks.

figure, the original *SRH* = 8'h00. Any change made to a block (leaf or parent) will result in a different root hash. Note that a change in a node's content only changes the hash values in the path from the node to the root node. Thus, only the contents of a small subset of blocks are affected when the content of a single block is updated.

There are many ways to organize an *n*-ary hash tree. For example, a hash tree with 32-B blocks may be organized as a 32-ary hash tree with 1-B hash values, as a 16-ary tree with 2-B hash values, as an 8-ary tree with 4-B hash values, or as a binary tree with 16-B (128-bit) hash values. In addition, the parent blocks may be stored with the leaf blocks in the same memory or maintained separately.

11.10 Secure Coprocessor Architecture

An SCP is as an embedded system and as it was discussed earlier in Sec. 11.4 includes a trusted firmware module (TFM). Because the instructions and data of firmware are not accessible from outside of the chip, TFM is not subject to spoofing, splicing, or replay attacks. However, because an SCP must communicate with the other components on the platform to exchange data, the SCP may be subject to physical attacks if an attacker is able to get a physical hold of the platform. For instance, any data that SCP must access from main memory is not secure.

The required cryptographic algorithms may be implemented in software as part of the TFM, but many are very time consuming and, therefore, for performance reasons, they would be implemented in hardware. Figure 11.27 shows the organization of an SCP containing a minimum set of required modules. It includes nonvolatile memory to store an secret root key (SRK), a random number generator, and encryption/decryption and hash algorithms implemented in hardware. The SRK may be a symmetric key [53] or the private key of a public/private key pair [47].

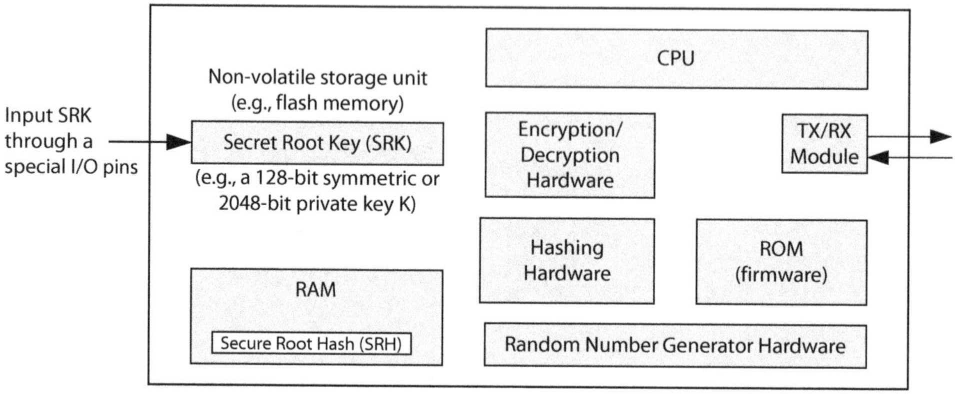

FIGURE 11.27 A secure coprocessor as an embedded system.

Random access memory (RAM) is used during the execution of the TFM and to temporarily store keys from a keychain (also see Sec. 11.8.2). The RAM may be also used to store an SRH if the TFM also manages a keychain protected by a hash tree.

The smart card [57] and trusted platform module (TPM) [47, 52, 58] are two examples of SCP. However, the smart card has very limited capabilities.

11.10.1 Trusted Platform Module

The TPM is designed to provide total platform security; it verifies the validity of the hardware and OS software components during startup. The OS and even application programs can use APIs to access secure services of the TPM.

TPM's specification is developed by a consortium of industry representatives, known as the Trusted Computing Group (TCG). Many companies, including AMD, HP, IBM, Intel, and Microsoft, are members of the TCG. Figure 11.28 illustrates a block diagram of the TPM chip as an embedded system. The nonvolatile storage is used to store an endorsement key (EK), an SRK, and flags to enable or disable certain functions. EK is a secret key embedded in the chip, typically by the manufacturer. SRK is used to protect TPM-generated keys. An Attestation Identity Key (AIK) is a

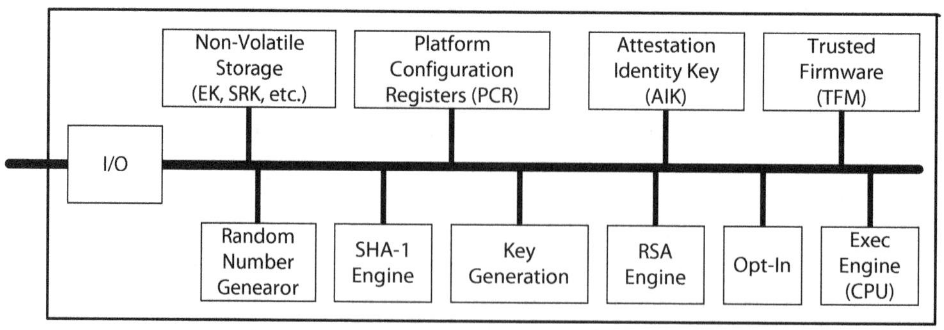

FIGURE 11.28 The architecture of the TPM [58].

private key generated based on SRK and is used for multiple purposes, including platform authentication.

The **random number generator** may use the thermal noise in the chip [59] to generate nonce as needed. The **RSA Engine** performs RSA encryption and decryption. The **RSA Key Generator** is used to generate asymmetric RSA keys. The **SHA-1 Engine** is used for hashing purposes.

The Opt-In module allows a user to opt in or opt out according to the privacy guidelines of the manufacturer of the platform. With the opt-in mechanism enabled, a user is prompted before a feature or service is provided. The opt-in mechanism is disabled by default. With the opt-out mechanism enabled, the user is prompted to either keep or disable a particular feature or function. By default, the opt-out mechanism is enabled.

11.11 Secure Processor Architecture

An SP may implement multiple secure execution modes (SXMs), as was discussed in Sec. 11.4, to create the desired secure execution environment for a given TSM. An SP may provide software developers with the option to choose the protection level of a program as integrity only, confidentiality only, or both and whether the protection should apply to program code (instructions and static data), to program dynamic data, or both to program code and data [30]. While program static data never changes, dynamic data is generated during program execution, including dynamically generated code produced by just-in-time compilers.

11.11.1 Program Code Integrity

Programs (i.e., trusted software modules, TSMs) executing in code integrity secure execution mode (CI-SXM) are protected from spoofing and splicing attacks. Replay attacks are not an issue because program code (including static data) does not change during execution. Hash values are used to verify the validity of instructions and static data during execution. However, because modern processor chips contain cache memories and cache transactions are in blocks, one hash value per block (i.e., cache line) is sufficient.

There are two ways the hash values of program code blocks may be organized in memory [31, 55]. One option is to include a hash value embedded within each block with instructions and static data, as illustrated in Fig. 11.29. For example,

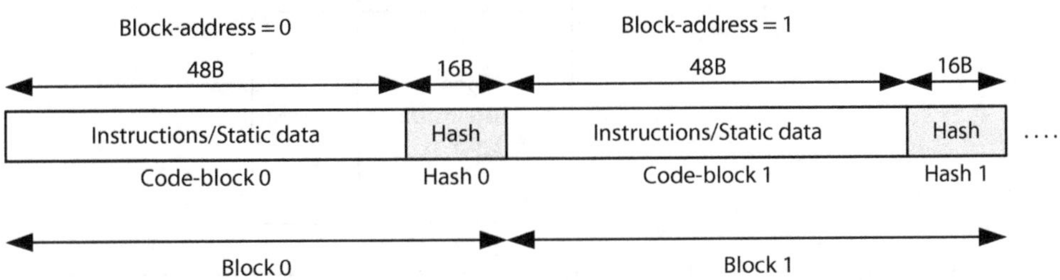

FIGURE 11.29 Program blocks with embedded hash values, assuming 64B blocks [55].

assuming that SP is a 32-bit RISC processor and the lowest (e.g., L2) cache lines are 64 B each, the figure illustrates a 128-bit (16-B) keyed hash (e.g., HMAC) embedded in each block. A hash is computed for every 48-B program code, named a **code block**, that could contain 12 4-B instructions or static data words. A **block** refers to 64 B of memory content.

Each code block that is loaded into the cache must be authenticated by computing and comparing each code block's hash with the hash embedded in the block. If the two hashes match, the block is considered valid and the hash bytes are changed to NOP (no operation) instructions before the block is stored in the cache. On the other hand, if the two hash values do not match, the block is marked invalid in cache, and an exception is raised that terminates the execution of the program.

Alternatively, the hash values can be kept separately and not embedded within the blocks. One way to do this is to build the memory from 72-bit ECC (error correcting code) synchronous dynamic random access memory (SDRAM) modules. In this case, 72-bit memory content is made of 64-bit program code and an 8-bit hash value in place of an 8-bit ECC [31]. The SP, however, would need to use the lowest-level cache with 128-B cache lines. The program binary would be divided into blocks of 128 B each, consisting of 16 64-bit content. A 128-bit (or 16-B) keyed hash is computed for each 128-B block and is stored as 16 8-bit hash quantities in the 16 ECC fields reserved for each block, as illustrated in Fig. 11.30. During a cache miss, 16 72-bit memory contents are transferred from memory into the SP. Each 72-bit memory content contains 64-bits program code and one of the 16 8-bit hash quantities. If the SP-computed 16-B hash value matches the 16-B hash read from memory, the cache line is marked valid. A 16-B hash is not stored in the cache. Although, in this case, a 16-B hash value cannot be used for error correction, it can, however, be used to detect multiple bit errors.

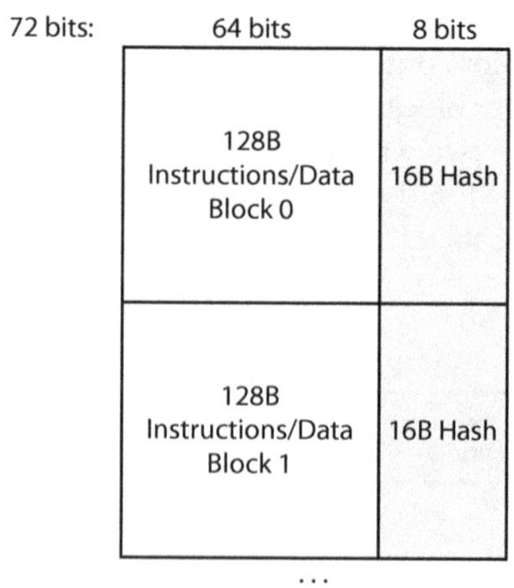

Figure 11.30 Program instructions and static data blocks in memory designed from ECC SDRAMs; the hash values are stored in the space reserved for the ECC bits.

Program Compilation

Two different memory block organizations for code integrity were discussed earlier. Consider the block organization shown in Fig. 11.29 where a hash is embedded in each block. In this case, each memory block only partly contains instructions or data. Therefore, the compiler would need to take into account the location of the hash bytes when computing jump/branch addresses.

11.11.2 Operational Security Mechanisms

The binary of the SXM program must not only be securely distributed and installed, but also securely loaded in memory during execution. The set of procedures similar to those outlined in [55, 60, 61] must be followed to install and load the program. In the following sections, a set of security mechanisms for software distribution and installation, as well as loading binary into memory for execution, are discussed.

Secure Binary Distribution

If the TSM is developed to be delivered for public distribution, a program plaintext binary ($binary_{ptxt}$) would need to be hashed in case an unauthorized modification is made to the program. The hash of the binary file is encrypted to create a header record using the private key ($PR_{company}$) of the developer. The record would be attached to the binary file for secure delivery. An example of public distribution is illustrated next, using E_{asym} to indicate an asymmetric cipher (e.g., RSA), H to indicate a (standard) hashing algorithm such as SHA-256, "PU" to indicate a public key.

```
//binary package includes both header and program binary
General distribution: {header_company, binary_ptxt};
//secret header record for code integrity
header_company = E_asym (PR_company, {H(binary_ptxt), "SHA-256"});
```

Program binary may also be delivered to a specific device using instead the public key (PU_{device}) of the device as follows:

```
header_device = E_asym (PU_device, {H(binary_ptxt), "SHA-256"});
Target device distribution: {header_device, binary_ptxt};
```

For device-specific delivery, the header record can only be encrypted by the device using its private key (PR_{device}); the program can only be executed by the target device.

Secure Program Installation

For secure installation of a TSM, the installer program would need to have access to the SRK of the SP and possibly a password provided by an authorized party (a person) who installs the program [61]. The installer program generates a secret program-specific signing key (e.g., $K_{sym\text{-}prog\text{-}sign}$) within the SP and uses the key to hash each of the program code blocks during installation.

One way to do this is to use an installer TFM (i.e., an embedded installer trusted firmware) to ensure that the installer program remains secure from attacks. In this

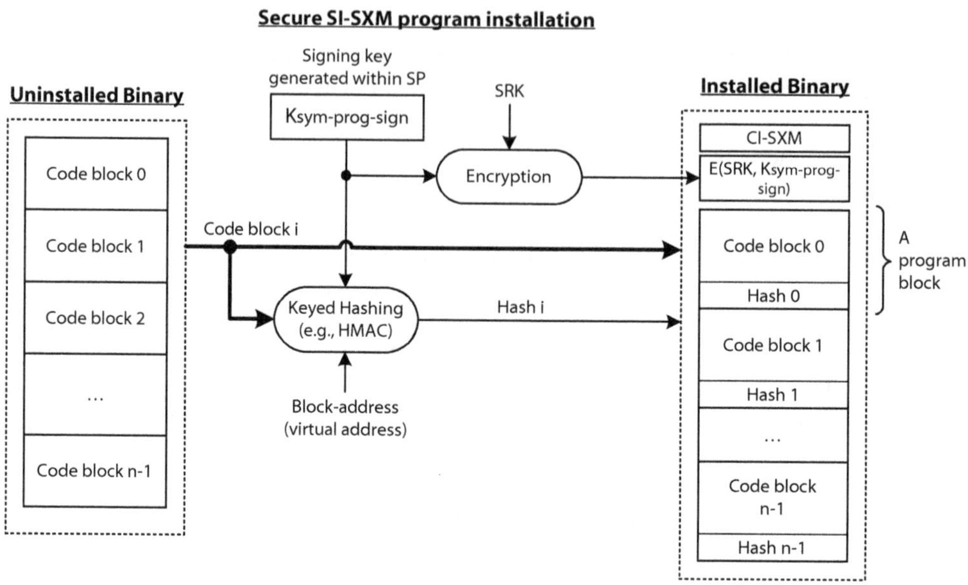

FIGURE 11.31 Secure installation for program code integrity SXM [30].

case, the installer input arguments, such as the size and location of uninstalled TSM binary in memory and the installation key information (e.g., $PU_{company}$) are stored by the OS in a known memory space before the firmware can install the program.

Figure 11.31 is an illustration of installation steps for a CI-SXM program using the block organization in Fig. 11.29. The hash of each code block is stored with the code block to create a program block.

The hashing of just each code block will detect spoofing attacks during execution. However, in order to also detect splicing attacks, the starting address of each block, referred to here as a **block address**, is also used to compute the hash of each code block. The installer-generated signing key $K_{sym\text{-}prog\text{-}sign}$ (e.g., a random number) is also encrypted using the processor SRK before it is stored (on the hard disk) with the installed binary, as illustrated in the figure. The following outlines the required steps to install a program for CI-SXM using the block organization in Fig. 11.29. Also, H_{keyed} indicates a keyed hash, $code\_block_j$ indicates program code (instructions and static data) sections within $block_j$, n is the number of blocks, and HV stands for hash value.

Steps to install a program for CI-SXM:

```
block_ptxt-j = {code_block_ptxt-j, HV_j}; //do this for all blocks,
//assume organization in Figure 11.29
//compute hash to detect spoofing and splicing attacks
HV_j = H_keyed (K_sym-prog-sign, {code_block_ptxt-j, block_address_j});
binary_installed = {"code integrity, HMAC-128", E(SRK, K_sym-prog-sign),
{{ block_ptxt-j } for j = 0 to n -1}};
```

Secure Loading Executable Binary

The loader program must also have access to the SRK within the SP in order to decrypt and store the installer-generated signing key $K_{sym\text{-}prog\text{-}sign}$ in a special register inside the SP before program execution can start. Therefore, like the installer program, the basic loader program is a TFM. Both the installer and loader firmware must not leak processor secrets. The OS accesses and stores loader input arguments in memory so the loader can extract the signing key before the execution of the TSM can be started.

Depending on the TSM protection level, the loader program may need to perform additional initialization, which will be discussed later, before program execution can begin.

11.11.3 Program Code Confidentiality

TSMs compiled to execute in code (instructions and static data) confidentiality secure execution mode (CC-SXM) must be kept confidential on the hard disk as well as in memory. The protection of instructions and static data integrity, however, is not required in this mode; thus, no hash values are needed. The installer firmware (see Sec. 11.11.2) generates an encryption symmetric key $K_{sym\text{-}prog\text{-}enc}$ to individually encrypt program code blocks. In addition, in order to prevent information leak, the block address is included in the encryption of each block in case two blocks have the same content. This is illustrated next using a block address to create an IV for a symmetric-key cipher (E_{sym}), such as AES.

Steps to install a program for CC-SXM:

```
IV_j = {block_address_j, 00...0}; //IV is padded with 0 to make up,
//for example, a 128-bit IV for 128-bit AES
block_ctxt-j = E_sym (K_sym-prog-enc, IV_j, block_ptxt-j); //encrypt each block
binary_installed = {"code confidentiality, AES-128", E(SRK, K_sym-prog-enc),
{{block_ctxt-j} for j = 0 to n -1}};
```

11.11.4 Program Code Integrity and Confidentiality

This secure execution mode, indicated as CICC-SXM, implements the protection of both integrity and confidentiality of a TSM's code (instructions and static data). It is the combination of CI-SXM and CC-SXM discussed earlier.

Steps to install a program for CICC-SXM:

```
IV_j = {block_address_j, 00...0}; //create a unique IV for each block
//The following two steps may be combined using AES-CBC-MAC
code_block_ctxt-j = E_sym (K_sym-prog-enc, IV_j, code_block_ptxt-j) ; //encrypt
//code block
HV_j = H_keyed (K_sym-prog-sign, {code_block_ctxt-j, block_address_j}) ; //hash
//each code block
```

```
blockctxt-j = {code_blockctxt-j, HVj} ; //combine to create
                                        //blocks as in Figure
                                        //11.29
binaryinstalled = {"code integrity and confidentiality,
                  AES-CBC-MAC", E(SRK, Ksym-prog-enc), E(SRK,
                  Ksym-prog-sign), {{blockctxt-j} for j = 0 to n-1}};
```

11.11.5 Program Data Integrity

Dynamic data is generated during execution, and contrary to static data, values change in memory. TSMs that are compiled to execute in data integrity secure execution mode (DI-SXM) must be protected from replay attacks in addition to spoofing and splicing attacks. Recall that a replay attack can replace an updated value in memory with a saved older value. Therefore, a hash value alone computed for each block, as discussed for CI-SXM, will not detect replay attacks. A hash tree is necessary to detect replay attacks of data blocks. Figure 11.32 illustrates a binary hash tree with data blocks as leaf nodes and hash blocks as parent nodes. It is assumed that each block is 32 B and each parent block can hold two 128-bit hash values computed from each of its children blocks.

In this case, the loader firmware (see Sec. 11.11.2) generates a **session signing key** (e.g., $K_{sym\text{-}session\text{-}sign}$) and then creates an initial hash tree using the session key before TSM execution in the DI-SXM can start. The figure also shows the tree organization in memory. The number shown below each block is a block address. Leaf blocks are in the high-address section of the memory and parent blocks in the

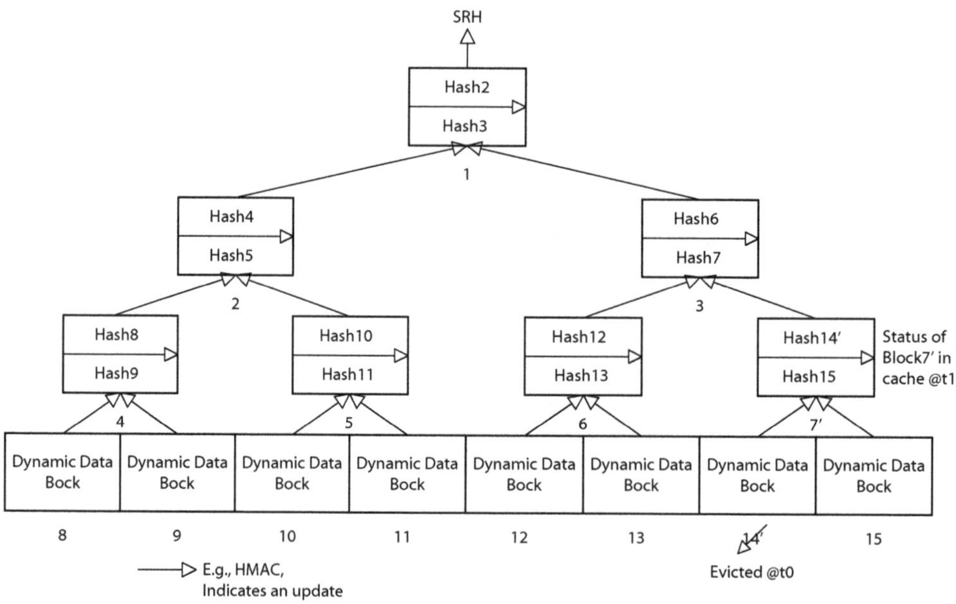

Figure 11.32 Illustrating an update to the hash tree of dynamic data blocks; numbers are block addresses.

low-address region. The hash of block 1 (the root block) is stored as an secure root hash (SRH) inside the SP; note that block 0 is not used.

In theory, each time a modified block is evicted from the lowest-level cache and leaves the secure perimeter of the SP, a new SRH must be computed. This requires that parent blocks in the path from the evicted block all the way to the root block must also be updated, as was illustrated in Fig. 11.26. However, in practice, because cache memories are inside SP and thus are considered secure, this updating of the parent blocks can be stopped as soon as a parent block in the path from the leaf block to the SRH is found in cache.

For example, suppose both leaf (i.e., data) block 14 and its parent block 7 are in cache, and block 14 is updated, shown as 14′ in the figure. Now suppose block 14′ is evicted from cache (shown crossed out) at time t0. Because block 7 is in cache and considered valid, the hash of block 14 (*Hash14*), which is stored in block 7, is replaced with the newly computed *Hash*14′ at time t1. There is no need to continue and update parent blocks 3 and 1 and the SRH. The next time that block 14′ is read from memory and loaded into cache, block 7′, assuming it is still in cache, would still contain *Hash*14′, the most recent hash of block 14′.

Now, suppose block 15 is loaded next from memory and block 7′ is still in cache. *Hash*15 must be compared with the hash stored in block 7′. Because *Hash*15, originally computed from the data in block 15, is still contained in block 7′, block 15 would be considered valid if its hash matches with *Hash15*. The process is the same; block 3, the parent of block 7′, is updated when block 7′ is evicted from cache; block 1 is updated if block 3′ is evicted from cache; and SRH is updated if block 1′ is evicted from cache. This reduces the overhead of maintaining a hash tree, which will be discussed in more detail in Sec. 11.12.

The tree organization of Fig. 11.32 works if the size of the dynamic data space is declared in advance in the program and the space is pre-allocated in physical memory. On the other hand, in order to allocate data memory space dynamically during run time, a different mechanism using paging (also see Chap. 9) is needed. In this case, the hash tree is a virtual tree and its nodes are blocks from virtual address space. One way to organize such a hash tree is to construct a two-level tree consisting of only root pages and leaf pages. A page consists of several blocks. For example, a 4-KB page would contain 1 to 64 64-B data blocks. Such a hash tree may be constructed as follows [30]:

1. Organize each dynamic data block in each leaf page with an embedded hash value, as was illustrated for code blocks in Fig. 11.29. For example, assuming that each hash value is 16 B, each 64-B dynamic data block would contain 48-B data and 16-B hash. A 4-KB leaf page would contain 3072 B (48 B $*$ 64) dynamic data and 1024 B (16 B $*$ 64) hash values.
2. Compute a checksum (using a bitwise XOR) of all the embedded hash values of each leaf page and store it in a block in a root page. A 4-KB root page would store a maximum of 256 16-B checksums in 64 blocks—four 16-B hash values in each block. In addition, a root page can hold checksums for 1 to 256 leaf pages.
3. Compute an accumulative checksum (again using bitwise XOR) of all the checksums in all the root pages and store it as an *SRH* inside SP.

As needed, more leaf and root pages are dynamically allocated. A virtual hash tree also protects leaf and root pages that migrate back to the hard disk. If there is an unauthorized change made to a page on the disk, the change can be detected the next time a modified page is copied back to memory and blocks from this page are accessed by SP.

Because the integrity of every data block that is loaded into cache must be verified, the secure loader firmware (discussed earlier) must create an initial hash tree for those blocks that are allocated prior to the start of program execution.

11.11.6 Program Data Confidentiality

DC-SXM is similar to CC-SXM, except that because data blocks change in memory during program execution, it may be possible for an adversary to find out, for example, that a data block in memory has the same value at different times. In order to prevent such information leaks, a **randomized encryption** of data blocks is needed [33]. Each time that a modified data block in cache gets evicted, in addition to the block-address a unique number is also used to encrypt the block using a session encryption key ($K_{sym\text{-}session\text{-}enc}$) generated by the loader. The loader firmware also performs the initial randomized encryption of any allocated data blocks before the execution of the TSM can begin. The session key, like the other keys, remains securely inside the SP.

In this case, even if the content of the block at times remains the same, the encrypted copy of the block would be different. Also, because dynamic data is generated at run time, the most recent unique number assigned to each block is saved in memory and then is accessed to decrypt the block the next time that the block is loaded from memory. For randomized encryption, there are two options to generate unique numbers for each data block:

1. **Random sequence.** A sequence of unique numbers for each block is randomly generated. The following illustrates a randomized encryption of $data\_block_j$ using a random sequence. RN_j indicates a random number assigned to $data\_block_j$ and Y_j is a memory location used to save RN_j. The IV of the cipher (e.g., AES) is created from the block address and its assigned random unique number. An $IV = \{block\_address, RN\}$ may be padded with 0's to create the right size IV for the encryption. Each time that there is a cache miss for $data\_block_j$, its current RN, stored in location Yj, is read to create the IV used to decrypt the block. A new RN is generated each time a block (modified or not) is evicted from the cache. Note that, using an initial RN (e.g., RN0) for each block j during initialization by the loader is not necessary. However, using an initial RN simplifies the architecture of the processor.

```
//Initialization by loader; RN0_j is saved at memory
//location Y_j
data_block_ctxt-j = E(K_sym-session-enc, {block_address_j, RN0_j},
data_block_ptxt-j});

//1st time evicted from cache, save RN1_j at location Y_j
```

```
data_blockctxt-j  = E(Ksym-session-enc, {block_addressj, RN1j},
data_blockptxt-j});
//2nd time evicted from cache, save RN2j at location Yj
data_blockctxt-j  = E(Ksym-session-enc, {block_addressj, RN2j},
data_blockptxt-j});
//3rd time evicted from cache, save RN3j at location Yj
data_blockctxt-j  = E(Ksym-session-enc, {block_addressj, RN3j},
data_blockptxt-j});
etc.
```

One must make sure a random number assigned to each block is indeed unique; but because there is no way to know this in advance, it is possible that, in some cases, some random numbers might not be unique for some blocks. If, for instance, short 32-bit random numbers are used, then there is a higher chance that some numbers may be repeated for a given block if the block is accessed many times. On the other hand, although using large random numbers for each block may reduce the probability of repeats, more memory space is needed to store large random numbers.

2. **In-order sequence.** A sequence of unique numbers, such as 0, 1, 2, etc., is sequentially generated for each data block. The following illustrates randomized encryption of $data\_block_j$ using an in-order sequence, assuming that the initial in-order number 0 is assigned by the loader and is stored in memory location Y_j.

```
//Initialization by loader; 0 is assigned and saved at
//location Yj
block_blockctxt-j  = E(Ksym-session-enc, {block_addressj, 0},
data_blockptxt-j}
//1st time evicted from cache; Yj + 1 = 1; 1 is saved at
//location Yj
block_blockctxt-j  = E(Ksym-session-enc, {block_addressj, 1},
data_blockptxt-j}
//2nd time evicted from cache; Yj + 1 = 2; 2 is saved at
//location Yj
block_blockctxt-j  = E(Ksym-session-enc, {block_addressj, 2},
data_blockptxt-j}
//3rd time evicted from cache; Yj + 1 = 3; 3 is saved at
//location Yj
block_blockctxt-j  = E(Ksym-session-enc, {block_addressj, 3},
data_blockptxt-j}
etc.
```

While using in-order sequences will guarantee unique numbers are assigned to a block each time the block is evicted from cache, one must make sure that each sequence is not exhausted while the program is still executing. For example, if a dynamic data item is updated once every 100 ns in memory, a 32-bit in-order sequence (i.e., 0, 2, 3, …, $2^{32} - 1$) will overflow in about 429 ($2^{32} * 100$ ns/10^9 ns) seconds, or about 7.16 minutes. A 64-bit in-order sequence, on the other hand, will overflow in about 58.5K years. However, like large random numbers, more memory space would be needed to store large in-order sequence of unique numbers.

In general, there are two options to minimize the required memory storage space using in-order sequences:

1. **Using short in-order sequences.** In this case, each time the in-order sequence for one of the blocks overflows while the program is still executing, the SP stops the execution of the program, generates a new session key ($K_{sym\text{-}session\text{-}enc}$), and encrypts all the data blocks using the new key and the initial unique number (e.g., 0) in each sequence before the execution of the program can resume [30, 33]. However, the time required to re-encrypt all the data blocks can be long if this scheme is used for TSMs that operate on a large number of data blocks.
2. **Using split in-order sequences.** In this case, blocks are organized into small groups, for example, 256 blocks in each group. A16-bit in-order unique number is assigned to each block and a longer (e.g., 48-bit) in-order unique number is shared with all the blocks in one group [62]. To implement randomized encryption of each block within a group, the concatenation of the shared 48-bit unique number of the group with the block's 16-bit private unique number creates a long 64-bit in-order unique number for the block. However, each time one of the short sequences in one group overflows, the corresponding shared 48-bit unique number is incremented, the private short in-order sequences in that group are initialized, and all the blocks in that group are re-encrypted. Because no new session key is required and the number of blocks in each group is relatively small as compared to a TSM's total number of data blocks, with split in-order sequences, the length of time required to re-encrypt only the blocks in one group is much shorter as compared to the time required in Option 1. This is illustrated next using $data\_block_{i,j}$ to indicate a data-block j in group i. Each time $data\text{-}block_{i,j}$ is evicted from cache, its assigned short 16-bit private unique number is incremented. After 65,536 evictions, the block's shared 48-bit unique number is incremented and is used with an initial 16-bit unique number (e.g., 0) to encrypt all the (256) blocks in group i. Assuming that the loader stores the initial 48-bit unique number (e.g., 0) assigned to group i in memory location X_i and the initial short private 16-bit unique number (e.g., 0) assigned to $data\_block_{i,j}$ in Y_j, the following illustrates the randomized encryption of $data\_block_{i,j}$ using split in-order sequences:

```
//Initialization by the loader; X_i = 0; Y_j = 0.
block_block_ctxt-i,j = E(K_sym-session-enc, {block_address_j, 0,
0}, data_block_ptxt-i,j}
//1st time evicted from cache; X_i = 0; Y_i + 1 = 1; save 1
//at Y_j
```

```
block_block_ctxt-i,j   =  E(K_sym-session-enc,  {block_address_j,  0,
1}, data_block_ptxt-i,j}
//2nd time evicted from cache; X_i = 0; Y_i +1 = 2; save 2
//at Y_j
block_block_ctxt-i,j   =  E(K_sym-session-enc,  {block_address_j,  0,
2}, data_block_ptxt-i,j}

//3rd time evicted from cache; X_i = 0; Y_i +1 = 3; save 3
//at Y_j
block_block_ctxt-i,j   =  E(K_sym-session-enc,  {block_address_j,  0,
3}, data_block_ptxt-i,j}

. . .

//65535th time evicted from cache; X_i = 0; Y_i +1 =65535;
//save 65535 at Y_j
block_block_ctxt-i,j   =  E(K_sym-session-enc,  {block_address_j,  0,
65535, data_block_ptxt-i,j}

//65536th time evicted from cache; X_i + 1 = 1; Y_i = 0; 1
//is saved at X_i; 0 is saved at Y_j
//A firmware in SP encrypts all the blocks in group i and
//then resumes program execution
for(j = 0; j < number_of_blocks_in_group_i; j++)
    block_block_ctxt-i,j  =  E(K_sym-session-enc,  {block_address_j,
    1, 0}, data_block_ptxt-i,j});
```

The unique numbers assigned to the most referenced blocks may be saved in a specialized cache inside the SP to improve performance.

The re-encrypting task of all (Option 1) or some (Option 2) blocks will be managed as part of the TSM process. If the TSM is interrupted, its corresponding re-encrypting task will be also stopped, and it will resume when execution of the TSM resumes.

11.11.7 Program Data Integrity and Confidentiality

In DIDC-SXM, each dynamic data block must be encrypted, as in DC-SXM, and must be hashed and a hash tree must be maintained, as in DI-SXM. In addition to protecting the integrity of each data block, each block's assigned unique number, used for implementing randomized encryption, must be protected as well. However, it has been shown that there is no need to maintain a hash tree for data blocks and another hash tree for the assigned unique numbers in order to detect replay attacks [63, 64]. A single hash tree for the unique numbers, which would be smaller than the hash tree for data blocks, is sufficient to detect replay attacks. Furthermore, the unique numbers need not be encrypted [65]. Note that, the data blocks are still hashed, as illustrated next, but maintaining a hash tree for data blocks as in Fig. 11.32 is not needed. The hash value of each data block may be embedded within each block or stored separately, as was illustrated for code blocks in Fig. 11.29 or Fig. 11.30, respectively.

Figure 11.33 illustrates the organization of dynamic data blocks with embedded hash values and the hash tree for the data blocks' assigned unique numbers. In the

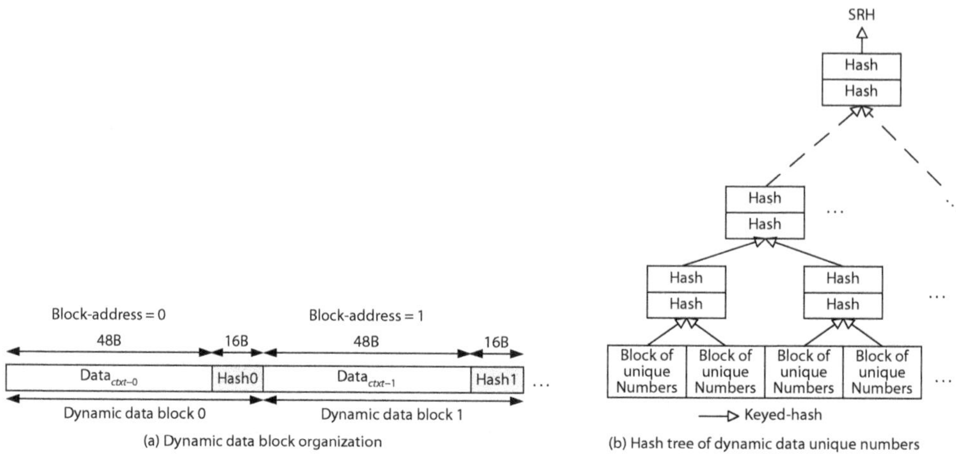

FIGURE 11.33 Protecting dynamic data: (*a*) dynamic data block with embedded hash; (*b*) hash tree for dynamic data unique numbers.

figure, each cache block is assumed to be 64 B and contains a 48-B data block and a 16-B (128-bit) hash. Assuming that a long (64-bit) unique number is used for each dynamic data block, eight such numbers can be stored in each 64-B cache block (8 = 64 B/64 bits).

Figure 11.34 illustrates a sophisticated replay attack using the data block organization shown in Fig. 11.33 with arbitrary unique numbers assigned to each block. The attacker replays both the cache block (data and embedded hash) and its assigned unique number. The attacker replaces both block 1 and its assigned unique number 23 at time t2 with an older copy saved at time t1. However, because the hash tree of the unique numbers will detect 17 as an invalid number at time t2, the execution of the process will be stopped, preventing the attacker from achieving the intended goal.

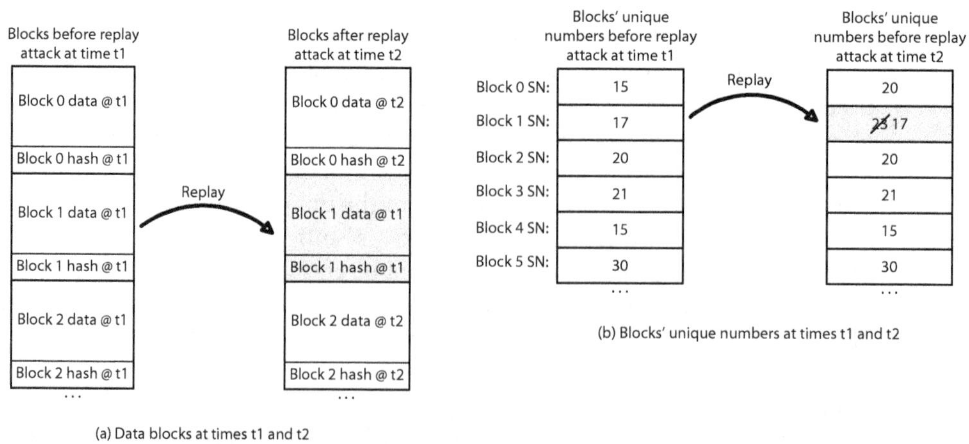

FIGURE 11.34 Illustrating a replay attack of a block and its sequence number.

11.11.8 Program Code and Data Integrity and Confidentiality

The CICC-SXM and DIDC-SXM combination provides maximum program protection. This combined execution mode requires four secret keys as $K_{sym\text{-}prog\text{-}enc}$ and $K_{sym\text{-}prog\text{-}sign}$ generated by the installer firmware and $K_{sym\text{-}session\text{-}enc}$ and $K_{sym\text{-}session\text{-}sign}$ generated by the loader firmware (Sec. 11.11.2). The first two keys are used for the encryption and hashing of program code blocks (including static data). The latter two keys are used for the encryption and hashing of program data blocks and maintaining a hash tree for the data blocks' assigned unique numbers used for randomized encryption of data blocks. This requires that the two types of blocks must be distinguishable so that the SP can use the right keys with each type of block during execution. There are two possible solutions, as follows:

1. **Physical Memory Organization.** One option is to split the main (physical) memory space into two regions: a non-SXM, including DMA access region, and an SXM region, which also includes a hash tree region. This option, however, may require a security kernel (part of operating system, OS) or a DMA transfer initiated by a trusted routine.

   If the SXM memory space reserved for code and data are further partitioned into code and data regions, then if the address indicates a code region (Fig. 8.5 in Chap. 8), the SP would use program keys to decrypt and authenticate an incoming cache block. On the other hand, if the memory address indicates a data block, the SP would use session keys to decrypt and authenticate an incoming cache block.

   Alternatively, if the SXM virtual pages are mapped to anywhere in the reserved SXM memory region (outside the hash tree region) and additionally randomized encryption of SXM data blocks is used, then number 0 could be assigned to each code block and a non-zero unique number to each data block [33]. Each time that there is a miss at the lowest-level cache, if the block's unique number is 0, the block is considered a code block, and program keys would be used to decrypt and authenticate the block; otherwise, if the unique number is non-zero, indicating a data block, the SP would use session keys to decrypt and authenticate the block. As stated earlier, zero and non-zero unique numbers need not be encrypted. However, a hash tree is used to authenticate the zero and non-zero assigned numbers.

2. **Virtual Memory Organization.** Another option is to use an additional virtual memory space, separate from the SXM and non-SXM virtual spaces, for the hash tree. Using a separate virtual space will allow an entire SXM virtual address space, divided into code and data regions, to be used by an SXM program (i.e., TSM). For any hash tree block that is cached, its virtual block address is also saved in the lowest cache so it can be used to determine the virtual address for the corresponding parent hash node, assuming physically addressed caches (Chap. 10). Furthermore, a separate translation look-ahead buffer (TLB) may be used for translating a hash tree virtual page number to its corresponding physical page number. Hash tree blocks may be saved in a separate cache memory for efficiency. If the virtual memory

address (stored in the cache) indicates a code block, the SP would use program keys to decrypt and authenticate an incoming cache block. On the other hand, if the virtual memory address indicates a data block, the SP would use session keys to decrypt and authenticate an incoming cache block. Again, in addition, if randomized encryption of data blocks is used, the unique numbers assigned to each data block, which would be saved in the virtual memory space reserved for the hash tree and its leaf blocks, need not be encrypted.

11.11.9 Handling Interruption

Interrupts require the state of the CPU (i.e., register contents and interrupt return address) to be saved upon interruption and then restored when the control is returned to the interrupted program (Chap. 9). In SXM, the register contents and the return address must be securely saved to detect attacks. The amount of extra resources required inside an SP depends on whether the SP is designed to execute only one SXM program (i.e., single TSM process) at a time [55] or multiple TSMs (i.e., SXM multiprocessing) concurrently [31–33].

Single Secure Execution Environment

In this case, only one TSM at a time can execute in SXM, indicated as SXM-OP (one process). Therefore, only one SXM state needs to be protected upon interruption. This can be done by encrypting and hashing SXM register contents using the SP's SRK. In addition, the encryption of register contents may be randomized to protect against any register information leak by using a unique number (e.g., a nonce randomly generated). The encrypted register contents are then stored back into their respective registers so they can be saved in memory by the interrupt handler (IH). The hash value, the interrupt return address, and the nonce (if any) are securely kept inside the SP. The required cryptography keys and their key materials, as well as the *SRH* (if any), also remain inside the SP.

Within the SP, registers and cache lines are tagged—for example, with 1 for SXM and 0 for non-SXM. Any read/write of a tag-1 register or tag-1 cache line by a non-SXM process or any read of a tag-0 register or tag-0 cache line by an SXM process will result in an exception. An SXM process can write any register or data block, thereby changing its tag to 1. Upon interruption, the SP clears all SXM-tagged registers, flushes SXM-tagged data blocks from caches, and changes the SP from SXM to non-SXM before turning control to the IH.

The SP may execute non-SXM programs in addition to one TSM (SXM program), all in a time-sharing environment. While a TSM is running, the OS cannot start to execute another SXM process. Upon returning from an interruption, the SP compares the return address with the one stored internally. If the two addresses match, the SP switches to SXM, decrypts the restored encrypted register contents, and resumes the execution of the SXM process. Otherwise, the return address indicates either the resumption of an interrupted non-SXM process if the address indicates non-SXM (Sec. 11.11.8) or an attack, in which case an exception would be raised.

Multiple Secure Execution Environments

In this case, the SP is designed to execute multiple SXM and non-SXM processes in a time-sharing environment. A **key table** would be used to store the interrupted SXM process state under the process ID. For example, a non-zero ID is used to identify an SXM process and $ID = 0$ to identify a non-SXM process. The SP resources (i.e., registers and cache lines) are also tagged by process ID. The table may be a **private key table** (embedded) within the SP or a **virtual key table** that resides outside the SP.

If a private key table is used, the SP can only execute a fixed number of SXM processes in a time-sharing environment. If there is an interruption, cryptography keys, return address, register contents, etc. are saved in the private table, SXM registers are cleared; and SXM data blocks are cleared from caches before the control is turned to the IH. Alternatively, instead of saving register contents in the private table, which would require maintaining a larger table, register contents may be processed as in SXM-OP. Register contents are encrypted, hashed, and then stored back in the registers to be saved by the IH in memory, and the hash along with the other information is saved in the table.

On the other hand, because a virtual key table may be copied on the hard disk, its size can grow as needed to allow the execution of any number of SXM processes in a time-sharing environment. However, the processor state (cryptography keys, register contents, return address, etc.) must be encrypted using the SP SRK before being stored under the process ID in the virtual table. The pages of the virtual key table are mapped to physical memory pages by the OS, just like how program virtual pages are mapped to physical pages (Chap. 10). A hash tree, similar to the one discussed in Sec. 11.11.5, is used to authenticate memory pages associated with the virtual key table.

Maintaining a virtual key table adds a delay to the handling of interrupts when compared to having an on-chip private key table. This delay, however, can be reduced if the recently accessed physical page addresses associated with the virtual key table are kept in a specialized cache memory within the SP for quick access.

Resuming from an interruption when using a private or virtual key table is handled similar to that described for the SXM-OP. However, since there can be more than one running SXM process, the return address is compared with that saved in the key table under the process ID; if the two addresses match, execution of the SXM process is resumed.

11.12 Design Example: Secure Processor

This section presents the architecture, including data path and sample SXM instructions, of an SP. The data path includes a standard processor core and the modules required for secure execution. An example application of TSM is also presented.

11.12.1 SP Specification

The following list specifies the features and limitations of an example SP:

1. The SP supports the execution of only one SXM process at a time (i.e., SXM-OP; see Sec. 11.11.9).

2. The SP implements the CICC-SXM and DIDC-SXM combination for maximum program protection. Throughout this section, the term SXM will mean maximum program protection that includes maintaining confidentiality and integrity of a trusted software module (TSM) code and data.
3. The SP includes a set of SXM instructions used to enable or disable SXM.
4. The SP contains a set of SXM status bits that indicates the SP status as either SXM or non-SXM.
5. A TSM code blocks (including static data) are encrypted and hashed to detect code spoofing and splicing attacks (if any). The code blocks are organized by compiler in the format shown in Fig. 11.29. When necessary, the term "program block" is used to refer to a cache block with an embedded code block and hash value, as illustrated in the figure. Furthermore, the terms TSM and "SXM program" may be used interchangeably. "SXM process" refers to a running TSM.
6. SXM program data blocks, which dynamically change during execution, are encrypted and a hash tree is maintained to detect dynamic data spoofing, splicing, and replay attacks (if any).
7. The encryption of dynamic data blocks is not randomized.
8. An SXM program is entirely self-contained with library routines statically linked at compile time. The program does not call external library or systems routines.
9. Data blocks are statically declared in the SXM program and memory space is allocated during compile time; no memory space is allocated during run time.
10. A region in main (physical) memory is reserved for SXM. The region is also partitioned into program code (including static program data) and dynamic data regions. The most significant address bit identifies each region; 0 identifies the code region and 1 identifies the data region. The data region is further partitioned into data and hash tree regions, as shown in Fig. 11.35(*a*). A hash tree of dynamic data blocks with four leaf (dynamic) blocks is shown in Fig. 11.35(*b*).
11. An SXM program is considered small enough to fit in its entirety in the SXM code region. Therefore, no virtual-to-physical address translation is performed in SXM.
12. L2 is the lowest cache memory and uses a write-back coherency protocol such as the MESI protocol (Chap. 10).
13. The SP contains an **encryption/decryption and hashing engine** (EDHE) implemented in hardware as an embedded system within the SP. It is used to decrypt and hash an incoming SXM code block, to decrypt an incoming SXM dynamic data block, and to encrypt an outgoing modified SXM data block.

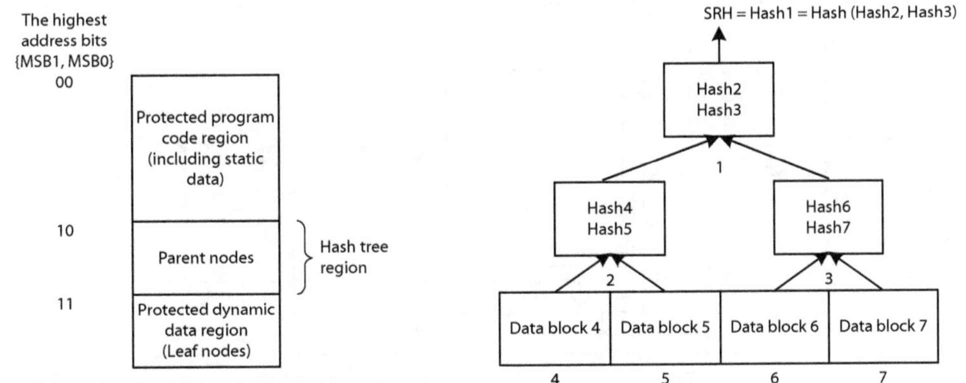

(a) Separation of code, dynamic data, and dynamic data hash tree regions.

(b) Illustrating a binary hash tree with only 4 leaf nodes as dynamic data blocks and parent nodes as hash blocks.

Figure 11.35 The organization of an SXM program code and data in memory: (*a*) memory map; (*b*) hash tree (shown as a binary tree).

14. The SP also contains a hash tree engine (HTE) implemented in hardware, which is also an embedded system within the SP. It is used to authenticate an incoming SXM data block using the hash tree and to update the hash tree when a modified data block is evicted from the L2 cache.
15. Both EDHE and HTE require trusted firmware where each is securely installed by the motherboard manufacturer.
16. The perimeter of the SP is the security boundary of the system. Therefore, caches are secure and contain instructions and data in plaintext.
17. The SP also includes the trusted program installer and loader firmware (Sec. 11.11.2). The loader firmware communicates with the OS to perform three tasks:
    a. The loader firmware extracts the two program cryptography keys $K_{\text{sym-prog-enc}}$ and $K_{\text{sym-prog-sign}}$ generated by the installer and stores them within the SP.
    b. The loader generates two session cryptography keys $K_{\text{sym-session-enc}}$ and $K_{\text{sym-session-sign}}$ and stores them within the SP. Session keys change each time the execution of an SXM program is started.
    c. The loader firmware creates the initial hash tree for TSM data blocks. The initial contents of the data blocks can be unknown.

    The OS starts the execution of the SXM program once the loader firmware completes its tasks.
18. Interruptions are handled the same way as was described for SXP-OP.
19. For simplicity, we will assume that the SP core includes the hardware to handle interruptions. Here, we will focus on the EDHE and HTE data paths.

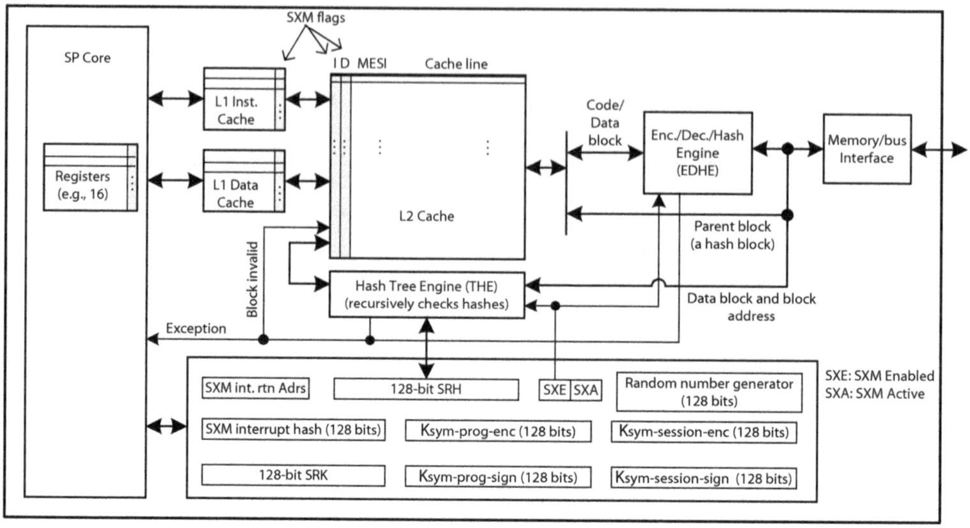

FIGURE 11.36 The data path of the example SP. The SXM implements CICC and DIDC secure execution modes.

11.12.2 Processor Architecture

Figure 11.36 illustrates the data path of the SP. It includes a processor core, L1 and L2 caches, and the modules required to implement the SXM. A set of registers is used to store cryptography keys, $K_{sym\text{-}prog\text{-}enc}$ and $K_{sym\text{-}prog\text{-}sign}$, generated by the installer firmware and extracted by the loader firmware, and two session keys, $K_{sym\text{-}session\text{-}enc}$ and $K_{sym\text{-}session\text{-}sign}$, generated by the loader firmware to protect the TSM's (dynamic) data blocks during execution.

The EDHE is responsible for decrypting and hashing an incoming SXM program block or SXM data block and encrypting a modified outgoing data block. An SXM program block contains a code block and an embedded hash. A code block contains instructions and/or static data. Because the contents of code blocks are not expected to change, these blocks are deleted from caches when they are replaced. The HTE is responsible for maintaining a hash tree for SXM data blocks. EDHE and HTE are discussed later in this chapter.

The SP data path also includes a 2-bit SXM status register [55]; the two register bits are called secure execution enable flag (SXEF) and secure execution active flag (SXAF). The SXEF enables both EDHE and HTE. The SXAF is used to ensure there is only one SXM process currently running. When SXAF is active, it prevents the OS from starting another SXM process as long as one is still running. The SP can be in one of three valid modes outlined in Table 11.10. An interruption of an SXM process resets SXEF, making it a 0, and a return from an SXM interruption sets SXEF, making it a 1.

As shown in Fig. 11.36, register contents and cache blocks used by the current process are tagged as 1 (SXM) or 0 (non-SXM). An SXM process can only read SXM-tagged register contents and cached blocks. The process, however, can write any register or data block, changing its tag to 1 (SXM). A non-SXM process, on the

SXAF	SXEF	State of the SP
0	0	The SP is in the non-SXM state and currently executing a non-SXM process. The OS may spawn an SXM process.
0	1	Invalid state (not used).
1	0	An SXM process is interrupted and not running. The currently executing program is a non-SXP process. The OS may not spawn another SXM process; otherwise, an exception is raised.
1	1	The SP is in the SXM state and currently executing an SXM process.

Table 11.10 The SP State Based on Values of SXM Status Bits SXAF and SXEF

other hand, can only read or write a non-SXM-tagged register content or non-SXM-tagged cache block. All the blocks in the combined L2 cache are also marked with non-SXM or SXM tags. In addition, instruction blocks in L2 cache are tagged "I" and data blocks as "D." This prevents an SXM process from accessing a data block in cache as instructions.

Table 11.11 presents a set of SXM instructions. Similar instructions are also defined elsewhere [31, 33, 55].

Application Example: Secure Encryption Service

Consider a TSM that implements an encryption API. Application software, as well as commodity OS, may use the API to encrypt application- or OS-specific data. For example, consider an application software that uses the API and specifies an encryption key number in a keychain, the starting memory address of its plaintext data in memory, and the starting memory address for the destination ciphertext (refer to Sec. 11.8.2 for an example). The following outlines the steps the TSM uses to securely encrypt the application's data:

1. The TSM executes instruction "SXM-ENTER." If SXAF is 0, and thus switches SP to SXM. Otherwise, if SXAF = 1, the application must wait until a currently executing (different) SXM process terminates, resets SXAF (making it a 0), and returns control to OS, which then can start the secure encryption TSM.
2. Once the TSM is invoked, it copies the application's plaintext, a non-SXM data, into the SXM data region in main memory using "LD" (Chap. 8) and "ST_TO_SXM" instructions.
3. The TSM securely extracts the encryption key from the application-provided keychain.
4. Using the "SXM_LD" and "SXM_ST" instructions, the TSM securely encrypts the plaintext, now stored in the SXM data region in memory, and stores the ciphertext, also in the SXM data region.
5. Finally, using the "LD_FROM_SXM" and also "ST" (Chap. 8) instructions, the TSM copies the generated ciphertext from the SXM data region to the application's ciphertext (non-SXM) data region in memory.

Instruction	Description
SXM_ENTER	If SXAF = 0, this instruction sets both SXAF and SXEF to 1 (active) causing the SP to enter the SXM state. The SXEF = 1 activates both the EDHE and HTE. The two session keys $K_{sym\text{-}session\text{-}enc}$ and $K_{sym\text{-}session\text{-}sign}$ that were generated by the loader firmware are used for encrypting/decrypting and hashing the SXM data blocks. An exception is raised if SXAF was 1.
SXM_EXIT	If SXAF = 1 and SXEF = 1, the SP exits the SXM state. The instruction clears the SXAF and SXEF setting them both to 0, resets the SXM tagged registers, flushes L1 data cache, and invalidates the SXM tagged data blocks in the L2 cache. L1 instruction cache and the instruction blocks in the L2 cache are not flushed in case the exit is temporary.
SXM_LD	If SXAF = 1 and SXEF = 1, the instruction loads from an SXM dynamic data block. If the instruction causes a cache miss, the block received from the main memory is decrypted before it is loaded into the L2 cache and is tagged as SXM. The HTE is invoked to authenticate the SXM data block. An exception, which terminates the SXM process, is raised if the block is not authentic. An exception is also raised if SXEF is 0.
SXM_ST	If SXAF = 1 and SXEF = 1, the instruction securely stores data in memory via caches. Specifically, the instruction writes and updates a block in cache and sets its tag to SXM. If the instruction causes a cache miss, the block is first loaded into the cache, tagged as SXM, and then it is updated. The HTE is invoked to authenticate the incoming SXM data block. A block's SXM tag in the cache refers to all the words in the block; the SP does not handle partially tagged blocks. An exception is raised if SXEF is 0.
LD_FROM_SXM	If SXAF = 1 and SXEF = 1, the instruction loads from an SXM data block to a register and changes the register tag to non-SXM allowing the register content to be used by a non-SXM instruction. If the instruction causes a cache miss, the block is decrypted before it is loaded into the L2 cache. The HTE is invoked to authenticate the incoming SXM data block. An exception, which terminates the SXM process, is raised if the block is not authentic. An exception is raised if SXEF is 0.
ST_TO_SXM	If SXAF = 1 and SXEF = 1, the instruction stores non-SXM register content into a SXM data block. If the instruction causes a cache miss, the block is decrypted before it is loaded into the L2 cache and then tagged SXM. The HTE is invoked to authenticate the incoming SXM data block. An exception, which terminates the SXM process, is raised if the block is not authentic. An exception is raised if SXEF is 0.

Table 11.11 The SXM Instruction Set

11.12.3 Encryption Decryption Hashing Engine

The EDHE contains encryption, decryption, and hashing functions implemented as an embedded system within the SP. It performs multiple tasks when SXEF is set to 1. The EDHE decrypts, hashes, and authenticates an incoming program cache block, which includes a code block and an embedded hash. It also decrypts an incoming data block when executing the "SXM_LD," "SXM_ST," "LD_FROM_SXM," or "ST_TO_SXM" instruction and encrypts an outgoing SXM-tagged modified data block. Note, no randomized encryption of SXM data blocks is implemented in the example SP. An unmodified SXM-tagged data block is deleted from caches when the block is replaced. The block address of an SXM program or data block is also used in the encryption, decryption, and hashing of the block. The inclusion of the block address in the hashing is used to detect splicing attacks (if any).

Cache Line Authentication: Code Blocks

The most significant two address bits (MSB1 and MSB0) identify the two different types of blocks; $(00)_2$ identifies a program block and $(11)_2$ identifies a data block (Fig. 11.35(*a*)). $K_{\text{sym-prog-enc}}$ is used to decrypt each SXM program block, and $K_{\text{sym-prog-sign}}$ is used to hash each SXM code block. The session encryption key $K_{\text{sym-session-enc}}$ is used to decrypt/encrypt an SXM data block.

For an incoming SXM program block (a code block plus an embedded hash), if the computed hash of the code block matches the embedded hash, the hash values in the program block are replaced with NOP instructions before the program block is stored in the L2 cache. The block is marked valid in the cache and its tag is set to 1 (SXM). Otherwise, the cache line is marked invalid (I) and an exception is raised, which terminates the SXM process (to prevent an attack).

If an incoming SXM data block is the result of executing an "SXM_LD" or "LD_FROM_SXM" instruction, the block is considered an SXM data block. The block is loaded from memory and decrypted (using $K_{\text{sym-session-enc}}$) before it is stored in the L2 cache. The data block and its block address are also loaded to the HTE for authentication. However, the block in the cache is considered valid and program execution continues as normal unless the HTE raises an exception, signaling an attack. The handling of an incoming SXM data block due to a write miss as a result of executing the "SXM_ST" or "ST_TO_SXM" instruction is the same—the block is decrypted and loaded into caches, HTE is invoked to authenticate the block, and the block is updated in the L1 data cache and marked modified in both the L1 data cache and the L2 cache.

Figure 11.37 illustrates the EDHE and HTE data paths for a read cycle, and Fig. 11.38 illustrates the data paths for a write cycle.

11.12.4 Hash Tree Engine

The HTE is responsible for both parsing the hash tree to authenticate an incoming SXM data block and for updating the hash tree when a modified SXM data block is evicted from the L2 cache and leaves the secure perimeter of the SP. However, because tree parsing takes time, the authentication is performed in the background; no data speculative execution is necessary. A data speculative execution, in this case, means the processor continues to execute the program, but does not commit the computed results to the registers until the data used in the speculative execution is validated. However, because there are no known universal techniques to recover from an attack other than to restart the program [66], the SP can commit the computed results to registers. That is, on a cache miss, an SXM data block that is loaded into the L2 cache is marked valid (e.g., the state E or S in the MESI protocol), and program execution continues as usual until the HTE raises an exception, which signifies an attack. The exception causes the SP to terminate the process and return control to the OS; the program nevertheless may be restarted.

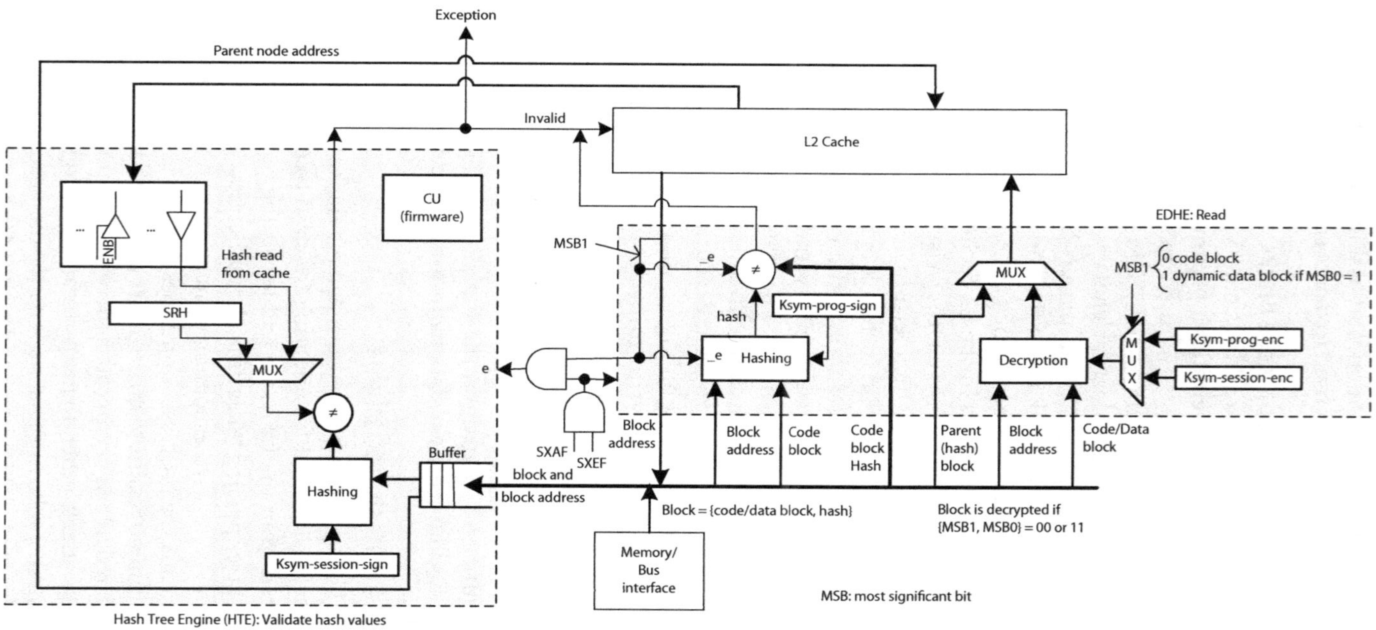

FIGURE 11.37 Loading an SXM block from memory; a code block decrypted and authenticated with embedded hash in the program block; a data block decrypted and authenticated by the HTE (some parts from [66]).

Cache Line Authentication: Data Blocks

The data path of the HTE for authenticating a data block is also illustrated in Fig. 11.37. The HTE recursively operates on the nodes of the hash tree, as illustrated by examples in Table 11.12 using the hash tree shown in Fig. 11.35(*b*).

Suppose caches are initially empty and the first data item accessed by the SP is in SXM data Block5. When memory supplies Block5, the block is decrypted and stored in the L2 cache by the EDHE. From there, the block is copied to the L1 data cache. At this time, the state of Block5 is assumed to be valid in all the caches. Block5 and its block address are also loaded into a buffer in the HTE when the block is loaded into the SP. Block5 is then authenticated by the HTE. Using the block address, the HTE determines that Block2 is the parent block of Block5 and attempts to access Block2 from the L2 cache. However, because the caches were initially empty, Block2 not being in the cache causes a miss. When memory supplies Block2, the block is stored in the L2 cache and, along with its block address, is also entered into the buffer in the HTE. Note that the parent blocks contain hash values and thus need not be decrypted before being loaded into the L2 cache, as illustrated in Fig. 11.37.

This time, the HTE tries to access Block1, the root and the parent block of Block2. Again, Block1 not being in the cache causes a miss. It is read from memory and loaded into L2, and along with its block address, is also entered into the buffer in the HTE. Recall that all the blocks, including the hash blocks, are marked valid in the L2 cache upon loading. Because Block1 is the root block, the HTE computes and compares its hash (*Hash1*) with the stored *SRH* inside the SP. If *Hash1* matches the *SRH*, Block1 is considered authentic and is removed from the buffer, leaving Block2 next in line in the buffer to be authenticated.

Data Block Read	Blocks in the Buffer	Target Parent	Cache Action	Action	Blocks in Cache
	5: authenticate	2	Read miss 2	Need Block2, enter 2	5
	2: authenticate	1	Read miss 1	Need Block1, enter 1	5, 2
Enter 5	1: authenticate	SRH		SRH =? Hash1, remove 1	5, 2, 1
	2: authenticate	1	Read hit 1	Block1.Hash2 =? Hash2, remove 2	5, 2, 1
	5: authenticate	2	Read hit 2	Block2.Hash5 =? Hash5, remove 5	5, 2, 1
Enter 4	4: authenticate	2	Read hit 2	Block2.Hash4 =? Hash4, remove 4	5, 2, 1
	7: authenticate	3	Read miss 3	Need Bock3, enter 3	5, 2, 1, 7
Enter 7	3: authenticate	1	Read hit 1	Block1.Hash3 =? Hash3, remove 3	5, 2, 1, 7, 3
	7: authenticate	3	Read hit 3	Block3.Hash7 =? Hash7, remove 7	5, 2, 1, 7, 3

"enter": Block and its block-address are entered into the buffer in the HTE.
"remove": Block and its block-address are removed from the buffer.

TABLE 11.12 Dynamic Data Block Authentication Examples Using a Hash Tree

This time, the HTE computes the hash of Block2 (*Hash2*) and compares it with Block1.*Hash2*. If the two hash values match, Block2 is considered authentic and is removed from the buffer, leaving Block5 still in the buffer to be authenticated. Finally, the hash of Block5 (*Hash5*) is computed and compared with Block2.*Hash5*. Again, if the two hash values match, Block5 is considered authentic and is removed from the buffer. This terminates the parsing of the hash tree to authenticate Block5. If at any time during the tree parsing any two compared hash values do not match, the HTE would raise an exception, which would cause the SP to clear all SXM-tagged registers and flush the L1 data cache and all the SXM-tagged L2 data blocks before terminating the SXM process and returning the control to the OS.

As illustrated in Table 11.12, since Block5 was the very first data block accessed from memory, it took the HTE several steps to authenticate the block. However, authenticating another data block, such as Block4, would take only one step. This is because Block2, the parent block of Block4, is already in the L2 cache (assuming not replaced) and was authenticated when the HTE was authenticating Block5, as indicated in the table. Therefore, there is no need to continue and verify the hash values in the path from Block2 all the way to the SRH as was done with Block5.

In the final example, SXM data Block7 is loaded into the cache and, along with its address, is also entered into the buffer in the HTE. Its hash value (*Hash7*) would need to be computed and compared with Block3.*Hash7*. However, because Block3 is not in the cache, Block3 is loaded from memory into the L2 cache and, along with its address, is also entered into the buffer in the HTE. Because Block1 is already authenticated and valid in the cache, as shown in the table, *Hash3* is computed and compared with Block1.*Hash3*. If the two hash values match, Block3 is considered authentic and is removed from the buffer, leaving Block7 in the buffer yet to be authenticated. The HTE computes and compares *Hash7* with the Block3.*Hash7*. If the two hash values match, data Block7 is considered authentic and is removed from the buffer.

Note that, while Block5 was being authenticated Block4 and Block7 may be entered into the buffer. The HTE authenticates data blocks in the first come first service (FCFS) order.

Hash Tree Update

The HTE also computes a new *SRH* when a modified SXM data block is evicted from the cache. This is illustrated by examples in Table 11.13 using the data path shown in Fig. 11.38 and the hash tree in Fig. 11.35(*b*).

Suppose SXM data Block5′, where ′ indicates the block is modified, is in the L2 cache. Also, suppose Block5′ is evicted from the cache, and thus will be encrypted and copied to memory. In order to update the hash tree, the encrypted Block5′, along with its block address, is copied to the buffer in the HTE as the block leaves the L2 cache. Because the caches are inside the SP and therefore are considered secure, the HTE only needs to update Block2, the parent block of Block5. Assuming that Block2 is not in the cache, this would cause a cache miss. Block2 would be copied from memory to the L2 cache and authenticated by the HTE, which also requires Block1 to be authenticated, as it was discussed earlier for the read cycle.

Data Block Evicted	Blocks in the HTE Buffer	Target Parent	Cache Controller	Action	Blocks in the Cache
Enter 5′	5′: update 2	2	Write miss 2	Need Block2, Enter 2	
	2: authenticate	1	Read-miss 1	Need Block1, Enter 1	2
	1: authenticate	SRH		SRH =? Hash1, remove 1	2, 1
	2: authenticate	1	Read-hit 1	Block1.Hash2 =? Hash2, remove 2	2, 1
	5′: update 2	2	Write-hit 2	Block2.Hash5 ← Hash5′, remove 5′	2′, 1
Enter 4′	4′: update 2	2′	Write hit 2	Block2′.Hash4 ← Hash4′, remove 4′	2″, 1
Enter 7′	7′: update 3	3	Write miss 3	Need Block3, push 3	2″, 1
	3: authenticate	1	Read-hit 1	Block1.Hash3 =? Hash3, remove 3	2″, 1, 3
	7′: update 3	3	Write hit 3	Block3.Hash7 ← Hash7′, remove 7′	2″, 1, 3′
Enter 2″	2″: update 1	1	Write hit 1	Block1.Hash2 ← Hash2″, remove 2″	3′, 1′
Enter 1′	1′: update SRH	SRH		SRH ← Hash1′, remove 1′	3′
Enter 3′	3′: update 1′	1′	Write miss 1′	Need Block1, enter 1′	3′
	1′: authenticate	SRH′		SRH′ =? Hash1′, remove 1′	3′, 1′
	3′: update 1′	1′	Write hit 1′	Block1′.Hash3 ← Hash3′, remove 3′	1″

′ and ″ indicate number of modifications, ′ (once) and ″ (twice); "enter" enters the block and block-address into the buffer in HTE; "remove" removes the block from the buffer.

Table 11.13 Hash Tree Update Examples

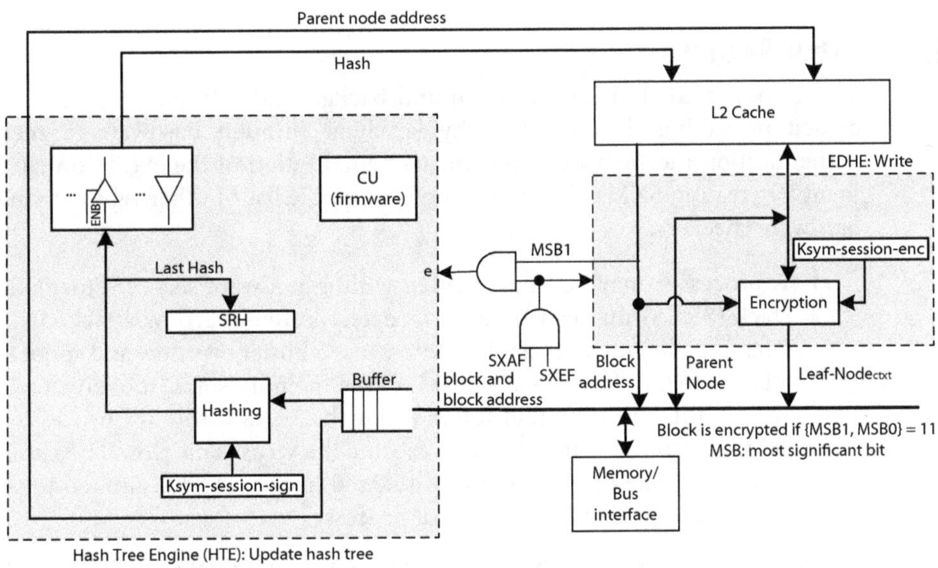

Figure 11.38 Write-back data block to memory; data path to evict a modified data block.

The content of Block2′ in the L2 cache changes to {*Hash4*, *Hash5′*}, where {} indicates concatenation, and Block5′ is removed from the buffer, completing the hash tree update. Next, suppose Block4′ is evicted from the L2 cache. Assuming that Block2′ is still in the cache, this update will not cause a cache miss and will be quick, changing Block2″ to {*Hash4′*, *Hash5′*}, where ″ indicates two updates. The eviction of Block7′ requires Block3 to be updated with *Hash7′*. However, assuming that Block3 is missing in the cache, it will be loaded from memory, authenticated, and then updated. This causes Block3.*Hash7* to be replaced with *Hash7′* in the cache and Block7′ to be removed from the buffer.

Next, suppose Block2″ that contains {*Hash4′*, *Hash5′*} is evicted from the cache. Block2″ and its block address are also entered into the buffer in the HTE. From its block address, it is determined that Block1, the parent of Block2, must now be updated. Assuming that Block1 is still in the cache, after the update, the content of Block1′ becomes {*Hash2″*, *Hash3*}, and Block2″ is removed from the buffer.

Table 11.13 also illustrates the eviction of Block1′, which causes the HTE to update the SRH with *Hash1′*, and then the eviction of Block3′, which causes Block1′ to be reloaded and authenticated and then modified with *Hash3′*. Note that hash blocks need not be encrypted when they leave the SP or decrypted when they are loaded into the cache.

The SP introduces additional overhead when compared to a standard processor. As illustrated in Fig. 11.37, each SXM program and data block is decrypted, requiring a cipher, before it is loaded into the L2 cache. In addition, hash tree parsing or updating generates additional cache traffic, which could slow down the execution of an SXM process. However, a separate cache may be used to store hash blocks to improve performance.

11.13 Further Reading

While we provided an introduction and background information, we also discussed protecting the cryptography keychain through hardware, a memory authentication mechanism, and compartmentalization of the execution environment by creating SXMs. The following is a sample list of other types of runtime hardware checkers:

- A processor may be implemented with a hardware secure return address stack (SRAS) that can be used to detect buffer-overflow attacks [67, 68]. Even if malicious software is able to cause a buffer overflow and spoof a new address (e.g., address of a virus) into the memory stack, the return address will be different from that stored inside the SRAS within the processor, and thus, no jump will take place to execute the virus. This would be similar to using a private or virtual table discussed in Sec. 11.11.9, limited to return addresses only, but for all types of processes.
- The hardware array bound checker uses the base address and size of the array to monitor out-of-bound errors [69].

- Hardware monitors can detect abnormal program behavior. This includes creating intra- and interprocedural control-flow monitors in hardware [70]. The monitor would be an FSM-based checker and would use program control-flow and data-flow graphs determined during compilation to dynamically monitor jump (intraprocedural) and call/return (interprocedural) addresses. An FSM and a table would be used to keep track of all permissible caller-callee relationships. The table stores the call/return addresses and is used to map an address to an FSM state. An invalid call/return address indicates an invalid behavior that causes the FSM to enter an invalid state.
- A program profiling–based checker verifies whether or not a program follows a normal execution path [71]. All possible program paths are recorded during some training runs, and the record is used by the checker to detect an invalid path. The training time must be long enough to reduce the number of false positives.
- Dynamic tracking of program information flow in hardware [72–75]. An integrity policy is implemented within the processor that prevents an OS-tagged low-integrity input data to be used as high-integrity data. An input is marked as low-integrity data if it enters the system, for example, through a device controller interface (DCI) such as a USB host controller interface. Tainted data values would be prevented from being used as instructions or memory addresses (pointers). Gate-level information flow tracking [74] hardware requires a shadow logic for every gate to track the trustworthiness of each bit. Each input bit and, thus each output bit, is marked as trusted (0) or untrusted (1). Therefore, simply using an untrusted bit does not always mean the result is also untrusted. That would depend on the gates used to process the untrusted bit. For example, using an AND gate with one trusted input $x = 0$ and one untrusted input $y = 0$ or 1, the output will be 0 and trusted. An instruction set architecture (ISA) data path using this methodology requires that the program pointer (PP) is never conditionally modified and there are no indirect memory load/store instructions. If the condition is untrusted, then the content of the PP would be untrusted, and this will lead to untrusted content for all registers and potentially all memory space. Therefore, all the instructions that depend on a condition must be converted to predicated instructions, and all unbounded loops must be converted into bounded loops (to prevent timing information leaks) using a counter with a termination condition where all the instructions within the loop are predicated with the negation of this termination condition. The counter is initialized by a special instruction ("countjump") and is decremented by 1 every iteration until it reaches 0. With this mechanism, a loop (including nested loops) is executed as an entity, and when it completes, it causes the PP to increment and thus exit the loop without branching. Untrusted information flow via hardware Trojans and physical attacks that may tamper with memory content is not considered. It has been shown that while shadow logic would increase the size of the circuit (e.g., by 70% in one study), it does not negatively affect clock frequency.

- Code and data replication to detect attacks by comparing the behavior of multiple copies of a running program [76]. With each replication, a different memory layout is used to detect memory access errors, a different hashing scheme is used with each copy to protect the integrity of critical data, and a different encryption scheme is used with each copy to better protect data confidentiality.
- Protect availability by implementing a better memory bandwidth allocation scheme. A memory controller (MC) typically employs a variety of algorithms to prioritize and schedule the outstanding cache misses supplied by memory. A first come first serve (FCFS) scheduler, for example, may also assign the highest priority to column accesses from the current active row (Chap. 7) to increase memory throughput and the next highest priority to the oldest nonactive row among the remaining outstanding requests [77, 78]. However, a malicious thread that flushes the MC with random (address) transactions could potentially increase the number of row accesses and thus increase stall times for other threads. A stall-time fair memory scheduler (STFM) uses **memory-slowdown** values to better schedule memory requests [79]. In this case, the memory controller computes a memory-slowdown (S) value for each thread that has a list of outstanding transactions as the ratio of the **average stall-time** if memory is shared with other threads (T_{shared}) and the **expected stall-time** if the thread is executed alone (T_{alone}). That is, $S = T_{shared}/T_{alone}$. An unfairness (U) parameter then is computed as $U = S_{max}/S_{min}$ where S_{max} and S_{min} are the maximum and minimum memory-slowdown values among all the outstanding requests. If U is less than some acceptable value (i.e., $U < a$), then the scheduler would use an algorithm to increase memory throughput—for example, by assigning a higher priority to burst transactions. On the other hand, if $U \geq a$ and thus there is unfairness in the processing of the outstanding memory requests, the requests from a thread with $S = S_{max}$ would be assigned top priority. The FCFS scheduling is then used to prioritize transactions among all the top priority requests.

References

1. M. M. Olama, J. J. Nutaro, V. Protopopescu, and R. A. Coop, Security concerns and disruption potentials posed by a compromised AMI network: risks to the bulk power system, The 2012 International Conference on Security and Management (SAM'12), Las Vegas, 2012, pp. 133-137.
2. A program example illustrating buffer overflow attack, http://www.cse.scu.edu/~tschwarz/coen152_05/Lectures/BufferOverflow.html.
3. Champagne David, "Scalable security architecture for trusted software," a Ph.D. dissertation, Princeton University, 2010.
4. Markus G. Kuhn, Cipher instruction search attack on the bus-encryption security microcontroller DS5002FP, *IEEE Transactions on Computers*, vol. 47, no. 10, October 1998, pp. 1153-1157.

5. Huang Andrew, *Hacking the Xbox: An Introduction to Reverse Engineering*, No Starch Press, San Francisco, 2003.
6. Butler Lampson, Martín Abadi, Michael Burrows, and Edward Wobber, Authentication in distributed systems: theory and practice, SOSP '91: *Proceedings of the Thirteenth ACM Symposium on Operating Systems Principles*, pp. 165-182
7. Hoglund, Greg. and Butler, James, *Rootkits: Subverting the Windows Kernel*, Addison-Wesley Professional, 2005.
8. Elias Levy, Approaching zero, *Security & Privacy*, IEEE Volume: 2 , Issue: 4. pp. 65-66.
9. Tal Garfinkel et al., Terra: a virtual machine-based platform for trusted computing, *Proceedings of the Nineteenth ACM Symposium on Operating Systems Principles* (SOSP '03), 2003, pp. 193-206.
10. Michael Fey, Brian Kenyon, Keven Readon, Brandon Rogers, and Charles Ross, *Security Battleground: An Executive Field Manual*, Intel Press, 2012.
11. Thomas A. Fuhrman, The new old discipline of cyber security engineering, *SAM '12*, 2012, pp. 547-553.
12. Ruby Lee, Simha Sethumadhavan, Edward Suh, and David Grawock, Tutorial on security for computer architects, ISCA Security Tutorial, San Jose, California, June 4, 2011.
13. Adam Waksman and S. Sethumadhavan, Tamper evident microprocessors, In: *Proceedings of the 31st IEEE Symposium on Security and Privacy*, 2010.
14. Mark S. Miller et al., *Capability Myths Demolished*, SRL, 2003, pp. 42-49.
15. Capability-Based Computer Systems, available from: http://www.cs.washington.edu/homes/levy/capabook/Chapter1.pdf.
16. D. E. Bell and L. J. LaPadula, *Secure Computer Systems*, Mitre Corporation, Bedford, MA, 1977.
17. K. J. Biba, *Integrity Consideration for Secure Computer Systems*, Mitre Corporation, Bedford, MA, 1977.
18. Timothy Fraser, *LOMAC: Low Water-Mark Integrity Protection for COTS Environments*, 2000 IEEE Symposium on Security and Privacy, 2000 (S&P 2000), pp. 230-245.
19. David D. C. Brewer and Michael J. Nash, The Chinese wall security policy. In *Proc. of the IEEE Symposium on Security and Privacy*, Oakland, IEEE Press,1989, pp. 206-214.
20. David D. Clark and David R. Wilson, A comparison of commercial and military computer security policies, *IEEE*, 1987, pp. 184-194.
21. Sally Adee, "The Hunt for the Kill Switch," IEEE Spectrum, May 2008, pp. 35-39.
22. K. Gandolfi et al., Electromagnetic analysis: concrete results, In: *Proceedings of 3rd International Workshop on Cryptographic Hardware and Embedded Systems (CHES)*, 2001, pp. 251-261.
23. D. Asonov and R. Agrawal, Keyboard acoustic emanations, In: *Proceedings of the IEEE Symposium on Security & Privacy*, May 2004, pp. 3-11.
24. Zhenghong Wang and Ruby B. Lee, A novel cache architecture with enhanced performance and security, In: *Proceedings of the 41st Annual IEEE/ACM International Symposium on Microarchitecture (Micro-41)*, 2008, pp. 88-93.
25. Waksman and S. Sethumadhavan, Silencing hardware backdoors, *SP '11 Proceedings of the 2011 IEEE Symposium on Security and Privacy*, pp. 49-63.
26. Craig Gentry, A fully homomorphic encryption scheme, Ph.D. dissertation, spring 2009, Stanford University.
27. M. Hicks, S. T. King, M. M. K. Martin, and J. M. Smith, Overcoming an untrusted computing base: detecting and removing malicious hardware automatically, In: *Proceedings of the 31st IEEE Symposium on Security and Privacy*, 2010.

28. Reouven Elbaz, David Champagne, Catherine Gebotys, Ruby B. Lee, Nachiketh Potlapally, and Lionel Torres, Hardware mechanisms for memory authentication: a survey of existing techniques and engines, *Trans. on Comput. Sci. IV*, LNCS 5430, 2009, pp. 1-22.
29. Champagne, David, Elbaz, Reouven, and Lee, Ruby B., The reduced address space (RAS) for application memory authentication, In *Proceedings of the 11th International Conference on Information Security* (Taipei, Taiwan, September 15-18, 2008, pp. 47-63.
30. Austin Rogers, Designing cost-effective secure processors for embedded systems: principles, challenges, and architectural solutions," a dissertation, University of Alabama in Huntsville, 2010.
31. D. Lie, C. Thekkath, M. Mitchell, et al., Architectural support for copy and tamper resistant software, *Proc. of the 9th Intl Conf. on Architectural Support for Programming Languages and Operating Systems (ASPLOS-IX)*, 2000, pp. 168-177.
32. G. Edward Suh, Dwaine Clarke, Blaise Gassend, Marten van Dijk, and Srinivas Devadas, AEGIS: architecture for tamper-evident and tamper-resistant processing, *Proceedings of the 17th Annual International Conference on Supercomputing (ICS '03)*, 2003, pp. 160-171.
33. G. Edward Suh, Dwaine Clarke, Blaise Gassend, Marten van Dijk, and Srinivas Devadas, AEGIS: architecture for tamper-evident and tamper-resistant processing, Computer Science and Artificial Intelligence Laboratory (CSAIL), MIT, 2004. (An extended version of [48]).
34. Qiong Liu, Reihaneh Safavi-Naini, and Nicholas Paul Sheppard, Digital rights management for content distribution, Proceedings of the Australasian information security workshop conference on ACSW frontiers 2003 - Volume 21, Australian Computer Society, January 2003.
35. Search for Extraterrestrial Intelligence (SETI), http://setiathome.ssl.berkeley.edu/.
36. Distributed.net, http://www.distributed.net/Main_Page.
37. Auguste Kerckhoffs, La cryptographie militaire, *Journal des Sciences Militaires*, http://www.petitcolas.net/fabien/kerckhoffs/.
38. National Institute of Standards and Technology (NIST), http://csrc.nist.gov/publications/.
39. AES-NI instruction set, http://software.intel.com/.
40. William Stallings, *Cryptography and Network Security*, Pearson Prentice Hall, 4th ed., 2006.
41. RSA calculator, https://www.cs.drexel.edu/~jpopyack/IntroCS/HW/RSAWorksheet.html.
42. Francis Crowe, Alan Daly ,and William Marnane, Scalable dual mode arithmetic unit for public key cryptosystems, *Information Technology: Coding and Computing*, Vol. 1, 2005, pp. 568-573.
43. Efficiency of ECC Cipher, http://www.certicom.com/index.php/the-basics-of-ecc.
44. R. Needham and M. Schroeder, Using encryption for authentication in large networked computers, *Communications of the ACM*, Volume 21 Issue 12, Dec. 1978, pp. 993-999.
45. SHA-1 calculator, http://www.sha1.cz/
46. Richard Spillman, *Classical and Contemporary Cryptology*, Pearson Prentice Hall, 2005.
47. Hans Brandl, *Trusted Computing: The TCG Trusted Platform Module Specification*, Infineon Technologies AG, Embedded Systems 2004.
48. Chu-Hsing Lin, Hierarchical key assignment without public-key cryptography, *Computers and Security*, Vol. 20, No. 7, 2001, pp. 612-619.
49. Blaise Gassend, Dwaine Clarke, Marten van Dijky, and Srinivas Devadas, Silicon physical random functions, *Proceedings of the 9th ACM Conference on Computer and Communications Security (CCS '02)*, 2002, pp. 148-160.
50. Yohei Hori, Hyunho Kang, Toshihiro Katashita, and Akashi Satoh, Pseudo-LFSR PUF: A compact, efficient and reliable physical unclonable function, *7th International Conference on Reconfigurable Computing and FPGAs (ReConFig '11)*, Cancun, Quintana Roo, Mexico, 2011, pp. 223-228.

51. G. Edward Suh, Charles W. O'Donnell, and Srinivas Devadas, AEGIS: a single-chip secure processor, *Information Security Technical Report* (2005) 10, pp. 63-73.
52. Sundeep Bajikar, Trusted platform module (TPM) based security on notebook PCs: white paper, Intel Corporation, June 2002.
53. Jeffry Dwoskin and Ruby Lee, Hardware-rooted trust for secure key management and transient trust, CCS'-07, Alexandria, Virginia, 2007, pp. 389-400.
54. Weiping Peng, Yajian Zhou, Cong Wang, Yixian Yang, and Yuan Ping, A new hierarchical key authdata management scheme for trusted platform, *International Conference on Multimedia Information Networking and Security*, 2010, pp. 463-467.
55. Ruby Lee et al., Architecture for protecting critical secrets in microprocessors, 32nd International Symposium on Computer Architecture, 2005 (ISCA '05), pp. 2-13.
56. Ralph C. Merkle, Protocols for public key cryptography, In: *IEEE Symposium on Security and Privacy*, 1980, pp. 122-134.
57. Smart card basics, http://www.smartcardbasics.com/.
58. TCG Specification Architecture Overview, Specification Revision 1.2 28 April 2004, http://class.ee.iastate.edu/tyagi/cpre681/papers/TCG_1_0_Architecture_Overview.pdf.
59. Y. Wang, H. Zhang, Z. Shen, and K. Li, Thermal noise random number generator based on SHA-2 (512), in *Proceedings of the 4th International Conference on Machine Learning and Cybernetics*, Guangzhou, China, 2005, pp. 3970-3974.
60. M. Milenković, A. Milenković, and E. Jovanov, A framework for trusted instruction execution via basic block signature verification, In: *Proceedings of the 42nd Annual ACM Southeast Conference*, 2004, pp. 191-196.
61. D. Kirovski, M. Drinic, and M. Potkonjak, Enabling trusted software integrity, In: *Proceedings of the 10th International Conference on Architectural Support for Programming Languages and Operating Systems (ASPLOS-X)*, 2002, pp. 108-120.
62. Chenyu Yan, Rogers B, Englender D, Solihin D, and Prvulovic, M, Performance and security of memory encryption and authentication, *Computer Architecture*, 2006. ISCA '06. 33rd International Symposium on Digital Object Identifier, 2006, pp. 179-190.
63. A. Rogers, Low overhead hardware techniques for software and data integrity and confidentiality in embedded systems, master's thesis, Electrical and Computer Engineering Department, University of Alabama in Huntsville, 2007.
64. B. Rogers, S. Chhabra, Y. Solihin, and M. Prvulovic, Using address independent seed encryption and bonsai Merkle trees to make secure processors OS- and performance-friendly, In: *Proceedings of the 40th Annual IEEE/ACM International Symposium on Microarchitecture (MICRO-40),* Chicago, IL, 2007, pp. 183-196.
65. Jun Yang, Lan Gao, and Youtao Zhang, Improving memory encryption performance in secure processors, *IEEE Trans on Computer*, 2005, pp. 630-640.
66. Blaise Gassend, G. Edward Suh, Dwaine Clarke, Marten van Dijk, and Srinivas Devadas, Caches and Merkle trees for efficient memory authentication, MIT-LCS-TR-857, 2002.
67. J. P. McGregor, D. K. Karig, Z. J. Shi, and R. B. Lee, A processor architecture defense against buffer overflow attacks, *Proc. IEEE Intl. Conf. on Information Technology: Research And Education (ITE 2003)*, August 2003, pp. 243-250.
68. R. B. Lee, D. K. Karig, J. P. McGregor, and Z. J. Shi, Enlisting hardware architecture to thwart malicious code injection, *Proc. Intl. Conf. on Security in Pervasive Computing* (SPC-2003), lecture notes in computer science, Springer Verlag, March 2003.
69. Joe Devietti, Colin Blundell, Milo M. K. Martin, and Steve Zdancewic, HardBound: architectural support for spatial safety of the C programming language, *Proceedings of the 13th International Conference on Architectural Support for Programming Languages and Operating Systems (ASPLOS-XIII)*, 2008, pp. 103-114.

70. Divya Arora, Srivaths Ravi, Anand Raghunathan and Niraj K. Jha, Secure embedded processing through hardware-assisted run-time monitoring, *Proceedings of the Conference on Design, Automation and Test in Europe (DATE '05)*, 2005, pp. 1530-1591.
71. Tao Zhang, Xiaotong Zhuang, Santosh Pande, and Wenke Lee, Anomalous path detection with hardware support, *Proceedings of the 2005 International Conference on Compilers, Architectures and Synthesis for Embedded Systems (CASES '05)*, 2005, pp. 43-54.
72. G. Edward Suh, Jaewook Lee, and Srinivas Devadas, Secure program execution via dynamic information flow tracking, *Proceedings of the 11th International Conference on Architectural Support for Programming Languages and Operating Systems (ASPLOS-XI)*, 2004, pp. 85-96.
73. Shashidhar Mysore, Bita Mazloom, Banit Agrawal, and Timothy Sherwood, Understanding and visualizing full systems with data flow tomography, *Proceedings of the 13th International Conference on Architectural Support for Programming Languages and Operating Systems (ASPLOS-XIII)*, 2008, pp. 211-221.
74. Mohit Tiwari, Hassan M. G, Wassel Bita, et al., Complete information flow tracking from the gates up, *Proceedings of the 14th International Conference on Architectural Support for Programming Languages and Operating Systems (ASPLOS-XIV)*, 2009, pp. 109-120.
75. Guru Venkataramani, Ioannis Doudalis, Yan Solihin, and Milos Prvulovic, FlexiTaint: a programmable accelerator for dynamic taint propagation, In: *14th International Symp. on High Performance Computer Architecture (HPCA)*, 2008, pp. 173-184.
76. Ruirui Huang, Daniel Y. Deng, and G. Edward Suh, Orthrus: efficient software integrity protection on multi-cores, *Proceedings of the 15th International Conference on Architectural Support for Programming Languages and Operating Systems (ASPLOS-XV)*, 2010, pp. 371-383.
77. Scott Rixner, Memory controller optimizations for web servers, *Proceedings of the 37th Annual IEEE/ACM International Symposium on Microarchitecture (MICRO-37 2004)*, pp. 355-366.
78. Scott *Rixner, William J. Dally, Ujval J. Kapasi, Peter Mattson, and John D. Owens, Memory access scheduling, Proceedings of the 27th Annual International Symposium on Computer Architecture (ISCA '00)*, 2000, pp. 128-138.
79. Onur Mutlu and Thomas Moscibroda, Stall-time fair memory access scheduling for chip multiprocessors, *Proceedings of the 40th Annual IEEE/ACM International Symposium on Microarchitecture (MICRO 40)*, 2007, pp. 146-158.

Exercises

11.1 List various security issues that users, organizations (e.g., military, banks), application programs, and systems (e.g., personal computers, cloud, handheld devices) might face.

11.2 Consider a government office issuing passports. Follow the SEM and the example in Table 11.1 to develop a security mechanism for "issuing a hard-to-forge passport."

11.3 Give a reason as to why a mandatory access control is needed in an organization.

11.4 Consider a bank with safe deposit boxes that customers can rent. The bank needs to select a secure scheme to allow only authorized access to safes. In addition, customers wish to have more freedom and occasionally allow their friends or relatives to access their boxes. For each of the following techniques, itemize what the bank and the owner of a safe need to do and what protections are needed to prevent an unauthorized person from accessing a safe or denying access to a legitimate owner.

a. Derive a secure mandatory ACL-based scheme that the bank can use. Hint: Each customer gives the bank a list of names that can also access the safe.
b. Derive a secure capability list-based scheme that the bank can use. Hint: Bank issues each customer n keys to give to a friend or a relative.

11.5 What is a multilevel security policy model?

11.6 What is a multilateral security policy?

11.7 Briefly explain the BLP's *-property.

11.8 Which policy model could be used to prevent the Stuxnet malware from changing the specifications of an industrial control system?

11.9 Stuxnet is designed to search for a specific control system known as a programmable logic controller (PLC). Which policy model could be used to prevent Stuxnet from transferring the control system information through the network?

11.10 Flame malware is designed to "suck" information (keystrokes, screenshots, audio, etc.) from a computer system and send it over the Internet to those who control it. Which policy model could be used to prevent Flame from transferring data through the network?

11.11 Why does the software that implements an access control mechanism and the system that runs it need to remain trustworthy?

11.12 Suppose a CPU uses an 8-bit carry look-ahead adder (CLA). Use two CLA modules, but purposely modify one of the CLAs so it outputs the wrong results. For example, change the correct expression $s_3 = p_3 \oplus c_2$ to $s_3 = \overline{p_3 \oplus c_2}$. With billions of transistors in a typical processor, it would be hard to detect there are two adders. Do the following:
   a. Design a single input triggering hardware Trojan that outputs wrong results after it is triggered using the input 0xAA.
   b. Suppose you apply 50 unique test vectors to test the adder. What are the chances of detecting the Trojan?
   c. Suppose the adder is a 32-bit CLA with the same exact Trojan with trigger input 0xAAAAAAAA. Determine the chance of detecting the Trojan with one million tests.

11.13 Suppose a MOD 24 counter is used to create a time bomb Trojan.
   a. How many test vectors do you need to trigger the Trojan?
   b. Suppose you applied 10M tests and the circuit worked correctly. How often do you need to reset power in order to prevent a time bomb trigger?
   c. Design a circuit that outputs 1 to reset the power to the module that may use a counter to create a time bomb Trojan. Also assume that the number of test vectors in part (b) includes tests applied during post silicon testing; that is, before mass production. Sample chips are manufactured for post-silicon testing purposes only in order to apply more tests using actual hardware instead of a simulation tool.

11.14 Design the following 16-bit encryption/decryption circuit shown in Fig. 11.39. It uses a multifunction 16-bit input register, a right-shift register as output register, and an 8-bit LSFR with a secret key. The circuit operates manually (no control unit is required). Validate your design.

11.15 What are the three sources of development threats?

11.16 What is the purpose of using homomorphic encryption, and in addition to requiring more hardware, what potential problem may exist that would make implementation of a homomorphic encryption more difficult? (Hint: Also refer to FP arithmetic in Chap. 3.)

11.17 Design an FSM-based control unit to encrypt/decrypt 16 bits of data at a time. The block diagram is shown in Fig. 11.40. The controller starts when signal *start* is asserted.

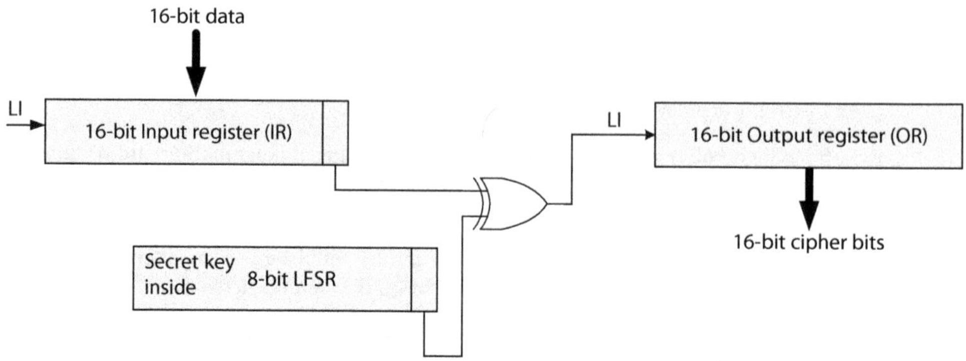

Figure 11.39 Data path for Exercise 11.14.

A user selects a 16-bit input and toggles the switch connected to the DFF to assert the *start* signal, as illustrated in the figure. Once the controller starts, signal *ack* (acknowledge) is used to reset the DFF. The encryption/decryption data path contains a counter for counting the number of times the multifunction registers must shift. In the block diagram, C.C. stands for combinational circuit. When the counter reaches the right number, the operation stops and the controller returns to standby, its initial state. The system should be reset only once and should work multiple times, each time encrypting/decrypting a 16-bit input. (Also see Exercise 11.14).

11.18 Given $e = 5$, $d = 29$, and $n = 91$, use RSA to encrypt $P = 6$ and decrypt $C = 41$.

11.19 Given $e = 5$, $d = 29$, and $n = 91$, use RSA to encrypt $P = 13$.

11.20 Suppose IV = 8'h77. Determine the ciphertext for plaintext = "HELLO" using RSA in CBC mode and use public key $e = 5$ and $n = 91$. For example, the first $P_0 = 72$ = 8'h48 is XORed with 8'h77 before it is encrypted to generate ciphertext C_0; then P_1 = 8'h45 is XORed with C_0 before it is encrypted to generate C_1, etc.

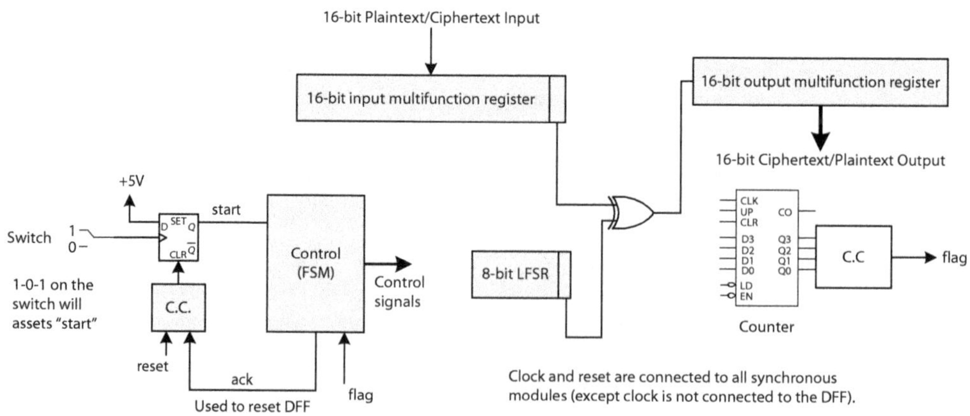

Figure 11.40 Data path and control unit for Exercise 11.17.

11.21 State why an MAC or HMAC is needed.

11.22 What is the advantage of an MAC? Also check the properties of AES-GCM online.

11.23 What is the advantage of an HMAC?

11.24 Consider a system with 8-B cache blocks and 256-B main memory. Organize this memory as a binary hash tree. Specifically, given the memory address of a child block, formulate a technique to identify the memory address of its parent block.

11.25 Consider a system with 8-B cache blocks and 128-B main memory. Organize this memory as a 4-ary hash tree. In a 4-ary tree, each parent node has four children nodes. Specifically, given the memory address of a child block, formulate a technique to identify the memory address of its parent block.

11.26 A hierarchical access control suitable for hardware implementation: Refer to reference [48] and write an HDL code to generate 8-bit R_{ji} and N_{ji} for given 8-bit ID_i, ID_j, *SRK*, K_i, and K_j. Use the following data to test your code. Then use each of the values for ID_i, ID_j, R_{ji}, and N_{ji} to compute secret K_i. Values of R_{ji}, and N_{ji} for all *i* and *j* would be public.

ID_i	K_i	ID_j	K_j	R_{ji}	N_{ji}
00000001	00000011	00000100	00001000	00000101	01011010
11110001	11100011	11010100	10111001	00100101	11011011
11111111	11100011	11111111	10111001	00000000	11110000

11.27 Explain in what ways a secure co-processor can improve the security of a computer.

11.28 Outline the basic functions an SP must be able to perform in a system that runs a TSM.

11.29 Explain in what ways an SP can improve the security of a computer.

11.30 Suppose for a DI-SXM program, a hash tree of sequence numbers is created for the program's data blocks. What would happen if an attacker performs a block spoofing, splicing, or replay attack? Will each type of attack be detected and why?

11.31 Suppose for DI-SXM programs, the sequence numbers of data blocks are stored inside the SP instead of on memory. Briefly state the advantage and disadvantage of storing the sequence numbers inside the SP.

11.32 Consider the design of a secure virtual memory management system. Suppose we would like to authenticate each virtual page using two-level hash trees. Each hash tree contains one root page and several leaf data pages. The hash (using bitwise XOR in this case) of each leaf page is stored in its corresponding root page. Assuming that pages are 4 KB, cache blocks are 64 B, and each hash value is 16 B, how many root pages are needed to authenticate 16 MB dynamically allocated virtual memory space?

11.33 Consider a hash tree of virtual memory data blocks (i.e., the hash tree uses virtual addresses). Suppose the SP implements physically addressed caches. In this case, a block's virtual address is also saved in the lowest-level cache. Briefly explain why saving the virtual address is necessary for processing the hash tree and detecting replay attacks.

11.34 Consider eight data blocks. Draw the hash tree and illustrate the authentication of blocks 8, 9, and 14 by the HTE. Also, determine how many cache misses will result due to accessing these blocks. Assume none of the cached parent blocks are replaced.

11.35 We would like to compare the memory space required to maintain a hash tree of sequence numbers assigned to memory data blocks. Two different sequence numbers are investigated: 64-bit vs. 64-bit split (48-bits long and 16-bits short) sequence numbers. Also, assume one long number is used for every 16 consecutive blocks, blocks are 64B, each hash value is 256 bits, and the maximum size of dynamic data memory space is 1MB. Determine how much memory space is needed to maintain the hash tree in each case.

11.36 Consider a SXM-OP system. Assume the CPU has four user-accessible registers and data and addresses are 8-bits each. Suppose during an interruption, the 8-bit register contents and an 8-bit return address are hashed using bitwise XOR and the hash value is saved inside the CPU. The register contents and the return address are then saved in memory by the interrupt handler. If the hash value is stored inside the CPU, show how it can be used to detect spoofing, splicing, and replay attacks. If any of the attacks cannot be detected, identify the reason and suggest a security mechanism. You may assume there are five registers numbered 0 to 4, where register 4 is used to store a return address.

11.37 An HTE is a microcontroller and executes a firmware located inside the SP. For simplicity, consider a hash tree of data blocks (as opposed to a hash tree of sequence numbers assigned to data blocks). Before a program can access its dynamic data in DI-SXM, the hash tree for the dynamic data blocks must already exist. Assuming the SP implements the MESI cache protocol, describe how the initial hash tree for a program's dynamic data blocks would be created. Also assume there are other SP state bits so that the OS can invoke the firmware when necessary and can choose to enable or disable the HTE read cycle, which, when disabled, causes data blocks that are loaded from memory to not be authenticated. The data memory space may be statistically declared during programming or allocated during run time.

11.38 Suppose split sequence numbers are used with each dynamic data block. Also assume that the starting address and the size of the dynamic data space are stored within the SP during the time a DI-SXM program is executing. Discuss/explain how a hash tree would be updated when one of the short sequence numbers overflows. Also, see Exercise 11.37.

11.39 Suppose an SP-based system uses separate virtual address spaces for an SXM process (code and data), a non-SXM process (code and data), a hash tree to protect SXM process data blocks, and a systems process (code and data). Outline a mechanism the SP could use to identify the right page table to use with each different virtual address.

Bibliography

Abd-El-Barr Mostafa and El-Rewini Hesham, *Fundamentals of Computer Organization and Architecture*, Wiley, 2005.

Agner Fog, "Branch prediction in the Pentium family," www.x86.org/articles/branch/branchprediction.htm.

Altera Quartus II, CPLD, FPGA design tool, http://www.altera.com/.

Anderson John A., *Foundations of Computer Technology*, CRC Press, 1994.

ATI Xenos GPU (for Xbox 360), www.amd.com.

Buchanan William J., *Introduction to Security and Network Forensics*, CRC Press, 2011.

Carpinelli John D., *Computer Systems Organization and Architecture*, Addison Wesley, 2001.

Christof P., Jan P., and Bart P., *Understanding Cryptography: A Textbook for Students and Practitioners*, Springer, 2010.

Ciletti Michael D., *Starting Guide to Verilog 2001*, Pearson Prentice Hall, 2004.

Clements Alan, *Principles of Computer Hardware*, Oxford, 2006.

Culler David, Singh Jaswinder, and Gupta Anoop, *Parallel Computer Architecture: A Hardware/Software Approach, Morgan Kaufman*, San Francisco, 1999.

Easttom William, *Computer Security Fundamentals*, 2nd ed., Pearson, 2011.

Gendrullis Timo, "Hardware-based cryptanalysis of the GSM A5/1 encryption algorithm," thesis, May 2008.

Hard drive interfaces, http://www.harddrivereport.com/.

Harris David and Harris Sarah, *Digital Design and Computer Architecture*, Morgan Kaufmann, 2007.

Harvey A. F., Data Acquisition Division Staff, "DMA Fundamentals on Various PC Platforms," National Instruments.

Hennessy John and Patterson David, *Computer Architecture: A Quantitative Approach*, 5th ed., Morgan Kaufman, Waltham, 2012.

Hwang Kai, *Computer Arithmetic Principles, Architecture, and Design*, Wiley, 1979.

Intel, "Optimization techniques for integer-blended code," http://download.intel.com/design/pentiumii/manuals/24281603.pdf.

Intel QuickPath, http://www.intel.com/technology/quickpath/introduction.pdf.

Katz R. and Borriello G., *Contemporary Logic Design*, Pearson, 2005.

King S. T., Tucek J., Cozzie A., Grier C., Jiang W., and Zhou Y. , Designing and implementing malicious hardware, In: *Proceedings of the 1st USENIX Workshop on Large-Scale Exploits and Emergent Threats*, April 2008.

Luebke David and Humphreys Greg, *How GPUs work?* IEEE Computer, February 2007, 96–100.

Mano Morris M. and Kime Charles R., *Logic and Computer Design Fundamentals*, 4th ed., Pearson Prentice Hall, 2008.

Mano Morris M. and Ciletti Michael D., *Digital Design*, 4th ed., Prentice Hall, 2007.

Marcovitz Alan B., *Introduction to Logic Design*, McGraw-Hill, 2005.

Microsoft Keyboard scan code specification, http://www.microsoft.com/.

Northbridge and Southbridge, http://www.nvidia.com/page/home.html, http://www.intel.com/products/chipsets/, http://www.amd.com/us/PRODUCTS/.

Null Lina and Lobur Julia, *Computer Organization and Architecture*, Jones Bartlett Learning, 2012.

NVIDIA GeForce GPUs, www.nvidia.com.

Osadchy M., Pinkas B., Jarrous A., and Moskovich B. , Scifi: a system for secure computation of face identification, In: *Proceedings of the 31st IEEE Symposium on Security and Privacy*, 2010.

Patterson David and Hennessy John, *Computer Organization and Design: The Hardware/Software Interface*, Morgan Kaufmann, San Francisco, 2005.

Saba A. and Manna N., *Digital Principles and Logic Design*, Jones and Bartlett, 2010.

Saltzer Jerome H. and Kaashoek M. Frans, *Principles of Computer System Design: An Introduction*, http://ocw.mit.edu/resources/res-6-004-principles-of-computer-systemdesign-an-introduction-spring-2009/online-textbook/protection_open_5_0.pdf.

Samsung hard drives, www.samsung.com.

Shen John P. and Lipasti Mikko H., *Modern Processor Design*, McGraw-Hill, 2005.

Smith James E. and Pleszun Andrew R., Implementing precise interrupts in pipelined processors, IEEE Transactions on Computers, 1988, 562–573.

Spansion Flash Memory, http://www.spansion.com.

Stallings William, *Computer Organization and Architecture*, Pearson Education, 2010.

Stallings William, *Cryptography and Network Security*, Pearson Prentice Hall, 4th ed., 2006.

Tanenbaum Andrew, *Structure Computer Organization*, Pearson, 2006.

Universal Host Controller Interface (UHCI) Design Guide, http://download.intel.com/technology/usb/UHCI11D.pdf.

Universal peripheral interface slave microcontroller (UPI-42), www.alldatasheet.com.

USB (universal serial bus), http://www.usb.org/home.

USB 3.0 specification, http://www.usb.org/developers/docs/.

Vahid Frank, *Digital Design with RTL Design, VHDL, and Verilog*, John Wiley and Sons Publishers, 2011.

Vray Jogn Shaley, *Interprocess Communications in UNIX*, Prentice Hall, 2003.

Wakerly J. F., *Digital Design: Principles and Practices*, 4th ed., Prentice Hall, 2006.

Index

www.ingramcontent.com/pod-product-compliance
Lightning Source LLC
LaVergne TN
LVHW062159200926
848484LV00006B/1284